D1126056

The Nalco Guide to Boiler Failure Analysis

The Nalco Guide to Boiler Failure Analysis

Nalco Company
James J. Dillon
Paul B. Desch
Tammy S. Lai
Daniel J. Flynn Editor

Second Edition

New York Chicago San Francisco
Lisbon London Madrid Mexico City
Milan New Delhi San Juan
Seoul Singapore Sydney Toronto

The McGraw·Hill Companies

Cataloging-in-Publication Data is on file with the Library of Congress.

The Nalco Guide to Boiler Failure Analysis, Second Edition

5 6 7 8 9 0 DSS/DSS 19 18 17 16

ISBN 978-0-07-174300-6
MHID 0-07-174300-6

The pages within this book were printed on acid-free paper.

Sponsoring Editor
Larry S. Hager

Editorial Supervisor
David E. Fogarty

Project Manager
Anupriya Tyagi, Glyph International

Copy Editor
Patti Scott

Proofreader
Surendra Nath Shivam, Glyph International

Indexer
Daniel J. Flynn

Production Supervisor
Richard C. Ruzycka

Composition
Glyph International

Art Director, Cover
Jeff Weeks

Cover Concept
Tammy S. Lai

About Nalco Company

Nalco Company is the leading integrated water treatment and process improvement company in the world. It provides essential expertise for water, energy, and air, delivering significant environmental, social, and economic performance benefits to its customers. Nalco Company helps customers reduce energy, water, and other natural resources consumption; enhance air quality; minimize environmental releases; and improve productivity and end products while boosting the bottom line. Together their comprehensive solutions contribute to the sustainable development of customer operations.

To our families, both professional and personal.
Robert D. Port and Harvey M. Herro, we are indebted to your laborious efforts to produce the 1st edition; it will forever be a classic work in the field. We truly appreciate the sacrifices that you made to make the guide a renowned success, and for your guidance and inspiration.
Dawn, Shane, Kevin, and Tyler Dillon; Gretchen, Jenny, and Steven Desch; and Brian Ingram, you have our sincere gratitude for your understanding, patience, and support during the process; it provided an environment that allowed for completion of our work.

Contents

About the Authors for the Second Edition

James J. Dillon received a B.S. degree in Metallurgical Engineering from the Illinois Institute of Technology in 1983. He joined Nalco Company in 1983 in the Metallurgy department specializing in failure analysis for industrial water and steam systems, as well as for refinery, oil production, and various chemical process industry systems. He has conducted over 5000 failure analysis investigations and hundreds of on-site inspections. He made numerous presentations, and provided training in a wide variety of industries. He has authored and coauthored numerous papers involving corrosion and failure analysis. He is an active member of the National Association of Corrosion Engineers (NACE) where he authored several papers, was chair for TEG-163X (Boiler Failure Analysis), and has been a member of several committees. He also was a major contributor to the writing of two books, including the first edition of *The Nalco Guide to Boiler Failure Analysis*, and The *Nalco Guide to Cooling Water Systems Failure Analysis.*

Paul B. Desch obtained a Ph.D. degree in Metallurgical Engineering in 1996 from the Illinois Institute of Technology. Prior to joining Nalco, he held positions at the Center for Material Science and the Metallurgy Group in the Material Science Technology Division of Los Alamos National Laboratory. Dr. Desch joined Nalco in 1997 as a member of the Metallurgy Group. He has authored or coauthored over 25 papers, including many on topics related to failure analysis of boiler components. He has made presentations at professional societies such as the National Association of Corrosion Engineers and the International Water Conference (IWC).

Tammy S. Lai obtained a B.S. degree in Chemical Engineering from Rice University, followed by a Ph.D. in Materials Science and Engineering from Northwestern University in 2006. Dr. Lai joined Nalco in 2007 as a member of the metallurgy group and has previous experience in the semiconductor industry and in teaching materials science at the university level. She is an active member in ASM International.

Foreword

Learning from our past helps guide us today and into the future as we apply lessons learned to avoid pitfalls similar to those we have experienced. We can look at failure analysis in much the same way. It is no longer acceptable to just repair a failed component. Knowledge of why the component failed is critical to prevent the situation from recurring in the future. From this understanding, modifications can be made to prevent recurrence, which can include changes in design, operations, automation, or the application of innovative chemistry. These modifications are undertaken to prevent future failures, assist in optimal equipment performance and protection, increase asset life, and reduce our overall environmental impact. With the increasing focus on sustainable use of our natural resources, and the market-driven need for continued cost productivity improvement, the understanding of the mechanisms of failure has become more critical today than in the past.

This second edition of *The Nalco Guide to Boiler Failure Analysis* is a comprehensive, authoritative field guide to boiler system failures. Contained within is information gathered during 60 years of failure analysis at Nalco Company. The authors of both editions of this book collectively have participated in excess of 12,000 failure analyses, including consultative services and field inspections of failed equipment. We believe that by gaining a thorough understanding of failure mechanisms, the knowledge can be applied in order to take preventive measures within a boiler and steam system. Instituting corrective actions or preventative measures will assist you in maintaining clean equipment, reducing corrosive conditions, increasing heat transfer, reducing cost, and lengthening asset life. We trust that this book will assist you in these endeavors.

J. Erik Fyrwald
Chairman, President and Chief Executive Officer
Nalco Company

Preface

During our discussions about the changes between the first and second editions of this book, we centered on several areas of improvement for the benefit of the reader. We restructured the book into three separate sections, Waterside Corrosion and Damage, Fireside Corrosion and Damage, and Material Defects. In each of these sections we grouped content where it made sense to do so, and we expanded the discussions including additional case histories as we looked deeper into each topic area. Our expanded discussions include material new to this edition. The chapters covering boiler types, phosphate corrosion, stress-assisted corrosion, steam and condensate damage, as well as flow-accelerated corrosion are all new.

Our chapter construction remains the same. Each chapter includes the following topics which we discuss in detail.

General description: Failure mechanisms are described. The causes and sources of attack are discussed.

Locations: Failure locations are pinpointed.

Critical factors: The most important or especially critical factors influencing failure are listed. These may include chemistry, pH, temperature and the like.

Identification: Typical attack morphologies in visual and microscopic inspection.

Elimination: Necessary preventive and/or corrective steps to eliminate or reduce damage are listed and explained.

Cautions: Comparisons are drawn between failure mechanisms that produce damage with a similar appearance.

Related problems: Related problems and damage are described.

Each chapter finishes with a selection of case histories. These examples are chosen to illustrate both similarities and differences between actual and "classical" failures. These histories provide diverse perspectives on how failures actually occur.

We have made liberal use of graphics, tables, and photographs to illustrate the concepts and failures. The majority of illustrations and photographs used are new to this edition. The changes we have made to the content and presentation should make the book appealing across a broad section of industries and steam systems.

While this book will help you in diagnosing failures, it is not a substitute for rigorous laboratory failure analysis. Many mechanisms are only properly identified by this type of in-depth evaluation. This book will, however, assist in the understanding of the mechanisms and locations of failures, allowing for steps to be taken in prevention.

DANIEL J. FLYNN
Editor

Acknowledgments

Undertaking the writing of the second edition of this book involved a great many individuals beyond the primary authors. When we started, we realized we would be requesting assistance from a good number of people. As the project progressed, the numbers grew as we cast our net farther into the many disciplines and departments throughout Nalco Company. We would like to take this opportunity to acknowledge the participation and efforts of the individuals involved with the second edition of *The Nalco Guide to Boiler Failure Analysis.*

The authors of the first edition of this book, Robert D. Port and Harvey M. Herro, deserve special recognition. Their efforts, insight, and invaluable tutelage over the years have made this extension of their initial work possible. The authors of the second edition, James J. Dillon, Paul B. Desch, and Tammy S. Lai, worked diligently at updating existing materials; writing new case histories; and expanding the depth, breadth, and scope of the book. Together these authors have completed in excess of 12,000 failure analyses.

We thank Anthony C. Biondo, Manian Ramesh, and David Slinkman for managerial support throughout the process, with additional thanks to Dr. Biondo for contributions to case histories included in this edition.

We thank Anton Banweg, Deborah Bloom, Kevin Gehan, Martin Godfrey, Peter Hicks, George Tortura, and Steef Vrijhoeven, for many helpful discussions over the years. Additional thanks are extended to Anton Banweg, Joshua Bishop, Robert Faedtke, Daniel Flynn, Kevin Gehan, Michael Martin, George Tortura, and Steef Vrijhoeven for review of various chapters in the book.

We acknowledge Richard W. Cloud and Joan B. Lewis for exceptional, timely scanning electron microscope analysis, especially for development of procedures and techniques for novel analysis of specimens. Support from the Deposit and Water Analysis groups in the Global Customer Analytical Group was also very much appreciated. We thank Michael S. Gagliano and Robert A. Gagliano, whose efforts and quality work provided multiple contributions of documentation used in the book. Michael J. Danko and Hayes O. Shackelford prepared samples and provided photos. Karen Baumann, Greg Coy, Nancy Limberg-Meyer, Susan Molloy-Vesley, Laura Pilling, and Christine Tokarz gave assistance with images to be used in the manuscript. We thank Kevin A. Smith for undertaking additional job responsibilities that provided increased time for participation in this project and Jacob Powers for obtaining various metallographic specimens for use in the book.

The Nalco Guide to Boiler Failure Analysis

SECTION 1

Waterside Corrosion and Damage

Introduction to Boiler and Steam Systems

A boiler and steam system is very complex and requires proper function of the various components to ensure successful operation. Mechanical, operational, and chemical treatment factors are all important. It is important to be familiar with the various components that are used within a boiler and steam system, since understanding their function is necessary to properly diagnose the causes of damage.

A schematic diagram of a typical industrial boiler and steam system is shown in Fig. S1.1. The ultimate goal of a boiler system is to boil water by heating it through use of appropriate heat sources to produce steam. The water typically flows through a suitable component, most commonly a tube (through the tube side). However, the water may also flow across the tube external surface (shell side) in certain boiler designs, such as firetube boilers. The most common heat sources in industrial boilers are provided by combustion of fossil fuels such as coal, oil, and natural gas. Many other fuel types or heat sources are also used. Steam has many possible uses. The steam can be used as a source of energy, for instance, to drive a turbine to produce electricity or pump fluids. Steam can also be used as a

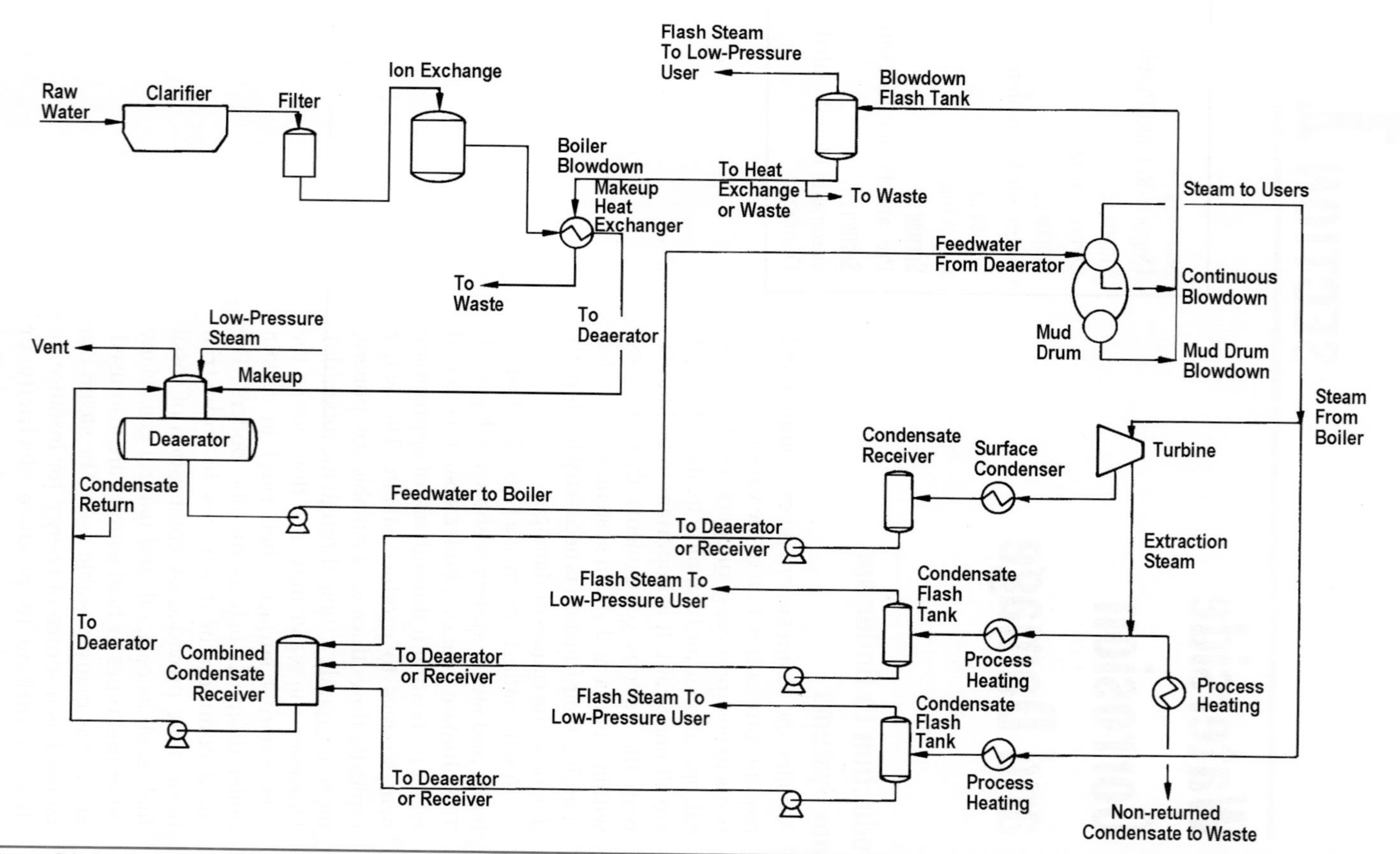

FIGURE S1.1 Boiler and steam circuit diagram. (*Courtesy of McGraw-Hill, The Nalco Water Handbook, 3d ed., New York, 2009.*)

heat source to aid production in various manufacturing processes. Temperature control for reactors, heating of air for drying or comfort heat in buildings, and sterilization processes are a few examples.

Preboiler

At the initial stages of boiler operation, and due to consumption or loss of steam within or by the system, it is necessary to provide a source of makeup water for the boiler. The beginning of the diagram (Fig. S1.1) highlights the processes and components used to produce makeup water that is suitable for use in the boiler from a raw water source. Essentially, since the water is to be boiled, it is necessary to reduce the concentration of agents that are suspended or dissolved in the water, such as calcium, magnesium, and silicon. This is necessary since these agents will produce deposits and/or will cause corrosion unless they are removed or reduced in concentration to appropriate levels. Boiler pretreatment processes that remove these agents are well described in *The Nalco Water Handbook*, third edition. The feedwater and boiler water quality requirements will vary according to pressure of the boiler, as described in Chap. 1, Water-Formed and Steam-Formed Deposits.

Downstream of the makeup water pretreatment system, it is necessary to remove dissolved gases in the water, particularly oxygen, to control corrosion of various components. This is achieved through use of a deaerating system. Various designs of deaerators are described in *The Nalco Water Handbook*, third edition. A common design is the use of trays within an enclosed vessel. The water cascades down a series of trays while being exposed to a countercurrent of steam (Fig. S1.2). The heating process drives dissolved oxygen and other gases out of the water, which is vented from the deaerator. Since the deaerator cannot remove dissolved oxygen completely, oxygen scavenging chemicals are added to the water to react with the residual dissolved oxygen prior to the pumping of the feedwater to the boiler.

Low-pressure boilers may use the deaerator or an open feedwater heating system to heat the feedwater prior to its introduction to the boiler. Following deaeration, however, in units such as high-pressure utility boilers, high-pressure feedwater heaters are used to preheat the water prior to introduction to the boiler. These feedwater heaters are typically of shell-and-tube design (Fig. S1.3). Feedwater flows through the tubes, and is heated by steam that flows across the shell sides. The steam is typically supplied by extraction from the turbine.

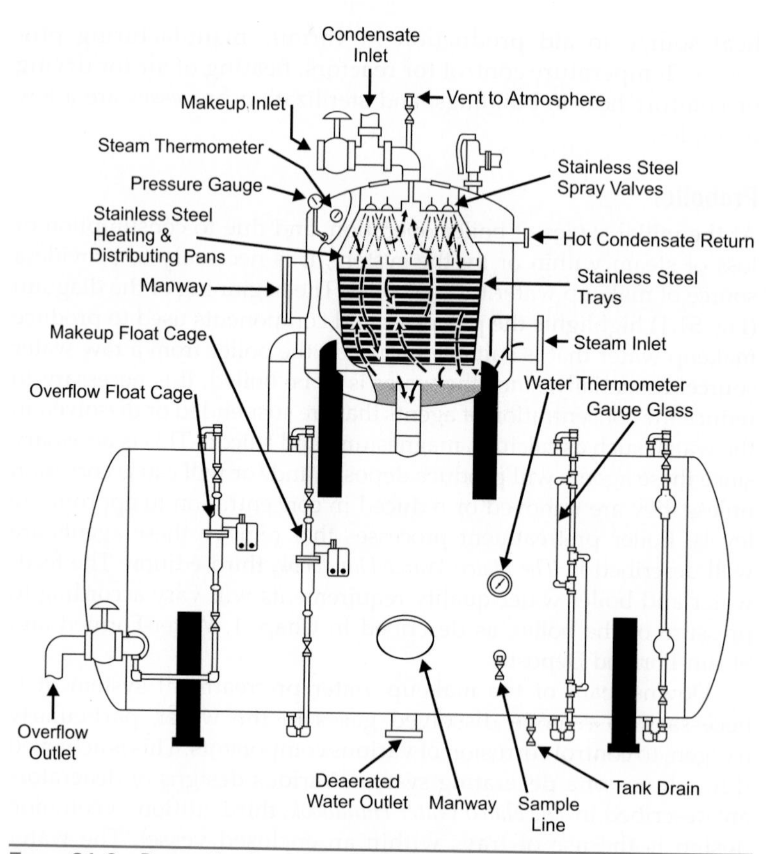

FIGURE S1.2 Deaerating heater (top) supplies deaerated water to the storage tank (bottom). (*Courtesy of McGraw-Hill, The Nalco Water Handbook, 3d ed., New York, 2009.*)

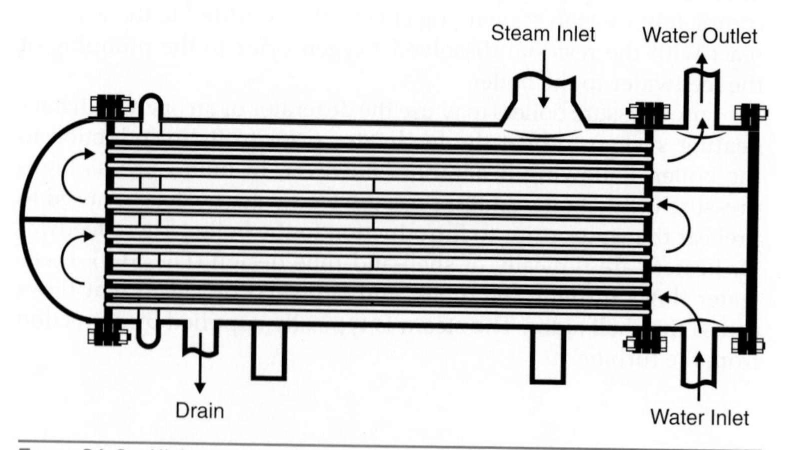

FIGURE S1.3 High-pressure feedwater heater.

Boilers

There are many different boiler designs. The appropriate boiler design is based upon many considerations. The choice depends upon factors such as the intended service, available space, fuel that is available or most economical to use, cost of materials and construction, expected service life, safety, reliability, and other considerations.

The material used for most water contacting boiler components, including tubing, is carbon steel. Higher-grade alloys are used when needed, such as when higher corrosion and thermal oxidation resistance is required. Such alloys are more costly. The most common industrial boilers employ the watertube design. Boiler water flows through the tube sides, and is heated on the shell sides of the tubes.

Within certain boilers, particularly high-pressure units, the first component within the boiler that the feedwater contacts, is the economizer. The economizer is composed of carbon or low-alloy steel tubes and is located in the low-temperature zone of the flue gases near the stack. The function of the economizer is to capture heat from the outgoing flue gases to heat the boiler feedwater. Some boiler designs do not include economizers, particularly low-pressure boilers.

Many industrial boiler designs employ two drums, or large, cylindrical, closed-ended vessels with rows of tubes connecting them on opposite sides (Fig. S1.4). The top drum is called the steam drum. Feedwater typically passes from the economizer to the steam drum. Internal treatment chemicals may be fed directly to the steam drum in some boilers to condition the boiler water. The bottom drum is called the mud drum. In the distant past, sludge tended to accumulate substantially in the mud drum. The sludge was removed or was blown down from this drum to maintain proper boiler water quality. Modern boiler operating practice tends to produce much less sludge in the mud drum.

The tubes that extend from the mud drum to the steam drum have different functions depending upon the heat fluxes inherent in the design. In some designs, heat flux is higher in one set of tubes than in the other. Flow in the tubes that experience the higher heat flux will rise from the mud drum to the steam drum. These are called riser tubes (Fig. S1.5). Where the boiler water within the circuit is cooler, it will flow downward through tubes from the steam drum to the mud drum. These tubes are called downcomers. The difference in temperature provides natural circulation through this portion of the system, which is called the generating bank, since steam is generated in the riser tubes. The locations of riser and downcomer tubes within a boiler are dependent upon the design. The produced steam collects from the riser tubes in the steam drum.

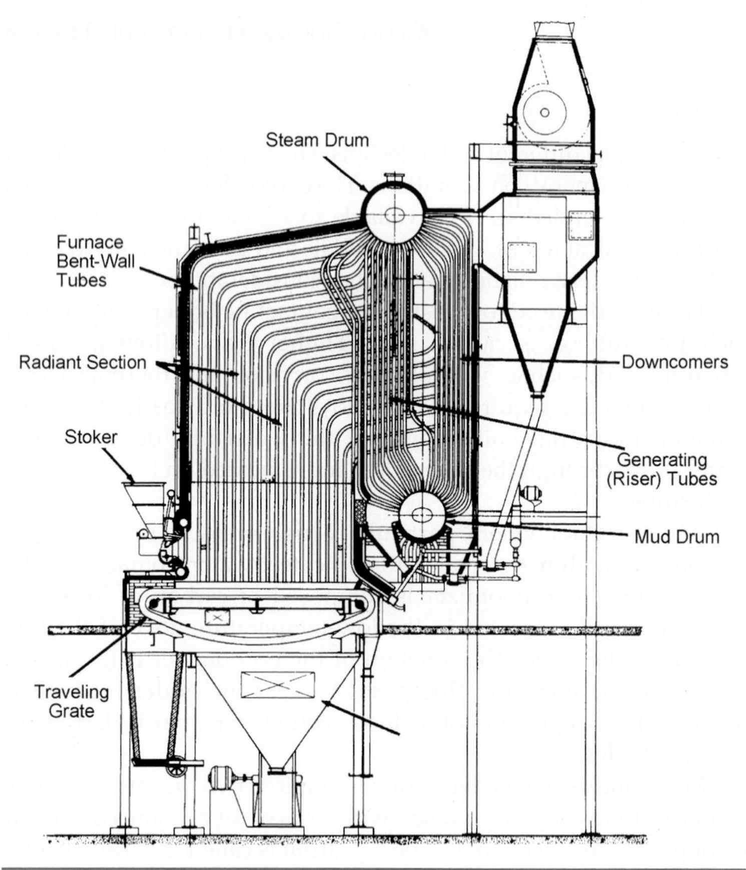

Figure S1.4 Generating bank. (*Courtesy of McGraw-Hill, The Nalco Water Handbook, 3d ed., New York, 2009.*)

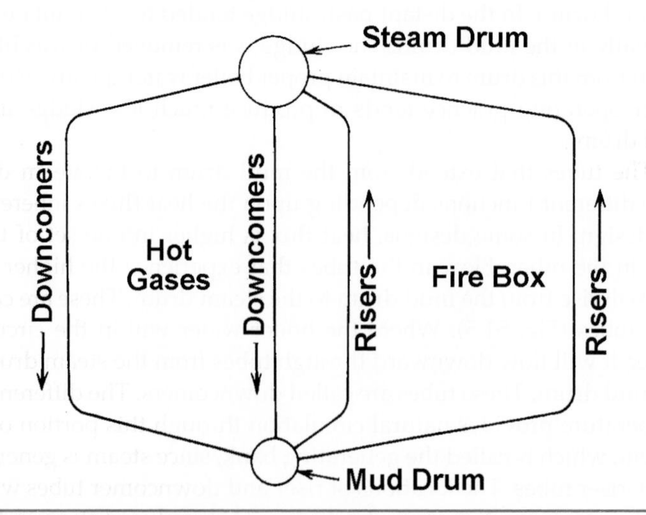

Figure S1.5 Riser and downcomer tube natural circulation. (*Courtesy of McGraw-Hill, The Nalco Water Handbook, 3d ed., New York, 2009.*)

The heat of fuel combustion within the firebox of the boiler is contained by water-filled tubes (Fig. S1.4). The tubes that serve this purpose are appropriately named wall tubes, floor tubes, and roof tubes. Some locations on the walls are exposed to radiant heat. Steam generation occurs within these tubes. The wall tubes are typically supplied by and flow into larger-diameter collection tubes called headers. The steam from the wall tubes is collected in the steam drum as well. It is common for the firebox to be sealed to prevent gas and heat leakage by attaching adjacent wall tubes to one another. Strips of metal, called membranes, are longitudinally welded to the external surfaces, bridging the tubes to form panels. The side of the tube facing the firebox is called the *hot side,* while the side facing the outside of the boiler, where no heat is input, is called the *cold side.*

The collected steam in the steam drum is prepared for discharge to the steam system. The steam drum contains components such as cyclone separators, screens, or other devices that separate water from the steam (Fig. S1.6). The separation is achieved by mechanically forcing the steam/water mixture that is present in the steam drum to follow a circuitous path. The heavier water droplets collect and flow back into the bulk water within the drum. This separation process controls carryover of boiler water into the steam to prevent deposition, corrosion, and overheating in downstream components.

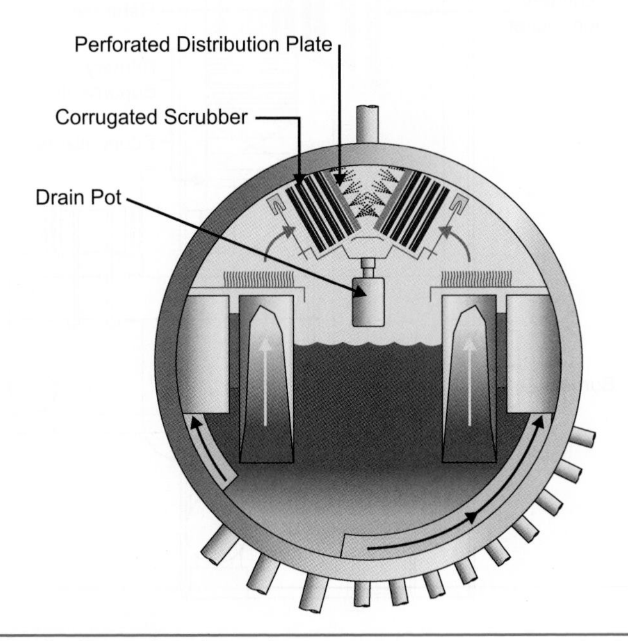

FIGURE S1.6 Steam drum internals.

Steam Systems

In high-pressure boilers, particularly those that supply steam to turbines for power generation, the steam passes from the steam drum to superheater sections where it is heated above the steam/water saturation temperature (Fig. S1.7). The superheater is located in a high-temperature zone in the firebox. There may be primary and secondary superheaters installed in series. Since superheater tubes operate at higher temperatures than the water-filled tubes upstream, it is common for more resistant alloys to be used such as low-alloy and stainless steels, and possibly higher alloys. Protection of superheater tubes, particularly in paper mill recovery boilers, is provided by screen tubes. Screen tubes are angled extensions of wall tubes that shield or screen the superheater. Superheated steam temperature may be reduced if necessary through use of attemperators, which supply pure water to the superheater (Fig. S1.8). The pure water used in attemperation may be supplied by a component called a sweet water condenser, which is typically a shell-and-tube heat exchanger that is constructed with stainless steel or other corrosion-resistant alloy

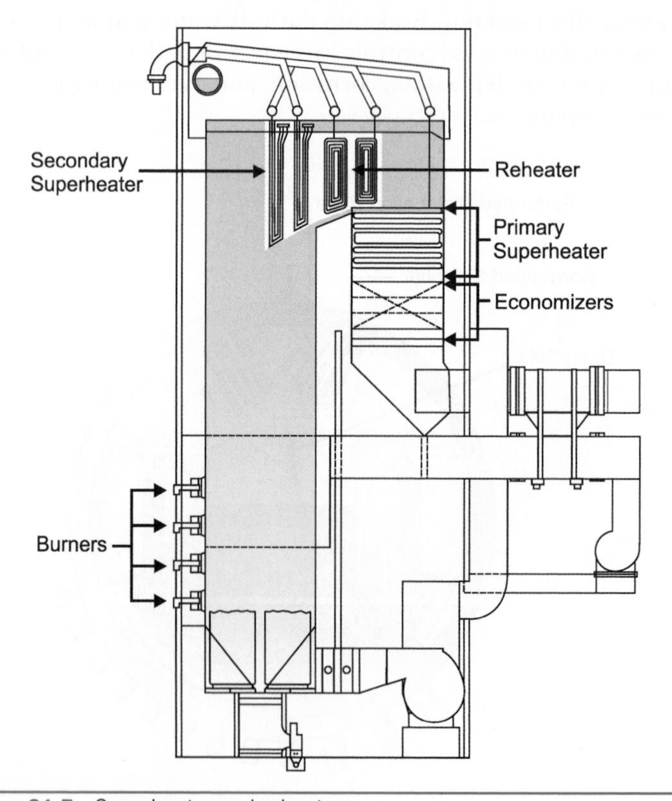

FIGURE S1.7 Superheater and reheater.

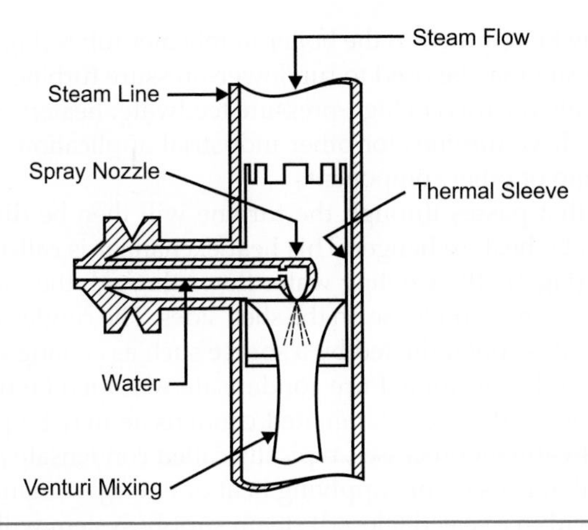

FIGURE S1.8 Attemperator.

tubes. Cool boiler feedwater flows across one side of the tubes to condense saturated steam, which is then used for attemperation. Alternatively, highly polished returned condensate may be used for attemperation as well.

For utility boilers, superheated steam passes to a turbine (Fig. S1.9), causing it to rotate to generate electric power. Steam extracted from the

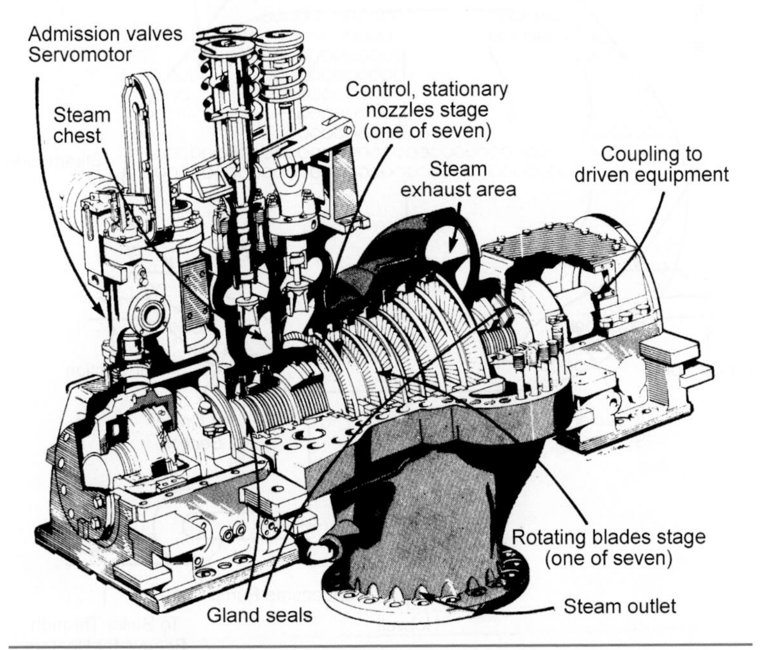

FIGURE S1.9 Steam turbine. (*Courtesy of McGraw-Hill, The Nalco Water Handbook, 3d ed., New York, 2009.*)

turbine may be returned to the boiler in reheater tubes (Fig. S1.7). The reheated steam may be used to run lower-pressure turbines. Extracted steam may also be used in high-pressure feedwater heaters. Steam may be used to drive turbines for other industrial applications such as to drive a pump or other components.

Steam that passes through the turbine will then be directed to a shell-and-tube heat exchanger. This heat exchanger is called a surface condenser (Fig. S1.10). Cooling water flows through the tubes, which causes the steam to condense on the shell sides. The condensate is very pure unless it is contaminated by a source such as cooling water from a leaking condenser tube. Pure condensate will then be returned to the feedwater system. Contaminated condensate may be purified by proper pretreatment processes, typically called condensate polishing.

Steam that is used for supplying heat or energy to manufacturing processes within a sealed (closed) steam supply system will condense as it cools. This condensate may also be returned to the boiler. Within such systems it is necessary to control the flow of steam and remove the condensate that forms along the path of the steam line. This may

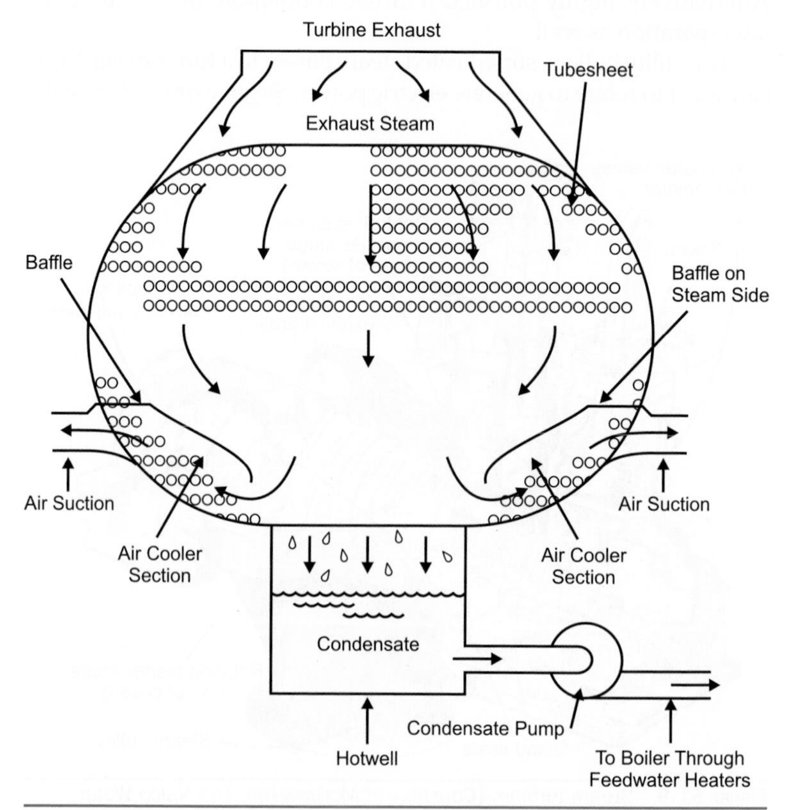

FIGURE S1.10 A view of a surface condenser showing a portion of the tubes.

be provided by the use of steam traps along the steam line. There are many types and designs for steam traps, but their main function is similar. As condensate builds up, or cooling occurs due to the presence of condensate within the steam trap, components within the trap will simultaneously block the flow of steam and drain the condensate. Once the condensate is cleared, steam flow will be reestablished.

Other Types of Boilers

Firetube boilers (Fig. S1.11) have a wide variety of applications. They are commonly used in heating, ventilating, and air conditioning (HVAC) systems for buildings. Such boilers may also be used in light industry. They operate at low pressure, typically ranging from 5 to 300 psi (34.5 kPa to 2.0 MPa), and are used for systems that have low steam demand. As the name *firetube boiler* suggests, the combustion gases flow through the tube sides of the steam-generating tubes, to generate steam on the shell sides.

Certain waste heat boilers used in refining applications also employ shellside steam generation. These boilers use shell-and-tube designs, and may be either horizontally or vertically oriented (Fig. S1.12). Very hot process gases that require cooling, flow through the tubes.

Other boilers have specific designs and functions. One boiler that was designed to provide increased efficiency for power generation is the heat recovery steam generator (HRSG). Such boilers use a turbine

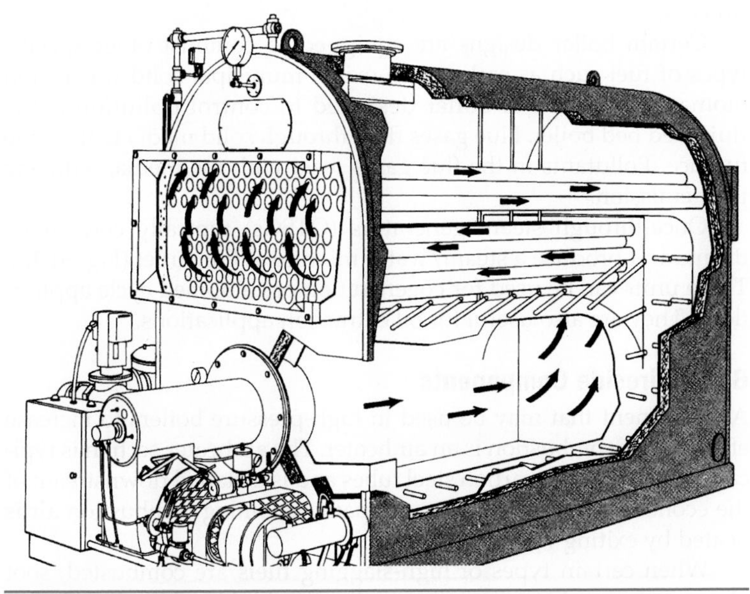

FIGURE S1.11 Firetube boiler. (*Courtesy of McGraw-Hill, The Nalco Water Handbook, 3d ed., New York, 2009.*)

FIGURE S1.12 Waste heat boiler. (*Courtesy of McGraw-Hill, The Nalco Water Handbook, 3d ed., New York, 2009.*)

that is driven by combustion gases. The heat from the combustion gases is then sent to a boiler to generate steam, which runs a steam turbine.

Recovery boilers used at paper mills combust a fuel that is generated during the papermaking process, typically black liquor. By-products from the combustion are then used in the papermaking process.

Certain boiler designs are equipped to combust other specific types of fuel such as carbon monoxide, municipal solid waste, and biomass. One type of boiler designed to control pollution is the fluidized bed boiler. Flue gases flow through solid media that fill the furnace. Pollutants in the flue gases react with the media, reducing their emission.

Once-through steam generators (OTSGs) typically contain no drums and produce a steam/water mixture at the outlet (Fig. S1.13). These units can be used for cogeneration or combined cycle applications. They are also useful for oil extraction applications.

Boiler Fireside Components

A component that may be used in high-pressure boilers to increase efficiency of combustion is an air heater. The air heater section is typically constructed of carbon steel tubes and is installed downstream of the economizer along the flue gas path. Incoming combustion air is heated by exiting warm flue gases.

When certain types of high-slagging fuels are combusted, soot blowers are used to remove excessive fireside deposits (Fig. S1.14). Soot blowers consist of tubes that contain ports through which

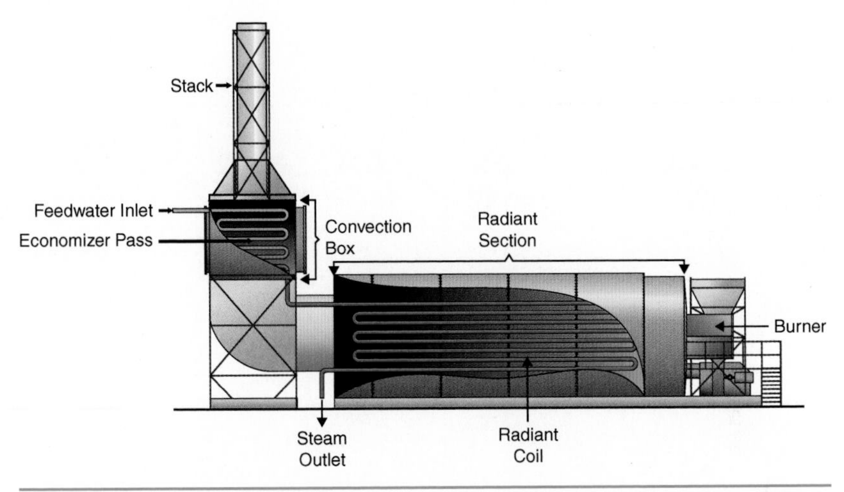

FIGURE S1.13 Once-through steam generators convert water to steam in a single pass. There is no recirculation of water in the system.

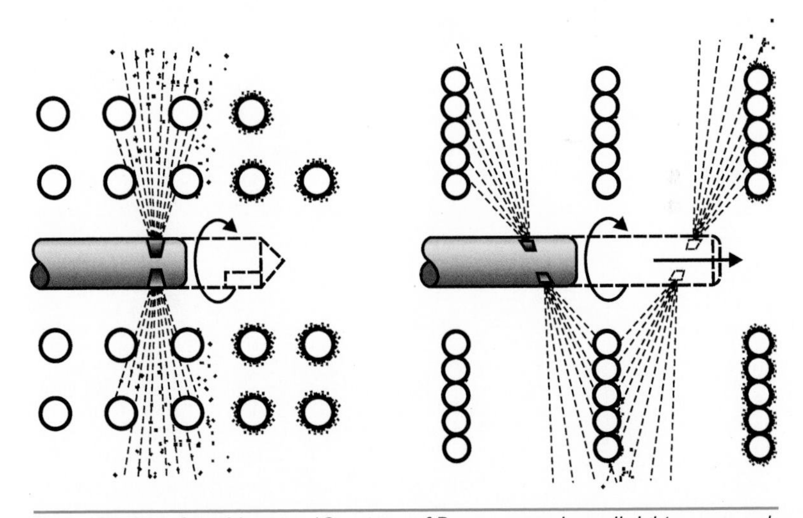

FIGURE S1.14 Soot blowers. (*Courtesy of Power magazine, all rights reserved.*

high-velocity steam is emitted. Some soot blower tubes retract from the boiler when they are not in service. During service, the soot blower is inserted into the firebox between tubes and is spun around to direct the emitted steam across the tube surfaces, in order to blast away deposits. This increases boiler efficiency and controls flue gas channeling.

More detailed information about the boiler designs and components described above, as well as many other aspects of steam systems, is provided in *The Nalco Water Handbook*, third edition.

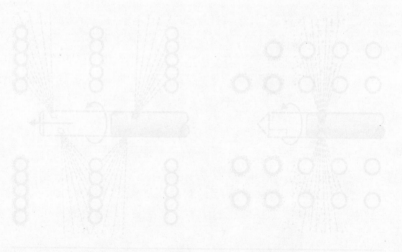

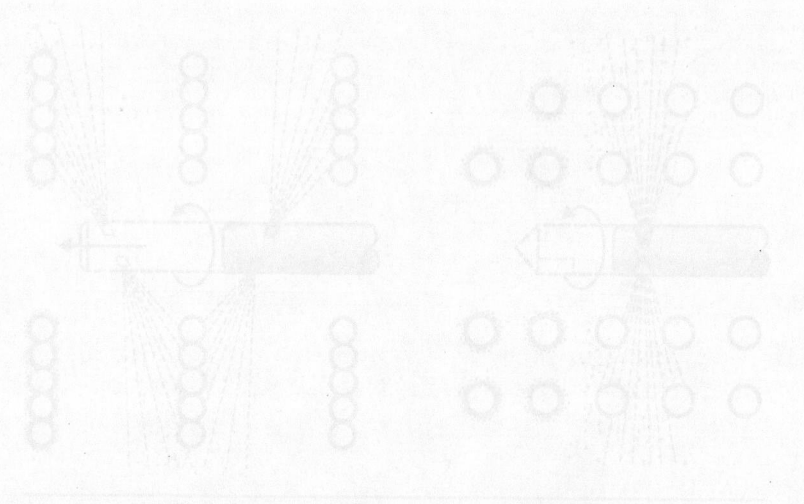

SECTION 1.1

Deposit-Related Problems

CHAPTER 1
Water-Formed and Steam-Formed Deposits

CHAPTER 1

Water-Formed and Steam-Formed Deposits

General Description

The term *deposit* generally refers to materials that originate elsewhere and are conveyed to a deposition site. In water-filled tubes, deposits may also form at a deposition site due to decreased solubility of species in boiler water solution. Deposits do not include corrosion products formed in place. Corrosion products that are formed in other locations and are transported and deposited elsewhere do qualify as deposits under this definition. Oxides formed from boiler metal are not deposits unless they have been moved from their origination sites. This distinction is fundamental.

Boiler deposits come from four sources:

- Waterborne minerals
- Compounds formed due to reactions of waterborne minerals with nonvolatile treatment chemicals
- Transported corrosion products from elsewhere in the system
- Contaminants

Deposits from these sources may interact to increase deposition rates, to produce a more tenacious layer, and to serve as nucleation sites for further deposit formation.

Boiler deposits in general may include (but are not limited to) metal oxides, copper, phosphates, calcium and magnesium compounds, carbonates, silicates, sulfates, and contaminants as well as a variety of organic and inorganic compounds. Specific common compounds that form in deposits will be described in the section entitled "Identification" in this chapter. The deposit components will depend upon the mechanism of deposition and the internal chemical treatment program used in the boiler. Boiler pressure also influences deposition. All the contributing factors are described in the following sections.

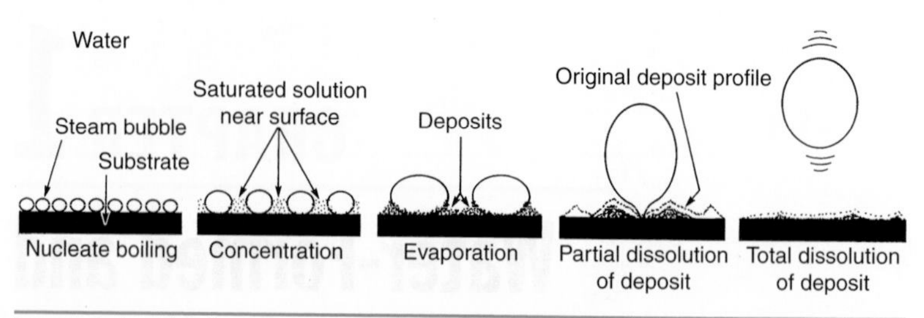

FIGURE 1.1 Five instants in the life of a steam bubble.

Deposit Mechanisms

One mechanism for deposition in steam-generating tubes involves the concentration of soluble and insoluble species in a thin film bordering the metal surface during steam-bubble formation (Fig. 1.1). The term *nucleate boiling* refers to a condition in which discrete bubbles of steam nucleate at points on a metal surface. When a steam bubble becomes dislodged from a tube wall under normal nucleate boiling conditions, soluble deposits redissolve in the water. Insoluble material will form a permanent deposit layer. Inverse-temperature solubility of certain materials, such as calcium-based compounds, leads to deposition where heat transfer is great. A deposit weight, measured in units of grams per square foot, is used to gauge the relative amount of deposition on the internal surfaces of tubes in high heat flux areas. Under normal nucleate boiling conditions, deposition rates will tend to be low, typically 5 g/ft^2 (5.4 mg/cm^2) or less per year of service. More detailed descriptions of deposit weight determination methods and interpretation will be provided in the section entitled "Critical Factors" in this chapter.

The tendency to form deposits is influenced by localized heat input, water turbulence, and water composition at or near the tube wall. The rate at which the deposit builds depends on the rate of bubble formation and the effective solubility of the deposit. In cases of high heat input relative to boiler water flow rate, a departure from nucleate boiling (DNB) condition is produced. A stable steam layer can cause concentration of even highly water-soluble material (Fig. 1.2). Deposits that form beneath stable steam layers do not redissolve, since the surface cannot be rinsed while covered with steam (Fig. 1.3). Insulation provided by deposit layers may promote or cause the production of the steam, which in turn will result in further deposition. Steam layers may also result from surface irregularities that disturb water flow, such as intruding welds or weld backing rings. Downstream of such irregularities, localized pressure drops favor steam bubble formation and consequently deposit formation.

Aside from deposition under normal nucleate boiling or even DNB conditions, other deposit mechanisms involve settling of particulate

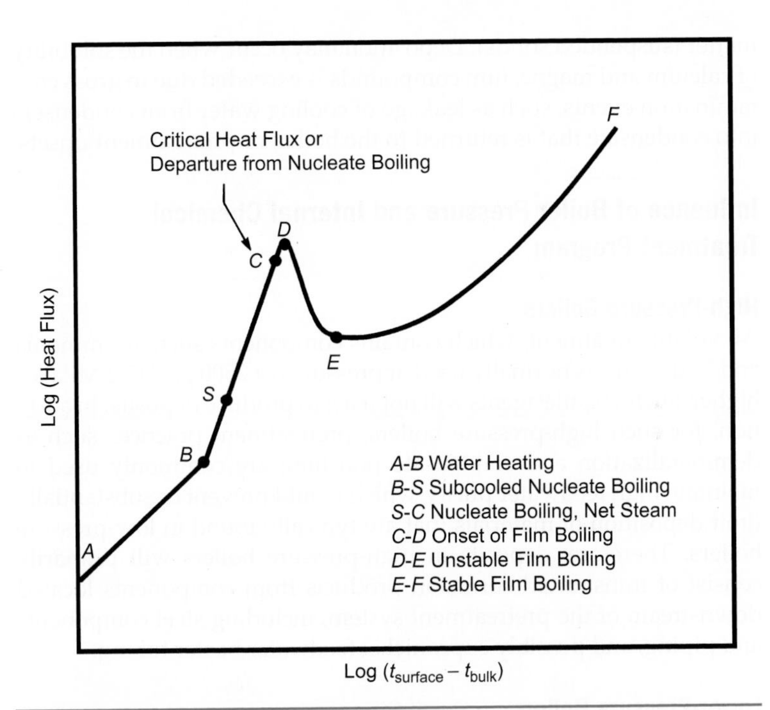

Figure 1.2 Heat transfer to water and steam in a heated flow channel. Relation of heat flux to temperature difference between channel wall and bulk water or steam temperature. (*Courtesy of The Babcock & Wilcox Company, Steam/Its Generation and Use, 38th ed., New York, 1972.*)

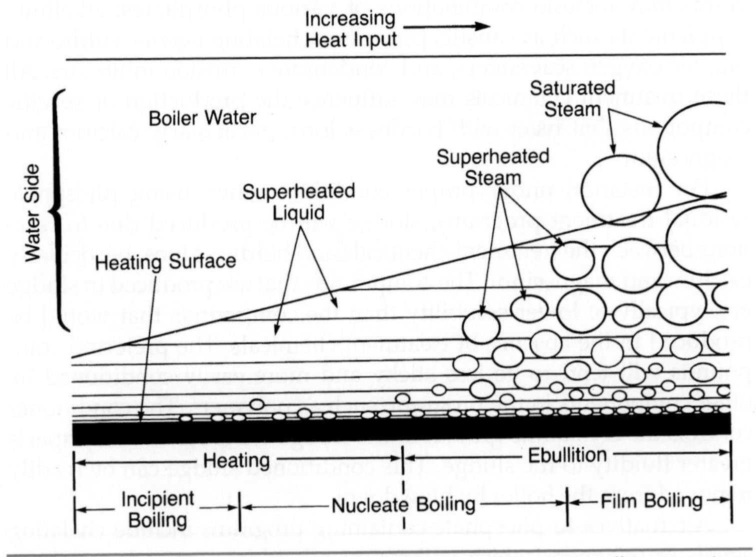

Figure 1.3 Transition from heating to boiling (ebullition) as wall temperature increases.

matter (suspended solids). Deposition may occur when the solubility of calcium and magnesium compounds is exceeded due to gross contamination events, such as leakage of cooling water from condensers into condensate that is returned to the boiler, or pretreatment upsets.

Influence of Boiler Pressure and Internal Chemical Treatment Program

High-Pressure Boilers

All volatile treatment, which contains components such as ammonia and hydrazine, is normally used at pressures of 2500 psi (17.2 MPa) or higher. Such volatile agents will not react to produce deposits. In addition, for such high-pressure boilers, pretreatment practices such as demineralization and condensate polishing are commonly used to maintain high feedwater purity, which would prevent or substantially limit deposition of materials that are typically found in low-pressure boilers. Therefore, deposits in high-pressure boilers will primarily consist of transported corrosion products from components located downstream of the pretreatment system, including steel components and piping, and possibly cupronickel feedwater heater tubing.

Lower-Pressure Boilers

In boilers below 2000 psi (13.8 MPa), pretreatment is not required to produce high-purity feedwater. Various nonvolatile chemical treatment programs may be used, and a wider variety of compounds may be found in deposits. The more common chemical treatment programs may include combinations of various phosphates; alkalinity supplements such as caustic, polymers, chelating agents, sulfite and organic oxygen scavengers; and condensate corrosion inhibitors. All these treatment chemicals may influence the production of specific compounds that react with hardness ions, particularly calcium and magnesium.

For instance, under proper conditions when using phosphate residual treatment programs, sludge will be produced due to reactions between the treatment chemicals and hardness ions, particularly calcium and magnesium. The compounds that are produced in sludge are typically of lower solubility than the compounds that would be produced in the absence of treatment chemicals. The preferred compounds will tend to be less sticky and more easily conditioned by other components in the program, such as polymers. The conditioner controls the crystalline growth of the sludge compounds and imparts greater fluidity to the sludge. This conditioned sludge can be readily removed from the boiler by blowdown.

Alternatives to phosphate-containing programs include chelating agents or polymers, which will complex hardness ions adequately in boiler water solution, thus preventing them from reacting with any other components in the boiler water. Under proper operating conditions, both sludge and complexed hardness will be removed sufficiently in the blowdown. As such, hardness ions and sludge components are reduced to levels that are low enough to limit deposition.

However, under upset conditions, certain compounds that can promote deposition may form, particularly at locations where tube metal temperatures are high. Upset conditions may occur when treatment program control is insufficient or becomes overwhelmed with contaminants or water of inappropriate quality. Under upset conditions, adherent deposits may form. Scale may deposit on surfaces at high rates during upsets, promoting under-deposit corrosion and overheating (see Chaps. 2 through 7). Typical scale deposits include calcium carbonate, calcium sulfate, and magnesium hydroxide. More in-depth descriptions of various internal chemical treatment programs for boilers are presented in *The Nalco Water Handbook*, third edition. These descriptions include the specific treatment chemicals used as well as the chemical reactions with calcium, magnesium, silica, and other dissolved solids to produce compounds that will be present in deposit layers.

Locations

Deposits can occur anywhere water or steam is present in a boiler. In economizers, where heat flux is low, steaming does not typically occur. Deposits in economizer tubes usually consist primarily of corrosion products that are transported from their upstream origination sites in the condensate and feedwater system, particularly if condensate polishing is not conducted properly (Fig. 1.4). In addition, iron

FIGURE 1.4 Thick layer of friable iron oxide that deposited in an economizer tube. Most iron was carried into the economizer from the preboiler system.

FIGURE 1.5 Fragmented iron oxide layers and tubercles on internal surface of an economizer tube. Most oxide was formed in place due to oxygen corrosion.

oxide corrosion products may form directly on surfaces in the economizer due to oxygen corrosion, resulting from exposure to feedwaters containing excessive dissolved oxygen concentration (Fig. 1.5) (see Chap. 9, Oxygen Corrosion). Such iron oxide corrosion products may break away from surfaces and produce deposits at downstream locations.

Wall and screen tubes located in high heat flux areas are usually the most heavily deposited. Roof and floor tubes often contain deposits as well. Tube orientation can influence the location and amount of deposition. Deposits are usually heaviest on the hot side of steam-generating tubes. Deposition can be substantial during steam generation under DNB conditions. Under DNB conditions, deposit accumulations are often heavier on top portions of horizontal and slanted tubes than on vertically oriented tubes. This is due to the lower critical heat flux for DNB conditions in horizontal and slanted tubes. The hottest regions of steam generator tubes, such as in areas near the bottom rear wall of boilers using chain-grate stokers, and screen tubes are particularly susceptible to deposition.

Coarse particulate matter is likely to be found in horizontal runs and where the flow velocity is low, producing conditions that promote settling. Pieces of deposited material may dislodge from steam-generating tube surfaces and collect at low points within the boiler. Such deposits are commonly called *chip scale*. Chip scale may dislodge during cleaning or during idle periods when deposits dry, thus reducing their strength. If the chip scale is not removed, it can disrupt flow to promote overheating. In addition, deposition will tend to increase in tubes that experience poor water circulation due to the design or operating practice of the boiler. The center section shown in Fig. 1.6 was removed from a tube located in the generating bank of a boiler, which had experienced substantial standby operation where boiler water circulation was poor.

Mud and steam drums may also contain deposits. Because drums are readily accessible, a visual inspection and deposit analysis can provide many details about water chemistry and deposition processes.

Superheater and reheater tubes may contain deposits that are primarily formed due to the carryover of boiler water with the steam. Carryover may occur due to foaming, caused by boiler water contamination with organic materials and/or due to high dissolved solids. Mechanical problems such as high water levels in steam drums or malfunctioning steam separation equipment may also promote carryover. Dissolved nonvolatile species in the boiler water, such as sodium, phosphate, and calcium, will boil out of solution to produce deposits. In addition, suspended deposit and corrosion product particles in carried over boiler water will produce deposits. Such deposits will usually be concentrated near the superheater inlet or in nearby pendant U bends. Contaminated attemperation water can also induce deposits, primarily in locations immediately downstream from the attemperation point.

FIGURE 1.6 Tube sections virtually plugged with deposits. The center tube contains silicates, phosphates, and other compounds; fouling occurred on standby service. The tube on the right is from a low-pressure boiler and is fouled with almost pure calcium carbonate. The section on the left is reddened by almost 20% elemental copper.

Figure 1.7 Severely corroded buckets from high-pressure condensing stage of a turbine. Deposits were removed to reveal attack.

In high-pressure boilers, volatile silicates and copper can be carried through superheaters and deposited on turbine components. If chloride- or sulfate-containing compounds are present in turbine deposits, hydration can cause severe corrosion due to hydrolysis (Figs. 1.7 and 1.8). More in-depth descriptions of such corrosion are given in Chap. 16, Steam and Condensate System Damage.

Figure 1.8 Deposits on turbine bucket before cleaning, as in Fig. 1.7.

Critical Factors

The rate at which deposits form on heat-transfer surfaces is controlled mainly by the solubility and physical tenacity of the deposit and by the amount of water washing that occurs where steam is generated. Solubility, tenacity, and water washing, in turn, depend on other factors such as dissolved-solids concentration, pH, alkalinity, chemical treatment residuals, temperature, agglomeration morphology, and boiler water flow rate.

However, the predominant factor for significant deposit formation is usually steam formation. In fact, deposits can form even when steaming is slight (Fig. 1.6). As long as nucleate boiling is occurring, heat transfer is controlled by tube-wall and deposit thermal conductivities and by the gas-side temperature. The thermal conductivities of several common metals, alloys, and boiler deposits are shown in Table 1.1. Note that the thermal conductivities vary with temperature.

Alloys	Thermal Conductivity	
	W/(m · K)	Btu/(h · ft · °F)
Carbon steel (SA-178)	46.0	26.6
Alloy steel (SA213-T11)	40.0	23.1
304 Stainless steel	14.0	8.1
Copper	415.1	240.0
Aluminum	235.1	136.0
Deposits		
Aluminum oxide, fused (Al_2O_3)	3.6	2.1
Analcite ($Na_2O \cdot Al_2O_3 \cdot 4SiO_3 \cdot 2H_2O$)	1.4	0.8
Calcium carbonate ($CaCO_3$)	0.9	0.5
Calcium phosphate [$Ca_3(PO_4)$]	3.6	2.1
Calcium sulfate ($CaSO_4$)	1.3	0.8
Basic magnesium silicate, lizardite, serpentine [$Mg_3Si_2O_5(OH)_4$]	1.0	0.6
Ferric oxide, hematite (Fe_2O_3)	0.6	0.4
Magnesium oxide (MgO)	1.2	0.7
Magnesium phosphate [$Mg_3(PO_4)_2$]	2.1	1.2
Magnetite (Fe_3O_4)	2.9	1.7
Silica (SiO_2)	1.6	0.9

All thermal conductivities are at room temperature.
Note: Btu/(h · ft · °F) = 1.73 W/(m · K)

TABLE 1.1 Thermal Conductivities of Alloys and Deposits

In addition, increased porosity in a deposit layer will decrease the thermal conductivity.

Unlike soluble salts such as caustic which increase in solubility as a function of increasing water temperature, certain salts in boiler water solution have inverse-temperature solubility. As such, they are particularly prone to deposit on heat-transfer surfaces as temperatures increase above a certain level. Calcium sulfate, calcium carbonate, and calcium phosphate have inverse-temperature solubility and may deposit preferentially in hotter areas. If DNB occurs, evaporation to dryness causes deposition of species having normal-temperature solubility, such as sodium and potassium salts as well as chlorides. Frequently, the deposits of the most insoluble materials are found in water-cooled tubes having the highest heat flux. When evaporation to dryness occurs, both soluble and insoluble deposits are usually found together.

Depending upon the pressure of the boiler, even a relatively small amount of deposit can cause wall temperatures to rise considerably in high heat-transfer locations. As wall temperature rises, the tendency for DNB increases (Fig. 1.3). The presence of a stable steam layer decreases the heat flow, potentially causing under-deposit corrosion and/or overheating that can result in tube failures (Fig. 1.9) (see Chap. 3, Long-Term Overheating; Chap. 4, Caustic Corrosion; Chap. 5, Low pH Corrosion during Service; Chap. 6, Hydrogen Damage; and Chap. 7, Phosphate Corrosion).

FIGURE 1.9 Thick deposit layers on hot side of wall tube from 600-psi (4.1-MPa) boiler. Note the bulge due to long-term overheating.

Pressure (psi)	Pressure (MPa)	Silica Range (ppm)	Total Hardness	Maximum (ppm)		
				Oxygen	Iron*	Copper*
100	0.69	15–25	75.00	–	–	–
200	1.38	10–20	20.00	–	–	–
300	2.07	7.5–15	2.00	–	–	–
500	3.45	2.5–5.0	2.00	0.030	–	–
600	4.14	1.3–2.5	0.20	0.030	–	–
750	5.17	1.3–2.5	0.10	0.030	0.050	0.020
900	6.21	0.8–1.5	0.05	0.007	0.020	0.015
1000	6.89	0.2–0.3	0.05	0.007	0.020	0.015
1500	10.34	0.3 max.	0.00	0.005	0.010	0.010
2000	13.80	0.1 max.	0.00	0.005	0.010	0.010
2500	17.24	0.05 max.	0.00	0.003	0.003	0.002
3200+	22.06+	0.02 max.	0.00	0.002	0.002	0.001

*In modern industrial boilers, which have extremely high rates of heat transfer, these concentrations should be essentially zero. Similarly, total hardness should not exceed 0.3 ppm $CaCO_3$, even at the lower pressures; suspended solids in the feedwater should be zero if possible.

Note: 1 psi = 0.006895 MPa

Source: Courtesy of Chemical Publishing Company, James W. McCoy, *The Chemical Treatment of Boiler Water*, New York, 1981.

TABLE 1.2 Recommended Feedwater Quality Limits

Water quality also has a significant influence on deposition. Suggested acceptable feedwater quality guidelines are shown as a function of boiler pressure in Table 1.2. This table indicates that fewer contaminants, particularly silica and hardness, can be tolerated at high boiler pressures. This is mainly due to steam purity requirements. In addition, the insulating effects of deposits become less tolerable as pressures rise, since overheating and under-deposit corrosion may occur more readily. Maximum acceptable concentrations of boiler water total dissolved solids and some individual species are shown in Table 1.3.

One of the key factors in the table is the recommended dissolved-solids concentration, which decreases by a factor of 100 when pressure rises from 100 psi (0.69 MPa) to 2000 psi (13.8 MPa). In addition, allowable silica levels decrease by a factor of 250, and suspended solids decrease by a factor of 500.

Pressure (psi)	Pressure (MPa)	Steam Saturation Temperature [°F (°C)]	Maximum (ppm)						Range (ppm)		
			Dissolved Solids	Suspended Solids*	Total Alkalinity†	Silica	Sludge Conditioners		Residual Phosphate	Residual Sulfite	Residual Hydrazine
							Natural	Synthetic			
100	0.69	328 (164)	5000	500	900	250	150	15	NR‡	90–100	NR
200	1.38	382 (194)	4000	350	800	200	150	15	40–50	80–90	NR
300	2.07	417 (214)	3500	300	700	175	100	15	30–40	60–70	NR
500	3.45	467 (241)	3000	60	600	40	70	15	25–30	45–60	NR
600	4.14	486 (252)	2500	50	500	35	70	10	20–25	30–45	NR
750	5.17	510 (266)	2000	40	300	30	NR	10	15–20	25–30	NR
900	6.21	532 (278)	1000	20	200	20	NR	5	10–15	15–20	0.10–0.15
1000	6.89	545 (285)	500	10	50	10	NR	3	5–10	NR	0.10–0.15
1500	10.34	596 (313)	150	3	0	3	NR	NR	3–6	NR	0.05–0.10
2000	13.80	636 (336)	50	1	0	1	NR	NR	1–3	NR	0.05–0.10
2500	17.24	668 (353)	10	0	0	0.5	NR	NR	NR	NR	0.02–0.03
3200	22.06	705 (374)	0.02	0	0	0.02	NR	NR	NR	NR	0.01–0.02

*Guidelines for pressures from 100 to 900 psi apply to conventional field-erected boilers with moderate rates of heat transfer, say 50,000 Btu/(ft^2 · h). At high rates characteristic of package boilers, large amounts of insoluble material cannot be managed effectively by a dispersant presently available.

†Zero alkalinity refers to hydroxide ion, i.e., P and M alkalinities (2P-M). There is always some alkalinity produced by ammonia, hydrazine, morpholine, or other bases.

‡NR = not recommended.

Source: James W. McCoy, *The Chemical Treatment of Boiler Water*, Chemical Publishing Company, New York, 1981.

Table 1.3 Recommended Concentrations of Boiler Salines

Effects of Deposition on Tube Metal Temperature

Online monitoring methods to assess the influence of deposition may include the use of thermocouples. Thermocouples may be used to measure tube-wall temperatures in high heat flux areas, where internal deposition is expected to be most significant, such as in furnace wall tubes adjacent to the combustion zone. Figure 1.10 shows tube-wall temperatures that were measured on walls of tubes in both clean and heavily deposited conditions. Tube-wall temperatures on the hot side of the heavily deposited tube were excessive and in an overheated condition. In addition, under-deposit corrosion may occur at such elevated metal temperatures. In general, waterside deposition is considered to be significant when tube metal temperatures exceed about 100°F (38°C) of an established normal tube-wall temperature.

Monitoring by Deposit Weight Determination

Another monitoring method that can be used to indicate conditions that are conducive to overheating or under-deposit corrosion is a deposit weight determination. Tube sections are removed on a periodic basis to measure and trend deposit accumulation. Deposit weight determinations are conducted on tube sections that are removed from radiant heated zones in the boiler. Typically, deposit weight guidelines are applicable to tubes removed from radiant zones. The deposit and iron oxide layers are removed from the internal surface using appropriate cleaning methods. Mechanical removal and chemical cleaning methods have been used in the past (see ASTM D3483), but a well-accepted National Association of Corrosion Engineers (NACE) approved method employs the use of glass bead blasting for material removal. The most common unit of measurement is grams. The surface area where the deposits were removed is measured, and a deposit weight is calculated by dividing the mass of the material removed by the surface area. The most common deposit weight unit is grams per square foot (g/ft^2), but the closely matching metric unit, which is milligrams per square centimeter (mg/cm^2), is also used (1.07 mg/cm^2 = 1.0 g/ft^2). Deposit weight measurements can vary depending upon the sample location, deposit thickness variations within sampled areas, material removal methods, and surface roughness.

Deposit weight guidelines published in the literature and by the manufacturer of the boiler from which the section was removed, are used to gauge the cleanliness of the tube. Based upon the pressure (see Table 1.4), the guidelines indicate deposit weight ranges where chemical cleaning is recommended. When deposit weights measure below the lower end of the range, the tube is considered to be clean. In some boilers, porous magnetite layers up to 10 g/ft^2 (11 mg/cm^2)

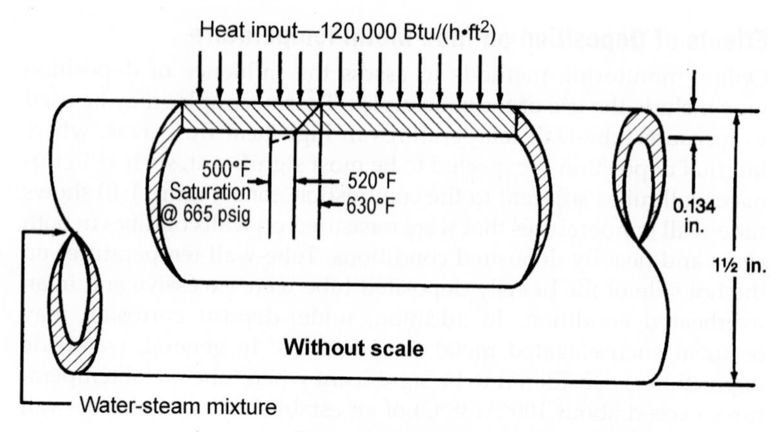

Temp. drop across water film = 20°F
Temp. drop across tube wall = 110°F

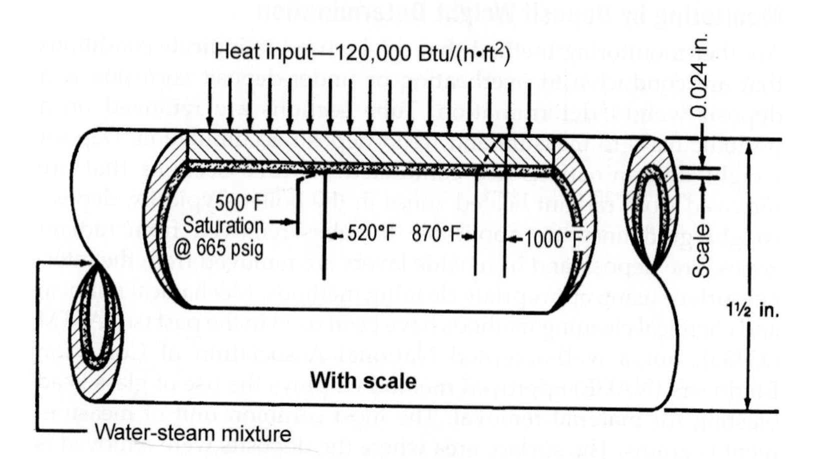

Temp. drop across water film = 20°F
Temp. drop across internal scale = 350°F
Temp. drop across tube wall = 130°F

Assumed $CaSO_4$ scale (0.024 in. thick)
Thermal conductivity 0.8 Btu/(h•ft•°F)

Maximum tube temp. (1000°F) is above allowable
Oxidation temp. limit for SA-210 carbon steel

FIGURE 1.10 Diagrams of tube-wall temperatures in the absence of internal deposits and with heavy internal deposits. (*Courtesy of McGraw-Hill, The Nalco Water Handbook, 3d ed., New York, 2009.*)

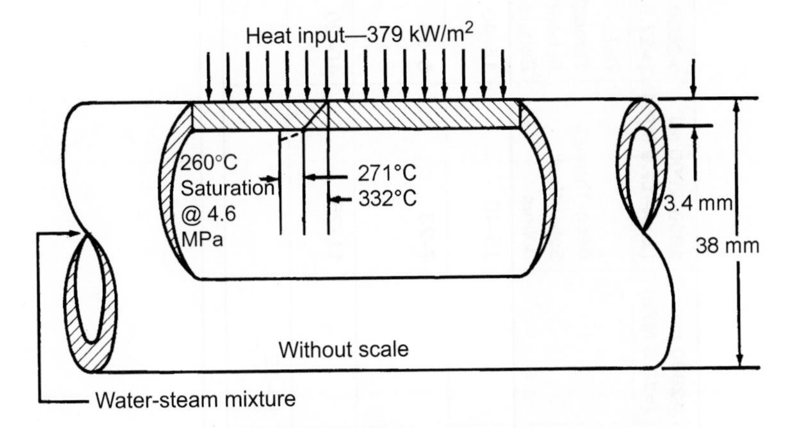

Temp. drop across water film = 11°C
Temp. drop across tube wall = 61°C

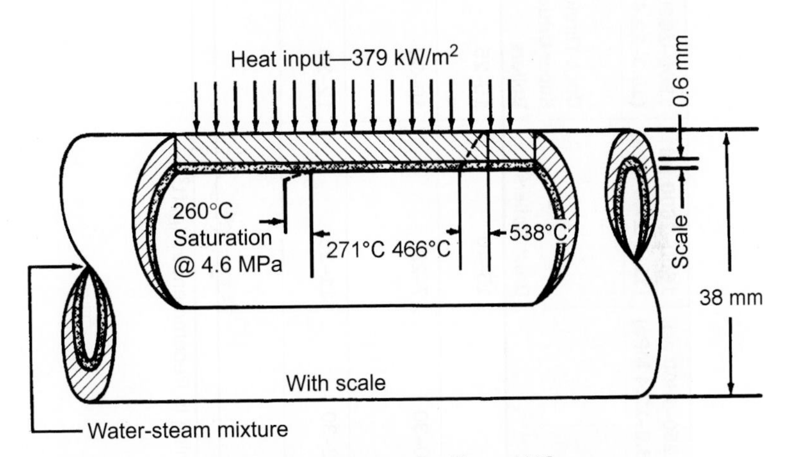

Temp. drop across water film = 11°C
Temp. drop across internal scale = 195°C
Temp. drop across tube wall = 72°C

Assumed $CaSO_4$ scale 0.61 mm
Thermal conductivity 1.44 W/(m•K)

Maximum tube temp. (538°C) is above allowable
Oxidation temp. limit for SA-210 carbon steel

FIGURE 1.10 (*Continued*)

Reference	<1000 psi (>6.9 MPa)	1000–2000 psi (6.9–13.8 MPa)	1250–1800 psi (8.6–12.4 MPa)	1800–2600 psi (12.4–17.9 MPa)	1500–1800 psi (10.3–12.4 MPa)	>2000 psi (>13.8 MPa)	1800–2600 psi (12.4–17.9 MPa)	>2600 psi (>17.9 MPa)
				Drum Boilers	**Once-Through Super-Critical Boilers**		**Once–Through Subcritical Boilers**	**Once-Through Subcritical Boilers**
Combustion engineering[1]				15–40	15–25		15–40	15–40
EPRI mechanical method[1]			10–30	7–23	10–30		7–23	5–15
EPRI solvent method[1]			12–30	11–25	12–30		11–25	10–20
Babcock & Wilcox[2]	20–40	12–20				10–12		

TABLE 1.4 Deposit Weight Ranges where Cleaning Is Recommended (g/ft^2)

produce no significant impairment of heat transfer. Within the range of the guidelines, cleaning is recommended as soon as possible, depending upon proximity to the upper threshold. At and above the upper end of the range, cleaning is necessary and should not be delayed. Heat transfer is severely reduced when deposit loading becomes excessive. Prolonged operation above the maximum deposit loadings may produce serious corrosion and overheating failures. However, the weight of deposits alone does not always accurately indicate the tendency to overheat. Deposit composition and morphology also influence heat transfer. Chemical cleaning is the preferred method of deposit and corrosion product removal in boilers, and it is described in greater detail in Chap. 10, Corrosion during Cleaning.

Identification

As stated earlier, boiler treatment chemicals, bulk feedwater and boiler water composition, and heat input affect deposition. The composition of material that forms in deposit layers may be identified through use of proper analytical testing techniques including x-ray fluorescence or scanning electron microscopy (SEM) for elements. X-ray diffraction may be used for inorganic compounds. Infrared analysis and carbon, hydrogen, and nitrogen analyses may be used for organic compounds. Individual layers within stratified deposits may be characterized by SEM using element mapping. Common boiler deposit compounds, their likely locations and characteristics, and their probable causes can be found in Table 1.5. Deposited material may be difficult to remove using mechanical scraping. Vibratooling or mechanical distortion of the tube wall may be helpful for removing tenacious material. Cross-contamination from other surfaces must be avoided.

Iron Oxides

A smooth, black, tenacious, dense, protective magnetite (Fe_3O_4) layer naturally grows on steel under reducing conditions found on boiler waterside surfaces (Fig. 1.11). Magnetite forms by direct reaction of water or steam with the tube metal. Particulate iron oxide can deposit on top of the magnetite layer if settling rates are high and/or if steaming is appreciable (Fig. 1.12). Usually, coarse, particulate magnetite will not tenaciously adhere to surfaces unless intermixed with other deposits.

Exfoliation refers to the loss of patches of thermally formed magnetite layers from surfaces of high-temperature components such as superheater and reheater tubes. Thermal stresses and multiple iron

Mineral	Formula	Nature of Deposit	Usual Location and Form	Cause of Formation
Calcium carbonate (aragonite)	$CaCO_3$	Hard, adherent	Tube scale, feed lines, sludge	Low or zero treatment chemical residuals. May be due to overwhelming hardness upset in the pretreatment system or inleakage of cooling water from condenser into returned condensate
Calcium silicate (xonotlite)	$5CaO \cdot 5SiO_2 \cdot H_2O$	Hard, adherent	Tube scale	Low or zero treatment chemical residuals, low alkalinity
Calcium phosphate (whitlockite)	$Ca_3(PO_4)_2$	Sludge	Tube scale and sludge	High phosphate concentration and/or low alkalinity
Basic calcium phosphate (hydroxyapatite)	$Ca_{10}(OH)_2\,(PO_4)_6$	Flocculant	Mud drum, waterwalls, sludge	Formed when treatment residuals are under proper control. If excessive, may need to increase sludge removal
Calcium sulfate (gypsum)	$CaSO_4 \cdot 2H_2O$	Hard, adherent	Tube scale, generating tubes	Low or zero treatment chemical residuals. May be due to overwhelming hardness upset in the pretreatment system or inleakage of cooling water from condenser into returned condensate
Calcium hydroxide	$Ca(OH)_2$	Hard, adherent	Tube scale, generating tubes	Low or zero treatment chemical residuals. May be due to overwhelming hardness upset in the pretreatment system or inleakage of cooling water from condenser into returned condensate

Magnesium hydroxide (brucite)	$Mg(OH)_2$	Flocculant	Sludge in mud drum and waterwall headers	Sludge that forms when there is insufficient silica concentration to form lizardite (serpentine)
Basic magnesium phosphate	$Mg_3(PO_4)_2$ $Mg(OH)_2$	Adherent binder	Tubes, mud drum, waterwalls	Low silica concentration, low alkalinity, excessive phosphate concentration
Magnesium phosphate	$Mg_3(PO_4)_6$	Adherent binder	Tubes, mud drum, waterwalls	Sticky sludge. Excessive phosphate and/or low alkalinity, low silica
Basic magnesium silicate (serpentine, lizardite)	$3MgO \cdot 2SiO_2 \cdot H_2O$	Flocculant	Sludge	Formed when treatment residuals are under proper control. If excessive, may need to increase sludge removal
Metallic copper	Cu	Metallic particles or layers	Boiler tubes and turbine blades	Transported corrosion products from copper-containing alloys in the water circuit such as condenser tubes and feedwater heater tubes. Plate from dissolved copper ions, typically in areas of corrosion
Cuprite	Cu_2O	Adherent layer	Turbine blades, boiler deposits	Transported corrosion products from copper-containing alloys in the preboiler such as condenser tubes and feedwater heater tubes
Iron oxide (hematite)	Fe_2O_3	Binder	Throughout boiler	Corrosion products formed due to oxygen corrosion. May be transported to boiler from sources upstream in the preboiler

TABLE 1.5 Components of Water-Formed Deposits (*Continued*)

Mineral	Formula	Nature of Deposit	Usual Location and Form	Cause of Formation
Iron oxide (magnetite)	Fe_3O_4	Protective film	All internal surfaces	Uniform, protective corrosion product formed on bare steel when treatment residuals are under proper control. Can be in the form of needles due to corrosion
Analcite	$Na_2O \cdot Al_2O_3 \cdot 4SiO_2 \cdot 2H_2O$	Hard, adherent	Tube scale under hydroxyapatite or serpentine	DNB condition. Aluminum in boiler water, low alkalinity, high silica. Indicates probable flame impingement or boiler water starvation
Silica	SiO_2	Hard, adherent	Turbine blades, mud drum, tube scale	Precipitates due to excessive silica concentration and low alkalinity in the boiler. Recommend 3:1 alkalinity-to-silica ratio. Silica deposits on turbine components result from excessive vaporous carryover. Silica should be limited to 20 μg/L in the steam
Acmite	$Na_2O \cdot Fe_2O_3 \cdot 4SiO_2$	Hard, adherent	Tube scale under hydroxyapatite or serpentine	Evaporation to dryness under DNB conditions. Also forms due to low alkalinity and high silica content. Combines with iron oxide corrosion products
Noselite	$3Na_2O \cdot 3Al_2O_3 \cdot 6SiO_2 \cdot Na_2SO_4$	Hard, adherent	Tube scale	DNB conditions. Aluminum in boiler water, low alkalinity, high silica. Indicates probable flame impingement or boiler water starvation

Pectolite	$Na_2O \cdot 4CaO \cdot 6SiO_2 \cdot H_2O$	Hard, adherent	Tube scale	DNB conditions. Low alkalinity, high silica. Indicates probable flame impingement or boiler water starvation
Sodalite	$3Na_2O \cdot 3Al_2O_3 \cdot 6SiO_2 \cdot 2NaCl$	Hard, adherent	Tube scale	DNB conditions. Aluminum in boiler water, low alkalinity, high silica. Indicates probable flame impingement or boiler water starvation
High loss on ignition (LOI)	Carbon, other organic compounds	Waxy, oily, graphite	Contaminant	Indicates oil or other organic contaminants. The tolerable concentration depends upon the pressure

Source: Courtesy of Chemical Publishing Company, James W. McCoy, *The Chemical Treatment of Boiler Water*, New York, 1981; J. N. Tanis, *Procedures of Industrial Water Treatment*, Ltan Inc., Ridgefield, Conn., 1987.

Table 1.5 Components of Water-Formed Deposits (*Continued*)

FIGURE 1.11 Smooth, tenacious, black, magnetite layer on a well-protected boiler tube.

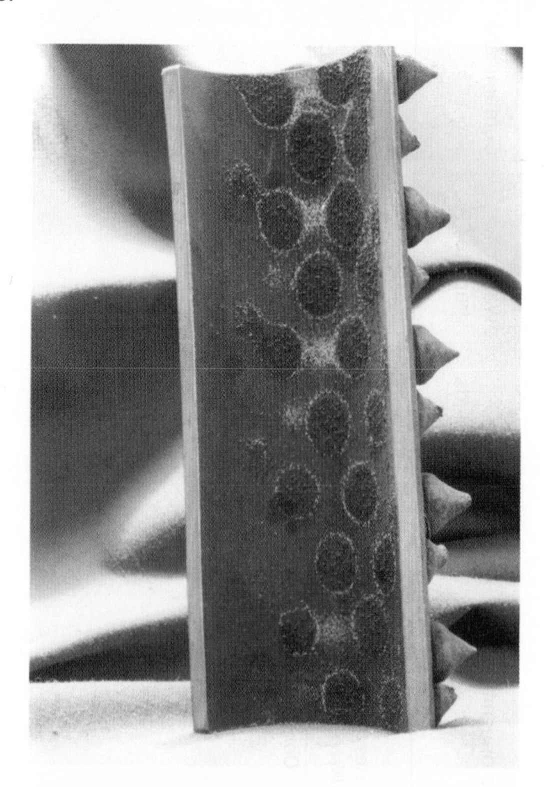

FIGURE 1.12 Spots of particulate iron oxide deposit mirroring stud locations on a wall tube of a recovery boiler. The studs, if exposed, concentrate heat, causing higher rates of steam generation and deposition. Note the white deposits of sodium hydroxide encircling each powdery magnetite spot.

FIGURE 1.13 Exfoliated magnetite patches in a tube from the primary superheater of a utility boiler. Magnetite chips can be carried into turbines and cause severe damage.

oxide layers promote their exfoliation from surfaces. Exfoliation may cause accumulation of iron oxide chips in U bends and long horizontal runs (Figs. 1.13 and 1.14).

Exfoliated iron oxide particles can be carried through the superheater into turbines to promote erosion damage.

FIGURE 1.14 Fine-mesh turbine inlet screen. Pieces of exfoliated oxide from superheater tubing are wedged into screen openings.

Figure 1.15 Sparkling black magnetite crystals on internal surface of a wall tube. Crystals are formed by precipitation on surfaces or are transported from preboiler areas.

Sparkling, black, highly crystalline magnetite needles will often be present near caustic corrosion sites. Crystalline magnetite needles form from concentrated iron solutions. Magnetite crystals can sometimes produce a sparkling surface coating on steam drums and tubes (Figs. 1.15 and 1.16). Scraping small amounts of this material from surfaces and exposing these particles to a magnet usually indicate whether iron oxide is present.

Figure 1.16 Patch of sparkling magnetite crystals at caustic corrosion site. Such crystals are often found near sites of caustic corrosion.

Hematite (Fe_2O_3) deposit formation, on the other hand, is favored at somewhat lower temperatures and higher oxygen concentrations. Hematite is a binder species that tends to accumulate and hold other materials in the deposit. Hematite can be red if formed where oxygen concentrations are high. Hematite usually is present in economizers in conjunction with oxygen corrosion.

Other Metals and Their Oxides

Metallic copper is deposited either by plating from dissolved copper ions in the water onto steel surfaces and/or by reduction of copper oxide by hydrogen evolved during corrosion. It is common to see large, reddish stains of metallic copper intermixed with corrosion products such as magnetite and hematite near caustic and acid corrosion sites due to the hydrogen generation associated with attack (Fig. 1.17). The reddish color superficially resembles hematite. Metallic copper can be easily discriminated from other material by a silver nitrate test. A single drop of silver nitrate will precipitate white silver crystals almost immediately if elemental copper is present. Galvanic corrosion of carbon steels associated with metallic copper deposits is extremely rare in well-passivated boilers. In particular, when electrically insulating deposits separate the copper from the steel surface, no coupling of dissimilar metals will exist to promote galvanic corrosion. Copper oxide formed under boiler conditions is black and nonmagnetic.

FIGURE 1.17 Metallic copper on wall tube from a high-pressure utility boiler.

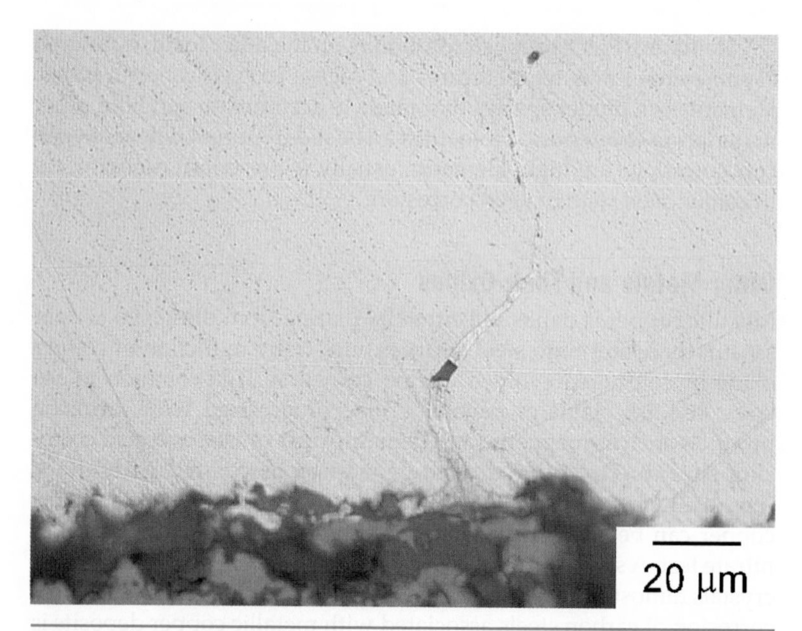

FIGURE 1.18 Metallic copper penetrating along grain boundaries due to liquid metal embrittlement caused by melting of copper deposits during welding.

Melting of metallic copper deposits during weld repairs may cause cracking of steel components by liquid metal embrittlement (Fig. 1.18).

Zinc and nickel oxides will sometimes be found in conjunction with copper deposits. This is to be expected, since zinc and nickel are often present in brass and cupronickel alloys used in condensers and feedwater heaters. Usually, these elements are deposited at lower concentrations than copper. Nickel oxide has been indicated as a "binder" compound promoting tenacious deposition. Nickel and zinc, though present at low concentrations, can be detected by chemical analysis.

Salts

Some compounds form due to reaction of dissolved hardness salts with proper concentrations of treatment chemicals. If these compounds are adequately removed by blowdown, deposition may be adequately controlled. Other deposits may form rapidly due to insufficient concentration or improper control of treatment chemicals. The least soluble compounds deposit first when boiling occurs. In particular, inversely soluble salts are most susceptible. If chemical treatment residuals are insufficient or become overwhelmed due to hardness upset conditions, calcium carbonate may form a white, friable deposit (Fig. 1.19). Calcium sulfate, which also may form due to insufficient

FIGURE 1.19 Calcium carbonate deposit layers.

treatment residual, requires a higher degree of concentration to deposit than calcium carbonate.

Magnesium phosphate is a sticky binding material that can produce very hard, adherent deposits. These deposits form due to low alkalinity or low silica content, as well as excessive phosphate concentration in the boiler water. Insoluble silicate deposits may also form. Many silicates are very hard and are almost insoluble in most acids used for chemical cleaning, except for hydrofluoric acid. Complex silicates such as analcite ($Na_2O \cdot Al_2O_3 \cdot 4SiO_2 \cdot 2H_2O$), acmite ($Na_2O \cdot Fe_2O_3 \cdot 4SiO_2$), or sodalite ($Na_2O \cdot 3Al_2O_3 \cdot 4SiO_2 \cdot 2NaCl$) may form due to DNB conditions, since intermediate sodium compounds are soluble. Analcite may form if alum is carried over from clarifiers (Fig. 1.20).

In general, water-soluble deposits can be retained only if localized concentration mechanisms are severe. Therefore, the presence of species such as sodium hydroxide, sodium phosphate, and sodium sulfate should be considered proof of evaporation to dryness.

Hydrolyzable salts such as $MgCl_2$ can gain entrance to boiler water through condensate that is returned to the boiler contaminated with cooling water from failed condenser tubes. Chlorides will concentrate beneath porous deposits and hydrolyze to form hydrochloric acid. Decomposition of oil, grease, and other organic contaminants can also produce organic acids. More detailed information about acid corrosion during service is described in Chap. 5, Low pH Corrosion during Service.

FIGURE 1.20 Analcite formed by carryover of alum from clarifiers. Note how the deposit has spalled from the bulge.

Deposit Morphology

Much of the preceding discussion involves the identification of deposits based only on individual chemical compositions. However, most real boiler deposits contain many intermixed compounds. The deposit morphology can influence the species that are found and also gives clues about the heat input, water chemistry, and corrosion mechanisms during the period in which they were formed.

Insoluble, permeable deposits such as coarse particulate matter are harmful. They permit the concentration of more soluble species in their pores. Hence, sodium hydroxide may be found intermixed with porous, insoluble species. Deposition of the porous species often predates the concentration of the more soluble species.

In general, the older the deposit material, the harder it is and the more tenaciously it adheres to tube walls. As deposits age, they fill their interstices with solid material, resulting in increased bulk, density, and hardness. There are exceptions to this rule, and these judgments are naturally somewhat subjective. Nevertheless, the presence of hard, tenacious deposits often indicates that deposition has been occurring for a long time and/or they formed at high metal temperatures.

Deposit stratification indicates changes in water chemistry, heat input, etc. It is often possible to identify deposits formed during different water treatment programs through careful inspection and analysis of stratified scales. Metallographic and scanning electron microscopy may be used to more clearly identify individual deposit and corrosion product layers. The order in which the deposits appear usually can provide a chronological deposition history.

Elimination

All deposits are undesirable and may become problematic due to changes from appropriate water chemistry or proper boiler operation. Proper water treatment can minimize deposition rates. Water chemistry upsets and operation changes should be minimized.

Deposition can be avoided by operating at or slightly below design loads and ensuring that all boiler components are functioning properly. The most important boiler operating characteristic influencing deposition is firing practice. Firing practice may cause excessive heat input in places or disrupt boiler water circulation. Also, elimination of hot spots, correct monitoring of water levels and flow, and maintenance of a constant load are necessary to avoid deposition. In addition, correct burner position, well-considered fuel adjustments, and appropriate blowdown practices contribute to reduced deposition.

Excessive deposition should be removed through suitable cleaning practice. Cleaning may be conducted on a periodic basis, depending upon the pressure of the boiler. Chemical cleaning is discussed in Chap. 10, Corrosion during Cleaning.

Cautions

Deposits rarely contain only a single compound. Chemical analysis is often necessary to determine the amount and variety of each chemical species. Soluble species such as sodium compounds will be washed away when the boiler cools and steam generation ceases. Washing sometimes results in laboratory analysis that does not accurately reflect in-service compositions. It is usually safe to assume that concentrations of highly soluble species are underreported in laboratory analysis. The presence of any highly soluble material strongly indicates that evaporation to dryness occurred.

The amount, composition, and stratification of deposits are often altered near a rupture site. Escaping fluids may remove deposits near the rupture. Occasionally, fireside combustion products may find their way onto internal surfaces near the rupture. Deposit weight determinations near bulged or ruptured surfaces may be lower than actual in-service values due to spalling of layers. Corrosion products can be confused with deposits. This is especially true in superheaters suffering idle-time oxygen corrosion, where tubercular growth is often confused with carryover of boiler water solids (see Chap. 9, Oxygen Corrosion). Metallographic examination may be used to distinguish between deposited materials and formed in place corrosion product and deposit layers.

Related Problems

See also Chap. 2, Short-Term Overheating; Chap. 3, Long-Term Overheating; Chap. 4, Caustic Corrosion; Chap. 5, Low pH Corrosion during Service; Chap. 9, Oxygen Corrosion; and Chap. 10, Corrosion during Cleaning.

Case History 1.1

Industry:	Pulp and paper
Specimen location:	Economizer tube from recovery boiler near upper header
Orientation of specimen:	Curved
Years in service:	10
Water treatment program:	Chelating agent
Drum pressure:	600 psi (4.1 MPa); feedwater pressure, 820 psi (5.7 MPa)
Tube specifications:	2-in. (5.1-cm) outer diameter
Fuel:	Black liquor

During removal of economizer tubes, heavy internal surface deposits were found unexpectedly. Internal surfaces of some economizer tubes were partially lined with an irregular layer of soft, flaky iron oxide. Underlying oxides were black. A 1-in.- (2.5-cm-) thick layer of friable material that accumulated at the tube bend is shown in Fig. 1.21. Close observation revealed that the accumulated material consisted of anthracite particles and resin beads (Fig. 1.22). The deposit had been transported from a disintegrated resin bed from the pretreatment system.

The accumulated deposit retarded coolant flow, though overheating could not occur because flue gas temperatures were too low (about 550°F, or 288°C).

FIGURE 1.21 Anthracite and resin beads beneath stratified layers of iron oxide deposits.

FIGURE 1.22 Anthracite and resin beads.

Departure from nucleate boiling beneath the deposit led to a concentration of sodium hydroxide, resulting in shallow corrosion (see Chap. 4, Caustic Corrosion).

It is extraordinary that this deposit did not cause failure. The resin-bed disintegration was not discovered until the economizer deposits were found. Although these deposits must have been carried into other parts of the boiler, no failures were traced to their presence.

Case History 1.2

Industry:	Pulp and paper
Specimen location:	Superheater outlet
Orientation of specimen:	Vertical (stub end down)
Years in service:	3½
Water treatment program:	Phosphate residual
Drum pressure:	900 psi (6.2 MPa)
Tube specifications:	2-in. (5.1-cm) outer diameter, stub end, low-alloy steel (1½% Cr)
Fuel:	Coal

A 2-ft section of superheater tubing with a stub end was plugged solid with deposits that were still soaked with water when the tube was cut. Sections were gently heated for 3 days to dry the deposit in situ. After drying, white deposits became visible at the end of the stub, which was the lowest point of the tube (Fig. 1.23).

The tube contained about 2 lb (0.9 kg) of deposit for every 2 in. (5.1 cm) of its length. Away from the stub end, the deposit contained about 80% magnetite, 7% sodium hydroxide, and 7% sodium carbonate by weight. Small concentrations of sulfur, chlorine, phosphorus, chromium, and manganese were also detected. At the stub end, about one-half of the material consisted of sodium hydroxide and sodium carbonate.

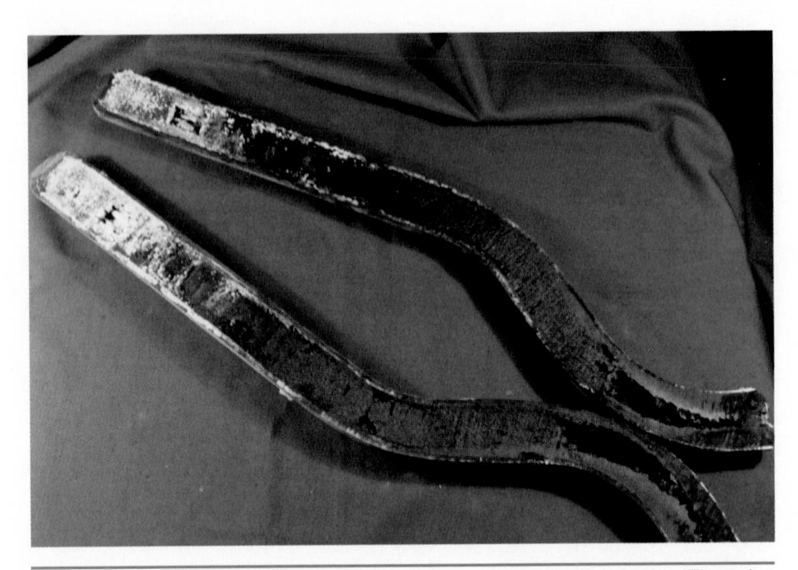

FIGURE 1.23 Longitudinally cut stub end from a superheater outlet. The tube is plugged with deposits. The white material is sodium hydroxide and sodium carbonate concentrated at the bottom of the stub end. The black material is mostly magnetite.

Deposits were caused by chronic carryover of boiler water into the superheater header from the steam drum. Water was converted to steam and left solids behind. Solids accumulated over a long period of time in the stagnant stub ends.

Cleaning stub ends is a very difficult task. However, cleaning or removal of the ends is necessary to prevent caustic corrosion and overheating.

Case History 1.3

Industry:	Pulp and paper
Specimen location:	Crossover tube
Orientation of specimen:	Horizontal
Years in service:	20
Water treatment program:	Phosphate
Drum pressure:	800 psi (5.5 MPa)
Tube specifications:	2½-in. (6.4-cm) outer diameter
Fuel:	Black liquor and fuel oil

The crossover tube contained a small bulge and rupture. A white, friable deposit layer covered the hot side internal surface. A deposit weight determination indicated the presence of material having a weight of 16 g/ft^2 (17 mg/cm^2) and 1 g/ft^2 (1.1 mg/cm^2), respectively, on the hot and cold side internal surfaces. Deposits were bounded on both sides by sharp borders (Fig. 1.24). The borders indicate the steam-blanketing terminus. Deposition took place during a brief episode of high fireside heat input. High heat input resulted in overheating and eventual failure.

Deposits consisted of basic calcium phosphate (hydroxyapatite), sodium aluminum silicate hydrate, magnetite, magnesium silicate, and small amounts

FIGURE 1.24 Deposit layer on the internal surface of the hot side of a crossover tube. Some deposit spalled off at a shallow bulge. Note the sharp boundaries indicating the terminus of steam blanketing.

of other materials. Some alum was carried over from clarifiers, producing aluminum-bearing deposits. Sodium-containing deposits indicated that evaporation to dryness occurred. Firing practices and water clarification were reviewed and changed appropriately.

Case History 1.4

Industry:	Building heating system
Boiler type	Heat recovery steam generator
Specimen location:	Riser tube externally finned
Orientation of specimen:	Vertical
Years in service:	9
Water treatment program:	Polymer
Drum pressure:	900 psi (6.2 MPa)
Tube specifications:	2-in. (5.1-cm) outer diameter
Fuel:	Natural gas

A failure occurred in a single riser tube during service. The tube was located within one of the inner rows of tubes. The failure occurred immediately after a bend that was located downstream of the inlet header. A long tube section was removed that was mostly free of internal surface deposits (Fig. 1.25), except within a 9-in.- (22.9-cm-) long portion at the location of the failure (Fig. 1.26). Very thick, highly stratified deposit layers covered the internal surface at that location occluding about three-quarters of the cross-sectional area of the tube bore. Though the deposit layers were very thick, the tube experienced no overheating since the heat flux at the location was low. Components of the deposits were highly alkaline in a distilled water solution. Analysis of the layers revealed that they consisted primarily of iron oxide, as well as metallic copper and a small percentage of sodium-containing compounds. A perforation occurred at the base

FIGURE 1.25 Example of the fairly clean internal surface found in most locations away from the failed area.

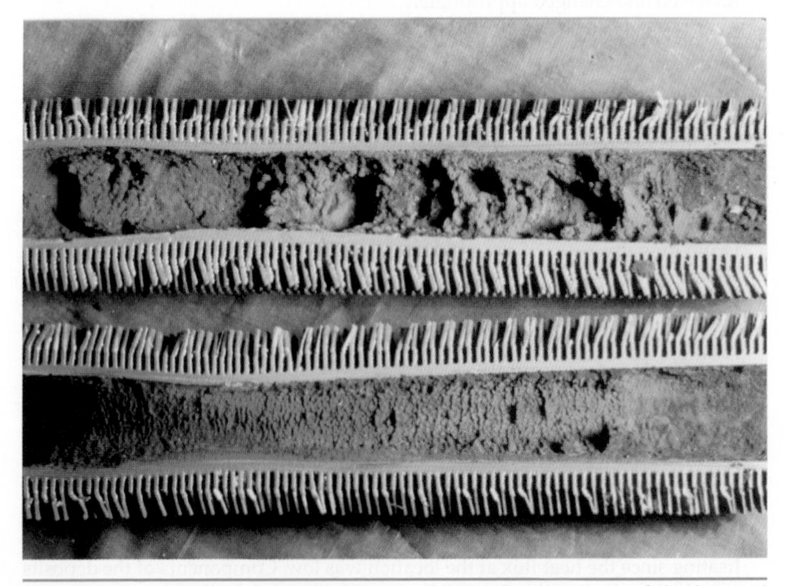

FIGURE 1.26 Very thick, stratified layers of deposits that covered the internal surface only within a 9″ long segment at the location of the failure. Up to ¾ of the cross-sectional area was occluded.

FIGURE 1.27 Very thick, highly stratified deposit layers consisting primarily of iron oxide cover the internal surface. A perforation occurred at the base of a deep crater that formed beneath the layers due to caustic corrosion.

of a deep crater that formed beneath the deposit layers (Fig. 1.27). The crater formed due to corrosion resulting from the concentration of caustic beneath the deposit layers; see Chap. 4, Caustic Corrosion.

The deposition occurred only within the highly localized portion of the tube due to poor boiler water circulation. Essentially, boiler water flow through the tube was stagnant. The water level within the tube floated up and down within the 9-in.- (22/9-cm-) long portion of the tube where the thick deposit layers formed at the steam/water interface. The boiler water evaporated to dryness at the water level, causing all dissolved and suspended material to deposit, including aggressive agents in the boiler water, in this case caustic. Internal inspection revealed that only this tube experienced localized deposition and corrosion. Poor circulation of boiler water through that tube resulted from the design and operating conditions of the boiler. Such circulation problems may not be easily rectified. For instance, installation of plugs to bypass flow to the affected tube may not prevent flow disruptions within other tubes. In some cases, isolated tubes that experience such flow disruptions will require periodic inspection and cleaning to control deposition and under-deposit corrosion.

Case History 1.5

Industry:	Utility
Specimen location:	Fine turbine screen
Orientation of specimen:	Vertical
Years in service:	8
Water treatment program:	Coordinated phosphate
Drum pressure:	2200 psi (15.2 MPa)
Tube specifications:	0.035-in. (0.89-mm) mild steel wire; screen opening size of 0.040 in. (1.01 mm)
Fuel:	Coal

After routine superheater inspections, it was revealed that spalling and exfoliation of magnetite on internal surfaces occurred. Sections of both coarse and fine turbine-inlet screens were removed for inspection.

Approximately 10% of the fine screen openings were plugged with magnetite chips. Small, exfoliated chips had been blown through the superheaters and had lodged in the screen (Fig. 1.14). The coarse screen caught magnetite chips as thick as 0.010 in. (0.25 mm) and as wide as 0.15 in. (3.8 mm). The side of the coarse screen facing steam flow was hammered and peened by impact with flying projectiles.

A slight decrease in turbine efficiency had been noticed. It was suspected to have been caused by hard-particle erosion from exfoliated oxide-damaged buckets. Subsequent turbine inspections revealed only minor damage.

Changes in tube metal temperature may promote exfoliation of magnetite layers. Fireside chemical treatments to reduce slagging were suggested, and soot-blowing practices were reviewed to help limit temperature fluctuations in service. Chemical cleaning of the superheater was deemed unnecessary.

References

1. S. Heifner and E. P. Thurston, "Collecting and Using Boiler Tube Deposit Loading Data," Paper 05442, National Association of Corrosion Engineers, Houston, Tx., 2005.
2. D. Frey, "Chemical Cleaning of Boilers," 54th Annual Meeting, International Water Conference, Pittsburgh, Pa., 1993.

SECTION 1.2

Overheating

Chapter 2
Short-Term Overheating
Chapter 3
Long-Term Overheating

Overheating is one of the most common failure mechanisms for boiler tubing, especially in components with relatively high operating temperatures, such as superheaters and reheaters. Degradation due to overheating increases as a function of increasing temperature, stress, and time. Failures can occur when the combination is excessive for the particular alloy of the component. Boiler and pressure vessel codes include maximum allowable stress values for different metal alloys.[1] These values decrease with increasing temperature, so proper design of components can limit overheating failures in a typical service life.

When a metal is subjected to appreciable stress levels, especially at elevated temperatures, deformation can result. Stresses on boiler tubes mainly arise from hoop (circumferential) stresses that develop from internal pressurization of the tube. Other applied stresses can also be appreciable in some cases and should also be considered. The hoop stress S is dependent on the internal pressure P, the tube diameter D, and the tube-wall thickness t, as indicated in Eq. (S1.2.1) for a thin-walled cylinder:[2]

$$S = PD/2t \qquad (1.2.1)$$

Stresses will increase with increasing pressure and diameter and decreasing tube-wall thickness. Decreasing tube-wall thickness may significantly promote overheating failures by increasing stress, and numerous corrosion and erosion mechanisms can result in wall loss.

Investigating overheating failures requires careful analysis of failed components. This analysis should include microscopic examination for determination of microstructural changes to the alloy (typically steel) and identification of other forms of degradation associated with high-temperature exposure. A thorough analysis of a failed component can provide essential information associated with the failure, such as estimates of the time and temperature of the overheating event or events. This information often needs to be compared to the operational history of the boiler, as well as boiler inspection findings, to verify the root cause.

Overheating failures have been generally divided into two relative types, typically referred to as *short-term overheating* and *long-term overheating*.[3] Specific characteristics associated with these types of overheating are presented in detail in the following chapters.

Microstructural evidence is a primary source of information regarding overheating conditions. Therefore, recognition of steel phase transformations and associated steel microstructures is necessary for diagnosis. It is important to note that the following discussion is intended to give an overview of microstructural changes for plain carbon steel and low-alloy steels with temperature and time, rather than detailed explanations of the changes, which can be found in other texts.[4]

Phase diagrams are useful to help understand the phases that form in alloy systems, such as steel, and the transformations that can occur with increasing or decreasing temperatures. They graphically represent the boundaries of phases in a system for a range of temperatures and compositions. Typical plain carbon steel can be considered an alloy of iron and carbon. Graphite and iron carbide are carbon-containing phases that can form in the system. Although graphite is more thermodynamically stable than iron carbide, the iron carbide phase (also known as cementite, Fe_3C) usually forms preferentially during typical steel manufacturing because of slow reaction kinetics to form graphite. Thus, the phase diagram typically shown for steel includes the iron carbide phase. A simplified version of the steel portion of the iron–iron carbide phase diagram is shown in Fig. S1.2.1.[5] This is not considered an equilibrium diagram, as the iron carbide will decompose to graphite over long periods (called *graphitization*). Detailed discussion of phase diagrams, phases, and crystal structures in the iron-carbon system can be found elsewhere.[6]

For typical plain carbon boiler tubing alloys (such as ASME SA-178, SA-192, and SA-210), the carbon contents in the alloys range from 0.06 to 0.35 wt %. For these steels, a normal and typical microstructure of as-fabricated boiler tubing consists of pearlite colonies in a matrix of ferrite grains (Fig. S1.2.2). Ferrite, also called alpha (α) iron, has a body-centered-cubic crystal structure and a low solubility limit for carbon. Pearlite colonies are lamellar aggregates of ferrite and iron carbide, Fe_3C. In some of the low-alloy steels containing

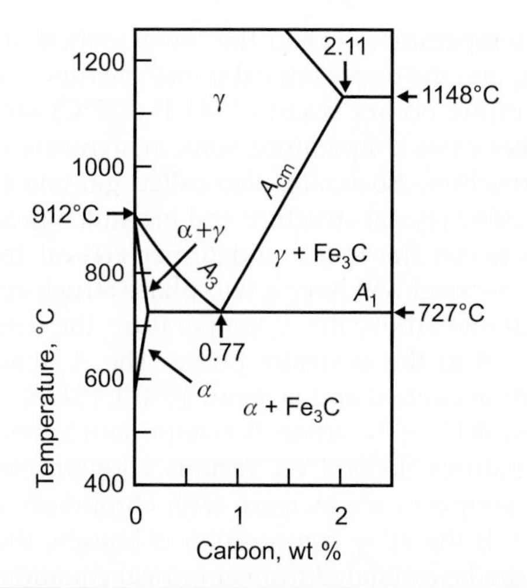

FIGURE S1.2.1 Steel portion of the iron-carbon phase diagram, showing the α–Fe (ferrite) and γ–Fe (austenite) phase fields and the A_1 and A_3 lines.

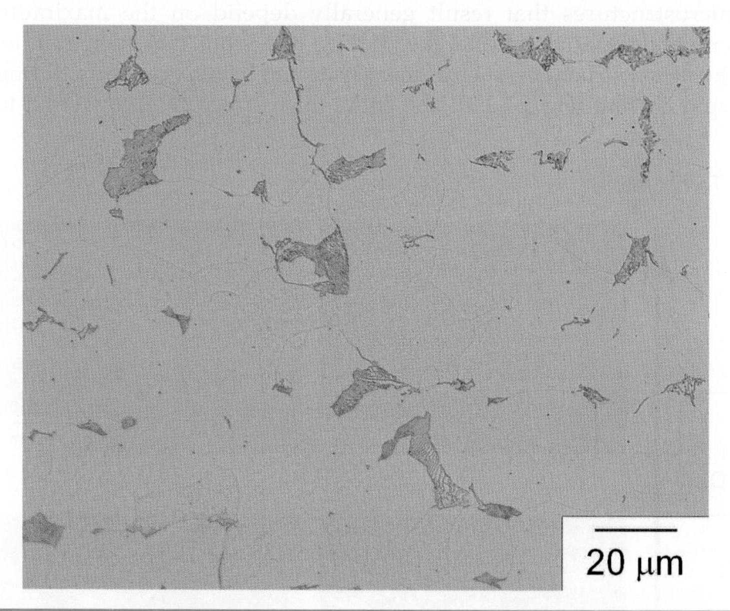

FIGURE S1.2.2 Microstructure normal and typical of plain carbon boiler tube steel consisting of pearlite colonies (the dark etching constituent having a lamellar appearance) in ferrite.

chromium and molybdenum, such as ASME SA-213 T11 and T22, the aggregate structure is often not well defined, consisting of indistinct pearlite colonies rather than lamellar morphology.

As typical boiler tube steel is heated, it will undergo a few phase transformations prior to melting. The first two-phase transformations occur at two temperatures called the lower critical transformation temperature A_1 and the upper critical transformation temperature A_3. The A_1 temperature occurs around 1341°F (727°C) for plain carbon steel alloys. Above this temperature, some austenite will start to form in the microstructure. Austenite, also called gamma (γ) iron, has a face-centered-cubic crystal structure and has much greater solubility for carbon. Between the lower and upper critical transformation temperatures, the steel will have a two-phase structure consisting of ferrite and austenite. Above the A_3 temperature, the steel will be completely converted to the austenite phase. The A_3 temperature will depend on carbon content and is about 1550°F (843°C) for plain carbon steel having 0.15 wt % carbon. It is important to note that both A_1 and A_3 temperatures depend on various alloying elements in the steel, and the temperatures increase with chromium and molybdenum additions. If the alloy composition is known, then the critical temperatures can be estimated from empirical equations.[7]

For steel tubing that has attained temperatures over A_1, high-temperature transformation products will form during cooling. The microstructures that result generally depend on the maximum temperature attained and the cooling rate from that temperature. Quenched structures, such as martensite and bainite, can form during rapid cooling (Fig. S1.2.3). The entire microstructure is converted to

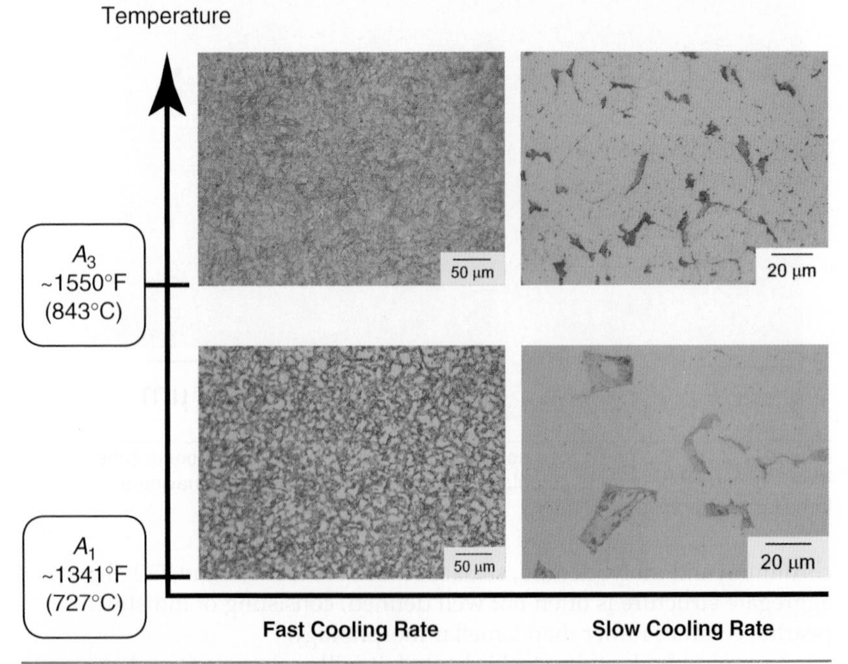

FIGURE S1.2.3 Microstructures that can form in plain carbon steel with exposure to elevated temperatures, with a comparison of fast and slow cooling rates.

quenched high-temperature at fast rates transformation products if temperatures exceed A_3 before cooling. If the tube metal is heated above A_3 and allowed to cool slowly, then the microstructure will consist of reformed pearlite in a ferrite matrix. Reformed pearlite colonies often have a different size and interlamellar spacing than the original pearlite colonies. The colonies may also be distributed differently (Fig. S1.2.3).

If the metal temperature was between A_1 and A_3 at the time of failure and then cooled rapidly, the temperature at failure can be estimated. The approximate temperature can be calculated by measuring the fractions of ferrite and quenched high-temperature transformation products in the microstructure and applying the lever rule in the two-phase (ferrite and austenite) field of the iron-carbon phase diagram. Examples of using the lever rule to estimate temperature are presented elsewhere.[8]

If the metal is at elevated temperatures below A_1 for periods of time that allow for carbon diffusion, then pearlite decomposition can occur. Depending on temperature and exposure time, the pearlite in plain carbon steels can decompose into spheroidized carbides (spheroidization) and/or graphite nodules (graphitization) (Fig. S1.2.4). Low-alloy steels containing more than 0.5% chromium, an element that has a strong attraction to carbon, do not experience graphitization. Graphitization is the typical mode of pearlite decomposition from about 850 to 1025°F (454 to 552°C). Graphitization also generally requires very long exposure periods, typically several months to years. Spheroidization dominates from about 1025 to 1341°F (552 to 727°C).[9] The degree of spheroidization, from partially spheroidized to completely spheroidized, can be used to estimate time and temperature ranges for overheating. Microstructure catalogs showing the effect of different times and temperatures for plain carbon and low-alloy steels can be found in other sources on boiler tube failure analysis.[10, 11] At metal temperatures below A_1, the cooling rate typically has little to no effect on the microstructure.

Lastly, extended times at elevated temperatures can result in decarburization, or loss of carbon from the microstructure. The surface conditions of the metal must be suitable for decarburization, such as an oxidizing environment. Decarburization requires diffusion of carbon, and diffusion rates increase with temperature. Decarburization due to overheating must be distinguished from decarburization that can occur during normal tube manufacture. In many boiler tube steels, narrow decarburized zones may be adjacent to the internal and external surfaces. These decarburized zones are normally present from tube manufacture and do not represent an in-service condition. When a decarburized microstructure develops during service, grain growth can also tend to occur. This is usually strong evidence that elevated temperatures, most likely above the lower critical transformation temperature A_1, were experienced for

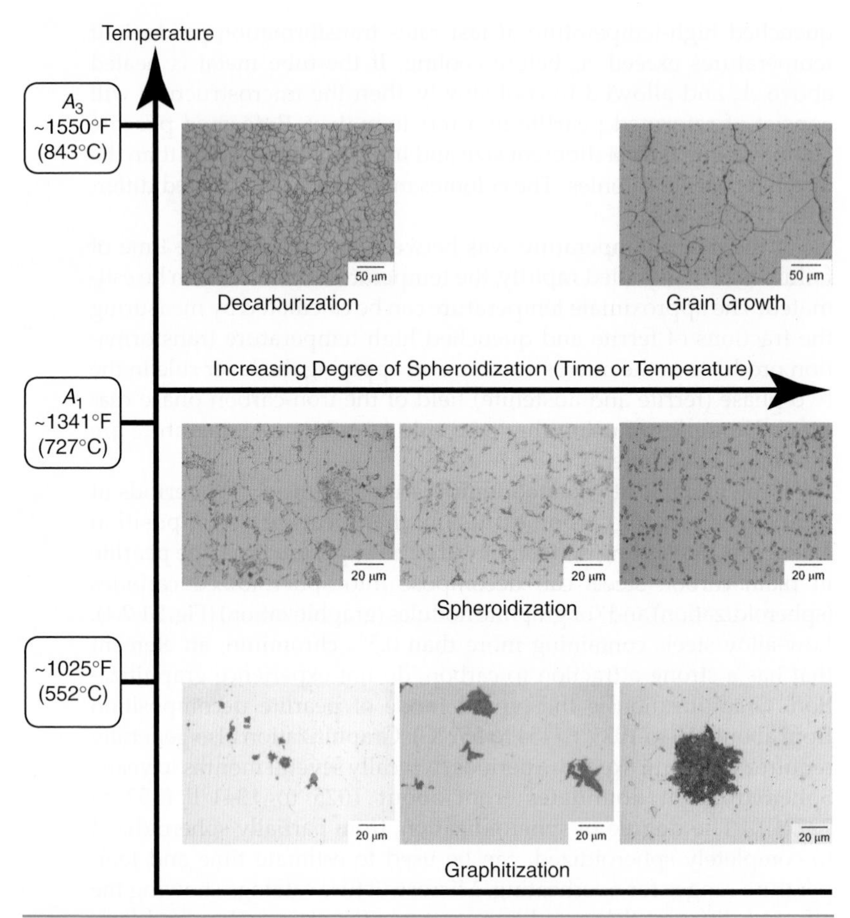

FIGURE S1.2.4 Microstructures that can form in plain carbon steel with exposure to elevated temperatures and extended times.

extended periods. In some cases carburization, which is an increase in carbon concentration of the metal, may occur. However, it is rare and requires that carburizing conditions exist at suitably high temperatures.

In addition to microstructural changes to the metal, overheating generally causes some thermal oxidation of the surfaces. *Thermal oxidation* refers to the conversion of metal to oxide at elevated temperatures, generally above where the oxidation rate becomes excessive. The thickness of the oxide layer depends on temperature and time; thus, oxide thickness measurements are additional evidence to help to assess thermal history. As thick layers of thermally formed oxide are generally associated with long-term type of overheating failures, detailed discussion of thermal oxidation is presented in Chap. 3, Long-Term Overheating.

The overheating discussion in this book is divided into long-term and short-term chapters. However, the terminology *severe* or *mild* is also sometimes used to characterize overheating failures. *Long-term* and *short-term* refer to time, whereas *mild* and *severe* refer to temperature. Mild and severe overheating failures are generally associated with metal temperatures below and above the A_1 temperature for the alloy, respectively. For instance, the presence of high-temperature transformation products indicates that temperatures well above design limits were reached in service. Regarding time, there is no clear-cut, established time value that separates short-term and long-term overheating. At the extreme ends of likely overheating time frames (e.g., minutes to hours for short-term overheating and months to years for long-term overheating), the appropriate terms can obviously be used. However, the distinction is not always clear for time ranges in the middle. It is also difficult to estimate the time frame associated with overheating to precise ranges. In addition to time and temperature, the amount of tube deformation (such as expansion or necking at the rupture edge) and microstructural features at the rupture are also used to distinguish between failure mechanisms (creep/stress rupture or yielding and tensile overload). It is important to note that time, temperature, deformation, and failure mode are generally related, as shown in Fig. S1.2.5. As temperature increases, the time to failure typically decreases and the amount of deformation associated with the failure increases.

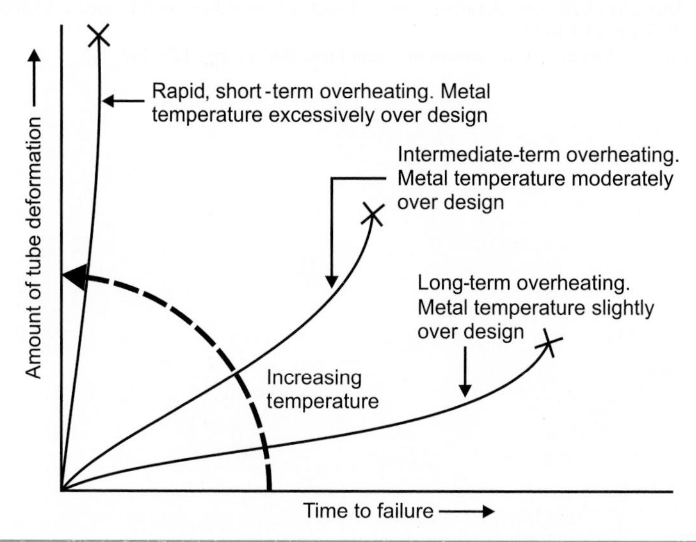

FIGURE S1.2.5 Relations between time, temperature, and the amount of deformation associated with overheating failures. (*Courtesy of Electric Power Research Institute.*)

In summary, careful sampling and examination of metallographic specimens is required as part of a thorough analysis. Specimens should be taken from locations at and away from the failure on both sides of the tube (where applicable) for comparison. Microstructural evidence and characteristic features obtained from these specimens can provide valuable information for estimating likely time and temperature exposure that occurred during the overheating event or events.

References

1. *2007 ASME Boiler & Pressure Vessel Code, Section II—Materials* (includes Addenda for 2008), American Society of Mechanical Engineers, 2008.
2. E. A. Avallone and T. Baumeister, III (Eds.), *Marks' Standard Handbook for Mechanical Engineers*, 10th ed., McGraw-Hill, New York, 1996, pp. 5–45.
3. EPRI Final Report CS-3945, *Manual for Investigation and Correction of Boiler Tube Failures*, Electric Power Research Institute, Palo Alto, CA, April 1985.
4. Leonard E. Samuels, *Light Microscopy of Carbon Steels*, ASM International, Materials Park, Ohio, 1999.
5. Ibid., p. 29.
6. George Krauss, *Principles of Heat Treatment of Steel*, American Society for Metals, Materials Park, Ohio, 1980, pp. 1–16.
7. K. W. Andrews, "Empirical Formulae for the Calculation of Some Transformation Temperatures," *J. Iron Steel Inst.*, 1965, vol. 203, pp. 721–727.
8. D. N. French, *Metallurgical Failures in Fossil Fired Boilers*, 2d ed., John Wiley & Sons, New York, 1993, pp. 85–87.
9. D. N. French, "Failures of Boilers and Related Equipment," *ASM Handbook*, 9th ed., vol. 11: *Failure Analysis and Prevention*, American Society for Metals, Materials Park, Ohio, p. 613.
10. EPRI TR-102433-2, *Boiler Tube Failure Metallurgical Guide*, vol. 2: *Appendices, Appendix D, Microstructural Catalogs*, Electric Power Research Institute, October 1993, pp. D-1 to D-108.
11. French, *Metallurgical Failures in Fossil Fired Boilers*, pp. 125–147.

CHAPTER 2

Short-Term Overheating

General Description

Short-term (sometimes referred to as *severe*) overheating occurs when the tube metal temperature rises significantly above design limits, generally for a brief period of time. In all instances, metal temperatures are at least 850°F (454°C) and typically exceed 1200°F (649°C). Some guides classify short-term overheating failures into three types, depending on the temperature attained at failure.[1] The three types are categorized as

- *Subcritical*: metal temperatures below the lower critical transformation temperature A_1
- *Intercritical*: metal temperatures between the lower and upper critical temperatures A_1 and A_3
- *Upper critical*: metal temperatures above the upper critical transformation temperature A_3

At elevated temperatures metal strength is markedly reduced (Fig. 2.1). When the yield strength of the metal is exceeded, the tube wall will experience plastic deformation, often in the form of bulging. This is accompanied by thinning of the tube wall. Wall thinning, in turn, increases the stress level of the tube. At higher temperatures and stresses, the tensile strength may be exceeded, resulting in rupture of the tube. This is the most common form for short-term overheating failures. The time to failure decreases with increasing temperature.[2] If temperatures rise to very high levels, then failure will occur quickly. At extremely high metal temperatures above the upper critical temperature, short-term creep mechanisms can also significantly contribute to tube failure.

Increases in metal temperature due to short-term overheating are related to insufficient coolant flow and/or excessive fireside heat input. Failure is usually caused by a boiler operation upset condition.

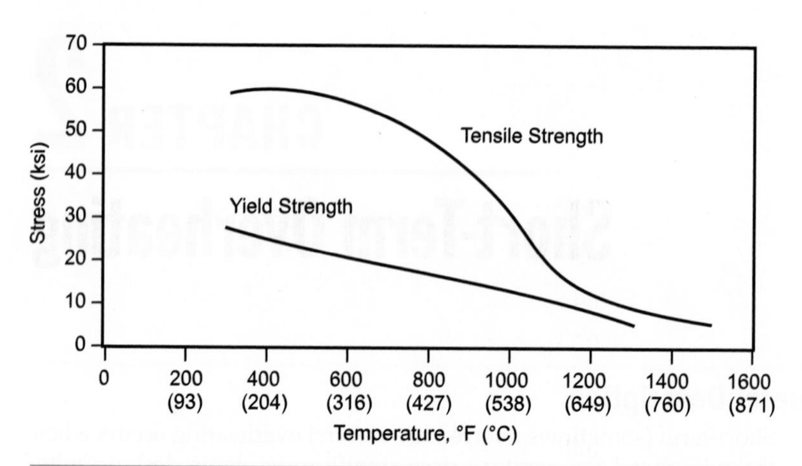

FIGURE 2.1 Yield and tensile strength for a plain carbon steel (SA-192) as a function of temperature.

Locations

Failures caused by short-term overheating are generally confined to steam- and water-cooled tubes including risers, waterwalls, roofs, screens, superheaters, and reheaters. Because of their high operating temperatures, superheaters and reheaters are common failure sites. Failures due to short-term overheating almost never occur in economizers, where temperatures are limited and steam generation does not typically occur.

Failure locations of components are often related to the upset conditions that lead to failure. For instance, a single ruptured tube in the midst of other apparently unaffected tubes suggests isolated pluggage or other flow-related problems. Failures can occur downstream from bends in pendant superheaters and reheaters if steam flow is limited. Short-term overheating failures may occur downstream from another failure due to reduction of coolant flow. When low water level is the cause, failures will often occur toward the top of waterwalls near the steam drum.

Critical Factors

A short-term overheating failure is caused by abnormal circumstances, such as an upset, generally occurring over a brief period of time (typically on the order of seconds to minutes). Therefore, pinpointing unusual events immediately preceding failure is extremely important in identifying the root cause of failure. Since short-term overheating frequently has little to do with water chemistry, efforts should be concentrated on review of operating procedures and system design that could result in reduced coolant flow or increased heat input.

Many factors can be responsible for short-term overheating failures. Some typical conditions that can lead to failures may include

- Rapid start-ups and shutdowns
- Large load swings for the boiler
- Low drum water levels
  - Equipment malfunction, such as feedwater pumps or valves
  - Operator error
  - Excessive blowdown
- Tube blockage
  - Repair debris not removed prior to start-up
  - Material accumulated from improper chemical cleaning
  - Exfoliated or transported oxides or deposits
  - Gross carryover into superheater tubing
  - Condensate that collects in bends of superheaters during idle time
- Low-flow or standby operation
- Excessive firing rate
- Changes in firing pattern
- Improper burner operation

Identification

Several factors that are often present in failures caused by short-term overheating, especially below the upper critical temperature, are uniform tube expansion, thinned rupture edges, violent rupture, and absence of significant internal deposits or large amounts of thermally formed oxide. Although some features may suggest a short-term overheating failure, it should be confirmed by metallographic examination. Examination of the microstructure can determine the likely temperature range for the overheating event. This information can be used to indicate the likelihood of plastic deformation due to yielding and subsequent tensile overload for the stress and temperature levels.

Short-term overheating is often accompanied by uniform tube swelling. Measurement of tube diameters at or near the failure will generally show appreciable expansion from the initial tube dimensions

FIGURE 2.2 Ring sections removed from near the rupture (left) and away (right) from a superheater tube that experienced a severe overheating event. Note the appreciable expansion of the tube diameter near the failure.

(Fig. 2.2). The tube circumference can be roughly measured at the rupture by using a piece of string or measuring tape. The circumference can then be used to estimate the outer diameter of the tube. Some sources suggest that increases of tube diameters of 5% or greater are indicative of short-term overheating.[3] However, microstructural changes and rupture characteristics should also be consistent to diagnose a short-term overheating failure. In some instances, severe overheating may cause the tube to be uniformly expanded, but without a rupture.

In some cases, severe overheating may produce bulging. Multiple bulges are usually absent, although a single bulge containing a rupture may occur, especially if long-term overheating has occurred previously (Fig. 2.3).

If failure happens rapidly, bulging may be absent and the rupture can be violent, sometimes flattening the tube wall or bending the tube severely and causing secondary metal tearing (Fig. 2.4). Ruptures are longitudinally oriented and can have a fish-mouth appearance. Rupture edges will generally be thinned to some degree, and frequently will taper to knifelike or chisel-like edges (Fig. 2.5). Thick-walled rupture edges may have a cup and cone type of appearance, typical of ductile fracture. Microscopic examination will generally show appreciable grain deformation in the microstructure adjacent to the rupture edge (Fig. 2.6). Elongated voids are often present at and near the rupture edge.

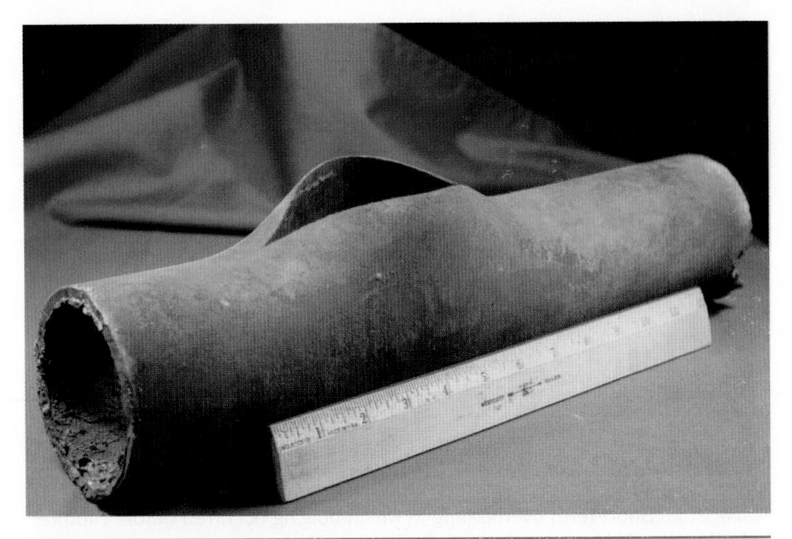

FIGURE 2.3 A rupture at a single bulge. The tube had experienced long-term overheating, followed by a brief episode of severe overheating.

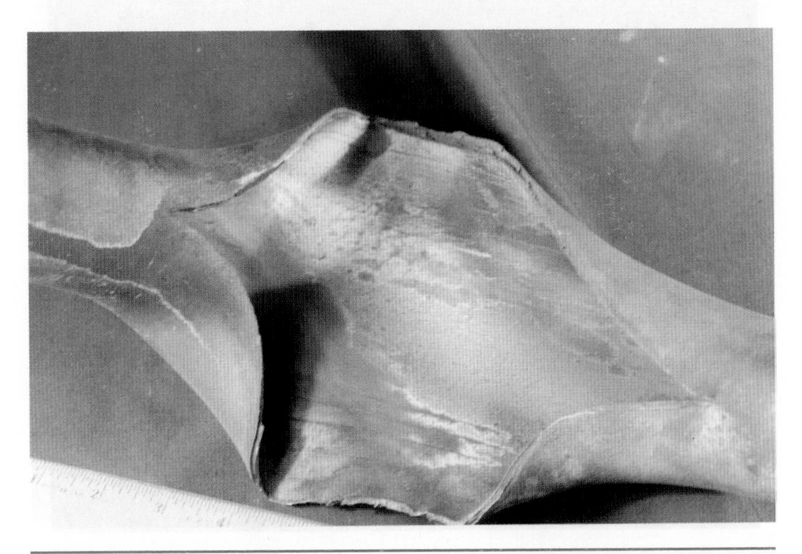

FIGURE 2.4 Violent rupture caused by short-term overheating. The tube is bent almost to a right angle, caused by the severity of the burst.

Some different features may be seen in failures that occur at significantly elevated temperatures, generally above the upper critical temperature A_3. Failures may be in the form of thick-walled ruptures (Fig. 2.7). In addition there may be no significant tube swelling, as the tube circumference at the rupture is sometimes nearly exactly equal to the tube circumference away from the rupture. The rupture

FIGURE 2.5 Short-term overheating in which bulging occurred before rupture. Note the chisel-like rupture edges.

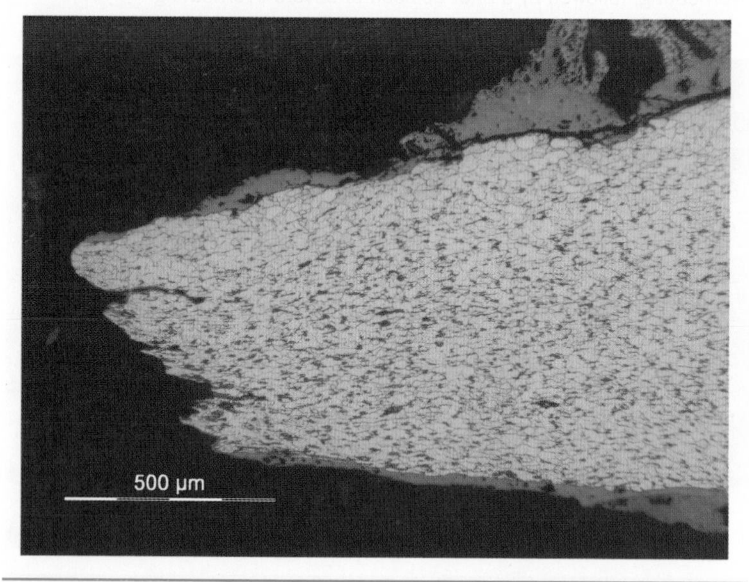

FIGURE 2.6 Microscopic cross section through a rupture edge, showing significant tube-wall thinning and elongation of the grains.

edges may be blunt and retain most of their original wall thickness (Fig. 2.8). Rupture edges may show some evidence of intergranular fracture.

In general, heavy internal deposits will not be present on a water-cooled tube that experiences a short-term overheating failure since

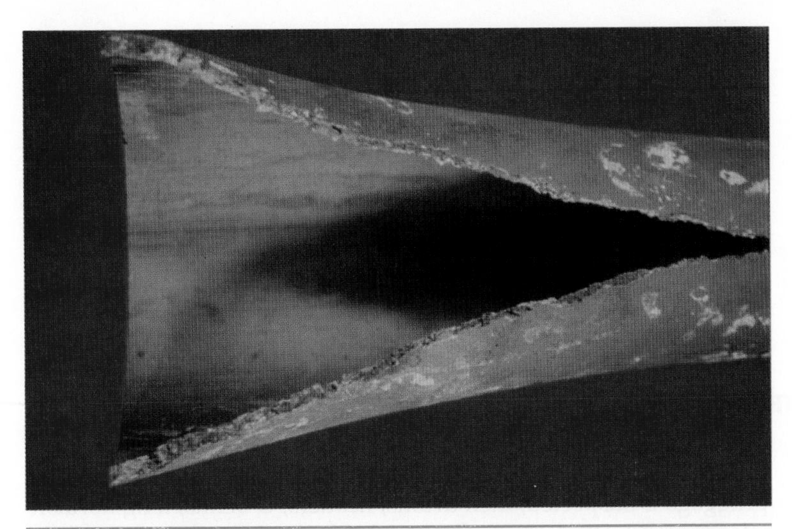

FIGURE 2.7 Longitudinal rupture in a superheater tube caused by partial pluggage upstream of failure, which in turn caused short-term overheating. Note the thick-walled rupture edges. Virtually identical tube circumferences are present at the rupture and away from the burst. Such failures often occur when temperatures exceed the upper critical temperature for the metal.

FIGURE 2.8 Microscopic cross section through a thick-walled rupture edge, showing some shallow intergranular penetrations along prior austenite grain boundaries.

these deposits are not likely to be the cause of the rupture. Further, if deposits do occur, they usually will be friable and easily removed by gentle probing, rather than tenacious and baked onto the surface as is typical with long-term overheating. Thick accumulations of thermally formed oxide will be absent if the overheating event was very short. However, because metal oxidation rates increase significantly with increasing temperature, a thermally formed oxide layer of considerable thickness can be present on the tube surfaces, if the overheating event was associated with very high temperatures or appreciable time at the elevated temperature.

Elimination

The solution for short-term overheating is to prevent brief upset conditions from occurring. If restricted coolant flow due to tube pluggage is suspected, then drums, headers, U bends, long horizontal runs, and other areas where debris may accumulate should be inspected and cleaned. This is especially true if a failure occurs shortly after a boiler cleaning. In many instances, the blockage may be blown from the tube due to the force of the rupture. Measurement of adjacent tubing for signs of swelling may be useful to determine if other tubes may have experienced short-term overheating. Suspicion should be aroused if a short-term failure occurs immediately after another failure. A failure may dislodge deposits or corrosion products that can be transported downstream to disrupt flow, cause pluggage, or promote deposition at other locations. A failure may also reduce flow downstream. All these circumstances, in turn, may promote subsequent overheating. The drum water level, firing procedures, blowdown frequency, and start-up and shutdown procedures should be carefully monitored to ensure that proper boiler manufacturer guidelines are followed. Failures related to excessive heat input may require review of firing practices and evaluation of flame patterns to determine if flame impingement is possible. Monitoring metal and combustion gas temperatures may be beneficial to evaluate whether heat input may be significantly greater than expected.

Cautions

Thin- or thick-walled longitudinal ruptures alone are insufficient evidence to warrant a diagnosis of short-term overheating. Severe wall thinning due to metal wastage from internal or external surfaces can also result in thin-walled ruptures due to dramatically increased stress levels. Thick-walled failures due to short-term overheating can easily be confused with failures from long-term overheating involving creep (see Chap. 3, Long-Term Overheating), failures from hydrogen damage (see Chap. 6, Hydrogen Damage), and failures from certain tube defects (see Sec. 3, Material Defects). The absence of deposits near a rupture may be due to the scrubbing action of escaping fluids during rupture. Also, short-term overheating may occur after long-term overheating. Single or multiple mechanisms that contributed to failure may be properly diagnosed through metallurgical examination.

Microstructural changes occurring during short-term overheating do not always lead to failure. In addition, tubes that have experienced short-term overheating do not always have to be replaced. Mechanical properties are not necessarily altered significantly by the overheating.

Related Problems

See Chap. 3, Long-Term Overheating, and Chap. 17, Cavitation.

Case History 2.1

Industry:	Utility
Specimen location:	Waterwall, nose arch
Orientation of specimen:	Slanted
Years in service:	5
Water treatment program:	Coordinated phosphate
Drum pressure:	1800 psi (12.4 MPa)
Tube specifications:	2½-in. (6.4-cm) outer diameter, internally rifled
Fuel:	Pulverized coal

A large, thick-walled rupture formed in an internally rifled waterwall nose arch tube. Rupture occurred shortly after start-up. The rupture edges were long and jagged (Fig. 2.9). Both internal and external surfaces were smooth and covered with thin, tenacious, dark oxide layers. No significant deposits were present anywhere.

FIGURE 2.9 Large, fish-mouth rupture of rifled nose arch tube. Rupture edges are thick, blunt, and ragged. Note the absence of significant deposits.

Microstructural evidence indicated that the tube metal near the rupture exceeded 1600°F (871°C). No significant accumulation of thermally formed oxide was found anywhere on the received section.

Internal rifling is sometimes used to reduce steam channeling and inhibit steam blanketing. No steam blanketing or liquid/vapor interface was found on internal surfaces. There was no change in microstructure on the cold side, indicating that the tube contained water at the time of the rupture. Rather, the burst was caused by insufficient coolant flow on start-up.

Case History 2.2

Industry:	Utility
Specimen location:	Superheater from a waste-heat boiler
Orientation of specimen:	Vertical
Years in service:	10
Water treatment program:	Polymer
Drum pressure:	600 psi (4.1 MPa)
Tube specifications:	3-in. (7.6-cm) outer diameter, with spirally wound 7/16-in. (1.1-cm) by 3/16-in (0.5-cm) fins
Fuel:	Waste heat from gas turbines

A rupture formed at the apex of a bulge (Fig. 2.10). A thick, black layer of thermally formed iron oxide was present on both internal and external surfaces and was thicker along the failed side. Deposit layers containing sodium, calcium, silicon, and iron overlaid the thermally formed iron oxide layers on the internal surface. Deposition occurred because of carryover of boiler water into the superheater due to water-level excursions, and foaming was common. The boiler had never been cleaned.

Temperatures of waste heat gases ranged from 1200 to 1800°F (649 to 982°C). The tube failed concurrently with several other tube failures. Metallographic analysis revealed that some adjacent tubes failed as a result of long-term overheating (creep rupture), while others failed as a result of short-term overheating. It was determined that the section had been mildly overheated for

FIGURE 2.10 Large, wide-open rupture in a superheater tube of a waste heat boiler. The tube experienced long-term overheating followed by a brief episode of short-term overheating when nearby tubes ruptured.

a considerable period, but had failed during a brief episode of severe overheating when temperatures exceeded 1341°F (727°C). The short-term overheating probably occurred because of coolant starvation, caused by leakage from other failed tubes upstream.

Case History 2.3

Industry:	HVAC and domestic water heating for a building
Specimen location:	Many tubes in a D-type package boiler
Orientation of specimen:	Vertical, horizontal, slanted
Years in service:	39
Water treatment program:	Polymer treatment
Drum pressure:	200 psi (1.4 MPa)
Tube specifications:	Plain carbon steel
Fuel:	Natural gas, maximum temperature 1800°F (982°C)

One day the boiler was brought on-line from standby to a 75 to 100% firing rate. A short time afterward the boiler experienced a restriction of feedwater flow, which eventually produced low water levels in the drum. Unfortunately, the boiler safety controllers and alarms did not trip the boiler off-line. After a period of operation under the low-water condition, water was rapidly reintroduced into the drum, at which time boiler water leaked out of the drum between the bores and rolled tube ends. The total time from boiler start-up to failure was about 2 h 45 min.

Inspection after the incident revealed significant damage throughout the boiler. Some of the tubes along the roof of the firebox experienced slight sagging (Fig. 2.11). The tube to drum joints showed evidence of leakage, and the tube

FIGURE 2.11 Sagging of the roof tubes in the firebox of a package boiler operated with low water levels.

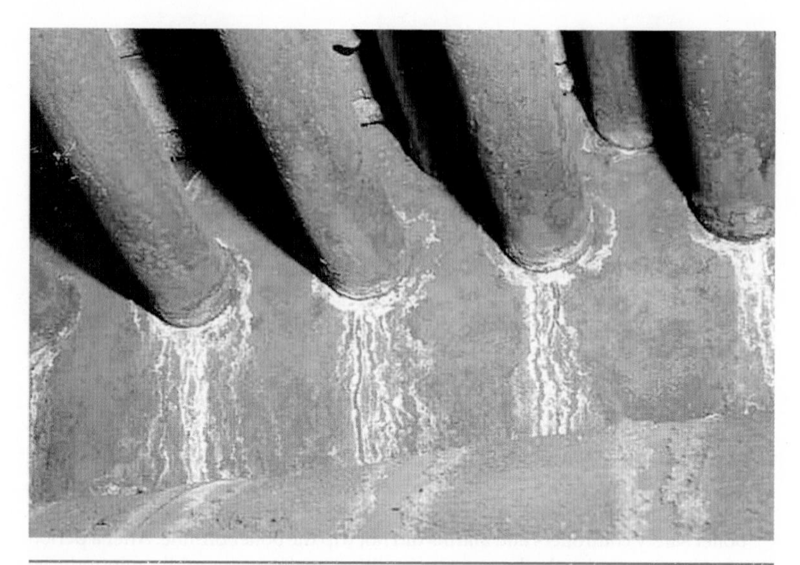

FIGURE 2.12 Deposits related to leaks that developed when the tube ends were pulled from the drum bores due to thermal contraction on cooling.

ends were pulled away from the drum (Fig. 2.12). The tubes thermally contracted at a faster rate than the drum wall when boiler water was reintroduced to the unit, allowing them to be pulled out of the bore under their own weight and other applied stresses. Tubes throughout the boiler experienced some thermal oxidation on the exposed surfaces, with blisters in the layers due to rapid thermal oxidation at very high temperatures.

Tube microstructure examinations revealed substantial grain growth and the formation of re formed pearlite, indicating tube metal temperatures exceeded 1341°F (727°C) and possibly were much higher. Thickness measurements were made of the thermally formed iron oxide layer, and the values suggested that the tubes were starved of boiler water for a significant length of time, probably at least 30 min or more. Periodic inspection was recommended to ensure that the level detectors, controllers, and safety alarms were operating properly.

Case History 2.4

Industry:	Pulp and paper
Specimen location:	Top rows in a firetube boiler
Orientation of specimen:	Horizontal
Years in service:	Unknown
Water treatment program:	Unknown
Drum pressure:	150 psi (1.0 MPa)
Tube specifications:	Plain carbon steel
Fuel:	Natural gas

A firetube boiler contained multiple collapsed tubes (Fig. 2.13). The collapsed tubes were localized to the top four rows of the unit. The tubes were deformed so severely that the opposite walls contacted. Collapsing occurred during a boiler start-up following a short idle period. Microscopic examination indicated

FIGURE 2.13 Three sections of collapsed tubing from a low-pressure firetube boiler.

reformed pearlite in the microstructure of the collapsed regions. The tube wall collapsed under the force of external pressurization when the metal strength was reduced significantly at elevated temperatures.

The severe overheating occurred when the external surfaces of the tubes were not submerged in water, and hot furnace gases flowed through the tubes during boiler operation. The top rows of tubes may have been exposed during rapid boiler start-up, when water levels were not maintained, as the boiler was fired up to maximum load.

Case History 2.5

Industry:	Chemical process
Specimen location:	Roof tubes in a D-type package boiler
Orientation of specimen:	Horizontal
Years in service:	Unknown
Water treatment program:	Polymer treatment
Drum pressure:	900 psi (6.2 MPa)
Tube specifications:	Plain carbon steel
Fuel:	Unknown

The roof tubes in a package boiler contained multiple bulges. One of the tubes contained a wide, thin-lipped rupture (Fig. 2.14). Some bulges in other tubes were directly adjacent to circumferential repair welds. No significant internal surface deposition was noted on the failed tube or the adjacent nonfailed tubes. The tube surfaces were smooth and covered by a thin, dark oxide layer. The microstructure of the ruptured tube consisted of relatively equal amounts of quenched high-temperature transformation products and ferrite, suggesting

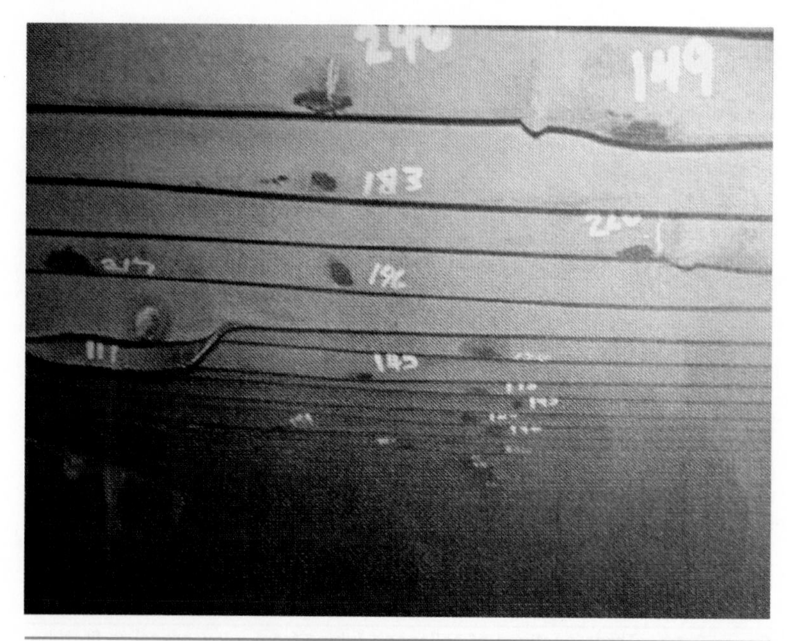

FIGURE 2.14 Bulged and ruptured roof tubes in a D-type package boiler.

metal temperatures of about 1500°F (816°C) were attained at the time of rupture. A microstructure of reformed pearlite was observed on the hot side of the nonfailed tubes, indicating these tubes exceeded 1341°F (727°C) and the severe overheating conditions were not localized to the failed tube.

Severe overheating occurred due to tube starvation that was related to circulation problems occurring during start-ups.

Case History 2.6

Industry:	SAGD oil recovery
Specimen location:	First row economizer section of a once-through steam generator (OTSG)
Orientation of specimen:	Horizontal
Years in service:	2
Water treatment program:	Polymer treatment
Drum pressure:	1500 psi (10.3 MPa)
Tube specifications:	ASME SA-106 plain carbon steel
Fuel:	Natural gas

The failed section of tubing contained multiple bulges. A short rupture occurred at one bulge apex (Fig. 2.15). Localized wall thinning due to deformation occurred at the rupture that reduced the tube-wall thickness by nearly 80%, relative to the nominal wall thickness measured away from the bulges. No significant metal wastage was observed along the internal and external surfaces. A cross section through the failure revealed rupture edges oriented at approximately 45° angles to the tube surfaces (Fig. 2.16). The rupture edges were generally free of appreciable oxide or deposit layers and had an irregular profile with many elongated microvoids in locations near the edges. The microstructure near the

Figure 2.15 Bulged and ruptured boiler tube from the economizer of a once-through steam generator.

Figure 2.16 Microscopic cross section through the rupture showing a cup and cone type of ductile fracture and elongated voids.

rupture consisted of large spheroidal carbides and elongated high-temperature transformation products in a matrix of ferrite grains (Fig. 2.17). These characteristics indicate that metal temperatures exceeded 1341°F (727°C) for a brief period at the time of rupture. The spheroidized carbides suggest that the short-term

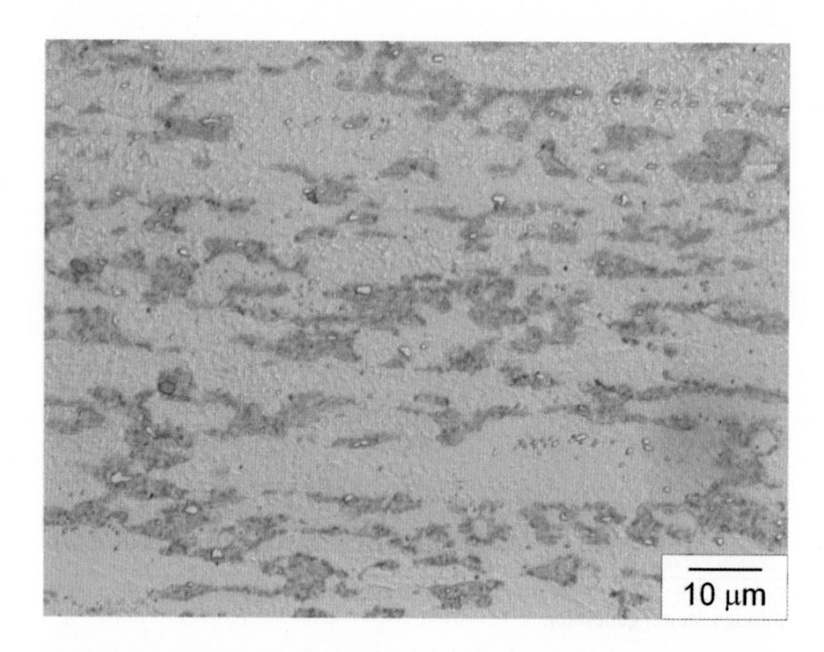

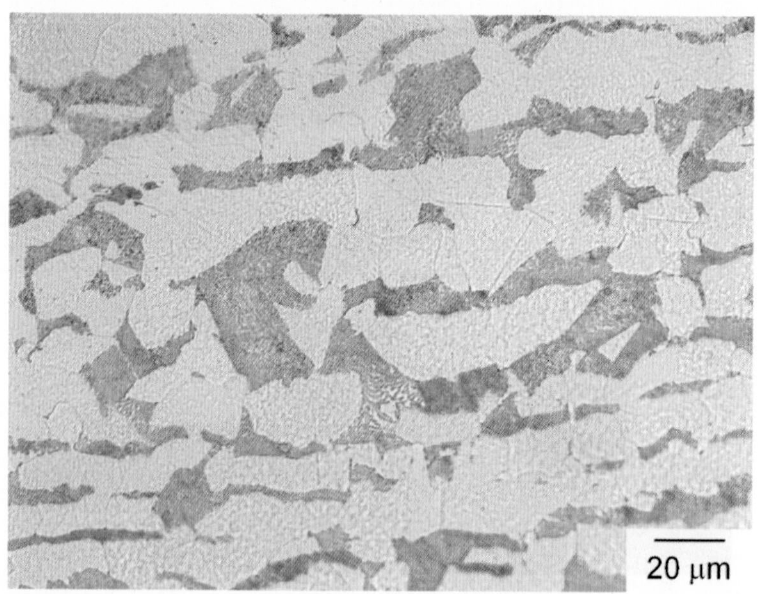

FIGURE 2.17 Microstructure near the rupture edge consisting of ferrite and high-temperature transformation products (top). Some spheroidal carbides are also present in places. Compare to the cold-side microstructure of pearlite in ferrite, which is normal and typical of plain carbon steel that has not experienced thermal alteration (bottom).

overheating event was preceded by overheating for an extended period of time in a lower-temperature range, likely between 1000 and 1341°F (538 and 727°C).

The internal surfaces were covered by moderately thick layers of deposits, which insulated the tube metal and contributed to overheating. However, the rupture was due to failure at temperatures that significantly exceeded typical temperature increases associated with the amount of deposition on the tubes. Thus, an upset condition related to coolant flow was suspected. Review of the boiler operation revealed a power loss event that tripped the boiler, shutting down the burner and boiler feedwater pumps, which likely resulted in a short period without coolant flow when the boiler was still at elevated temperatures.

References

1. EPRI TR-102433-1, *Boiler Tube Failure Metallurgical Guide*, vol. 1: *Technical Report*, Electric Power Research Institute, October 1993, Palo Alto, CA, pp. 3–166 to 3–175.
2. EPRI Final Report CS-3945, *Manual for Investigation and Correction of Boiler Tube Failures*, Electric Power Research Institute, April 1985, Palo Alto, CA, pp. 2–2 to 2–3.
3. EPRI Final Report CS-3945, *Manual for Investigation and Correction of Boiler Tube Failures*, Electric Power Research Institute, April 1985, Palo Alto, CA, pp. 2–7.

CHAPTER 3

Long-Term Overheating

General Description

Long-term overheating is a condition in which metal temperatures exceed design limits for days, weeks, months, or longer. Although the terms *long-term overheating* and *mild overheating* are often used interchangeably, it should be emphasized that the former refers to time, whereas the latter refers to temperature. That being said, the two often go hand in hand; however, there are cases in which severe overheating (at metal temperatures exceeding the A_1 temperature [1341°F (727°C)] for plain carbon steel) shows evidence of extended rather than brief elevated-temperature exposure.

Long-term overheating is the cause of more boiler failures than any other mechanism. Because of the extended times at elevated temperatures, creep deformation commonly contributes to tube rupture. Long-term overheating failures are due to a combination of factors that produce increased damage over extended time periods at elevated temperatures. This is different from short-term (severe) overheating, where failure typically results from exceeding the reduced yield and tensile strengths at elevated temperatures.

Locations

Failures resulting from long-term overheating generally occur in water- and steam-cooled tubes such as waterwalls, risers, screen tubes, superheaters, reheaters, and roof tubes. Almost 90% of failures caused by long-term overheating occur in superheaters, reheaters, wall tubes, and risers. Long-term overheating rarely occurs in economizers and floor tubes. Tubes that are subject to overheating conditions often contain significant deposits, have reduced coolant flow, experience excessive fireside heat input, or are near or opposite burners.

Some preferential locations that are more susceptible to overheating are specific to certain boiler types, and the locations are areas exposed to higher heat input. Tubes near chain-grate stokers commonly suffer long-term overheating. Failures related to excessive heat input due to burner misalignment usually occur in relatively broad areas and involve many tubes, e.g., a number of adjacent tubes in a wide region of a waterwall at about the same elevation. In heat recovery steam generators (HRSGs), areas in line with the locations of supplemental duct burners are subjected to higher heat input. For recovery boilers, waterwall tubes near the smelt bed that are not covered with solidified smelt and carbonaceous material are subject to overheating. Other tubes subject to overheating include sections in which refractory has been spalled. Channeling or laning of flue gases can result in tubes that are subjected to higher gas temperatures. Tubes at the inlet end of waste heat boilers are subjected to the highest gas temperatures, and these can be sites of overheating. Firetube boilers are rarely affected.

Slanted tubes, such as nose arches, are particularly susceptible to long-term overheating due to steam channeling. Locations downstream from flow disruptions, such as protruding welds, may be more likely to experience higher metal temperatures if departure from nucleate boiling occurs (see "Critical Factors" section). In superheater tubing, failures related to carryover of deposits are more frequent in the tubing immediately downstream from the steam drum or superheater inlet. Along superheater and reheater tubing circuits, tubing near the outlet will be exposed to the highest steam temperatures. In addition, tubing immediately upstream of a higher-grade alloy is more likely to experience overheating failures.

Critical Factors

Failure due to long-term overheating depends on temperature, length of time at temperature, stress level, and tube metallurgy. It is a chronic, rather than transient, problem that is associated with internal surface deposition and/or system operating problems that result in excessive heat input relative to coolant flow rate. These conditions result in metal temperatures that increase slightly over the design temperature. The maximum allowable design temperature is primarily a function of stress and tube metallurgy. As the amount of alloying elements is increased, particularly chromium and molybdenum, higher temperatures can be tolerated. Chromium and molybdenum provide greater resistance to thermal degradation that occurs at higher temperatures in unalloyed steel. Chromium is generally added in amounts of 0.5 to 2.5% to retard rapid, high-temperature oxidation of the metal and to enhance creep resistance, although alloys with higher chromium contents are used in some cases. Molybdenum is added in amounts of 0.25 to 1.25% to increase the creep resistance of the metal. Low-alloy steel tubes are therefore frequently used in superheaters and reheaters. Creep and thermal oxidation will be described in detail later in this chapter.

Furnace gas temperatures often exceed 2000°F (1093°C). Heat transfer into a boiler tube is controlled partly by the insulating characteristics of material near internal and external surfaces (refer to Chap. 1, Water-Formed and Steam-Formed Deposits). Heat transfer is markedly influenced by a thin gas film that normally exists on fireside surfaces. A temperature drop of over 1000°F (538°C) commonly occurs across this film.[1] Deposits such as slag or ash layers, corrosion products, refractories, and other materials on external surfaces also reduce temperatures between the furnace and the tube's fireside surface.

When one is considering heat transfer across the waterside surface, the effect of deposits is reversed. Steam layers and deposits insulate the metal from the cooling effects of the water, resulting in reduced heat transfer into the water and increased metal temperatures. This is illustrated in Fig. 3.1[2] which shows the effects of departure from nucleate boiling (film boiling) and thermally insulating waterside deposits that significantly raise metal temperatures relative to optimum operating conditions. Factors such as the amount, composition, and morphology of the deposits influence the thermal insulation of the deposit layers and the resulting increase in metal temperatures.[3]

Heavy internal surface deposition on the hot sides of steam-generating tubes insulates the tube wall from cooling effects of the water, contributing to overheating. Significant deposit weights (see Chap. 1, Water-Formed and Steam-Formed Deposits, for a detailed description) on the hot side but not on the cold side of the tube can indicate that excessive heat input promoted deposit formation. Experience has shown that when the ratio of deposit weights on hot to cold sides of water-cooled tubes exceeds three; heat input is substantially higher on the hot side. When this ratio approaches 10, heat input on the hot side relative to the cold side can be quite excessive.

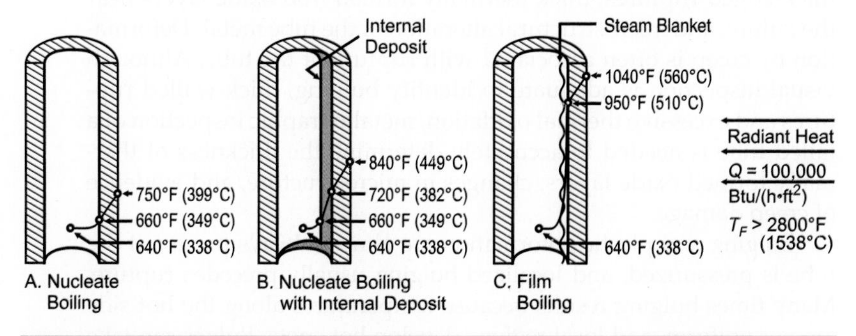

FIGURE 3.1 Schematic illustrating increased metal temperatures on a furnace wall tube experiencing radiant heat due to departure from nucleate boiling (DNB) conditions and thermally insulating deposits on the water side.

In cases when the deposit weights are high and the ratio is below 3, a few different conditions may have existed:

1. Heat input was not excessive on the hot side.
2. Chemical water treatment was deficient.
3. Water chemistry was overwhelmed by contaminants.

These values are rules of thumb, and other factors, such as circulation issues, also need to be considered with the deposit weight ratio.

Deposits on the internal surfaces of superheater tubes are caused by carryover and/or contaminated attemperation waters. These deposits can significantly increase metal temperatures and result in overheating. Deposits may collect in the bottom bends in pendants, which can cause flow restrictions that promote overheating in downstream tubing. Other sources of overheating include overfiring, incorrect flame pattern, restricted coolant flow, inadequate attemperation, and improper alloy composition. In most steam-cooled tubes, heat flux differences between the hot and cold sides are not pronounced, unless the tube is subjected to direct furnace radiation. In the latter case, the temperature on the front side (facing the gas flow and furnace) may be on the order of twice that of the opposite side.[4]

In addition to deposition, thermally formed oxide layers that develop on internal surfaces can increase metal temperatures appreciably, especially on superheater and reheater tubing. Metal temperatures will increase as the steamside oxide thickness increases, a condition very typical for long-term overheating.

Identification

Numerous features are generally associated with failures caused by long-term overheating. These features typically include bulging, thick-edged ruptures, thick thermally formed iron oxide layers near the failure, and microstructural alteration of the tube metal. Deformation by creep is often associated with rupture of the tube. Although visual inspection is adequate to identify bulging, thick-walled ruptures and excessive thermal oxidation, metallographic inspection of a failed tube is needed to accurately determine the thickness of thermally formed oxide layers, changes in microstructure, and evidence of creep damage.

Bulging and plastic deformation are almost always present if the tube is pressurized, and localized bulging usually precedes rupture. Many times bulging results because temperatures along the hot side are not uniform, and local regions develop hot spots. Bulges can take many forms; some are shallow and sloping, while others are abrupt (Fig. 3.2). A single bulge or many bulges may occur. When many bulges occur along the hot side, internal deposition is frequently appreciable.

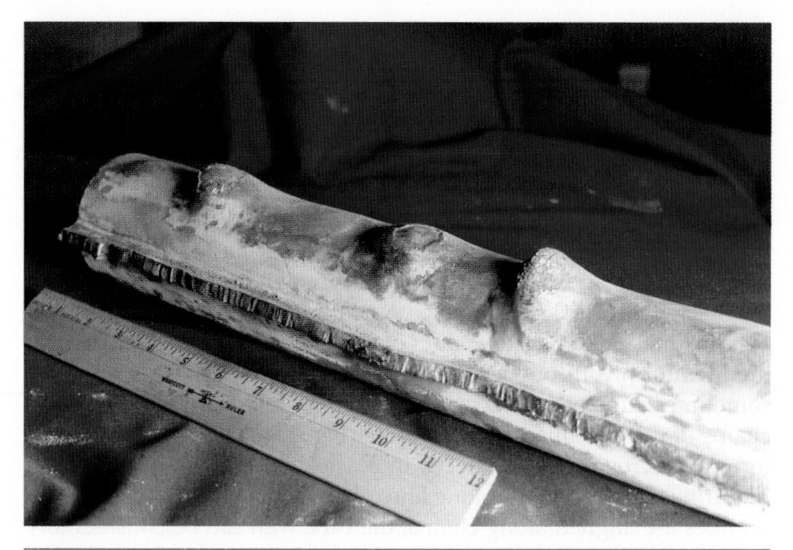

FIGURE 3.2 Multiple, abrupt bulges on the hot face of a waterwall tube.

Bulging usually causes spalling of deposits at the bulge site, initially reducing metal temperature locally. However, the bulge geometry produces a disturbance in the coolant flow that makes steam blanketing (in water-cooled tubes) more likely, thus raising the temperature. Also, bulging increases surface area, allowing a greater local fireside heat input. In general, metal temperatures become higher at bulges than in the surrounding metal. In many cases, the continued existence of bulges is sufficient to cause overheating failures, even if associated deposits are removed.

Waterside deposits will usually be present on steam-generating tubes. Deposits that formed over a long period of time prior to the failure will usually be hard and brittle. Multiple deposition events tend to result in stratified layers, which may be of different colors and textures, the innermost layers being hardest and most tenacious.

Microstructural Alteration

Microstructural changes that can occur to typical boiler tube steel, with exposure to elevated temperatures for extended periods of time, are presented in Section 1.2, Overheating. In the temperature range from 800°F (427°C) to the lower critical temperature [1341°F (727°C)], spheroidization of the microstructure can occur over time. For steel alloys that contain less than 0.5% chromium, graphitization generally will occur rather than spheroidization at temperatures between 800 and 1025°F (427 and 552°C). Spheroidization and graphitization are competing processes. At temperatures over 1025°F (552°C), spheroidization can occur relatively quickly, thus significantly reducing the chances for the slower process of graphitization. The extent of microstructural changes will depend on the time and temperature of the overheating event or events. A slight decrease in strength (and hardness) of the metal will generally accompany these microstructural changes.

Chain Graphitization

A specific form of microstructural alteration of particular concern is chain graphitization. Graphitization occurs when the iron carbide in the microstructure decomposes into graphite nodules after prolonged overheating. They are generally not a concern when they are distributed uniformly throughout the microstructure. However, nodules sometimes align along narrow bands in the microstructure, forming chains of cavities filled with graphite (Fig. 3.3). The nodules usually form at microstructural defects, in places where there are chemical impurities, and especially along lines of high stress. When the graphite nodules are aligned along chains, the tube wall can be considerably weakened. Stresses generated by normal internal pressurization cause tearing of the metal along chains of nodules, much as a postage stamp is torn from a sheet along the perforated edges. Rupture edges are often at an angle of about 45° to the tube surfaces.

Chain graphitization is usually found at welds (Fig. 3.4). Less commonly, damage occurs away from welds and forms helical cracks

FIGURE 3.3 Chain graphitization, showing alignment of graphite nodules along planes of high stress in the tube wall.

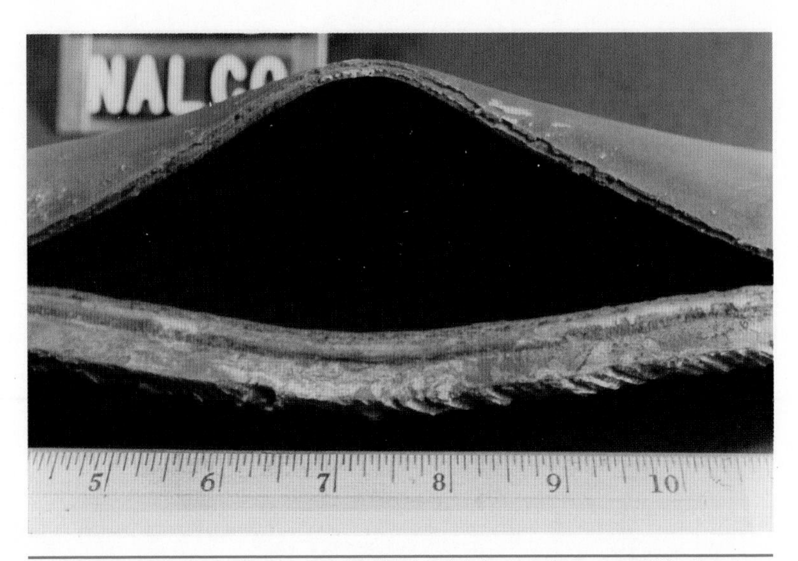

FIGURE 3.4 Rupture along a longitudinal weld due to chain graphitization. Note the blunt, rough rupture edge. Such failures occur after long-term mild overheating.

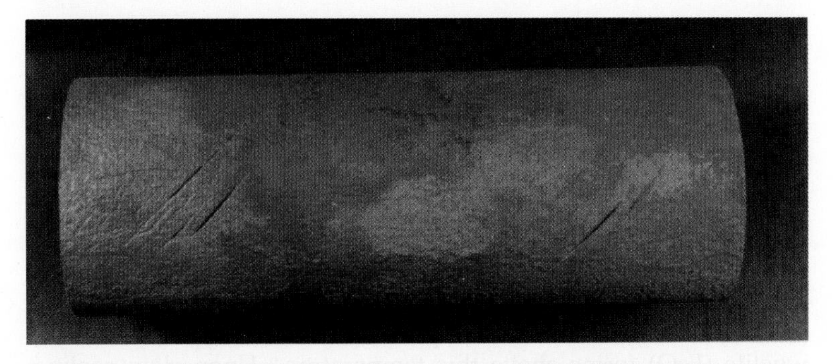

FIGURE 3.5 Spiral through-fissures due to chain graphitization. These fissures formed along lines of maximum shear stress.

spiraling around tube surfaces (Fig. 3.5). Such failures are sometimes confused with creep damage, but careful microscopic observation will reveal the presence of graphite nodules at or near fracture edges.

Thermal Oxidation

One sign of long-term overheating is usually a thick, brittle, dark iron oxide layer on both internal and external surfaces (Fig. 3.6). These oxide layers are often referred to as scale, but they should not be confused with deposition from constituents in the water.

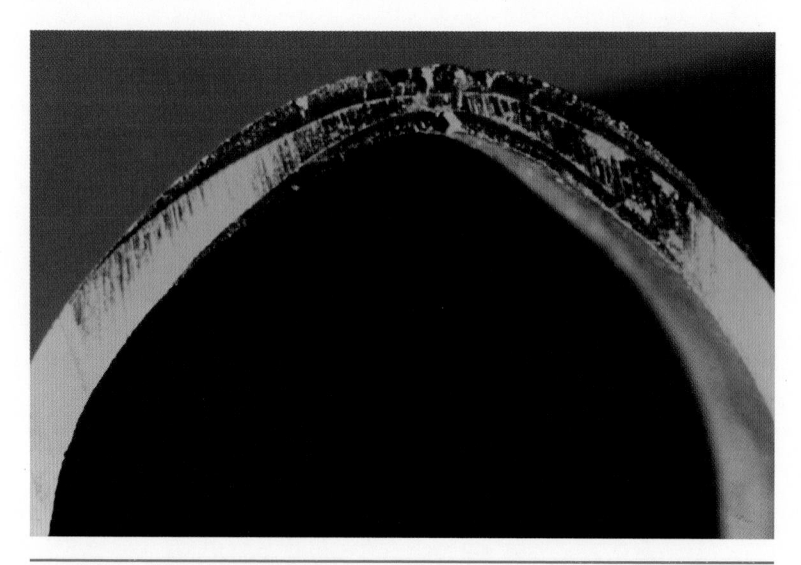

FIGURE 3.6 Thick, thermally formed oxide on internal and external surfaces of a bulged wall tube. Thinning of intact metal at bulge is due to both plastic deformation and thermal oxidation.

Typical boiler tube steels will generally form a thin, protective oxide layer under normal service conditions, depending on the environment. This layer will grow if the metal experiences elevated temperatures for extended times due to thermal oxidation (also known as metal burning). The weight loss associated with the conversion of metal to oxide as a function of time for carbon steel shows the increase of oxidation rate with increasing temperatures (Fig. 3.7).[5]

The type of oxide layers that form depends on temperatures and alloy composition. For plain carbon steel, magnetite (Fe_3O_4) is the primary oxide that forms at temperatures below 1060°F (571°C), and wustite (FeO) forms above this temperature.[6] Upon slow cooling, wustite can decompose into magnetite and elemental iron, with iron particles generally having a rounded morphology.[7] Because of slow transformation kinetics, wustite is often retained in the oxide layers on cooling. The presence of appreciable amounts of wustite in material overlying the tube surfaces can be an indication of overheating for extended periods above 1060°F (571°C). However, care should be taken to ensure that the wustite is not associated with tube cutting or welding debris.

If metal temperatures exceed a certain value, depending on the alloy, thermal oxidation can become excessive. Alloying the steel with chromium can dramatically increase the resistance to thermal oxidation,[8] as shown in Fig. 3.8.

Various references give temperatures considered to be the use limits for typical tube alloys. These limits account for excessive

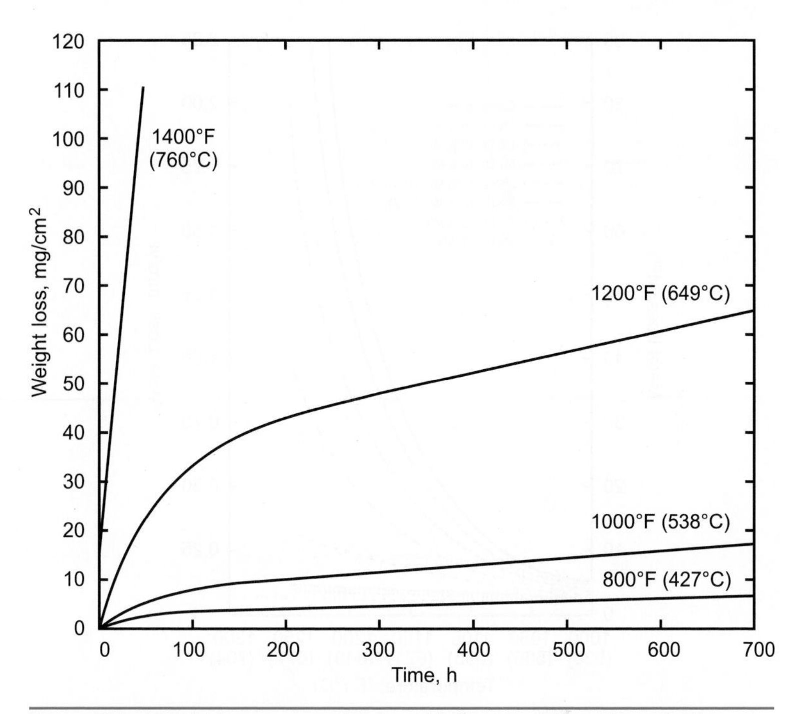

FIGURE 3.7 Weight loss curves due to oxidation on plain carbon steel as a function of time, showing increased conversion of metal to oxide with increasing temperatures. (*Reprinted with permission of ASM International. All rights reserved. www.asminternational.org*)

thermal oxidation as well as creep strength considerations. Temperature limits for some typical boiler tube alloys are compiled in Table 3.1. Note that there is considerable variation in the temperatures for some alloys, as the value depends on the criteria used to determine the limit. For instance, in the ASTM reference, the temperature at which oxidation becomes excessive is defined as that at which the area-normalized oxide weight reaches 10 mg/cm^2 in 1000 h.

Often, the thermally formed oxide layer contains longitudinal fissures and cracks. In some areas, patches of oxide may have exfoliated (Fig. 3.9). Cracks and exfoliated patches result from tube expansion and contraction caused by deformation during overheating and/or thermal stressing. Tube-wall thinning can result from cyclic thermal oxidation and spalling. Thermal oxidation can continue until the entire wall is converted to oxide, resulting in a failure (Fig. 3.10).

In many cases, tube surfaces will exhibit intergranular oxidation associated with long-term overheating. Intergranular oxidation generally has the form of numerous spherical iron oxide particles present intermittently along grain boundaries (Fig. 3.11). Although intergranular oxidation is often shallow, the depth of penetration and amount

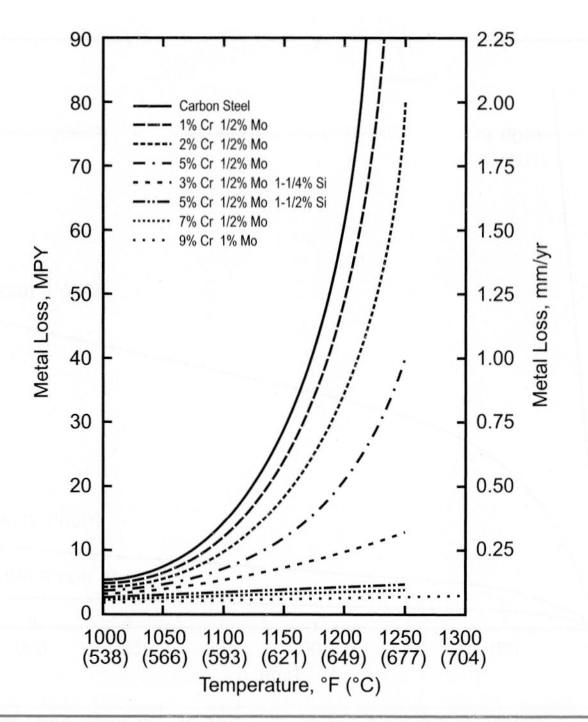

FIGURE 3.8 Effect of chromium on metal loss due to thermal oxidation in air as a function of temperature, showing increasing metal loss with temperature and decreasing metal loss with increasing levels of chromium for a given temperature. (*Reprinted with permission of ASM International. All rights reserved. www.asminternational.org*)

Steel Type	Typical Alloys	Combustion Engineering[9]	Riley Stoker[10]	Babcock & Wilcox[11]	ASTM[12]
Carbon steel	SA-178, SA-192, SA-210A, SA-106	850 (454)	850 (454)	950 (510)	1025 (552)
½ Mo steel	SA-209 T1, SA-209 T1a	900 (482)	900 (482)	975 (524)	1050 (565)
1¼ Cr, ½ Mo steel	SA-213 T11	1025 (552)	1025 (552)	1050 (566)	
2¼ Cr, 1 Mo steel	SA-213 T22	1075 (579)	1075 (579)	1125 (607)	1100 (593)
9 Cr, 1 Mo Steel	SA-213 T91	1175 (635)		1200 (649)	1420 (771)
18 Cr, 8 Ni (stainless) steel	SA-213 304H	1300 (704)		1400 (760)	1600 (871)

TABLE 3.1 Temperature Limits for Some Different Tube Alloys [°F (°C)]

FIGURE 3.9 Thermally deteriorated metal on failed wall tube of a utility boiler. Note the spalled and cracked oxide resembling tree bark caused by expansion of the tube during bulging and thermally induced stresses.

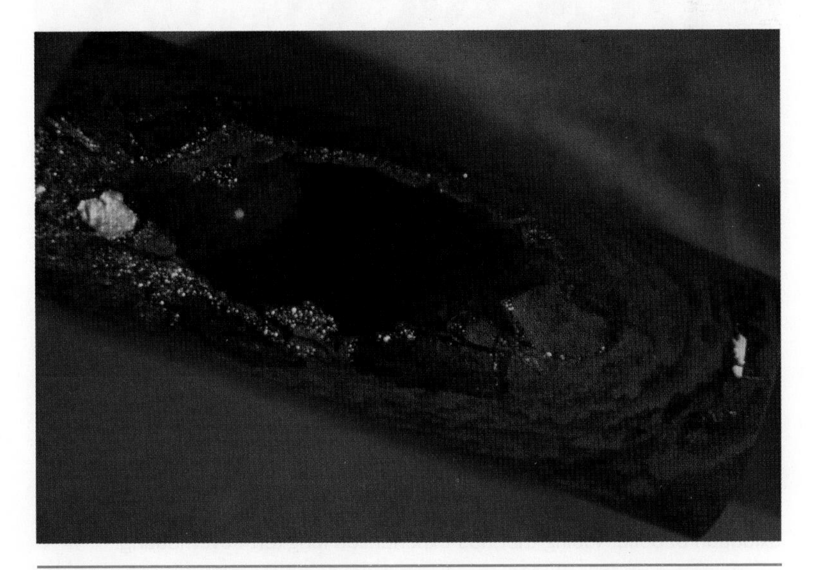

FIGURE 3.10 Through-wall oxidation of a tube from a low-pressure boiler. Brittle black magnetite patches still adhere to the rupture mouth. Elsewhere the oxide has spalled. Such through-wall oxidation is most common in low-pressure boilers, where internal pressures are not sufficient to cause premature failure.

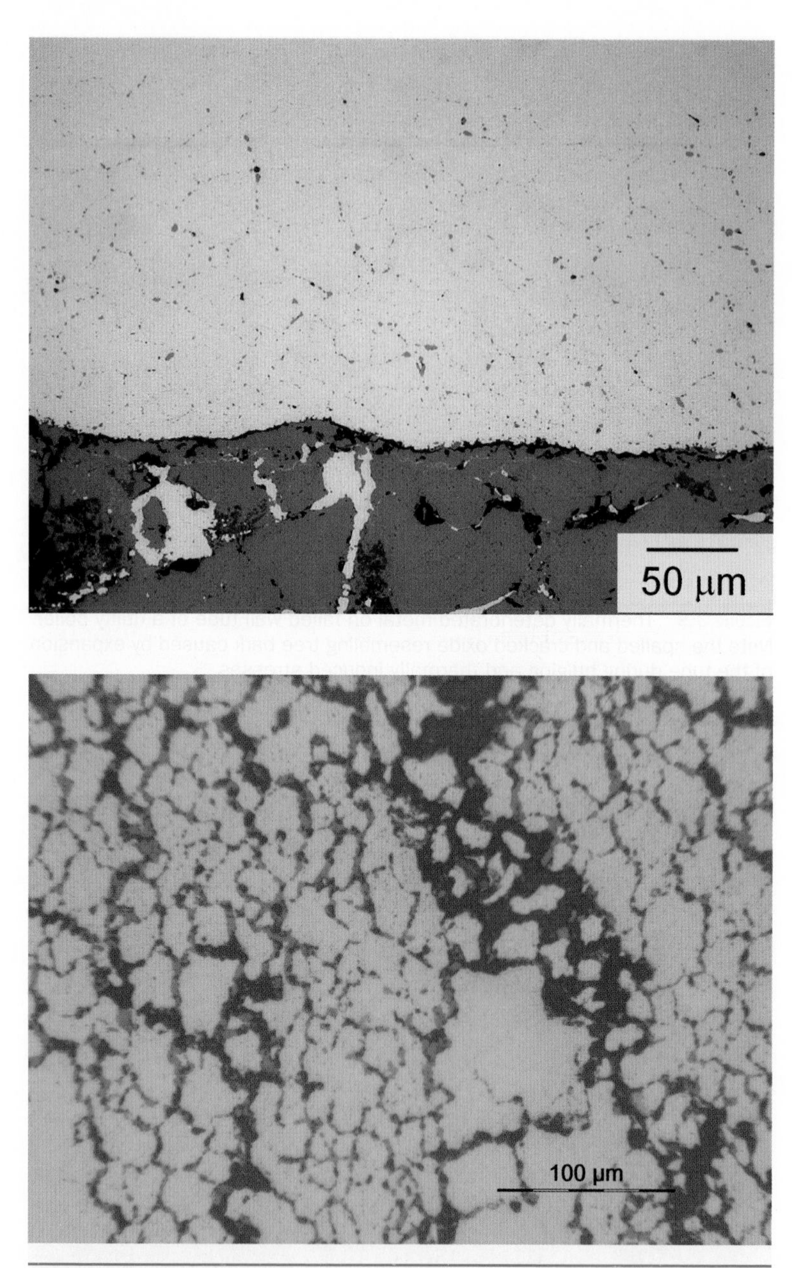

FIGURE 3.11 Examples of shallow (top) and severe (bottom) intergranular oxidation of boiler tubing that has experienced long-term overheating.

of oxide particles along the grain boundary tend to increase with increasing time and temperature. In rare cases, intergranular oxidation can be severe enough to significantly weaken the tube wall.

Creep Rupture

Creep is the plastic flow or deformation of metals at stresses lower than the yield stress, occurring at elevated temperatures over long periods. Creep deformation often produces microvoids and fissures at grain boundaries. Intergranular fissures can form when the microvoids link together, or when grain boundary sliding occurs (Fig. 3.12). Intergranular creep voids and fissures (referred to as *creep damage*) are often located in the areas adjacent to ruptures caused by long-term overheating. Evidence of creep deformation also tends to preferentially occur at locations of elevated stress, such as fissures on the surfaces. Failure occurs by rupture when the weakened tube wall cannot support the hoop stresses that develop from internal pressurization of the tube.

Creep rupture (also referred to as stress rupture) usually produces a thick-lipped rupture at the apex of a bulge. Often, a series of small, parallel, longitudinal fissures will be present at the apex of a heavily oxidized bulge (Fig. 3.13). In other cases, the rupture will be larger and have a fish-mouth shape (Fig. 3.14). The rupture edges are usually blunt and slightly ragged. Rupture edges may be slightly thinned

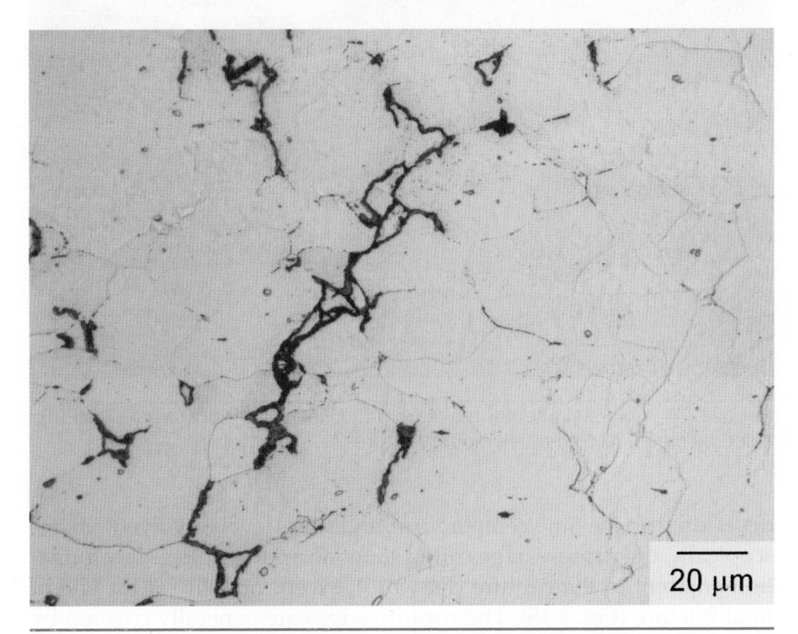

FIGURE 3.12 Close-up showing typical intergranular creep voids and fissures.

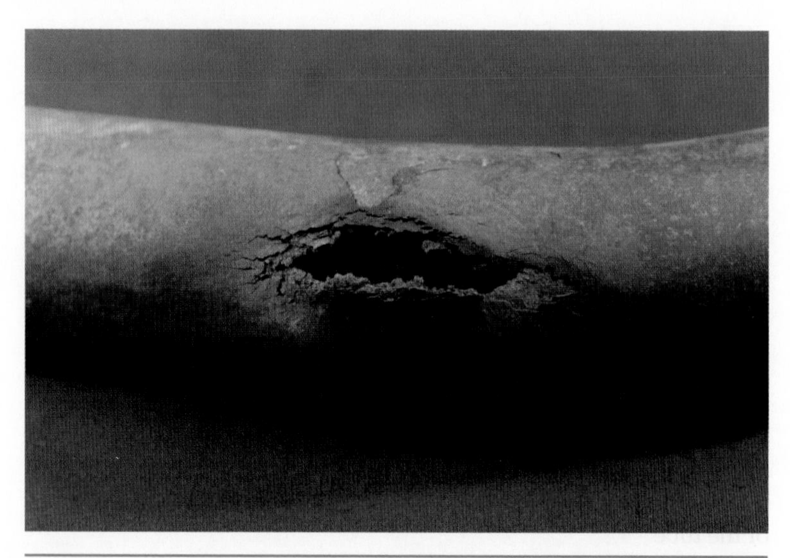

FIGURE 3.13 A small, ragged creep rupture at the apex of a bulge. Note the small secondary fissures and the relatively thick rupture edges.

FIGURE 3.14 Two superheater tubes from a utility boiler [1400 psi (9.7 MPa)], failed by creep. Note the fish-mouth ruptures.

depending on the time, temperature, and stress levels involved. Microscopically, the rupture edges often show no appreciable grain elongation adjacent to the rupture (Fig. 3.15), except possibly at the site of final fracture (Fig. 3.16). The rupture edges are generally covered by oxide layers that can be appreciably thick. Numerous secondary

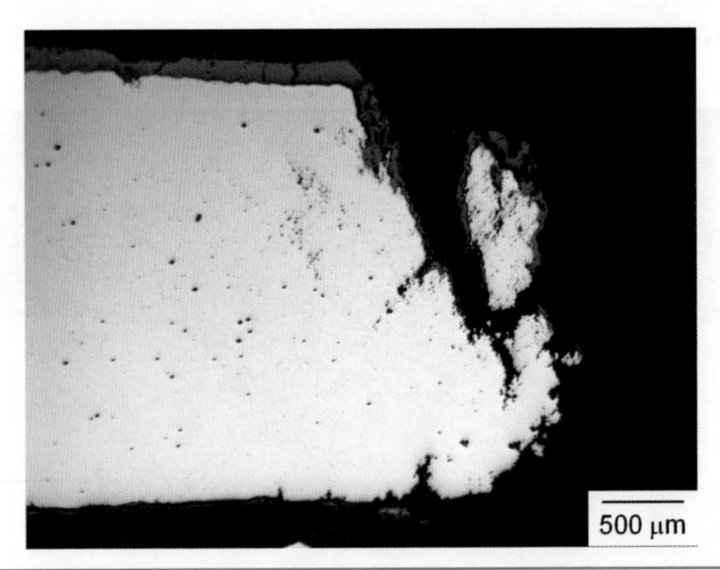

FIGURE 3.15 Microscopic cross section through a rupture that failed due to creep. Note the oxide layers covering the edges, especially near the external surface (top of image).

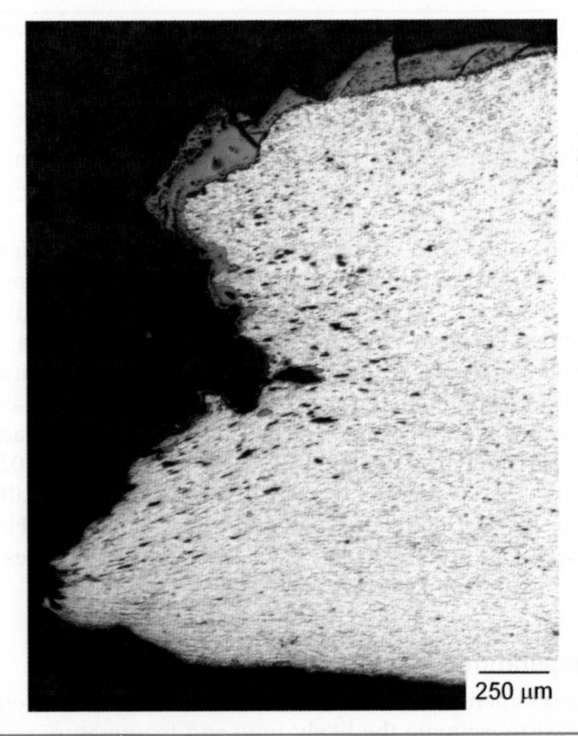

FIGURE 3.16 Microscopic cross section through a rupture that failed due to tensile overload after deep penetration of a creep fissure. Note the oxide-lined fissure penetrating from the external surface and the deformation and necking of a portion of the tube wall near the internal surface (bottom of image).

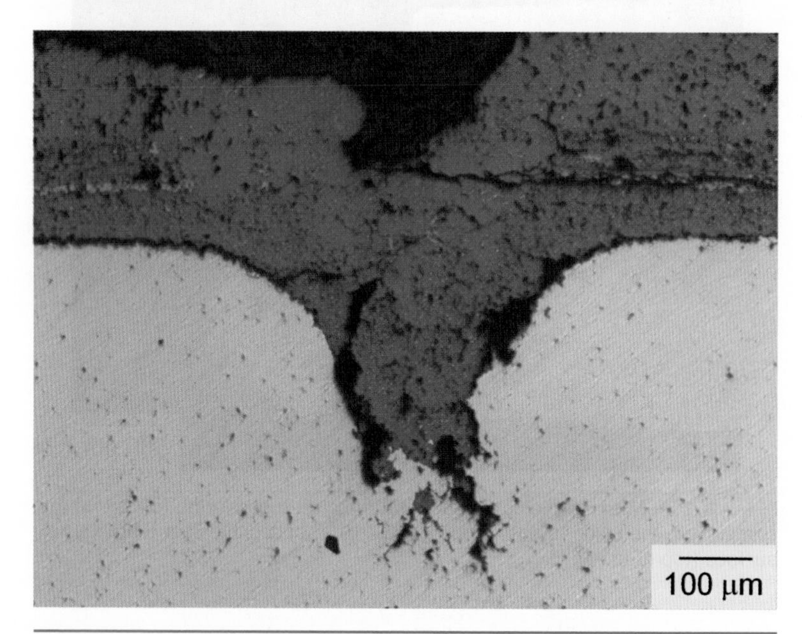

FIGURE 3.17 Oxide-filled creep fissure on the external surface of an overheated boiler tube showing intergranular creep voids and fisures at the tip.

oxide-filled fissures with intergranular penetrations are often present on the tube surfaces near ruptures, especially along the external surfaces (Fig. 3.17).

Alloying steel with certain elements can improve resistance to creep and stress-rupture properties. The *stress-rupture* (*creep-rupture*) *strength* is the stress that causes fracture in a creep test in a given time in a specified constant environment.[13] Stress-rupture data, specifically the temperatures to rupture in 100,000 h, for three typical boiler tubing alloys with different molybdenum levels, are shown as a function of stress and temperature in Fig. 3.18.[14] Molybdenum additions (SA-192 has nominally 0% Mo, SA-213 T11 has nominally 0.5% Mo, and SA-213 T22 has nominally 1% Mo) provide increased allowable stresses for a given temperature for some low-alloy steels. However, carbon steels are typically used where elevated temperatures are not expected, mostly due to lower costs.

Time-Temperature Estimation

Careful assessment of features associated with long-term overheating can be used to estimate the likely time and temperature ranges for overheating. For instance, stresses required to produce a certain amount of creep deformation or for stress-rupture failure have been correlated to a parameter known as the *Larson-Miller parameter* (LMP).[15]

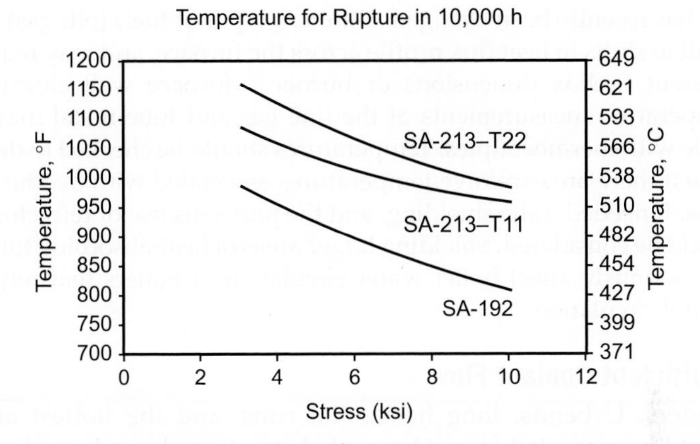

FIGURE 3.18 Stress rupture data for three boiler tube steels, showing an increase in temperature for rupture with increasing molybdenum content (SA-192 has nominally 0% Mo, SA-213 T11 has nominally 0.5% Mo, and SA-213 T22 has nominally 1% Mo).

The LMP incorporates both time and temperature into the following equation:

$$\text{LMP} = (T + 460)(20 + \log t)$$

where T = temperature, °F
t = time, h

In addition to rupture and creep stresses, steamside oxide thickness has been correlated to the LMP.[16] Rupture strength and oxide thickness data are compiled for some typical boiler tube alloys in multiple sources.[17, 18] These relationships provide a method to estimate likely time and temperature scenarios for overheating failures. The data are also useful for remaining tube life assessment calculations for superheaters and reheaters.

Elimination

Eliminating long-term overheating requires removal of a chronic condition that causes increased metal temperatures in the system. In some cases, the component is at the end of its expected service life. Some conditions and possible contributory factors that can reduce the expected service life of a component are discussed below.

Excessive Heat Input

The firing procedures and burner alignment for the boiler should be reviewed. An increase in the Btu value of fuels can be an issue if the fuel has recently been changed. Different types of fuels (oil/gas) will result in shifts in heat flux profile across the furnace, and may require different firebox dimensions or burner-to-furnace wall clearance. Temperature measurements of the flue gas and tube metal may be made with thermocouples. Temperatures should be checked to determine if there are excessive temperatures associated with overheated areas. If needed, tube shielding, and the judicious use of refractories, should be considered. Shielding larger areas of heat-absorbing tubing may seriously affect boiler water circulation in boilers that rely on natural circulation.

Insufficient Coolant Flow

Headers, U bends, long horizontal runs, and the hottest areas should be inspected for evidence of obstruction, deposits, and other foreign material that can restrict coolant flow. In addition, mechanical, operational, and design issues that may limit boiler circulation should be investigated. For instance, excessive blowdown rates can also result in insufficient coolant flow. Attemperation procedures should be reviewed.

Deposits

Excessive deposits should be removed by chemical or mechanical cleaning and prevented from recurring. The source of significant deposits must be identified and eliminated. Common causes of deposits include ineffective pretreatment, fluctuations in makeup water quality, improper water treatment, system contamination, poor circulation, and/or excessive heat input. Each potential cause must be addressed methodically.

Carryover of boiler water into superheater tubing should be prevented. Some common causes of mechanical carryover of boiler water include leakage around steam separating equipment, foaming caused by water chemistry or contamination, and priming generally related to high drum levels.[19] Significant boiler load swings can also promote carryover.

Improper Material Selection

Material deficiency occurs when a metal is used in an environment that exceeds its thermal stability with regard to oxidation and stress-rupture strength. Typically, this involves the use of unalloyed carbon steel in high-temperature superheater or reheater sections, where normal operating temperatures would require the use of low-alloy steels or even higher alloys, such as austenitic stainless steels. An examination of material specifications for the superheater or reheater sections will disclose the type of steel required for service. The primary responsibility for prevention of improper material use rests with the boiler manufacturer, boiler constructor, and purchaser of the boiler. Elimination requires close adherence to effective quality control procedures, beginning with equipment design and carrying through to equipment installation.

Cautions

It is incorrect to assume that long-term overheating automatically produces significant tube damage. While small microstructural changes may occur in the tube wall, these changes often do little to reduce service life or significantly weaken the tube. However, if overheating continues for a long period of time, failures will eventually result. Accumulation of other damage related to overheating, such as creep and thermal oxidation, can appreciably reduce the strength of the tube. Moreover, this damage is irreversible.

Long-term and short-term overheating failures may appear similar. Frequently, evidence of long-term overheating at temperatures slightly above the design limits (mild) and a brief period at severe temperatures will be present at the same failure. Because a brief episode of short-term overheating may follow long-term overheating, the sequence of thermal events can be difficult to diagnose without microscopic examination. Microscopic examination can also distinguish a thick-walled rupture that is associated with creep damage from other mechanisms that produce thick-walled failures, such as hydrogen damage, corrosion fatigue cracking, or stress corrosion cracking. In addition, careful analysis can determine if a combination of mechanisms may have occurred, such as long-term overheating and corrosion fatigue cracking or hydrogen damage. In these cases, metallographic analysis can also reveal which mechanism was predominately responsible for the failure, and ascertain which mechanism was likely active first.

Failures due to long-term overheating are sometimes associated with chemical attack and other significant metal wastage, while chemical attack in short-term overheating is rare. Many fireside corrosion mechanisms occur only at elevated temperatures or have corrosion rates that increase significantly with increasing temperature. Creep-rupture types of failures are promoted at higher stress levels, which can arise due to thinning of the tube wall by metal wastage such as corrosion or erosion.

Long-term overheating failures due to excessive oxidation and creep rupture of superheater and reheater tubes may occur both in properly specified and installed low-alloy steels and in improperly used plain carbon steel tubes. Therefore, the occurrence of a long-term overheating failure is not proof that a material has been incorrectly used in an application. If long-term overheating failures occur in properly operating superheater or reheater sections; however, it is advisable to confirm the correct use of the specified tube alloy. The time in operation should also be considered, as the tubing may be at the end of its useful service life.

Related Problems

See also Chap. 1, Water-Formed and Steam-Formed Deposits; Chap. 2, Short-Term Overheating; Chap. 4, Caustic Corrosion; Chap. 20, Waterwall (Furnace) Fireside Corrosion; and Chap. 21, High-Temperature Fireside Corrosion.

Case History 3.1

Industry:	Gas products
Specimen location:	Superheater, 3 ft (0.91 m) above firebox floor near center of boiler
Orientation of specimen:	Horizontal, immediately adjacent to a firebrick wall
Years in service:	12
Water treatment program:	Phosphate
Drum pressure:	Design is 700 psi (4.8 MPa) package boiler, but operated at 600 psig (4.1 MPa)
Tube specifications:	2-in. (5.1-cm) outer diameter, SA-213 T22
Fuel:	Natural gas

A brittle black magnetite layer covered both the external and internal surfaces of this superheater tube. The thermally formed oxide fractured and spalled, substantially reducing wall thickness (Fig. 3.19). Thinning was more severe along the side of the section abutting a firebrick wall.

Most metal was lost from external surfaces, and this was caused by thermal oxidation at temperatures between 1100 and 1341°F (593 and 727°C). Gas channeling and higher temperatures were present along the tube side abutting the firebrick. These factors accelerated the oxidation and spalling processes.

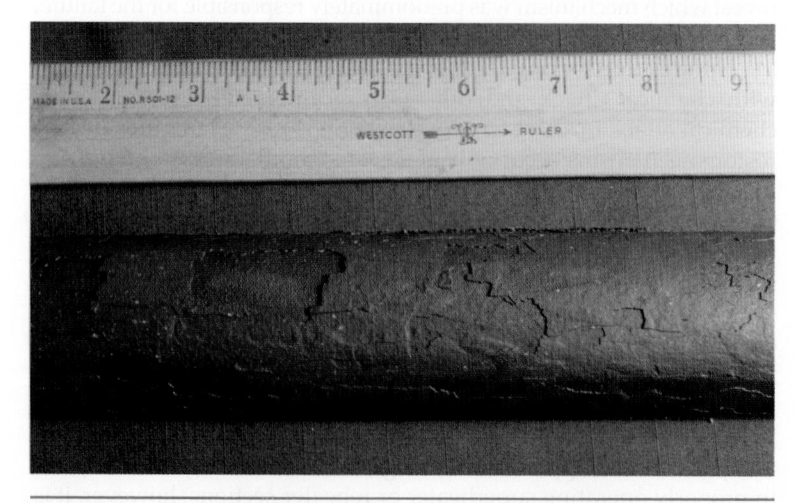

FIGURE 3.19 Severe thermal oxidation and spalling of magnetite on external surface. The laminated nature of the scale indicates multiple episodes of spalling and oxide reformation.

In the past, nearby tubes had ruptured as a result of thinning associated with thermal oxidation. No deposits were present on internal surfaces, and attemperation was not used. Fireside slagging was not present in this intermittently operated boiler.

Failures were a chronic problem associated with design and operation. The proximity of failures to the firebrick wall strongly linked the overheating to boiler design.

Case History 3.2

Industry:	Steel
Specimen location:	Center of waterwall
Orientation of specimen:	Vertical
Years in service:	8
Water treatment program:	Phosphate
Drum pressure:	1200 psi (8.3 MPa)
Tube specifications:	3-in. (7.6-cm) outer diameter
Fuel:	Blast furnace gas

Longitudinal fissures and thick-walled ruptures were present at bulges along the hot side of this section (Fig. 3.20). External surfaces are covered with an irregular tan slag layer, and the area around a rupture is checkered.

The rupture and fissuring were caused by very long-term exposure of the metal to temperatures between 850 and 1050°F (454 and 566°C). Evidence suggests that exposure to these temperatures may have been occurring for several years. Overheating was caused by excessive heat input relative to the coolant flow rate.

Nearby tubes showed similar, although less severe, evidence of overheating. Deposits on hot-side internal surfaces were less than 20 g/ft^2 (22 mg/cm^2). The boiler had been cleaned 3 years prior to failure.

FIGURE 3.20 Longitudinal fissures along hot side. Note the through-fissure at the bulge apex.

Case History 3.3

Industry:	Utility
Specimen location:	Rear wall, 30 ft (9.1 m) from front wall
Orientation of specimen:	Vertical
Years in service:	1½
Water treatment program:	Polymer
Drum pressure:	620 psi (4.3 MPa)
Tube specifications:	3-in. (7.6-cm) outer diameter
Fuel:	Pulverized coal

The tube had a series of several prominent bulges longitudinally aligned along the hot-side crown (Fig. 3.2). Thick layers of hard, tenacious iron oxide covered each bulge, except where spalling dislodged the oxide. Internal surfaces on the hot side were covered with spongy, porous deposits, which cover a hard, black magnetite layer. The back wall at the same elevation saw many failures. The boiler had frequent load swings and was operated intermittently.

The tube was overheated in a temperature range between 950 and 1150°F (510 and 621°C) at the bulges for a long time. Formation of deposits was due to an imbalance between coolant flow and fireside heat input. Deposit weights were about 5 g/ft^2 (5 mg/cm^2) on the cold side and 26 g/ft^2 (28 mg/cm^2) on the hot side. Deposits were caused by exceeding the solubility of inversely soluble species, by evaporative concentration, and by mechanical entrainment of particulate matter.

Because of the frequent load swings and intermittent operation, closer operational monitoring was suggested.

Case History 3.4

Industry:	Steel
Specimen location:	Primary superheater bank
Orientation of specimen:	Vertical pendant
Years in service:	16
Water treatment program:	Polymer
Drum pressure:	1200 psi (8.3 MPa)
Tube specifications:	2¼-in. (5.7-cm) outer diameter
Fuel:	Coke gas

Three large bulges are present on the pendant legs of a U-bend section. Each bulge is ruptured at its apex. The legs expanded to about a 2½-in. (6.4-cm) outer diameter from their 2¼-in. (5.7-cm) original value. A bulge is shown in Fig. 3.21. Internal surfaces are free of significant deposits, while external surfaces are covered with a tenacious, fragmented oxide layer. Eighteen tubes were similarly affected. This boiler operated on waste heat from a slab furnace.

The tube was overheated for several days or longer at temperatures above 1000°F (538°C), but below 1341°F (727°C). The wall strength was decreased at elevated temperatures, and tubes were thinned and weakened by thermal oxidation. As a result of these two forms of weakening, the legs bulged and then ruptured.

Operating records showed a significant increase of almost 100°F (38°C) in steam temperature approximately 2 months prior to failure. This increase was correlated with changes in firing associated with alteration of the slab mill's operation.

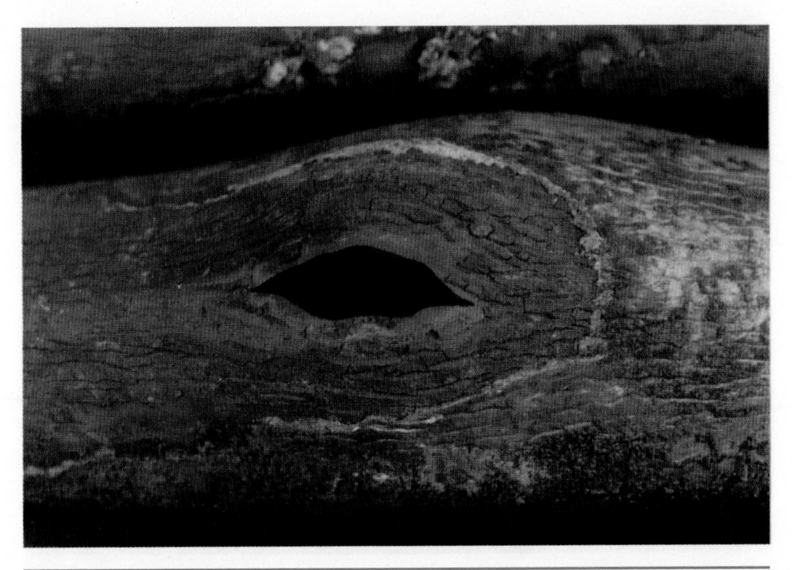

FIGURE 3.21 Fragmented thermally formed oxide surrounding a rupture.

Case History 3.5

Industry:	Utility
Specimen location:	Primary superheater inlet; section with hottest flue gas
Orientation of specimen:	Horizontal
Years in service:	20
Water treatment program:	Phosphate
Drum pressure:	1200 psi (8.3 MPa)
Tube specifications:	Circumferentially welded tubes, 2¼-in. (5.7-cm) outer diameter
Fuel:	Coal (1% S, less than 10% ash) 6 months prior to failure, previously oil (2% sulfur, 200 to 400 ppm vanadium) was used as fuel

The section contains a massive, thick-walled, longitudinal rupture just above a circumferential weld. The rupture occurred immediately downstream of the weld. The tube was bent into an L shape by the burst (Fig. 3.22). A thick, tenacious magnetite layer covers external surfaces, except at areas near the rupture, where the oxide has cracked and spalled. Elemental copper and spotty deposits are present on internal surfaces.

Failure was due to stress (creep) rupture caused by prolonged overheating at temperatures above 1050°F (566°C). Contributing factors included coolant flow irregularities immediately downstream of a partially intrusive circumferential weld and internal deposition, which reduced heat transfer. Additionally, a switch from oil to coal firing likely changed the fireside heat input.

The superheater had a history of boiler water carryover, and load swings were common. A previous failure had occurred in this region at least 2 years before this rupture.

FIGURE 3.22 Massive thick-walled fracture caused by creep. Note the circumferential weld just below the failure. The tube is torn and bent to a 90° elbow by the violence of the rupture.

Case History 3.6

Industry:	Utility
Specimen location:	Reheater
Specimen orientation:	Vertical
Years in service:	20
Water treatment program:	Coordinated phosphate
Drum pressure:	2875 psi (19.8 MPa)
Tube specifications:	2¼-in. (5.7-cm) outer diameter, plain carbon steel
Fuel:	Coal

The thick-walled rupture shown in Fig. 3.23 is one of many similar failures that occurred sporadically in this section of the boiler. The external surfaces are covered with thick layers of black iron oxide (thermally oxidized metal) overlaid with light-colored deposits. The internal surface is also covered with thick layers of iron oxide.

Microstructural examinations of the metal revealed a dense population of graphite nodules and spheroidal iron carbides in the tube wall. Intergranular oxidation was apparent at all surfaces.

The failure of the tube was a direct result of operation in an environment that exceeded the thermal stability of the plain carbon steel. A low-alloy steel tube (SA-213 T22) that had been butt-welded to the ruptured tube did not fail and had sustained only mild thermal deterioration. Replacing the plain carbon steel in the failing sections with a low-alloy steel such as SA-213 T22 was recommended to prolong service life.

Case History 3.7

Industry:	Food processing
Specimen location:	Waterwall tube in a D-type package boiler

FIGURE 3.23 Thick-walled rupture in a reheater tube. Note fissures on external surface aligned with the rupture.

Specimen orientation:	Vertical
Years in service:	2
Water treatment program:	Polymer, sulfite
Drum pressure:	300 psi (2.1 MPa)
Tube specifications:	2-in. (5.1-cm) outer diameter, carbon steel
Fuel:	Natural gas

A section of waterwall tubing contained multiple narrow, longitudinally oriented ruptures at slight bulges. The ruptures were located along one end of the section in an area covered by dark, thick oxide on the external surface (Fig. 3.24). This area abruptly terminated near the center of the section. A cross section through the tube near this location revealed a large amount of accumulated chip scale (Fig. 3.25), which partially blocked the tube's inner diameter over a length of the section upstream from the failures.

Microstructural evidence, including partially spheroidized high-temperature transformation products on the cold side, suggested that metal temperatures in all locations examined likely exceeded 1341°F (727°C) for a very brief period of time. However, spheroidization of the microstructure along the hot side indicated that metal temperatures in the range of 1000 to 1341°F (538 to 727°C) were likely reached for extended periods after the brief higher-temperature overheating event. Decarburization and grain growth in bulged regions indicated higher metal temperatures at these locations. The failures ultimately occurred due to creep rupture, as evidenced by the thick-edged ruptures as well as intergranular cracks and voids near the rupture edges.

Very thick, stratified layers of deposited material were present at locations away from the bulged region and blockage. Analysis of the deposit indicated that it consisted primarily of sodium calcium silicate (pectolite). The presence of appreciable amounts of sodium indicated evaporation to dryness conditions occurred in the tube.

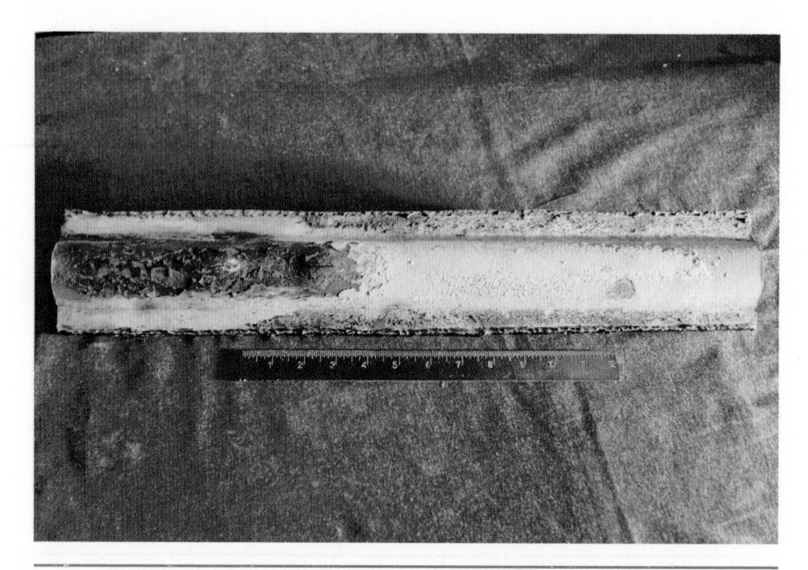

Figure 3.24 Waterwall tube from a package boiler containing multiple narrow ruptures in an area on the hot side that experienced appreciable thermal oxidation.

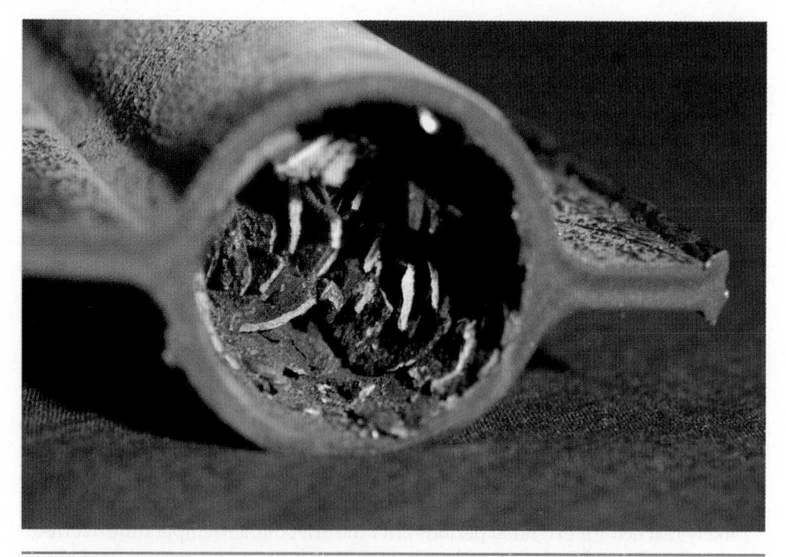

Figure 3.25 Cross section near a region of partial blockage, showing accumulated chip scale on the internal surface.

Overheating was caused by insufficient coolant flow due to partial blockage of the tube's inner diameter by chip scale. The chips of deposited material originated from other locations in the system, likely the same tube. The system was reportedly cleaned annually by hydroblasting, and the chip scale was likely related to this cleaning method.

Case History 3.8

Industry:	Facility powerhouse
Specimen location:	Riser tubes in a heat recovery steam generator (HRSG)
Specimen orientation:	Vertical
Years in service:	5
Water treatment program:	Polymer
Drum pressure:	825 psi (5.7 MPa)
Tube specifications:	2-in. (5.1-cm) outer diameter, externally finned, SA-178A plain carbon steel
Fuel:	Natural gas turbine exhaust, supplemental duct firing

Numerous failures occurred in the outer bank of tubes in two HRSG units that used the same boiler feedwater. The failures consisted of narrow, longitudinally oriented ruptures in bulged regions (Fig. 3.26). The fins in the bulged regions were consumed by thermal oxidation along the outer edge (sometimes referred to as "crisped fins"). The microstructure along the tubes' hot sides consisted of graphite nodules and spheroidized carbides, indicating metal temperatures from 850 to 1341°F (454 to 727°C) were attained for extended periods. Some locations in the bulged regions exceeded 1341°F (727°C). This temperature was above the reported upper gas turbine exhaust temperature of 1200°F (649°C), indicating that supplemental firing of the duct burners contributed to the overheating. The hot-side internal surfaces of the failed tube sections were covered by thick deposit layers that spalled from the bulge locations (Fig. 3.27).

Overheating was caused by excessive heat input relative to the coolant flow rate in combination with significant internal surface deposition. Deposit weight density measurements indicated hot-side values that ranged between about 120 and

FIGURE 3.26 Multiple bulges and ruptures along a hot side of externally finned HRSG tubing.

FIGURE 3.27 Internal surfaces showing thick deposit layers that spalled from a bulged region.

160 g/ft^2 (129 and 172 mg/cm^2). Significant deposition was also present on the cold side of the tubes, ranging from 60 to 120 g/ft^2 (65 to 129 mg/cm^2), suggesting that water chemistry was overwhelmed by contaminants and/or the chemical water treatment was possibly insufficient. The deposit layers were very distinctly stratified (Fig. 3.28), indicating that they formed during different periods of time and water chemistries. Analysis of the deposits indicated they contained major amounts of calcium and silicon and minor amounts of magnesium and iron in the form of magnesium silicate hydroxide (lizardite), $Mg_3Si_2O_5(OH)_4$; magnesium hydroxide (brucite), $Mg(OH)_2$; calcium carbonate (calcite), $CaCO_3$; and iron oxide (magnetite), Fe_3O_4. Review of the operator logs highlighted periodic upset conditions that occurred in the feedwater, including high hardness excursions, elevated silica levels, and insufficient chemical treatment levels.

Case History 3.9

Industry:	SAGD oil recovery
Specimen location:	Economizer tubes in a once-through steam generator (OTSG)
Specimen orientation:	Horizontal
Years in service:	6
Water treatment program:	Polymer
Pressure:	1500 psi (10.3 MPa)
Tube specifications:	3½-in. (8.9-cm) outer diameter, SA-106C plain carbon steel
Fuel:	Natural gas

FIGURE 3.28 Close-up of the stratified deposit layers on the hot-side internal surface.

The furnace pressure for the unit steadily increased over time, possibly many days, until the safeties for the unit tripped and it was brought down. Inspection revealed many tubes in multiple passes and rows of the economizer section that had experienced bulging (Fig. 3.29). Short, narrow, longitudinally oriented ruptures were present along many of the bulges. Microscopic examination through the ruptures revealed creep fissures (Fig. 3.30), and the microstructure consisted of spheroidal carbides in a ferrite matrix. These characteristics indicated long-term overheating in the temperature range 1025 to 1341°F (552 to 727°C).

FIGURE 3.29 Economizer tube section from an OTSG with multiple bulges along the hot side.

FIGURE 3.30 Cross section through a rupture showing deep oxide-lined fissures with surrounding intergranular creep damage.

Oxide thickness measurements suggested that the overheating event was on the order of many days to weeks. Evidence of overheating was localized to the bulged regions.

The internal surfaces of the failed tubes were covered by layers of thick deposits (Fig. 3.31). Deposit weight measurements on the bulged tubes ranged from 100 to 160 g/ft^2 (108 to 172 mg/cm^2) on the hot side to 70 to 110 g/ft^2 (75 to 118 mg/cm^2) on the cold side. The deposit morphology and consistency on the hot and cold sides were generally very similar. The deposit layers consisted primarily of carbon (generally greater than 40%) with minor amounts of silicon (generally between 15 and 25%) and magnesium (generally between 8 and 15%). Some sodium was also present in most samples, suggesting evaporation to dryness. The deposit was very stratified with layers rich in magnesium and silicon, and layers that were predominately carbon, suggesting periods of upsets in the feedwater to the unit. Review of data from the pretreatment facility indicated periods of high turbidity that were likely associated with multiple issues with the warm lime softener.

Overheating was primarily caused by the thick layers of internal surface deposits that significantly reduced heat transfer. In addition to the amount of deposits on the tube, the morphology of the deposit layers was a significant

FIGURE 3.31 Thick, stratified deposit layers along the entire tube circumference contain many blistered regions that created pockets between the deposit layers.

factor. The deposit layers contained many large blisters that created pockets where steam could have formed (Fig. 3.31). The combination of thick internal surface deposits and steam pockets in blistered deposits thermally insulated the tube metal from the cooling effects of the water. The blisters in the deposit layers were likely related to the high carbon levels in the deposit, which were due to bitumen contamination in the boiler feedwater. Removal of the internal surface deposition would increase heat transfer and reduce metal temperatures. Control measures related to pretreatment of the boiler feedwater and monitoring for upset conditions were needed.

Case History 3.10

Industry:	Chemical process
Specimen location:	Inlet tubesheet of a shell-and-tube waste heat boiler
Specimen orientation:	Horizontal tubes
Service time:	7 months
Water treatment program:	Polymer, sulfite
Unit pressure:	250 psi (1.7 MPa)
Tube specifications:	2-in. (5.1-cm) outer diameter, plain carbon steel
Fuel:	Waste heat gas

The unit was experiencing recurrent leaks at scattered locations on the waste heat inlet tubesheet, where the tubes were welded to the tubesheet. Waste heat from process gases typically ranging in temperature from 1500 to 1850°F (816 to 1010°C) entered the boiler and tubes on the inlet end. Other waste heat boilers reportedly operated for years with similar service conditions but did not experience any failures.

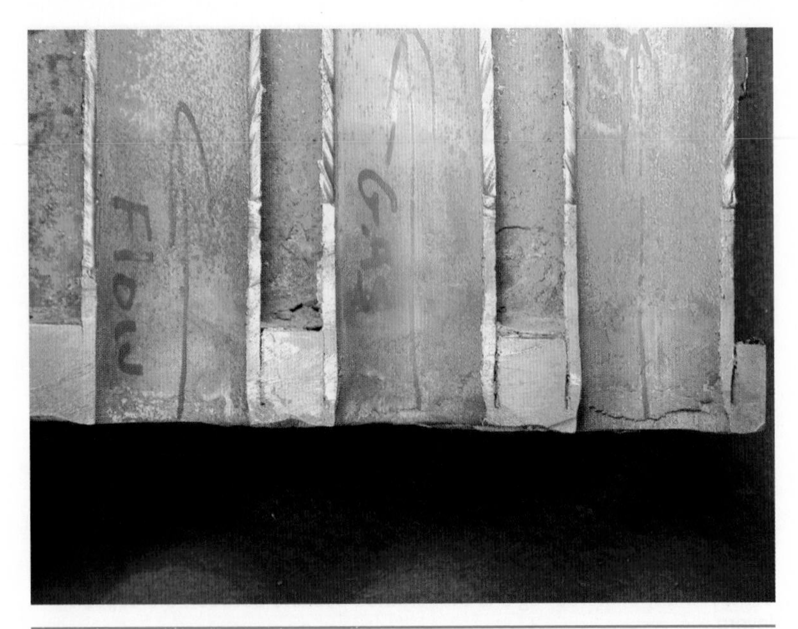

FIGURE 3.32 Cross section through some tube-to-tubesheet joints, showing significant thermal oxidation on the tubes at wide gaps between the tubes and tubesheet. (*© NACE International.*)

Cross sections revealed gaps between the tubes and the tubesheet. The gaps varied in size at the different locations (Fig. 3.32), but tended to be wider along the top sides of the tubes. Cross sections revealed appreciable tube-wall thinning at locations with wide gaps, especially near the welded end. Wall thinning was predominately caused by the formation of very thick layers of thermally formed iron oxide on both the waterside and fireside surfaces of the tube (Fig. 3.33). The thick, thermally formed oxide layers only extended a slight distance away from the tubesheet. In addition, only thin oxide layers were present on the tube surfaces in areas with narrow gaps between the tubes and tubesheet. The tube and tubesheet microstructures at locations with wide gaps consisted of spheroidal carbides and graphite nodules in a ferrite matrix (Fig. 3.34). However, the microstructures at locations with narrow gaps did not show evidence of significant thermal alteration, only slight spheroidization of the pearlite colonies in places.

Although evidence of a failure was not observed on the section, the leaks in the unit likely developed from wall thinning due to localized overheating at the wide gaps for extended periods and associated thermal oxidation. Because the gaps were open to the water side, water could enter the gap and form steam. However, the geometry of the wide gap restricted water flow and promoted formation of a steam layer, which resulted in significantly increased metal temperatures.

The waste heat boiler tube-to-tubesheet joints were supposed to be rolled, welded, and rerolled. However, appreciable variation was observed across

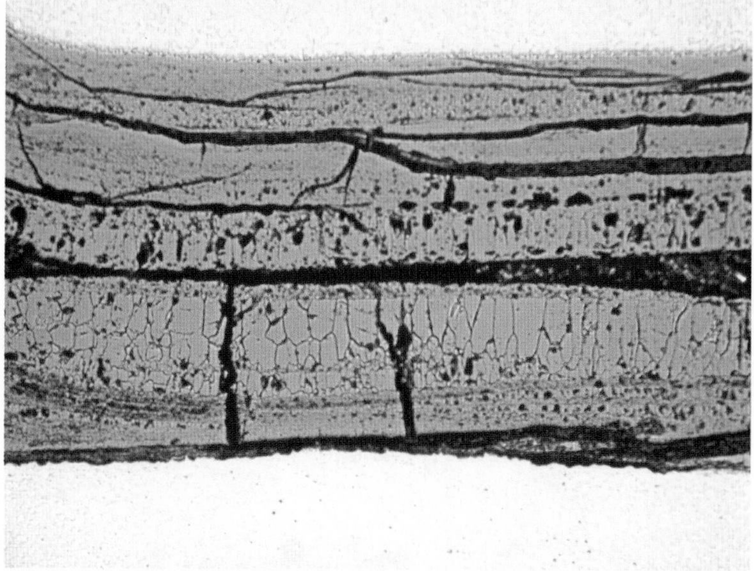

FIGURE 3.33 Thick layers of thermally formed oxide on the fireside (top) and waterside (bottom) surfaces at a location of a wide gap (unetched). (*© NACE International.*)

the tubesheet. In addition, tube expansion did not extend along the complete thickness of the tubesheet in some locations. The reported scattered failures strongly suggested that the rolling operations were insufficient. Because of the degradation from oxidation due to overheating, it was likely that the tubesheet had to be rebuilt.

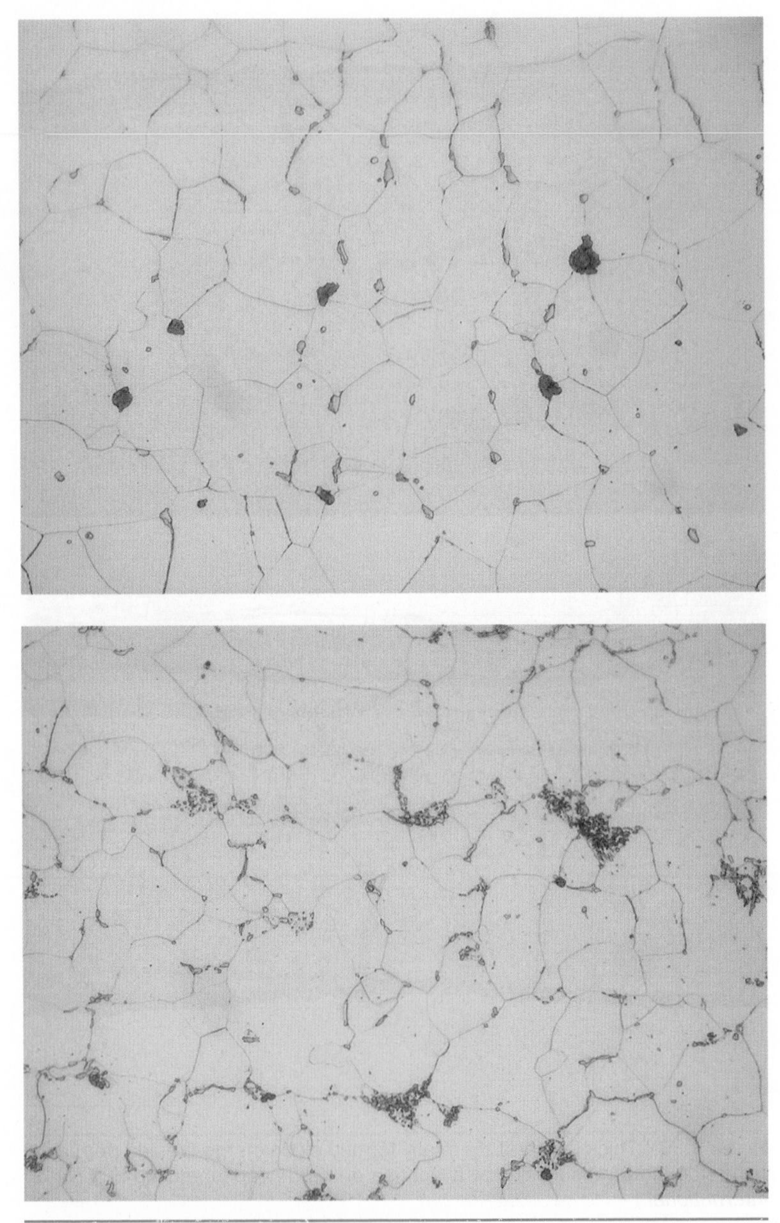

FIGURE 3.34 Comparison of tube microstructures at a wide gap (top) and a tight gap (bottom), showing graphitization and spheroidization at the location of a wide gap. Etchant Nital. (© *NACE International.*)

References

1. J. G. Singer (Ed.), *Combustion–Fossil Power Systems,* 3d ed., Combustion Engineering, Inc., Windsor, Conn., 1981, pp. 6–20.
2. Ibid., pp. 11–25.
3. D. J. Flynn (Ed.), *The Nalco Water Handbook,* 3d ed., McGraw-Hill, New York, 2009, pp. 11–17.
4. Singer, *Combustion–Fossil Power Systems,* pp. 6–41.
5. W. R. Patterson, *Designing for Automotive Corrosion Prevention, Proceedings,* Society for Automotive Engineers, Troy, Mich., November 8–10, 1978, p. 71.
6. *ASM Handbook,* vol. 13A: *Corrosion: Fundamentals, Testing, and Protection,* ASM International, Materials Park, Ohio, 2003, p. 92.
7. L. E. Samuels, *Light Microscopy of Carbon Steels,* ASM International, Materials Park, Ohio, 1999, p. 367.
8. A. W. Zeuthen, *Heating, Piping, and Air Conditioning,* vol. 42, no. 1, 1970, p. 152.
9. Singer, *Combustion Fossil Power Systems,* pp. 6–43.
10. D. N. French, *Metallurgical Failures in Fossil Fired Boilers,* 2d ed., John Wiley & Sons, New York, 1993, p. 496.
11. *Steam: Its Generation and Use,* 39th ed., Babcock & Wilcox Company, New York, 1978, pp. 29–11.
12. American Society for Testing and Materials, *Data on Corrosion and Heat Resistant Steels and Alloys—Wrought and Cast,* May 1950.
13. *ASM Handbook,* vol.11: *Failure Analysis and Prevention,* ASM International. Materials Park, Ohio, 2002, p. 1065.
14. EPRI TR-102433-2, *Boiler Tube Failure Metallurgical Guide,* vol. 2: *Appendices, Appendix C, Elevated Temperature Properties,* Electric Power Research Institute, Palo Alto, CA, October 1993.
15. F. R. Lawson and J. Miller, "A Time-Temperature Relationship for Rupture and Creep Stresses," *Transactions AIME,* July 1952, pp. 765–775.
16. I. M. Rehn, W. R. Apblett, Jr., and J. Stringer, "Controlling Steamside Oxide Exfoliation in Utility Boiler Superheaters and Reheaters," *Materials Performance,* June 1981, pp. 27–31.
17. EPRI TR-102433-2, *Boiler Tube Failure Metallurgical Guide,* vol. 2: *Appendices, Appendix C and Appendix D, Elevated Temperature Properties and Microstructural Catalogs,* Electric Power Research Institute, Palo Alto, CA, October 1993.
18. French, *Metallurgical Failures in Fossil Fired Boilers,* pp. 473–483.
19. Singer, *Combustion–Fossil Power Systems,* pp. 11–42 to 11–46.

SECTION 1.3

Waterside Corrosion

CHAPTER 4

Caustic Corrosion

General Description

Caustic (sodium hydroxide) is intentionally added at controlled concentrations to boiler water in many treatment programs to maintain proper pH. This ensures the formation and retention of passive, protective magnetite layers on boiler tube surfaces. Alkaline-producing salts may also be introduced to the boiler water unintentionally. If caustic or alkaline salt concentration becomes excessive, corrosion may result.

The susceptibility of steel to attack by sodium hydroxide is based on the amphoteric nature of iron oxides. That is, oxides of iron are corroded by both low-pH and high-pH environments (Fig. 4.1).

High-pH substances, such as sodium hydroxide, which is the most common alkaline salt in boiler water solution, may dissolve magnetite [Eq. (4.1)]:

$$4NaOH + Fe_3O_4 \rightarrow 2NaFeO_2 + Na_2FeO_2 + 2H_2O \quad (4.1)$$

When magnetite is removed, the sodium hydroxide may react directly with the iron, as in Eq. (4.2):

$$Fe + 2NaOH \rightarrow Na_2FeO_2 + H_2 \uparrow \quad (4.2)$$

Caustic corrosion may also be referred to as *caustic gouging* and *ductile gouging*. All terms refer to the corrosive interaction of sufficiently concentrated sodium hydroxide, with boiler tube steel, to produce distinct hemispherical or elliptical craters or gouges. The term *gouging* when applied to caustic corrosion may be considered to be somewhat misleading, since it tends to imply a mechanical form of damage. Caustic corrosion is rarely associated with mechanical damage. The craters may be filled with dense corrosion products that sometimes contain sparkling crystals of the iron oxide magnetite (Fe_3O_4). Frequently, a crust of hard deposits and corrosion products containing magnetite crystals will surround and/or overlie the attacked region. The affected metal surface within the crater generally has a smooth, rolling contour.

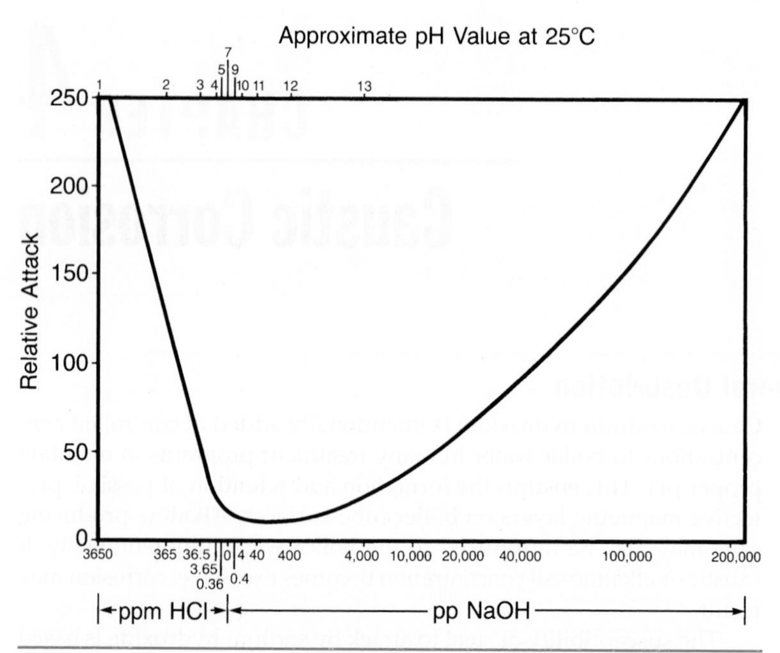

FIGURE 4.1 Attack on steel at 310°C (590°F) by water of varying degrees of acidity and alkalinity. (*Curve by Partridge and Hall, based on data of Berl and van Taack. Courtesy of Herbert H. Uhlig, The Corrosion Handbook, John Wiley & Sons, New York, 1948.*)

Locations

Generally, caustic corrosion is most common in boilers that operate above 1000 psi (6.9 MPa), or at lower pressures where high-purity makeup water is used. Higher operating pressures provide both higher saturated steam temperatures and higher tube metal temperatures, which promote concentrating mechanisms for caustic. In addition, corrosion rates by caustic increase as a function of increasing metal temperatures. Lower-pressure boilers that require high-purity boiler feedwater typically are of a challenging design that includes zones of high heat flux, poor circulation, or both. Challenging designs include externally finned tubes, which are used in heat recovery steam generators (HRSGs). Certain types of waste heat boilers provide challenging designs, particularly those that employ shell-and-tube design, where steam generation occurs on the shell side of the tubes. Crevices at tube-to-tubesheet joints in such boilers are also susceptible to corrosion. When high-purity water is used, metal oxides are the primary source of deposits. Such deposits are conducive to caustic corrosion. Caustic corrosion rarely occurs in lower-pressure boilers that use low-quality makeup water. Here deposits tend to consist of hardness scales. Caustic corrosion damage is confined to localized areas on water-cooled tubes in

1. Regions of high heat flux
2. Slanted or horizontal tubes
3. Locations beneath heavy deposits
4. Heat-transfer regions at or downstream from features that disrupt flow such as intruding welds, weld backing rings, or other devices

In some boiler designs, devices are installed within portions of tubes with the intention of directing or channeling flow. In some cases, such devices may fail to perform the intended function adequately and/or may even promote caustic corrosion. Superheaters may experience caustic corrosion if there is substantial boiler water carryover.

Critical Factors

Two critical factors are required for caustic corrosion. The first is the availability of sodium hydroxide or of suitable alkaline-producing salts (i.e., salts whose solution in water may produce a base and are capable of reacting with protective iron oxide layers and boiler tube steels). As stated previously, sodium hydroxide may be intentionally added to boiler water at noncorrosive levels. Sodium hydroxide is produced in boiler water solution by components of phosphate-based pH control programs, which are commonly used in high-pressure boilers. Detailed descriptions of typical phosphate chemistries are presented in *The Nalco Water Handbook*, third edition, page 33.23. Alkaline-producing salts may also be introduced unintentionally if chemicals from a caustically regenerated demineralizer, or condensate polisher, are inadvertently released into makeup water. Alkaline-producing salts may also contaminate condensate that is returned to the boiler. Contamination may result from inleakage of cooling water through failed surface condensers, or from process streams into the steam and condensate system. Poorly controlled or malfunctioning chemical feed equipment may also contribute excessive concentrations of alkaline salts. Boilers that use potassium hydroxide or phosphates that produce potassium hydroxide will experience higher corrosion rates than at equivalent concentrations of sodium hydroxide. As a consequence, the use of potassium salts is not recommended for boilers that operate at pressures over 580 psi (4 MPa) or in boilers that include superheaters and steam turbines.

The second contributing factor is the mechanism of concentration. Because sodium hydroxide and alkaline-producing salts are not present at corrosive levels in the bulk boiler water under proper control guidelines, a means of concentrating them must be present. Three basic concentration mechanisms exist:

1. *Departure from nucleate boiling (DNB):* The term *nucleate boiling* refers to a condition in which discrete bubbles of steam nucleate at points on a metal surface. Normally, as these steam bubbles form, minute amounts of boiler water solids will deposit at the metal surface, usually at the interface of the bubble and the water. As the bubble separates from the metal surface, the water will redissolve or rinse soluble solids such as sodium hydroxide (Fig. 1.1).

   At the onset of DNB, the rate of bubble formation exceeds the rinsing rate of the soluble solids. Under these conditions, sodium hydroxide, as well as other less soluble dissolved solids or suspended solids, will begin to deposit (Figs. 1.3 and 4.2). The presence of sufficiently concentrated sodium hydroxide and other concentrated corrosives will both compromise the thin film of protective magnetite and cause metal loss.

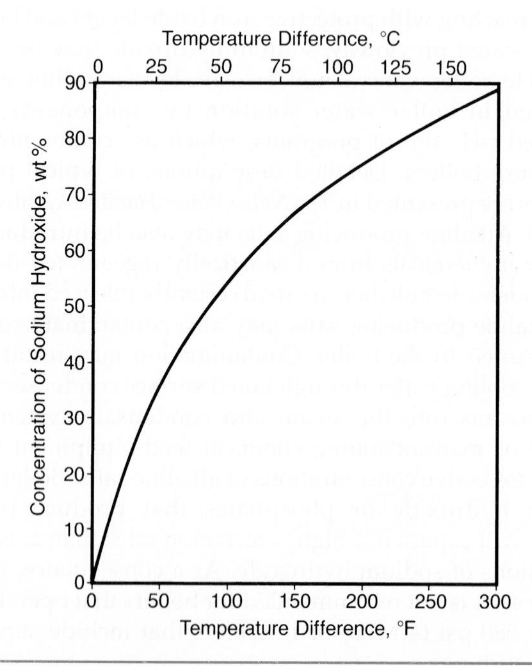

FIGURE 4.2 Sodium hydroxide content attainable in concentrating film of boiler water. [*Based on data from International Critical Tables, 3:370(1928). Courtesy of Herbert H. Uhlig, The Corrosion Handbook, John Wiley & Sons, New York, 1948.*]

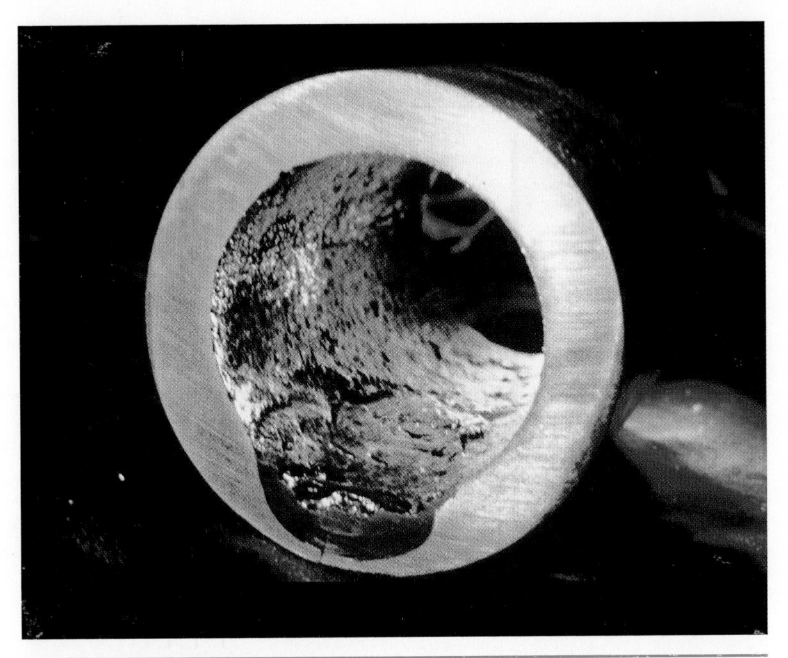

FIGURE 4.3 Deep caustic corrosion beneath insulating internal deposits. (*Courtesy of National Association of Corrosion Engineers.*)

Under the conditions of fully developed DNB, a stable layer of steam will form. Corrosives concentrate at the edges of the steam layer to cause corrosion.

2. *Deposition:* Another concentrating mechanism may occur when deposits shield the metal surface from the bulk water. Steam that forms under deposits escapes, leaving behind and entrapping a corrosive residue that can deeply corrode the metal surface (Fig. 4.3). Insulation provided by deposit layers may promote or cause the production of steam. This phenomenon is commonly referred to as *wick boiling* (Fig. 4.4). When caustic corrosion occurs in this manner, it is sometimes referred to as *under-deposit corrosion*. However, caustic corrosion is only one type of under-deposit corrosion. Other types of under-deposit corrosion will be described in Chap. 5, Low-pH Corrosion during Service, and Chap. 7, Phosphate Corrosion.
3. *Evaporation along a steam/water interface:* Horizontal or slanted tubes are more susceptible to steam/water stratification since the critical heat flux for DNB is significantly lower than in vertically oriented tubes. Under unfavorable conditions of

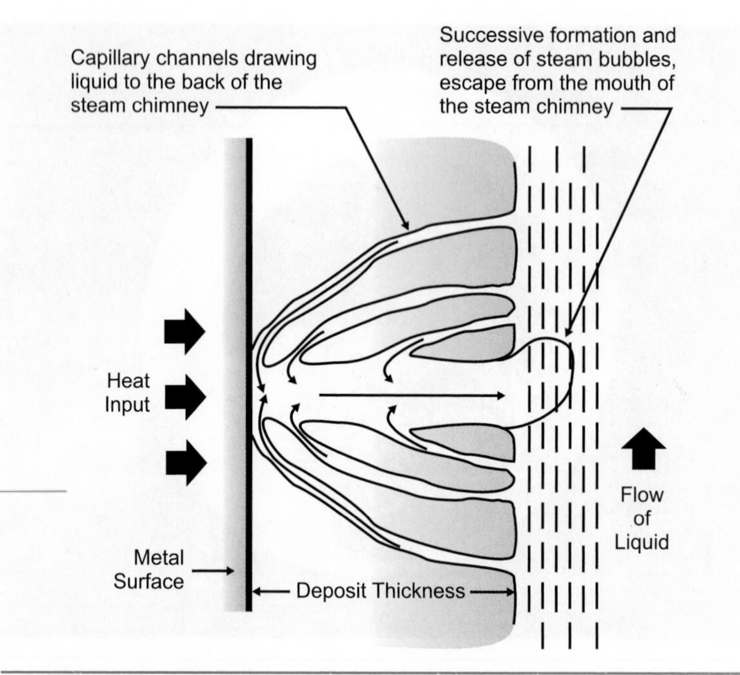

FIGURE 4.4 Wick boiling. (*Courtesy of McGraw-Hill, The Nalco Water Handbook, 3d ed., 2009.*)

heat flux relative to boiler water flow rate, steam/water stratification may occur. In addition, boiler operation at excessively low water levels, or excessive blowdown rates, may create waterlines. Waterlines may also be created by excessive load reduction when pressure remains constant. In this situation, water velocity in the boiler tubes is reduced to a fraction of its full-load value. If velocity becomes low enough in vertical, horizontal, or slanted tubing, steam/water stratification occurs, creating stable or metastable waterlines. Damaged or missing refractory layers on floor tubes may allow higher fireside heat input to cause steam/water stratification. Blowdown lines may also be susceptible to stratification under certain operating conditions. At locations of steam/water stratification, corrosives may concentrate by evaporation, resulting in corrosion.

Intruding welds, or welds employing backing rings, can create localized pressure drops immediately downstream. The pressure drop may promote localized DNB, which may cause deposition and under-deposit corrosion.

Identification

If affected waterside surfaces are accessible, visual examinations that reveal craters filled with hard corrosion products are potential sites of caustic corrosion damage. The extent of corrosion damage may be determined by removing the corrosion products.

In horizontal or slanted tubes, a pair of parallel or converging, longitudinal trenches may form (Fig. 4.5). If the tube is nearly full, the parallel trenches will coalesce into a single elongated, longitudinally oriented crater, along the crown of the tube (Fig. 4.6). In vertically oriented tubes, corrosive concentration at a steam/water interface will yield a circumferentially oriented crater (Fig. 4.7).

If the waterside surface is not accessible, nondestructive testing techniques such as ultrasonic testing may be required. Steam studies using a hydrogen analyzer may also function as an alert for the possibility of caustic corrosion.

Once possible caustic corrosion damage is identified, it is necessary for metallographic analysis to be conducted to confirm that caustic or alkaline salt corrosion occurred, since acid and phosphate corrosion during service may cause similar damage (see Chaps. 5 and 7). Supporting evidence for caustic corrosion may include the presence of

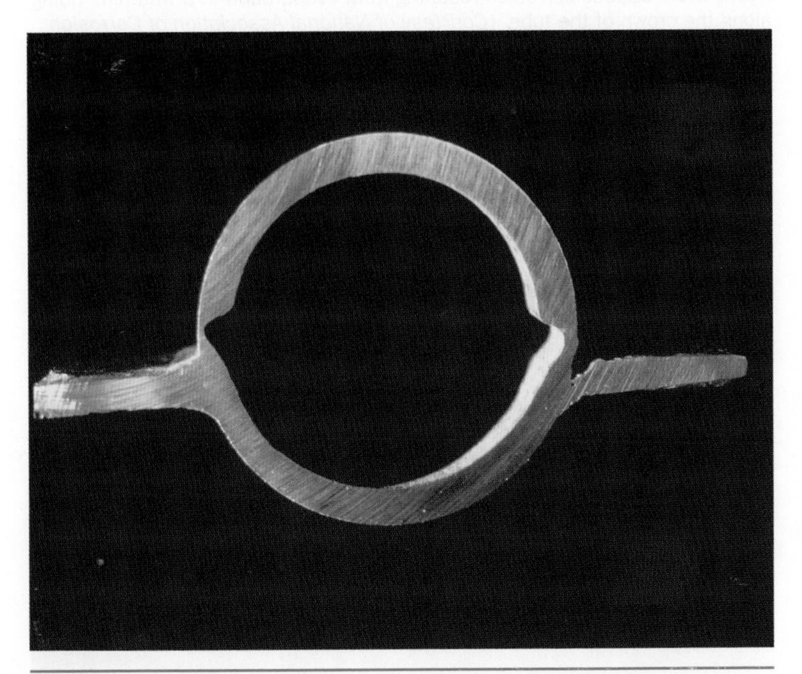

FIGURE 4.5 Caustic corrosion along a longitudinal waterline. (*Courtesy of National Association of Corrosion Engineers.*)

FIGURE 4.6 Caustic corrosion resulting from evaporation at a waterline riding along the crown of the tube. (*Courtesy of National Association of Corrosion Engineers.*)

FIGURE 4.7 Metal loss on a vertically oriented tube that penetrates around the entire circumference.

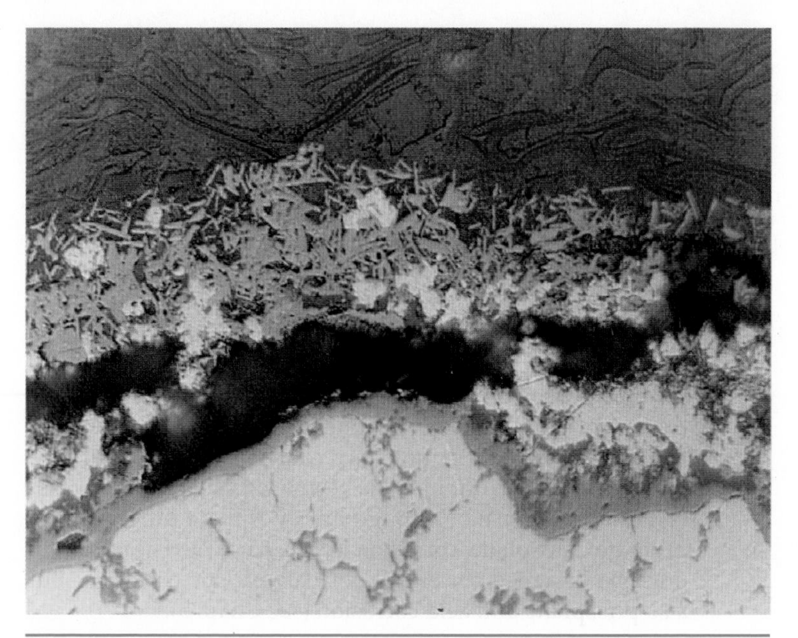

FIGURE 4.8 Metallographic cross section of a surface that experienced caustic corrosion. The surface is covered with corrosion products that contain magnetite needles and metallic copper particles.

alkaline material in deposits that cover the corroded surfaces. X-ray fluorescence and diffraction may be used to identify compounds that may cause corrosion, particularly sodium compounds. However, since such compounds are highly water soluble, they may wash away subsequent to the corrosion events. Magnetite needles may also be present in the corrosion products that form on the corroded surface, as shown in Fig. 4.8. Though the presence of magnetite needles provides strong supporting evidence for caustic and alkaline salt corrosion, such evidence should be used with caution since the needles may be formed by other environments as well. Metallic copper particles may also deposit in the crater. This is due to the evolution of hydrogen in the caustic corrosion reaction, which reduces dissolved copper in the water. Metallic copper typically will not promote galvanic corrosion of the boiler steel, since deposit and corrosion product layers provide an electrically insulating barrier layer. The chemical treatment program and treatment historical data should be reviewed to determine if conditions may have supported caustic or alkaline salt corrosion.

Elimination

Corrosion may occur when the boiler water contains dissolved sodium hydroxide or alkaline-producing salts and a concentrating mechanism exists simultaneously at a particular location.

The following remedies may be used to significantly control or eliminate corrosion by concentrated sodium hydroxide or alkaline-producing salts:

- *Reduce the amount of available free sodium hydroxide:* This is the underlying concept for various phosphate and caustic treatment programs implemented in high-pressure boilers (see *The Nalco Water Handbook*, third edition, pages 33.22 to 33.28). If a phosphate treatment program is used, it should be appropriate for the specific conditions of boiler operation. For instance, the congruent phosphate program was designed to prevent caustic corrosion. Below the congruent point, free sodium hydroxide is not produced, thus preventing caustic corrosion. However, the treatment program must consider the possible effects of not only caustic corrosion, but also other types of corrosion. It has been found that under certain conditions, the use of a congruent phosphate program may promote phosphate corrosion (see Chap. 7). The program must be designed to control not only caustic corrosion, but also phosphate and acid corrosion (see Chap. 5).
- *Prevent in-leakage of alkaline-producing salts by cooling water and process streams:* Contamination of steam and condensate by cooling water leaks at condensers, or by process streams that return to the boiler must be eliminated.

  For most industrial and utility boilers, the chemical treatment programs employ caustic or alkaline salts for pH and alkalinity control. Contaminants may contribute sodium hydroxide and alkaline salts that are outside of the treatment program control parameters. This may require significant adjustments to the program to control corrosion. Because of the powerful concentrating mechanisms that may operate in a boiler, the addition of only a few parts per million of contaminants to the boiler water may be sufficient to initiate or accelerate corrosion. In treatment programs that do not use caustic or alkaline salts, such as all volatile treatment (AVT), or oxygenated treatment (OT), contamination by alkaline salts cannot be tolerated. They may cause corrosion, even when present at small concentrations.
- *Prevent inadvertent release of caustic regeneration chemicals from makeup water demineralizers and condensate polishers.*

Prevention of localized concentration is the most effective means of avoiding caustic corrosion. However, in practice it is also the most difficult to achieve. The methods for preventing localized concentration include:

- *Prevent DNB:* This usually requires the elimination of hot spots, achieved by controlling the boiler's operating parameters. Hot spots are caused by excessive overfiring, misdirected burners, change of fuel, and gas channeling. Conditions that reduce flow below appropriate levels, such as reduced firing rates or excessive blowdown, should be avoided. In heat recovery steam generators (HRSGs), excessive heat input due to the use of duct burners may promote deposition and under-deposit corrosion.
- *Prevent excessive waterside deposition:* As mentioned previously, deposition should be controlled by maintaining consistently good feedwater quality. Tube sampling on a periodic basis may be performed to measure the relative amount of deposit buildup on tubes. This sampling is referred to as a *deposit weight determination, deposit loading,* or *gram loading.* Deposit weight determination practice is described in Chap. 1, Water-Formed and Steam-Formed Deposits. Susceptibility to under-deposit corrosion increases as a function of increasing deposit weight and boiler pressure. The boiler manufacturer should be consulted for chemical cleaning recommendations. Aside from boiler manufacturer's guidelines, associations such as the National Association of Corrosion Engineers (NACE) and Electric Power Research Institute (EPRI) provide cleaning guidelines. Proper monitoring will ensure that corrosion is controlled in the preboiler system, in order to reduce the amount of corrosion products that are introduced to the boiler.
- *Prevent steam/water stratification in tubes:* Proper boiler operating parameters must be followed to ensure proper steam production rates and boiler water circulation. If the design and operating conditions cannot be altered practically to prevent poor circulation, then consistent periodic inspection and cleaning practices may be required for certain tubes.
- *Ensure proper component design:* Intruding welds and other components within the fluid flow path on the water side should not promote DNB, entrap steam, or cause caustic and alkaline salts to concentrate.

Cautions

It is very difficult to distinguish localized attack by high-pH substances from localized attack by low-pH corrosion during service simply by visual examination. A formal metallographic examination is required. Evaluating the types of concentrateable corrosives that may be contaminating the boiler water will aid in the determination.

Because corrosion products may fill the depressions caused by caustic corrosion, the extent and depth of metal loss in the affected area—and even the existence of a corrosion site—may be overlooked. Probing a suspect area with a hard, pointed instrument may aid in the determination, but because the corrosion products are often very hard, a corrosion site may remain undetected.

Related Problems

See also Chap. 1, Water-Formed and Steam-Formed Deposits; Chap. 5, Low-pH Corrosion during Service; and Chap. 7, Phosphate Corrosion.

Case History 4.1

Industry:	Utility boiler
Specimen location:	Camera port, waterwall
Orientation of specimen:	Vertical and slanted, S shaped
Years in service:	25
Water treatment program:	Coordinated phosphate
Drum pressure:	2000 psi (13.8 MPa)
Tube specifications:	3-in. (7.6-cm) outer diameter
Fuel:	Ground coal

Visual examinations disclosed a thickened patch of hard corrosion products covering a crater adjacent to one bend (Fig. 4.9). Perforation of the wall had not occurred, but transverse cross sections cut through the site revealed substantial metal loss (Fig. 4.10).

FIGURE 4.9 Patch of hard iron oxides on internal surface.

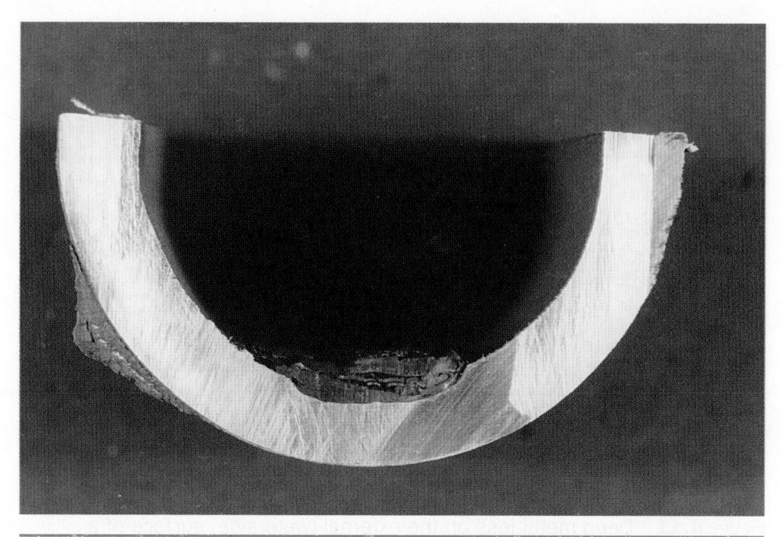

FIGURE 4.10 Cratered region beneath patch of iron oxides.

The crater was caused by sodium hydroxide that concentrated to corrosive levels due to wick boiling beneath thermally insulating deposits. Previous failures of this type had not occurred in this region of the boiler. The boiler had been cleaned 4 years previously, and was in peaking service. Closer control over the water treatment program was recommended. If control under the coordinated phosphate program could not be improved, alternate programs may be used to minimize the corrosion.

Case History 4.2

Industry:	Ethylene production
Specimen location:	Transfer line exchanger (TLE)
Orientation of specimen:	Vertical
Years in service:	25
Water treatment program:	Coordinated phosphate
Drum pressure:	1750 psi (12.1 MPa)
Tube specifications:	1¼-in. (3.2-cm) outer diameter
Fuel:	Ethylene process gas (tubeside)

Many transfer line exchangers have a shell-and-tube design, where hot process gases are cooled by boiler water that flows across the shell sides. They may be either vertically or horizontally oriented.

In this case, the unit was vertically oriented, with heat flux being highest at the bottom tubesheet where the hot process gases were introduced. Deep metal loss penetrated around the entire circumference of several tubes at the bottom tubesheet interface, beneath thick layers of deposits and corrosion products (Figs. 4.11 and 4.12). Above about 7 in. (17.8 cm) from the tubesheet interface, no significant external surface deposition occurred. The deposit layers within the corroded area contained components that were alkaline in a distilled water solution. Metallographic analysis revealed the presence of substantial amounts of magnetite needles in the corrosion products that covered the wasted surface

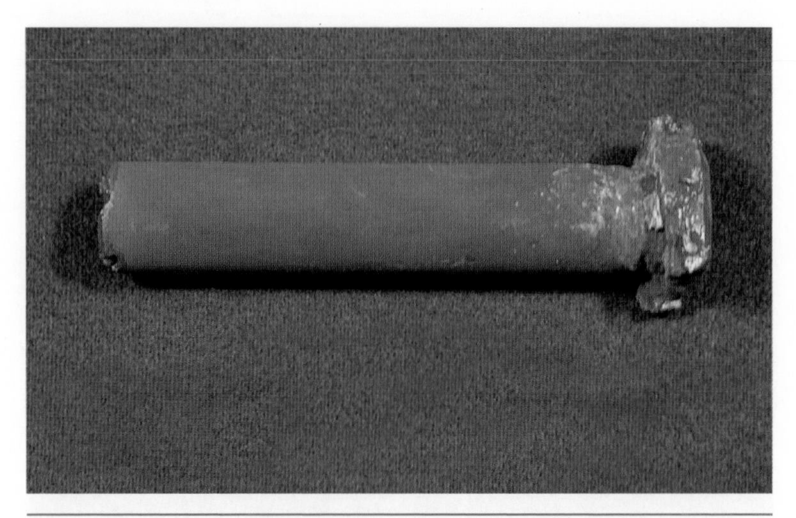

FIGURE 4.11 Deep metal loss on the external (waterside) surface of a vertically oriented tube from a shell-and-tube waste heat boiler at the bottom tubesheet interface.

FIGURE 4.12 Higher magnification view of wastage at tubesheet interface.

(Fig. 4.13). The metal loss was caused by caustic corrosion, primarily resulting from the concentration of alkaline salts beneath deposit layers. Suspended material in the water tends to settle along the bottom tubesheet. High heat flux also promotes deposition at that location. The deposit layers gradually build in thickness. This reduces heat transfer and promotes wick boiling, resulting in further deposition and corrosion. It was determined that periodic cleaning of deposits at the bottom tubesheet was not conducted on a timely basis.

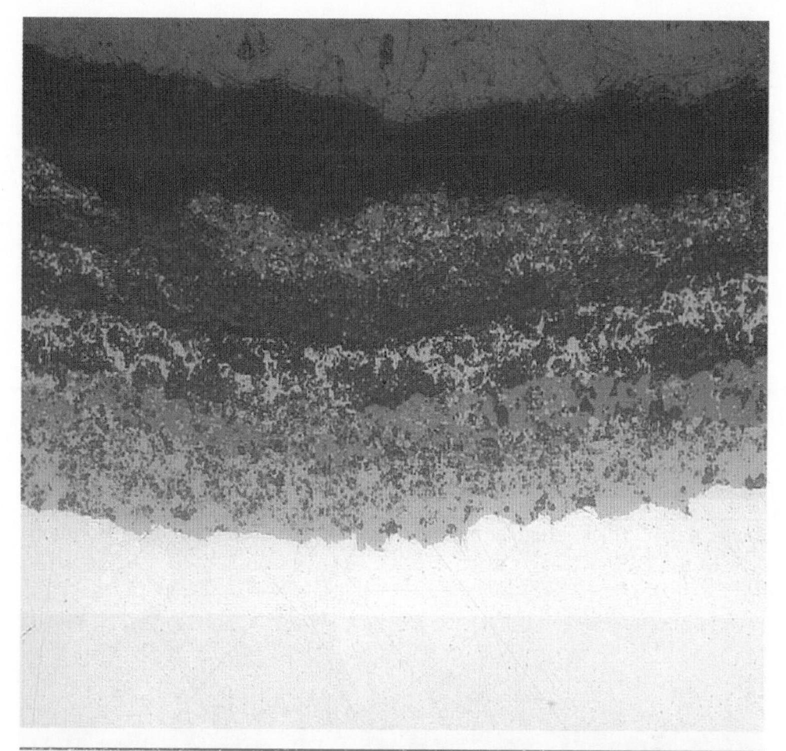

FIGURE 4.13 Metallographic cross section of corrosion products covering the corroded surface, which contain magnetite needles and metallic copper particles.

Deposition and corrosion may also happen at top tubesheets in shell-and-tube design boilers, if design and operating conditions promote steam stratification. In some cases, steam becomes entrapped along the top tubesheet. Proper design changes have been made to minimize or prevent steam entrapment in some units. If steam entrapment cannot be adequately controlled along the top tubesheet, periodic cleaning is recommended.

Case History 4.3

Industry:	Utility
Specimen location:	Bottom slag-screen tube
Orientation of specimen:	Slanted, 15° slope
Water treatment program:	Coordinated phosphate
Drum pressure:	2200 psi (15.2 MPa)
Tube specifications:	3-in. (7.6-cm) outer diameter

A growing number of small leaks were occurring in lower slag-screen tubes of this boiler. One of the leaking tubes was removed for examination.

Figure 4.14 illustrates the appearance of the internal surface in the area of leakage. A small perforation of this rifled tube was observed in the center of a large elliptical area, or crater, of metal loss (Fig. 4.15). The crater was covered with a

FIGURE 4.14 Thick, irregular mound of hard iron oxides covering perforation.

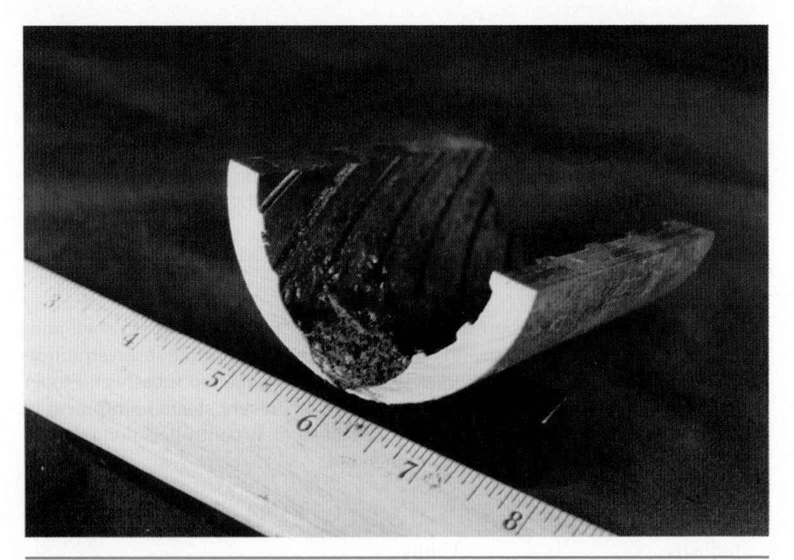

FIGURE 4.15 Perforation at bottom of crater.

thick, irregular mound of coarsely stratified iron oxides. Removing the mound revealed that the underlying crater had a smooth, rolling metal-surface contour (Fig. 4.16). The rest of the internal surface had suffered no metal loss.

Since deposits were not present and evidence of a waterline was not observed, it can be assumed that the concentration of the caustic material was caused by highly localized nonnucleate boiling (DNB). The rifling of the internal surface is designed to induce swirling of the water to prevent nonnucleate boiling and steam/water phase stratification. It is surprising, therefore, to find severe caustic

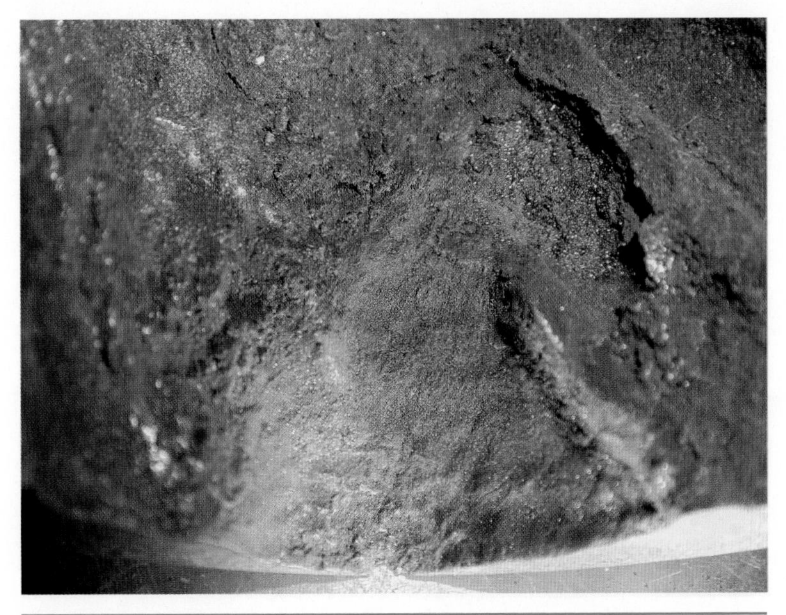

FIGURE 4.16 Corrosion products removed to reveal crater surface that has a smooth, rolling contour.

gouging in this tube design. However, this boiler was idle on weekends. It is possible that highly localized nonnucleate boiling occurred during start-up, before normal boiler water circulation was fully established.

Case History 4.4

Industry:	Ore refining
Boiler type:	Circulating fluidized-bed boiler
Specimen location:	Bottom leg of U bend in bed
Orientation of specimen:	Horizontal
Years in service:	0.75
Water treatment program:	Polymer
Drum pressure:	1280 psi (8.8 MPa)
Tube specifications:	2½-in. (6.4-cm) outer diameter
Fuel:	Coal

Failures recurred within the horizontally oriented bottom legs of U tubes at the 12 o'clock position. The bottom legs were located on the upstream end. No leaks occurred within the U bends or top legs of the tubes. Narrow strips of sheet steel that were bent into a helical shape were welded in place within the bottom legs of the tubes (Fig. 4.17). These strips reportedly were used to promote a preferred fluid flow pattern through the tubes to control or prevent steam/water stratification. The bottom legs of several different U tubes were longitudinally saw-cut to reveal their internal surfaces. This revealed repeating patterns of deposition and under-deposit corrosion at the 12 o'clock positions every 4 in. (10.2 cm), corresponding to locations where the sheet steel strips contacted the tube internal surface in a helical pattern (Figs. 4.18 and 4.19). Metallographic,

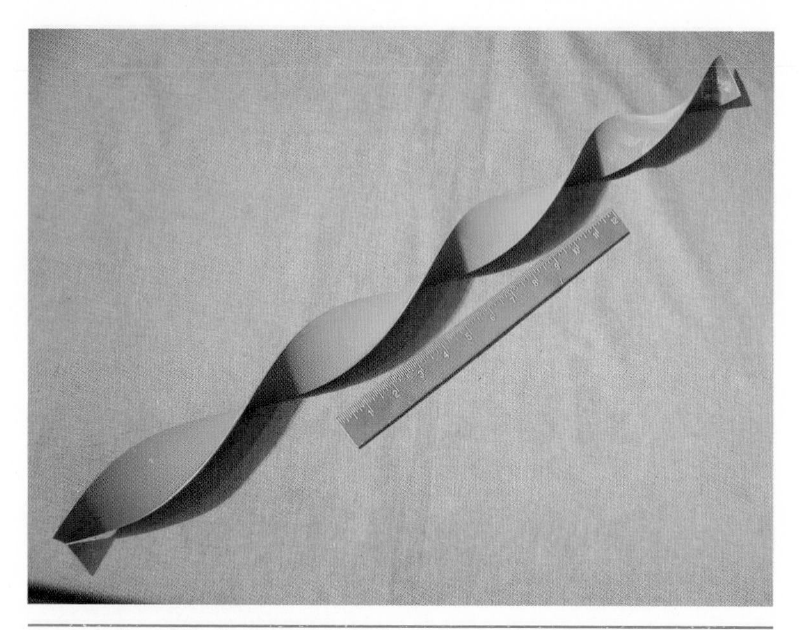

FIGURE 4.17 Helically shaped strip of sheet steel that was welded inside the horizontally oriented bottom leg of a U bend.

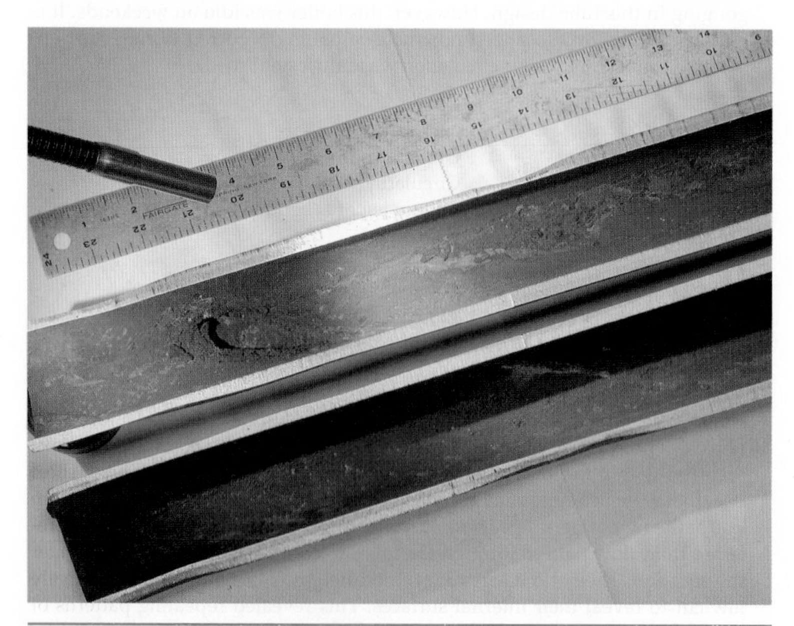

FIGURE 4.18 Internal surface deposition and metal loss within localized areas at 12 o'clock position where the helical strip shown in Fig. 4.17 was installed.

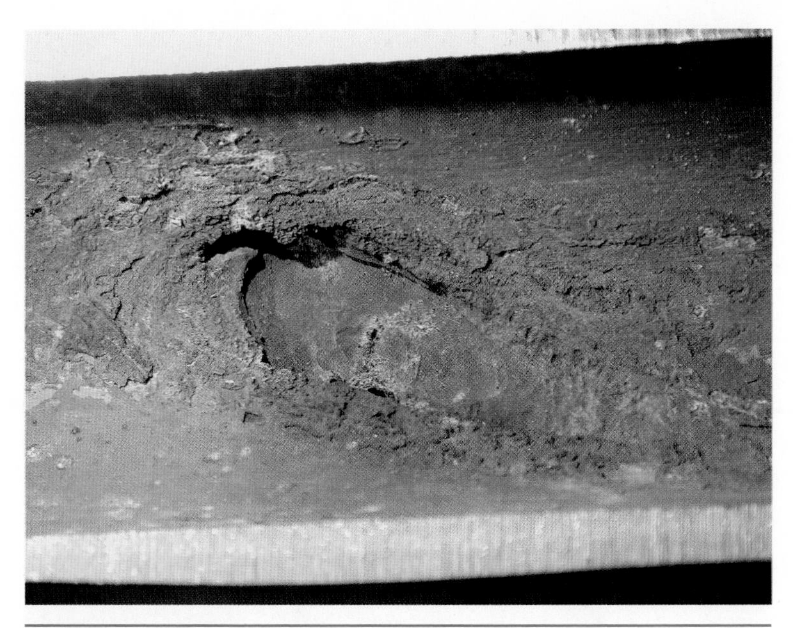

FIGURE 4.19 Close-up of crater shown in Fig. 4.18.

corrosion product, and deposit analyses revealed that the metal loss was caused by caustic corrosion that resulted from steam blanketing. Steam blankets became entrapped at the tops of the tubes along the helical strips due to insufficient flow. This allowed alkaline salts, which were dissolved at low and noncorrosive levels in the bulk water, to concentrate to corrosive levels. Methods to increase boiler water flow, such as installation of booster pumps, were considered. Increased flow would more effectively control steam blanketing in the lower legs of the U bend, since circulation under normal operating conditions was insufficient.

Case History 4.5

Industry:	Soap and detergent manufacturing
Specimen location:	Riser tube at tubesheet interface
Orientation of specimen:	Slanted
Water treatment program:	Polymer
Drum pressure:	1000 psi (6.9 MPa)
Tube specifications:	3-in. (7.6-cm) outer diameter
Fuel:	Natural gas

A substantial tube leak in a package boiler required an unscheduled shutdown for repair. On-site inspection of riser tubes within a steam drum revealed a single perforation in one of the tubes (Fig. 4.20). The perforation, which was located about 4½ in. (11.4 cm) from the tube end, was irregularly shaped and transversely oriented. A section containing the tube end was torch-cut for examination. Close visual inspection revealed that no significant metal loss occurred on the internal waterside surface in areas surrounding the perforation (Fig. 4.21). However, the perforation occurred at the base of a deep groove on the external surface at the interface of the rolled portion of the tube and the external surface

FIGURE 4.20 Perforation in tube found during internal inspection of steam drum. (*© NACE International, 2005.*)

FIGURE 4.21 No metal loss on internal surface surrounding the perforation. (*© NACE International, 2005.*)

FIGURE 4.22 Perforation at base of deep groove on external surface at tube/tubesheet interface. (*© NACE International, 2005.*)

of the steam drum (Fig. 4.22). Shallow metal loss and deposition occurred on the external surface within the adjacent rolled portion. Metallographic examination revealed the presence of corrosion product, deposit layers containing alkaline material, and agglomerated magnetite needles. The deposits and corrosion products reveal that the attack was not caused by erosion from escaping steam. Maintenance work was conducted a few months prior to the failure. During that time, the tube was improperly rolled into the tubesheet, allowing boiler water to seep through the joint and into the firebox. The water evaporated upon escape. This caused the dissolved caustic in the boiler water, which was at proper concentration, to concentrate to corrosive levels, eventually perforating the tube wall. Proper surface cleaning and tube rolling practice must be followed to prevent leaks that can result in such damage.

CHAPTER 5

Low-pH Corrosion during Service

General Description

Although it is relatively rare, a general depression of bulk water pH may occur if certain contaminants gain access to the boiler. Boilers using water of low buffering capacity can realize a bulk pH drop to less than 5, if contaminated with strong mineral acids such as hydrochloric acid or sulfuric acid.

The concern of this chapter, however, is with the more common creation of localized low-pH environments. Two conditions must exist simultaneously.

First, the boiler must be operated outside of normal recommended water chemistry parameters. Low concentrations of acid-producing salts such as chloride and sulfate may be introduced through contamination of boiler feedwater.

Second, to produce a localized low-pH environment, a mechanism must exist for concentrating acid-producing salts. This condition exists where boiling occurs and adequate rinsing is hindered by the presence of porous deposits. Dissolved acid-producing salts concentrate beneath the deposits. Hydrolysis of the salts produces localized low-pH conditions, while the bulk water remains alkaline. The hydrolysis of metal (M) chloride salts (such as magnesium, sodium, and calcium chloride) produces hydrochloric acid, shown by Eq. (5.1). Wherever low-pH conditions exist, the thin film of protective magnetic iron oxide is dissolved, and subsequently the metal is attacked [Eq. (5.2)]. Hydrogen is evolved during the acid corrosion reaction at the cathode [Eq. (5.3)]. In some cases, particularly in high-pressure boilers, acid corrosion may be accompanied by hydrogen damage. Hydrogen damage is described in Chap. 6.

Hydrolysis of chloride salts

$$M^+ + Cl^- + H_2O \rightarrow M(OH) \downarrow + H^+Cl^- \qquad (5.1)$$

Anodic reaction

$$Fe \rightarrow Fe^{2+} + 2e^- \tag{5.2}$$

Cathodic reaction

$$2H^+ + 2e^- \rightarrow H_2 \uparrow \tag{5.3}$$

Locations

Acid corrosion can occur in both low-pressure and high-pressure boilers. Acid corrosion damage is confined to localized areas on water-cooled tubes in the following four regions:

1. Regions of high heat flux
2. Slanted or horizontal tubes
3. Locations beneath heavy deposits
4. Heat-transfer regions at or downstream from features that disrupt flow, such as intruding welds, weld backing rings, or other devices

In some boiler designs, devices are installed within portions of tubes with the intention of directing or channeling flow. In some cases, such devices may fail to perform the intended function adequately and/or may even promote concentrating conditions required for acid corrosion.

Critical Factors

Two critical factors are required for low-pH corrosion during boiler operation. The first is the availability of free acid or acid-producing salts (i.e., salts whose solution with water may produce acid and are capable of reacting with protective iron oxide layers and boiler tube steels). The second contributing factor for low-pH corrosion is the mechanism of concentration.

Contamination may result from in-leakage of cooling water through failed surface condensers or from process streams into the steam and condensate system. Another source of contamination is the inadvertent release of acidic regeneration chemicals from a makeup water demineralizer or condensate polisher into the feedwater system. Note that sodium hydroxide, which causes caustic corrosion (as described in Chap. 4), must concentrate 10 to 100 times that of chloride ions to produce an equivalent amount of metal loss.[1] Poorly controlled or malfunctioning chemical feed equipment may also contribute excessive concentrations of acidic salts that are used in certain portions of the chemical treatment program.

It is important to note that acid may be produced in boiler water solution by components of phosphate-based pH control programs; which are commonly used in high-pressure boilers. Detailed descriptions of typical phosphate chemistries are presented in *The Nalco Water Handbook,* third edition. The acid corrosion mechanism that may occur when using phosphate/pH control programs is referred to as phosphate corrosion. This mechanism is described in Chap. 7.

Organic contaminants, such as sugar shots in boilers used at sugar mills, may decompose in boiler water to form organic acids that can significantly depress pH to cause corrosion. This is a particular concern for boilers that use high-purity, unbuffered waters. Organic acids are typically weaker than the mineral acids described earlier, and will tend to produce general metal loss instead of deep pitting. In addition, organics and volatile organic breakdown products can produce acidic condensate and may form carbonaceous deposits within the steam and condensate system.

Because acids and acid-producing salts are not present at corrosive levels in the bulk boiler water under proper control guidelines, a means of concentrating them must be present. Due to the powerful concentration mechanisms that may operate in a boiler, in-leakage of only a few parts per million of contaminant may be sufficient to cause localized low-pH corrosion. Residual contamination from chemical cleaning and poorly controlled or malfunctioning feedwater chemical equipment may also foster localized low-pH conditions. Three basic concentration mechanisms exist:

1. *Departure from nucleate boiling (DNB):* The term *nucleate boiling* refers to a condition in which discrete bubbles of steam nucleate at points on a metal surface. Normally as these steam bubbles form, minute amounts of boiler water solids will deposit at the metal surface, usually at the interface of the bubble and the water. As the bubble separates from the metal surface, the water will redissolve or rinse soluble acidic salts such as chloride (Fig. 1.1).

   At the onset of DNB, the rate of bubble formation exceeds the rinsing rate of the soluble solids. Under these conditions, acid salts, as well as other less soluble dissolved solids or suspended solids, will begin to deposit (Figs. 1.3 and 4.2). The presence of sufficiently concentrated acids or acid salts will both dissolve the protective magnetite layer and cause metal loss.

   Under the conditions of fully developed DNB, a stable, insulating layer of steam will form. Corrosives concentrate beneath the steam layer to cause corrosion.

2. *Deposition:* Another concentrating mechanism may occur when deposits shield the metal surface from the bulk water.

Steam that forms under deposits escapes, leaving behind and entrapping a corrosive residue that can deeply corrode the metal surface (Fig. 4.3). Insulation provided by deposit layers may promote or cause the production of steam. This phenomenon is commonly referred to as *wick boiling* (Fig. 4.5). When acid corrosion occurs in this manner, it is sometimes referred to as under-deposit corrosion. However, acid corrosion is only one type of under-deposit corrosion. Other types of under-deposit corrosion are described in Chaps. 4, 6, and 7.

3. *Evaporation along a steam/water interface:* Horizontal or slanted tubes are more susceptible to steam/water stratification, since the critical heat flux for DNB is significantly lower than in vertically oriented tubes. Under unfavorable conditions of heat flux relative to boiler water flow rate, steam/water stratification may occur. In addition, boiler operation at excessively low water levels, or excessive blowdown rates, may create waterlines. Waterlines may also be created by excessive load reduction when pressure remains constant. In this situation, water velocity in the boiler tubes is reduced to a fraction of its full-load value. If velocity becomes low enough in vertical, horizontal, or slanted tubing, steam/water stratification occurs, creating stable or metastable waterlines. Damaged or missing refractory layers on floor tubes may allow higher fireside heat input to cause steam/water stratification. Blowdown lines may also be susceptible to stratification under certain operating conditions. At locations of steam/water stratification, corrosives may concentrate by evaporation, resulting in corrosion.

Intruding welds, or welds employing backing rings, can create localized pressure drops immediately downstream. The pressure drop may promote localized DNB, which may cause deposition and under-deposit corrosion.

Identification

If affected waterside surfaces are accessible, visual examinations that reveal craters filled with hard corrosion products are potential sites of acid corrosion damage (Fig. 5.1). The extent of corrosion damage may be determined by removing the corrosion products. In some cases, particularly in lower-pressure boilers, thick corrosion product and deposit layers may not be found on the corroded surfaces. Rather, acid-corroded surfaces may be lined with thin corrosion product and deposit layers (Fig. 5.2). Occasionally in lower-pressure boilers, damage will not be in the form of a crater. Small, localized, undercut pits or depressions may be produced instead, particularly by strong acid

FIGURE 5.1 Crater formed by acid corrosion filled with dense corrosion products.

FIGURE 5.2 Acid-corroded surface that is covered with thin corrosion product and deposit layers.

solutions (Fig. 5.3). The corroded surfaces may exhibit straight, parallel, longitudinally oriented striations. Weak acid solutions will tend to produce more uniform, general metal loss.

In horizontal or slanted tubes, a pair of parallel or converging longitudinal strips of undercut pits may form (Fig. 5.4). If the tube is

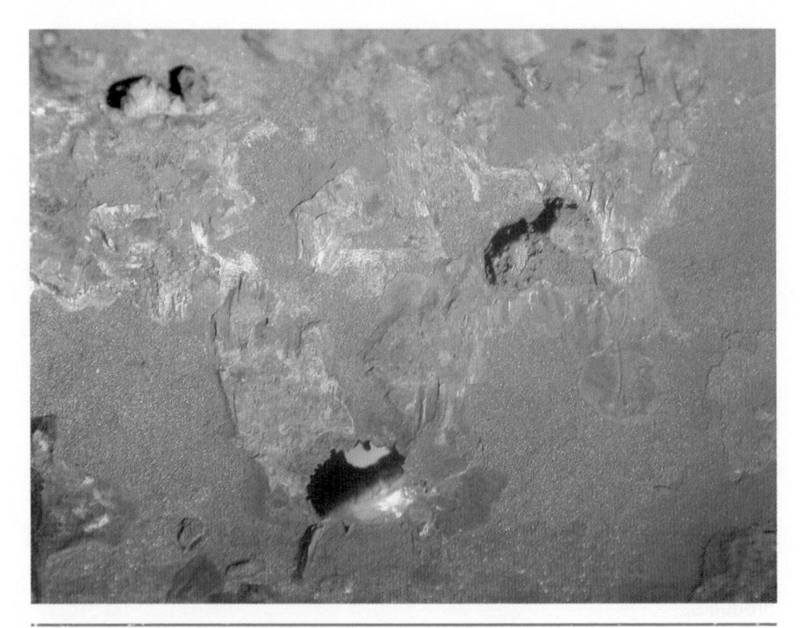

FIGURE 5.3 Undercut pits formed by acid corrosion.

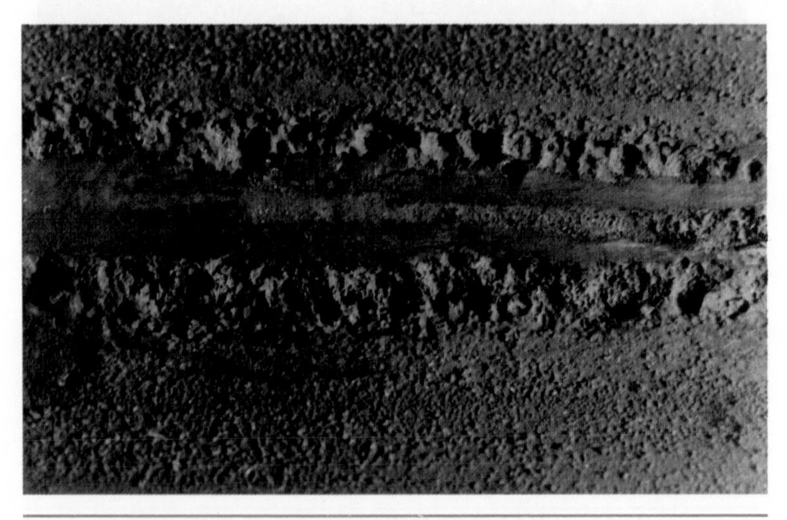

FIGURE 5.4 Metal loss coalescing to form a groove.

nearly full, the parallel trenches will coalesce into a single, elongated, longitudinally oriented crater, or coalition of pits along the crown of the tube. In vertically oriented tubes, corrosive concentration at a stable steam/water interface will yield a circumferentially oriented groove.

If the waterside surface is not accessible, nondestructive testing techniques such as ultrasonic testing may be required. Steam studies using a hydrogen analyzer may also function as an alert for the possibility of acid corrosion.

Once possible acid corrosion damage is identified, it is necessary for a metallographic analysis to be conducted. Since caustic and phosphate corrosion during service may cause similar damage, metallographic analysis is used to confirm that acid or acid salt corrosion occurred (see Chaps. 4 and 7). Supporting evidence for acid corrosion may include the presence of acidic material in deposits that cover the corroded surfaces. X-ray fluorescence, scanning electron microscope (SEM) analysis, and spot tests may be used to identify elements and compounds that may cause corrosion, such as chloride and sulfate. However, since such compounds are highly water-soluble, they may wash away subsequent to the corrosion events. Metallographic examination that reveals the presence of stratified layers of alternating dense and porous iron oxides and deposits provides strong evidence for acid corrosion (Fig. 5.5). Evidence of hydrogen damage may also indicate that acid corrosion occurred. Metallic copper particles may deposit in locations that experience acid corrosion. This is due to the evolution of hydrogen in the acid corrosion reaction, which reduces dissolved copper in the water. Metallic copper typically will not promote galvanic corrosion of the boiler steel, since deposit and

FIGURE 5.5 Metallographic cross section of stratified iron oxide layers that covered acid-corroded surface.

corrosion product layers provide an electrically insulating barrier layer. The chemical treatment program and historical data of water quality should be reviewed to determine if conditions may have supported acid or acid salt corrosion.

Elimination

Corrosion may occur when the boiler water contains dissolved free acid or acid-producing salts and a concentrating mechanism exists simultaneously at a particular location. The following remedies may be used to significantly control or eliminate corrosion by concentrated acid or acid-producing salts:

- *Prevent in-leakage of acid-producing salts from cooling water and process streams:* Contamination of steam and condensate by cooling water leaks at condensers, or by process streams that return to the boiler, must be eliminated.

  Because of the powerful concentrating mechanisms that may operate in a boiler, the addition of parts per billion (ppb) levels of contaminants to the boiler water may be sufficient to initiate or accelerate corrosion. Guidelines for allowable chloride concentrations are published by Electric Power Research Institute (EPRI). In treatment programs such as all volatile treatment (AVT) or oxygenated treatment (OT), contamination by acid or acid salts cannot be tolerated.
- *Prevent inadvertent release of acidic regeneration chemicals from makeup water demineralizers and condensate polishers.*

Prevention of localized concentration mechanisms is the most effective means of avoiding acid corrosion. However, it is also the most difficult to achieve. The methods for preventing localized concentration include:

- *Prevent DNB:* This usually requires the elimination of hot spots, achieved by controlling the boiler's operating parameters. Hot spots are caused by excessive overfiring, misdirected burners, change of fuel, and gas channeling. Conditions that reduce flow below appropriate levels, such as reduced firing rates or excessive blowdown, should be avoided. In heat recovery steam generators (HRSGs), excessive heat input due to the use of duct burners may promote deposition and under-deposit corrosion.
- *Prevent excessive waterside deposition:* Tube sampling on a periodic basis may be performed to measure the relative amount of deposit buildup on tubes. This sampling is referred to as a deposit weight determination, deposit loading, or gram loading. Deposit weight determination practice is described in Chap. 1. Susceptibility to under-deposit corrosion increases as a function of increasing deposit weight and boiler pressure. The

boiler manufacturer should be consulted for chemical cleaning recommendations. Aside from boiler manufacturer guidelines, associations such as the National Association of Corrosion Engineers (NACE) and Electric Power Research Institute provide cleaning guidelines. Proper monitoring will ensure that corrosion is controlled in the preboiler system in order to reduce the amount of corrosion products introduced to the boiler.

- *Prevent steam/water stratification in tubes:* Proper boiler operating parameters must be followed to ensure proper steam production rates and boiler water circulation. If the design and operating conditions cannot be altered practically to prevent poor circulation, then consistent periodic inspection and cleaning practices may be required for certain tubes.
- *Ensure proper component design:* Intruding welds and other components within the fluid flow path on the water side should not promote DNB, entrap steam, or cause acid salts to concentrate.

Cautions

It may be very difficult to distinguish localized attack by low-pH corrosion during service from localized attack by high-pH substances simply by visual examination. A formal metallographic examination is required. Evaluating the types of concentrateable corrosives that may be contaminating the boiler water will aid in the determination.

Because corrosion products may fill the depressions caused by acid corrosion, the extent and depth of metal loss in the affected area—and even the existence of a corrosion site—may be overlooked through simple visual or boroscope inspections. Probing a suspect area with a hard, pointed instrument may aid in the determination, but because the corrosion products are often very hard, a corrosion site may remain undetected.

Related Problems

See also Chap. 1, Water-Formed and Steam-Formed Deposits; Chap. 4, Caustic Corrosion; Chap. 6, Hydrogen Damage; and Chap. 7, Phosphate Corrosion.

Case History 5.1

Industry:	Utility
Specimen location:	Side wall
Orientation of specimen:	Vertical
Years in service:	30
Water treatment program:	Coordinated phosphate
Drum pressure:	2000 psi (13.8 MPa)
Tube specifications:	3-in. (7.6-cm) outer diameter, carbon steel

FIGURE 5.6 Crater formed at weld due to low-pH corrosion.

Numerous occurrences of the type of metal loss illustrated in Fig. 5.6 resulted in an extensive tube replacement program. Most replacements were in one of the boiler's side walls. This type of corrosion was a recurring problem, but the frequency suddenly increased. The boiler was in peaking service.

Figure 5.7 illustrates the depth and appearance of the metal loss. The crater was filled with thick, hard iron oxides and metallic copper. Figure 5.8 is a photomicrograph showing highly stratified iron oxide layers.

FIGURE 5.7 Cross section of deep crater shown in Fig. 5.6 filled with dense corrosion products.

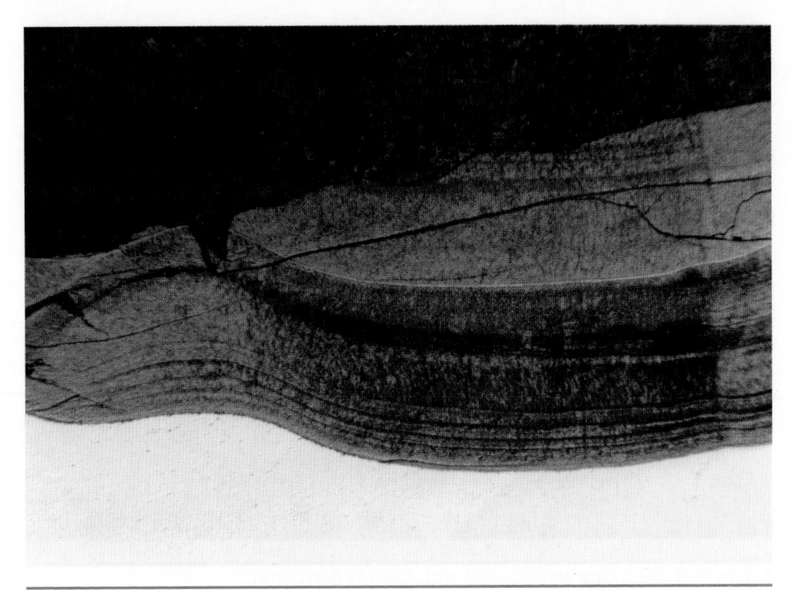

FIGURE 5.8 Metallographic cross section of stratified iron oxide layers that covered acid-corroded surface.

Analysis of corrosion products on the internal surface of the crater revealed significant amounts of chloride. The chloride was determined to have been introduced to the boiler through cooling water leaks from the condenser into returned condensate.

Most metal loss occurred immediately downstream of circumferential welds. This indicates that disrupted water flow across the welds was instrumental in establishing concentration sites for corrosive substances.

Case History 5.2

Industry:	Utility
Specimen location:	Wall tube
Orientation of specimen:	Vertical
Years in service:	26
Water treatment program:	Coordinated phosphate
Drum pressure:	1900 psi (13.1 MPa)
Tube specifications:	3¼-in. (8.3-cm) outer diameter, carbon steel

Figure 5.9 illustrates the massive rupture that occurred in a distinct zone of deep internal surface metal loss. Surrounding areas on the internal surface were unaffected and quite smooth. Microstructural examinations revealed extensive hydrogen damage in the tube wall immediately below the wasted surface.

The visual and microstructural appearance of the gouged region was consistent with that of low-pH exposure. The hydrogen damage that is associated with low-pH attack indicates that corrosion occurred during boiler operation. The boiler was exposed to returned condensate that was contaminated by frequent condenser cooling water leaks.

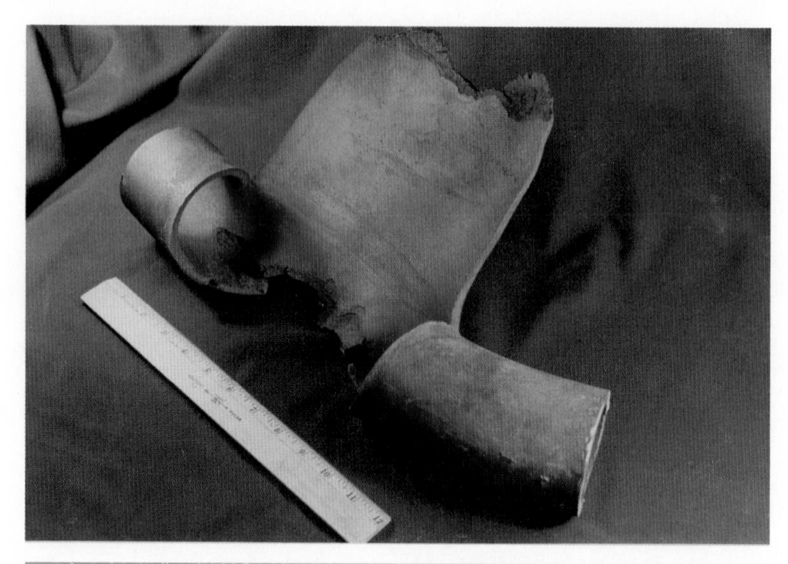

FIGURE 5.9 Rupture associated with low-pH corrosion.

Concentration of low-pH corrosives may have occurred beneath deposits that were dislodged from the internal surface at the time of rupture. It is also possible that concentration occurred as a result of a departure from nucleate boiling.

Case History 5.3

Industry:	Marine boiler
Specimen location:	Desuperheater tube in steam drum
Orientation of specimen:	Horizontal
Years in service:	20
Water treatment program:	Coordinated phosphate
Drum pressure:	870 psi (6 MPa)
Tube specifications:	1¼-in. (3.2-cm) outer diameter, carbon steel

The desuperheater was installed inside the steam drum. Superheated steam flowed through the tube sides, which was cooled (desuperheated or attemperated) by boiler water that flowed across the shell sides. The boiler makeup water was produced from seawater, processed through low-pressure evaporators. The chloride concentration in the makeup water was < 1 ppm, while the boiler water chloride concentration was typically < 10 ppm. After several years of service, a change in operating practice for the boiler required increased usage of the desuperheater. Several desuperheater tubes failed due to deep metal loss at the inlet ends (Fig. 5.10). Thick accumulations of iron oxide and metallic copper formed in the wasted areas. Acidic material containing chloride was also found in the layers. Metallographic examination revealed thick, highly stratified iron oxide layers covering the deeply wasted external surface (Fig. 5.11). Away from the inlet end, no significant external surface metal loss or deposition occurred.

Localized deposition at the inlet end resulted from a departure from nucleate boiling, which caused substantial chloride concentration and subsequent

FIGURE 5.10 Deep metal loss and buildups of thick iron oxide and metallic copper layers at inlet ends of desuperheater tubes.

FIGURE 5.11 Metallographic cross section of stratified iron oxide layers that covered acid-corroded surface.

acid corrosion. Increased usage of the desuperheater accelerated the corrosion. Reduction of chloride levels in the boiler water was considered to be impractical. Therefore, mechanical and operational control methods were required. Changes in operating practice to reduce use of the desuperheater were considered. Methods to increase water flow at the inlet end, if possible, were recommended. In addition, periodic inspection and cleaning was suggested in an effort to reduce corrosion over time. It was also suggested that alternate, more corrosion-resistant materials be considered as substitutes for the carbon steel desuperheater tubing.

Case History 5.4

Industry:	Chemical manufacturing
Specimen location:	Waste heat boiler
Orientation of specimen:	Horizontal
Years in service:	14
Water treatment program:	Polymer
Drum pressure:	400 psi (2.8 MPa)
Tube specifications:	1-3/8-in. (3.5-cm) outer diameter, carbon steel

The waste heat boiler had a shell-and-tube design. Hot salt solutions flowed across the shell sides of the tubes, which were cooled by boiler water that flowed through the tubes. Tube failures occurred following forced boiler shutdowns. Figure 5.12 shows deep undercut pits that were isolated to localized areas on the bottoms of tubes. Figure 5.13 shows a metallographic cross section of the deep, undercut pits, which formed due to acid corrosion. The pits were lined with thin layers of iron oxide.

During shutdowns, water flow to the boiler was stopped. Residual hot salt remained on the process side, which caused pools of residual water inside the tubes to evaporate. Aggressive acidic salts in the water concentrated to very high and corrosive levels to produce the metal loss. No remnants of the acidic salts that caused the corrosion remained on the corroded surfaces. Based upon the boiler water chemistry, it was determined that the most probable sources of acid salts were chloride or sulfate. The acid solution was so strong that metal

FIGURE 5.12 Deep undercut pits on bottom-side internal surface of a waste heat boiler tube.

FIGURE 5.13 Metallographic cross section of deep, undercut pits.

ions were dissolved and subsequently washed away when boiler water was reintroduced to the system. Changes were made to the operating practice to ensure that water flowed through the tubes during shutdowns, until the process side was cooled to suitably low temperatures.

Case History 5.5

Industry:	Building
Specimen location:	Firetube boiler
Orientation of specimen:	Horizontal
Years in service:	20
Water treatment program:	Phosphate, sulfite
Drum pressure:	50 psi (0.3 MPa)
Tube specifications:	2½-in. (6.4-cm) outer diameter, carbon steel

The boiler operated for only about 6 months per year during the heating season. During the remainder of the year, the boiler was in wet layup. No deaerator was used at the facility for removal of dissolved oxygen from the feedwater.

Figure 5.14 shows a tube that experienced deep metal loss in scattered locations on the top side. Deposits and corrosion products settled on the top side to promote under-deposit corrosion. The depressions were filled or lined with loosely adherent iron oxide corrosion products (Fig. 5.15). Material beneath the corrosion product layers produced an acidic distilled water solution and contained substantial concentrations of sulfate.

The localized depressions formed initially due to oxygen corrosion during idle periods. Aggressive sulfate ions from the boiler water concentrated beneath the occlusive iron oxide corrosion products, due to a wick boiling and/or ion concentration cell mechanism.[2] The sulfate concentrated to levels that formed corrosive, acidic environments, which substantially accelerated corrosion.

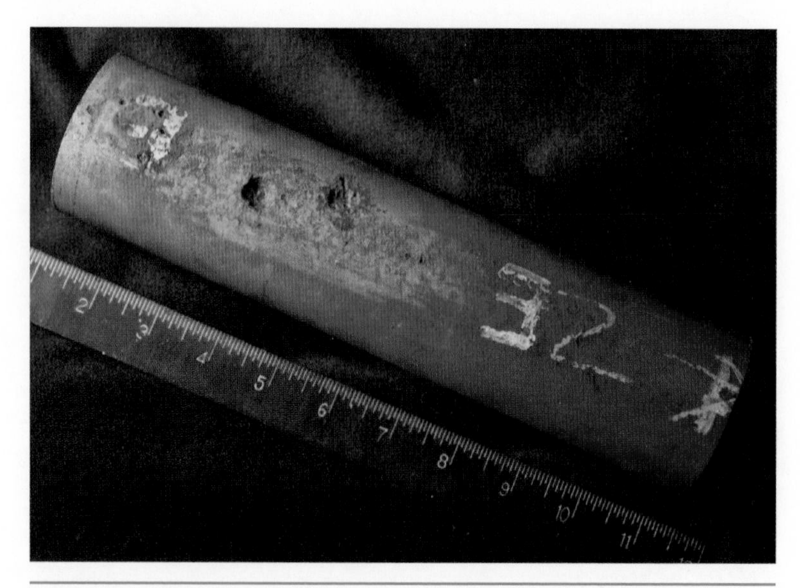

FIGURE 5.14 Localized depressions on tube from a firetube boiler.

FIGURE 5.15 Close-up of depressions that are lined with acidic deposits.

Sulfate was produced by sulfite that was used in the boiler to scavenge dissolved oxygen. Since a deaerator was not used in the system, high concentrations of sulfite were required to ensure that dissolved oxygen was reduced to proper levels to prevent oxygen corrosion. As the sulfate concentrated in the boiler water, under-deposit corrosion increased.

It is a common and recommended practice for boilers in building HVAC systems to use sulfite oxygen scavengers. Sulfite is preferred because it reacts more rapidly and is more cost-effective than other oxygen scavengers. For this reason, it is recommended that alternate methods of deaeration for the feedwater be investigated. In addition, the sulfite and sulfate concentrations in the boiler water should be maintained to proper levels.

References

1. H. G. Masterson, J. E. Castle, and G. M. W. Mann, "Waterside Corrosion of Power Station Boiler Tubes," *Chemistry and Industry*, Sept. 6, 1969, pp. 1261–1266.
2. H. M. Herro and R. D. Port, *The Nalco Guide to Cooling Water Systems Failure Analysis*, McGraw-Hill, New York, 1993, p. 15.

CHAPTER 6

Hydrogen Damage

General Description

Hydrogen damage may occur within the tube wall adjacent to locations where corrosion reactions at the internal surface result in the production of atomic hydrogen.* Some references describe hydrogen damage as an under-deposit corrosion mechanism that is specifically related to low-pH conditions.[1] Low-pH corrosion (refer to Chap. 5) beneath deposits and corrosion product layers produces the conditions that are most favorable for hydrogen damage. Larger amounts of atomic hydrogen are produced during the acid corrosion reactions than by other corrosion reactions that occur in boilers [Eq. (6.1)]. Since hydrogen is also produced during caustic corrosion reactions [Eq. (6.2)], hydrogen damage may occur. However, since the amount of atomic hydrogen produced is much less than that provided by acid corrosion, hydrogen damage will only occur during caustic corrosion reactions under extreme circumstances.

$$Fe + 2H_2O + 2H^+ + 2Cl^- \rightarrow Fe(OH)_2 \downarrow + 2HCl + 2H^+ \uparrow \qquad (6.1)$$

$$Fe + 2NaOH \rightarrow Na_2FeO_2 + 2H^+ \uparrow \qquad (6.2)$$

If atomic hydrogen is liberated, then it is capable of diffusing into the steel. Some of this diffused atomic hydrogen will combine at grain boundaries or inclusions in the metal to produce molecular hydrogen, or will react with iron carbides in the metal to produce methane [Eq. (6.3)].

$$Fe_3C + 4H \rightarrow CH_4 + 3Fe \qquad (6.3)$$

Since neither molecular hydrogen nor methane is capable of diffusing through the steel, these gases accumulate, primarily at grain boundaries. Eventually, gas pressures will cause separation of the metal at its grain boundaries, producing discontinuous, intergranular microcracks (Fig. 6.1). Portions of the iron carbides in the microstructure that were involved in the reaction will be reacted in locations

*_Hydrogen in the atomic form_ refers to uncombined atoms of hydrogen (H) as contrasted with molecules of hydrogen (H_2).

FIGURE 6.1 Discontinuous intergranular microcracks resulting from methane formation in the grain boundaries. Note decarburization of adjacent pearlite colonies (dark islands).

directly adjacent to the cracks. This loss of carbide is referred to as *decarburization*. As microcracks accumulate, tube strength diminishes until stresses imposed by boiler pressure exceed the tensile strength of the remaining intact metal. At this point a thick-walled longitudinal rupture may occur (Fig. 6.2). Deep fissures may form in locations within the corroded areas, adjacent to the primary rupture edges

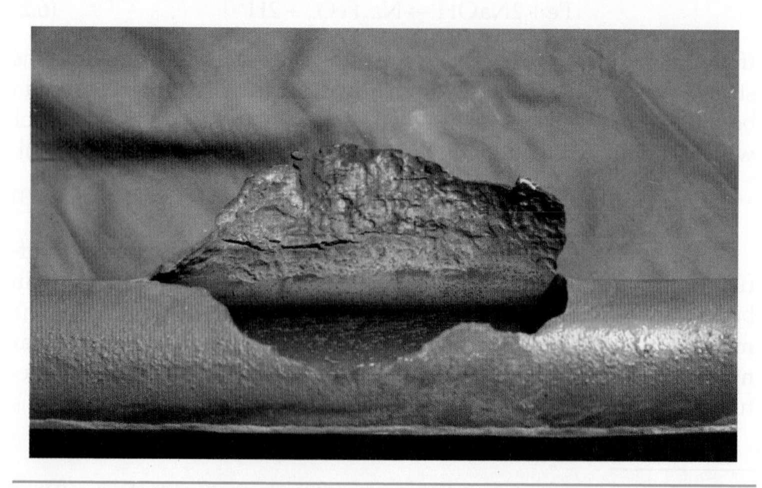

FIGURE 6.2 Thick-walled rupture resulting from hydrogen damage. Note the crater of metal loss adjacent to the rupture on the internal surface. Several deep fissures formed within the crater.

FIGURE 6.3 Deep fissures formed within corroded area adjacent to the primary rupture edge.

(Fig. 6.3). Such fissures will not be found on the external surface within the same portion of the tube wall. Depending upon the extent of hydrogen damage, large window sections of tube wall may be blown out, producing a gaping hole (Fig. 6.4). Serious injuries may occur due to expelling of fragments from hydrogen-damaged tubes.

FIGURE 6.4 Section of tube wall that was blown out of a hydrogen-damaged tube.

Locations

Generally, hydrogen damage is confined to water-carrying tubes. However, hydrogen damage may occur in steam-carrying tubes when they are exposed to substantial amounts of liquid water, such as during massive carryover events. Damage usually occurs in regions of high heat flux; beneath heavy deposits; in slanted or horizontal tubes; in heat-transfer regions at or adjacent to backing rings at welds; or near other devices that disrupt flow. Experience has shown that hydrogen damage rarely occurs in boilers operating below 1000 psi (6.9 MPa).

Critical Factors

The main factor governing hydrogen damage is an aqueous corrosion reaction in which hydrogen in the atomic form is liberated. The critical factors involved with hydrogen damage resulting from low-pH corrosion are identical to those outlined for on-line acid corrosion in Chap. 5, Low-pH Corrosion during Service. The critical factors involved with hydrogen damage resulting from high-pH corrosion are identical to those outlined for caustic corrosion in Chap. 4, Caustic Corrosion.

Identification

Locations where hydrogen damage was produced cannot be visually identified with certainty during boiler inspections prior to failure. In boilers operating at pressures greater than 1000 psi (6.9 MPa), areas that have sustained metal loss from either high-pH or low-pH corrosion should be considered suspect. However, the presence of deep internal surface fissures within the corroded areas with no external fissures formed within the same portion of tube wall provides compelling visual evidence for hydrogen damage.

Hydrogen damage may be detected by nondestructive testing methods, such as ultrasonic techniques. Typical ultrasonic thickness tests may reveal corroded areas where hydrogen damage should also be suspected. In some cases, more sophisticated ultrasonic testing techniques are capable of revealing hydrogen-damaged metal. Ultimately, hydrogen damage may only be confirmed through metallographic analysis.

Key microstructural features that can be used to positively identify hydrogen damage are

1. The presence of discontinuous, intergranular fissures within the tube wall extending from the corroded surface
2. Decarburization of the adjacent carbides in the microstructure

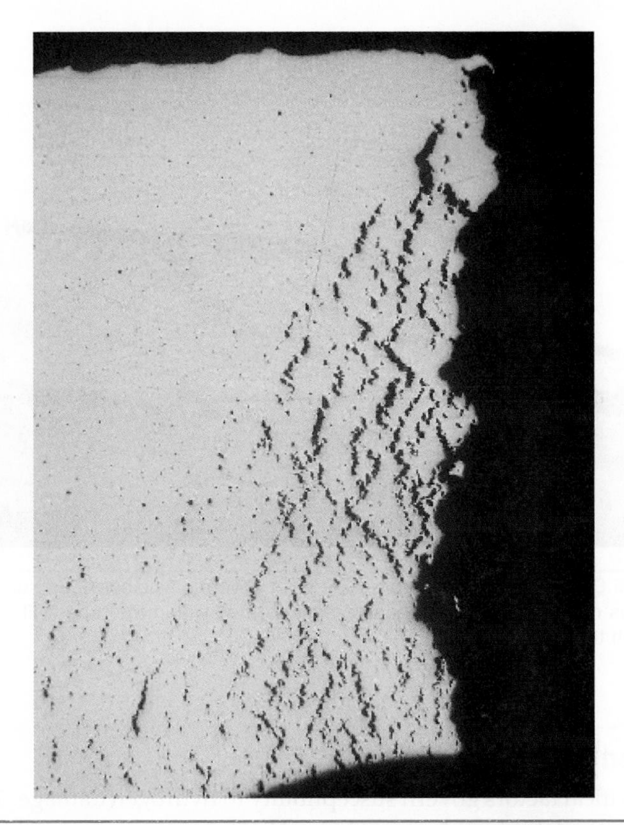

FIGURE 6.5 Large population of intergranular fissures extending almost completely through the tube wall from a corroded area on the internal surface (unetched).

Fissures will be most heavily concentrated near the corroded surface and may extend to a substantial depth (Fig. 6.5). Microstructural alteration other than decarburization may not be observed since corrosion is not always associated with overheating.

Corrosion and hydrogen damage depends upon the boiler water chemistry and the probable sources of contamination. For example, a common source of contamination of boiler water is condenser cooling water in-leakage. Seawater and recirculating cooling water from systems that incorporate cooling towers contain chlorides and other dissolved solids that may hydrolyze to form acidic solutions. Corrosion products and deposits may be studied through appropriate analytical techniques, such as x-ray fluorescence and diffraction or energy-dispersive spectroscopy, to indicate the possible corrodents responsible for corrosion and hydrogen damage. When acid corrosion is associated with hydrogen damage, metallographic analysis will reveal characteristic highly stratified iron oxide layers on the corroded surfaces (Fig. 6.6).

FIGURE 6.6 Highly stratified iron oxide layer covering acid-corroded surface. Fissures resulting from hydrogen damage formed within the tube wall beneath the layers (unetched).

Elimination

Two critical factors govern susceptibility to hydrogen damage. These are the availability of high- or low-pH substances and a mechanism of concentration that causes corrosion. Both must be present simultaneously for hydrogen damage to occur. Because of the powerful concentrating mechanisms that may operate in a boiler, the addition of parts per billion (ppb) levels of contaminants to the boiler water may be sufficient to initiate or accelerate corrosion. In treatment programs such as all volatile treatment (AVT) or oxygenated treatment (OT), contamination by acidic or alkaline salts cannot be tolerated.

The remedies described in the chapters concerning acid corrosion (Chap. 5) and caustic corrosion (Chap. 4) may be used to significantly control or eliminate corrosion. By controlling corrosion, hydrogen damage will also be controlled or prevented.

Cautions

Hydrogen damage typically produces thick-walled ruptures. Other failure mechanisms producing thick-walled ruptures include stress-corrosion cracking, corrosion fatigue, and creep (also known as stress rupture). It may be difficult to visually distinguish ruptures caused by hydrogen damage from thick-walled failures caused by other mechanisms. However, certain features will be more closely associated with hydrogen damage than with other mechanisms. For example, hydrogen damage is commonly located within deep and wide craters of metal loss. (Note cautions listed in Chaps. 4 and 5.) The other failure modes are not typically associated with gross corrosion. In addition, tube failures in hydrogen-damaged metal are often manifested as a blowout of a "window" of the tube wall. This is not a common feature of the other failure modes.

A definitive diagnosis of hydrogen damage requires metallographic examination.

Related Problems

See also Chap. 4, Caustic Corrosion, and Chap. 5, Low-pH Corrosion during Service.

Case History 6.1

Industry:	Utility
Specimen location:	Waterwall
Orientation of specimen:	Vertical
Years in service:	10
Water treatment program:	Coordinated phosphate
Drum pressure:	1800 psi (12.4 MPa)
Tube specifications:	2.5-in. (6.4-cm) outer diameter

Several similar failures occurred within a localized area in the waterwall of a utility boiler. A section containing a typical thick-walled rupture revealed that the failure occurred within a wide crater of metal loss on the internal surface (Fig. 6.2). Deep fissures, which paralleled the rupture edge, formed within the crater. No fissures were found on the external surface within the area where the crater formed on the internal surface. Metallographic analysis revealed no evidence of overheating. Corrosion products covering the crater contained highly stratified iron oxide layers and metallic copper (Fig. 6.6). No aggressive agents were found within the layers. Directly beneath the crater, a large population of discontinuous, intergranular fissures formed, some of which linked to form deep fissures (Fig. 6.7). The pearlite colonies in the microstructure that were directly adjacent to the fissures were decarburized (Fig. 6.8). The boiler was exposed to returned condensate that was contaminated with seawater which had leaked through failures in condenser tubes. Chlorides from the seawater concentrated beneath deposit and corrosion product layers to cause acid corrosion. Hydrogen that was evolved during the acid corrosion reaction produced the intergranular fissures. The condenser leaks were repaired, and proper monitoring was instituted to provide a warning system for future contamination to prevent similar failures.

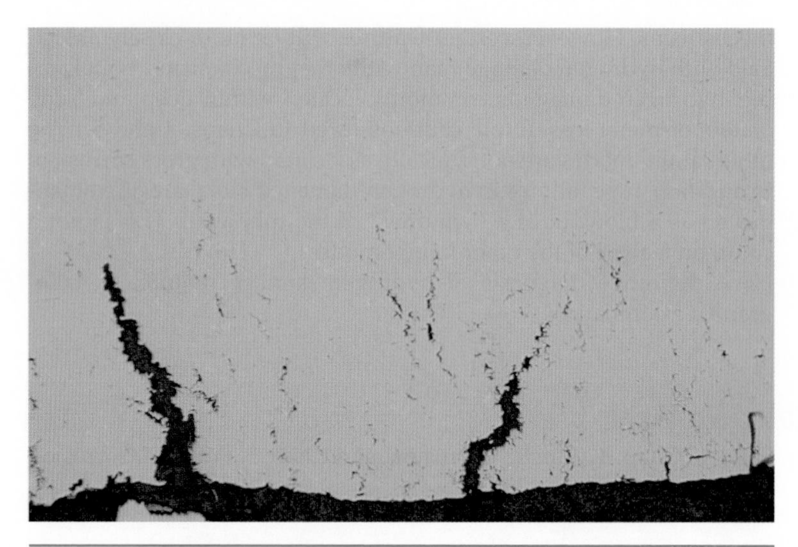

FIGURE 6.7 Numerous intergranular microcracks that formed within the tube wall at the internal surface within the crater. Fissures linked to form deep fissures.

FIGURE 6.8 Pearlite directly adjacent to fissures experienced decarburization.

Case History 6.2

Industry:	Ethylene production
Specimen location:	Transfer line exchanger (TLE) at bottom tubesheet
Orientation of specimen:	Vertical
Years in service:	11
Water treatment program:	All volatile treatment
Drum pressure:	1740 psi (12 MPa)
Tube specifications:	2-in. (5.1-cm) outer diameter

The TLE was a vertically oriented shell-and-tube heat exchanger. Cracked ethylene flowed through the tubes, while boiler water flowed across their shell sides. A thick accumulation of deposits formed at the bottom tubesheet. Deep metal loss occurred on the external surfaces of several tubes within the bundle. Figure 6.9 shows the typical deep metal loss that occurred, with a perforation at the base. Energy-dispersive spectroscopy revealed that the corrosion products and deposits covering the corroded surface consisted primarily of iron oxide and metallic copper. However, small concentrations of aluminum, zinc, sulfur, manganese, silicon, phosphorus, and chlorine were also present. Metallographic examination revealed no evidence of overheating. The deeply wasted surface was covered with highly stratified iron oxide layers (Fig. 6.10). Intergranular fissures formed within the adjacent tube wall. Acid corrosion was caused by the concentration of chloride and/or sulfate beneath the deposit and corrosion product layers. The acid corrosion reaction generated atomic hydrogen, which penetrated the metal to produce hydrogen damage.

FIGURE 6.9 Deep, localized groove that formed on the external waterside surface at the interface of the tube and bottom tubesheet.

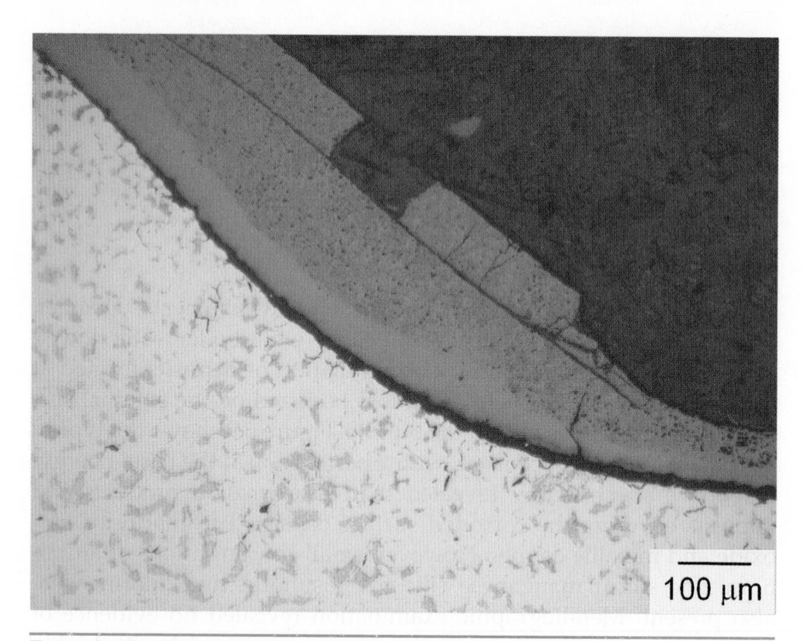

FIGURE 6.10 Highly stratified iron oxide layers covering the corroded surface. Intergranular fissures formed within the tube wall at and near the surface. Etchant: Picral.

Deposition and acid salt concentration resulted from settling of suspended solids in the water as well as departure from nucleate boiling (DNB). DNB resulted from poor water flow conditions and high heat flux at the bottom tubesheet. Improved control of boiler water contamination is imperative due to the all volatile treatment used in the boiler that has very limited ability to control corrosion by acid salts. Methods to improve cleaning of the external surfaces of the tubes and tubesheet were also instituted.

Case History 6.3

Industry:	Refinery
Specimen location:	Secondary superheater
Orientation of specimen:	Vertical U bend
Years in service:	2
Pressure	1000 psi (6.9 MPa)
Water treatment program:	Polymer
Tube specifications:	2-in. (5.1-cm) outer diameter, low-alloy steel (1% chromium, 0.5% molybdenum)

A superheater tube U bend bulged and ruptured (Fig. 6.11). The internal surface was heavily fouled with extremely thick, stratified layers of deposits, which formed due to chronic boiler water carryover (Fig. 6.12). Chemical analysis indicated that the deposit layers consisted primarily of sodium chloride, with moderate amounts of

FIGURE 6.11 Bulged and ruptured superheater U bend.

FIGURE 6.12 Thick, stratified layers of deposits covering the internal surface of superheater U bend.

FIGURE 6.13 Fissures formed adjacent to partially decarburized re-formed pearlite colonies.

magnesium oxide. Seawater leaked through failed condenser tubes and contaminated the returned condensate. Contamination caused foaming, resulting in severe carryover of boiler water to the superheater. Carryover in turn caused deposition and corrosion on the internal surfaces. Metallographic examination revealed that the tube experienced severe overheating with metal temperatures exceeding 1340°F (727°C). This was the primary cause of bulging and failure. In addition, shallow metal loss was caused by acid corrosion resulting from exposure to chlorides. Figure 6.13 shows intergranular fissures that formed adjacent to re-formed pearlite that was partially decarburized. The hydrogen that evolved during the acid corrosion reaction caused the cracks to form. Condenser leaks were repaired both to prevent chloride and hardness salt contamination of the boiler water as well as to control the carryover that caused the serious corrosion and hydrogen damage.

Case History 6.4

Industry:	Chemical industry
Specimen location:	HRSG first evaporator
Years in service:	20
Watertreatment program:	Phosphate treatment (PT)
Drum pressure:	1250 psi (8.6 MPa)
Orientation	Vertical
Tube specifications:	2-in. (5.1-cm) outer diameter, externally finned plain carbon steel tube

A thin-lipped rupture occurred on one side of an evaporator tube that was located within an outer row of the evaporator section near

FIGURE 6.14 Thin-lipped rupture through tube wall of externally finned HRSG evaporator tube section.

the hottest gas pass (Fig. 6.14). Very thick layers of deposits formed on the internal surface (Fig. 6.15). Deposit weights around the entire circumference were measured to be about 700 g/ft^2 (749 mg/cm^2). Feedwater quality was poor, with hardness levels averaging 100 to 500 ppb and iron levels ranging from 10 to 50 ppb. The boiler was never cleaned during its entire service life. A phosphate treatment program using only trisodium phosphate was applied to maintain a fairly high phosphate concentration and high pH, due to the known

FIGURE 6.15 Very thick internal surface deposit and corrosion product layers around the entire circumference.

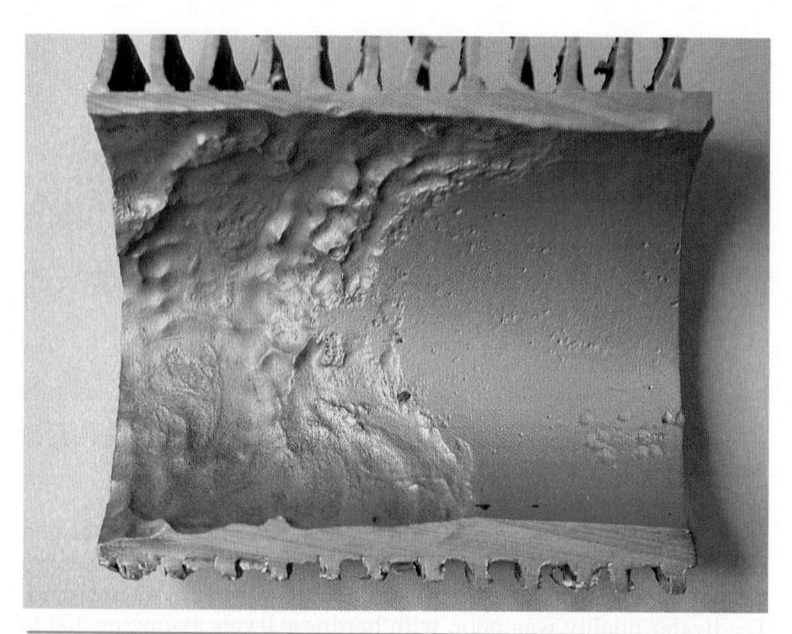

FIGURE 6.16 Internal surface cleaned to reveal deep metal loss that occurred beneath the deposit and corrosion product layers.

frequent contamination of the boiler water. The presence of hardness salts in the deposits, as well as chloride, indicates that condenser leaks were the primary source of contamination. The phosphate treatment program was used in an attempt to control corrosion by acid salts and hydrogen damage due to the known contamination. Unfortunately, deep acid corrosion occurred despite these measures, due to concentration of chloride beneath the deposit and corrosion product layers (Fig. 6.16). Hydrogen damage also occurred beneath highly stratified iron oxide layers, further weakening the tube wall (Fig. 6.17).

The deeply corroded and hydrogen-damaged tubes needed to be replaced. In addition, the boiler needed to be cleaned due to the known chronic contamination, in order to succeed in controlling corrosion and hydrogen damage going forward. It was imperative to control the ingress of contaminants into the boiler. Once contamination is brought under control, the phosphate program could be optimized.

Case History 6.5

Industry:	Utility
Specimen location:	Nose slope, waterwall
Specimen orientation:	45°
Years in service:	25
Water treatment program:	Coordinated phosphate
Drum pressure:	2000 psi (13.8 MPa)
Tube specifications:	3-in. (7.6-cm) outer diameter

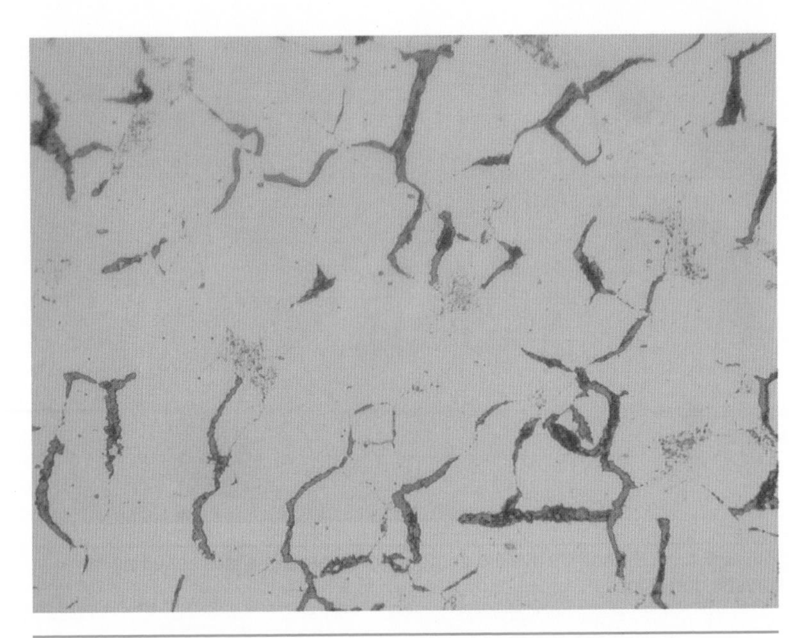

FIGURE 6.17 Intergranular fissures that formed within tube wall adjacent to deeply wasted internal surface that was covered with stratified iron oxide layers.

The thick-walled rupture shown in Fig. 6.18 was one of numerous similar failures recurring in both the nose-arch section and the roof tubes, requiring this area of the boiler to be rebuilt. The boiler was in peaking service. The rupture coincided with a distinct zone of deep metal loss on the internal surface (Fig. 6.19). The surface in the corroded region had a wavelike contour and was covered with black iron oxide. Examination of the microstructure revealed no thermal alteration. Magnetite needles covered the internal surface in the area

FIGURE 6.18 Thick-walled rupture.

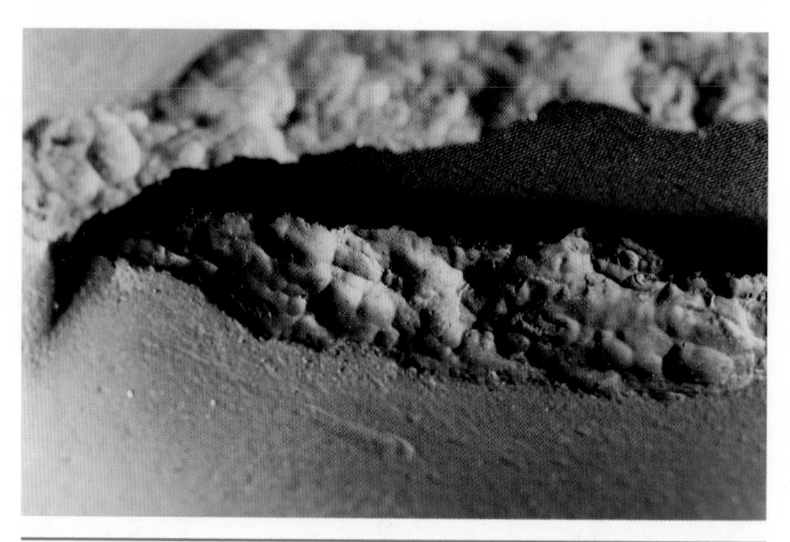

FIGURE 6.19 Smooth, wavelike contours of internal surface resulting from caustic corrosion.

where metal loss occurred. Numerous randomly oriented intergranular microcracks were present in the tube wall just below the corroded region. Localized DNB resulted in concentration of sodium hydroxide, which caused deep caustic corrosion at this site. The hydrogen damage and resulting fracture were a direct consequence of the caustic corrosion.

Reference

1. *EPRI HRSG Tube Failure Manual*, Electric Power Research Institute, Palo Alto, CA, 2002, pp. 12–5.

CHAPTER 7

Phosphate Corrosion

General Description

To describe the mechanism of phosphate corrosion, it is necessary to discuss in detail the phosphate treatment program and the history of phosphate programs in high-pressure boilers. Phosphate treatment of boiler water has been used since the 1950s. Sodium phosphate salts are intentionally added at controlled concentrations to boiler water as internal treatment chemistry in several boiler water treatment programs. In relatively low-pressure boilers, generally using softened makeup water, residual phosphate treatment programs are used primarily for scale control and may be used with or without synthetic polymers as a conditioner and dispersant. Traditional residual phosphate treatment involves controlling a specified orthophosphate concentration along with a controlled level of free hydroxide in the boiler water. These control ranges vary as a function of boiler drum operating pressure. Residual phosphate treatment is generally used as a precipitating treatment chemistry for any hardness contamination of the feedwater. The purpose is to intentionally precipitate the calcium as calcium hydroxyapatite and the magnesium as magnesium hydroxide. This treatment is generally used in boilers operating at <1000 psi (6.9 MPa), and it allows a significant concentration of free hydroxide ions in the boiler water to aid in the precipitation of the magnesium.

This treatment is also occasionally used in high-purity feedwater boiler systems. However, in higher-pressure boilers where higher feedwater purity is necessary, several other pH/phosphate control philosophies have been used. The primary function of these programs is for pH buffering and corrosion control. Proper pH control ensures the formation and retention of passive, protective magnetite layers on steel surfaces. The pH/phosphate control programs that have been used in high-pressure boilers include coordinated phosphate control, congruent phosphate control, equilibrium phosphate treatment, and finally what has been termed *phosphate continuum treatment.*

Free hydroxide in the boiler water can lead to several potentially corrosive conditions when highly concentrated. For example, caustic stress corrosion cracking of carbon steel drum materials may occur if there is a leak at a mechanical rolled tube joint (see Chap. 13, Stress Corrosion Cracking). Under-deposit caustic corrosion may also occur on waterside heat-transfer surfaces that are heavily fouled by relatively porous metal oxide deposits (see Chap. 4, Caustic Corrosion).

Several phosphate/pH control philosophies have been developed to try to minimize the presence of free hydroxide in the boiler water. The initial program was called *coordinated phosphate chemistry*, where the sodium-to-phosphate molar ratio of the boiler water was controlled to be less than 3.0:1. This was done to minimize the presence of free hydroxides in the boiler water. Unfortunately since it was found that sodium orthophosphates precipitate incongruently under dry-out conditions, this philosophy was not sufficient to prevent the generation of free hydroxide. The subsequent pH/phosphate control philosophy, called *congruent phosphate*, was developed to control the upper molar sodium-to-phosphate ratio in the boiler water between the congruent ratio 2.6:1 at high pressure (2.85:1 at lower pressures) and a lower sodium-to-phosphate ratio of about 2.3:1, again to minimize the presence of free hydroxide in the boiler water. The actual boiler water phosphate control range is chosen as a function of boiler drum operating pressure.

This treatment chemistry has been used for many years, but was found to be prone to a phenomenon known as *phosphate hideout*. As the term suggests, phosphate is removed from boiler water solution since it hides out on surfaces in high heat-transfer areas within the boiler. As boiler drum operating pressures increased with time, phosphate hideout became a progressively serious concern. Phosphate control ranges had to be reduced as a function of the operating pressure, to minimize the potential for phosphate hideout.

During a significant load increase in a conventional fuel-fired boiler on congruent phosphate treatment, the boiler water phosphate concentration would significantly decrease due to deposition or hideout of phosphate in high-temperature areas. Simultaneously, the boiler water pH would increase. This chemical hideout phenomenon is a unique temperature-dependent reaction between sodium orthophosphates and magnetite. There can be other reasons for the loss of phosphate from the boiler water that are unrelated. Following hideout events when the load is reduced without additional chemical feed, the boiler water phosphate concentration will increase, and the pH will decrease.

When the hideout phenomenon is not recognized, it is the natural tendency of the operator to try to correct the phosphate and pH with additional chemical feed. This correction would involve the feed of low-ratio sodium to phosphate chemistry to adjust the phosphate and reduce the pH. Monosodium phosphates and disodium phosphates have both been used. Unfortunately, incorrect adjustments will produce conditions that are conducive to phosphate corrosion, which can be also referred to as *acid phosphate corrosion*. The proper response during a hideout situation is to resist adjustment with these phosphates, since during a hideout event the accumulated phosphates will result in phosphate corrosion in high heat-transfer areas. Phosphate corrosion is most common when high-purity feedwater pH/phosphate control is used, particularly congruent phosphate control. Phosphate corrosion occurs when a localized concentrating mechanism raises the normally protective bulk boiler water concentrations of phosphate to levels that are destructive to carbon steel. These concentrating mechanisms under steam generator heat-transfer conditions typically are as follows:

- Wick boiling in porous metal oxide deposits
- Steam water stratification (steam blanketing)
- Phosphate hideout

The presence of maricite ($NaFePO_4$) and/or iron phosphate on a corroded surface is strong evidence that phosphate corrosion has occurred.

In an attempt to correct phosphate hideout behavior, potassium phosphates have been used in the past when congruent phosphate treatment was employed. Unfortunately serious under-deposit corrosion by concentrated alkaline potassium salts was experienced instead. Today potassium phosphate salts are rarely used because of this experience.

In high-pressure boilers with drum operating pressures approaching 2800 psi (19.3 MPa), problems with phosphate hideout on congruent phosphate treatment led to the development of a new treatment philosophy originally called *equilibrium phosphate* treatment. Recently this program was updated slightly and was renamed *continuum phosphate* treatment. The control boundaries for continuum phosphate treatment are shown in Fig. 7.1. This treatment philosophy has

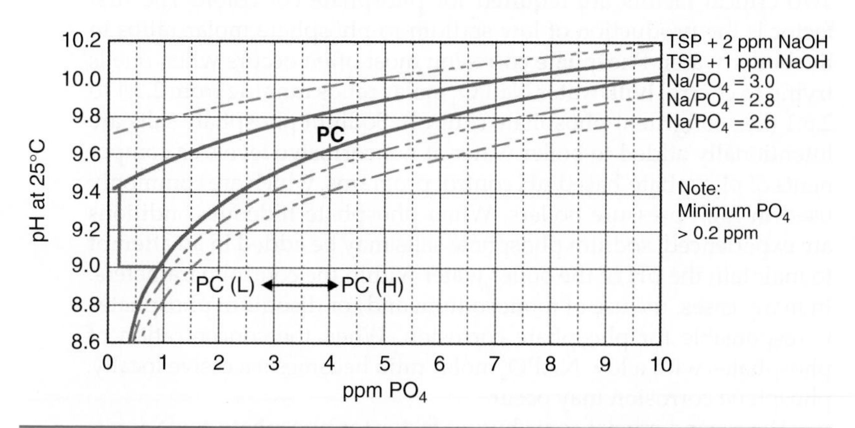

FIGURE 7.1 Phosphate continuum control chart. (*Courtesy of McGraw-Hill, The Nalco Water Handbook, 3d ed., New York, 2009.*)

generally eliminated the phosphate hideout behavior previously experienced by the same units that had used congruent phosphate treatment in the past, thus eliminating phosphate corrosion. This treatment philosophy allows up to 1 ppm NaOH in the boiler water along with a low concentration of phosphate in the boiler water.

Locations

Generally, phosphate corrosion is most common in boilers that operate above 2000 psi (13.8 MPa), or at lower pressures where high-purity makeup water is used. When high-purity water is used, metal oxides are the primary source of waterside deposits. Such deposits tend to promote wick boiling, which can provide the concentration mechanism necessary for phosphate corrosion. Phosphate corrosion damage is confined to localized areas on water-cooled tubes in

- Regions of high heat flux
- Slanted or horizontal tubes
- Locations beneath heavy deposits
- Heat-transfer regions at or downstream from features that disrupt flow such as intruding welds, weld backing rings, or other devices

In some boiler designs, devices are installed within portions of tubes with the intention of directing or channeling flow. In some cases, such devices may fail to perform the intended function adequately and/or may even promote phosphate corrosion (Case History 7.2).

Critical Factors

Two critical factors are required for phosphate corrosion. The first factor is the production of low sodium-to-phosphate molar ratios in the boiler water. Phosphate corrosion most often occurs when one is trying to control bulk water $Na:PO_4$ molar ratios ranging from 2.3:1 to 2.6:1 (i.e., congruent phosphate control). Sodium phosphate salts are intentionally added to boiler water at noncorrosive levels as components of phosphate-based pH control programs, which are commonly used in high-pressure boilers. When phosphate hideout conditions are experienced, sodium phosphate salts may be added in an attempt to maintain the pH of the boiler water within the control parameters. In many cases, the use of monosodium and/or disodium phosphates is responsible for phosphate corrosion. When the concentration of phosphates with a low $Na:PO_4$ molar ratio becomes excessive locally, phosphate corrosion may occur.

The second crucial contributing factor for phosphate corrosion is the mechanism of concentration. Because sodium phosphate salts are not present at corrosive levels in the bulk boiler water under proper control guidelines, a means of concentrating them must be present. Three basic concentration mechanisms exist:

1. *Departure from nucleate boiling (DNB):* The term *nucleate boiling* refers to a condition in which discrete bubbles of steam nucleate at points on a metal surface. Normally, as these steam bubbles form, minute amounts of boiler water solids will deposit at the metal surface, usually at the interface of the bubble and the water. As the bubble separates from the metal surface, the water will redissolve or rinse soluble solids, some of which are corrosive when concentrated, such as monosodium phosphate or disodium phosphate (Fig. 1.1).

   At the onset of DNB, the rate of bubble formation exceeds the rinsing rate of the soluble solids. Under these conditions, acid phosphate salts, as well as other less soluble dissolved solids or suspended solids, will begin to deposit (Figs. 1.3 and 4.2). The presence of sufficiently concentrated acid phosphate salts will both compromise the thin film of protective magnetite and cause metal loss.

   Under the conditions of fully developed DNB, a stable layer of steam will form. Corrosives will concentrate beneath the steam layer to cause corrosion.

2. *Deposition:* Another concentrating mechanism may occur when deposits shield the metal surface from the bulk water. Steam that forms under deposits escapes, leaving behind and entrapping a corrosive residue that can deeply corrode the metal surface (Fig. 4.3). Insulation provided by deposit layers may promote or cause the production of the steam. This phenomenon is commonly referred to as *wick boiling* (Fig. 4.4). When phosphate corrosion occurs in this manner, it is sometimes referred to as under-deposit corrosion. However, phosphate corrosion is only one type of under-deposit corrosion. Other types of under-deposit corrosion are described in Chaps. 4, 5, and 6.

3. *Evaporation along a steam/water interface:* Horizontal or slanted tubes are more susceptible to steam/water stratification since the critical heat flux for DNB is significantly lower than in vertically oriented tubes. Under unfavorable conditions of heat flux relative to boiler water flow rate, steam/water stratification may occur. In addition, boiler operation at excessively low water levels, or excessive blowdown rates, may create steam/water stratification. Steam/water stratification may also be created by excessive load reduction when pressure remains constant. In this situation, water velocity in the boiler tubes is reduced to a fraction of its full-load value. If velocity

becomes low enough, steam/water stratification occurs, creating stable or metastable waterlines. Damaged or missing refractory layers on floor tubes may allow exposure to higher fireside heat input, causing steam/water stratification. At locations of steam/water stratification, corrosives may concentrate by evaporation, resulting in corrosion.

Intruding welds or welds employing backing rings can create localized pressure drops immediately downstream. The pressure drop may promote localized DNB, which may cause deposition and under-deposit corrosion.

Identification

If affected waterside surfaces are accessible, visual examinations that reveal craters filled with hard corrosion products are potential sites of phosphate corrosion damage (Fig. 7.2). Frequently, a crust of hard deposits and corrosion products containing stratified iron oxide layers will surround and/or overlie the attacked region. The extent of corrosion damage may be determined by removing the corrosion products. The affected metal surface within the crater underlying the deposits and corrosion products generally has a smooth, rolling contour.

FIGURE 7.2 Thick corrosion product and deposit layers covering crater at location of phosphate corrosion damage.

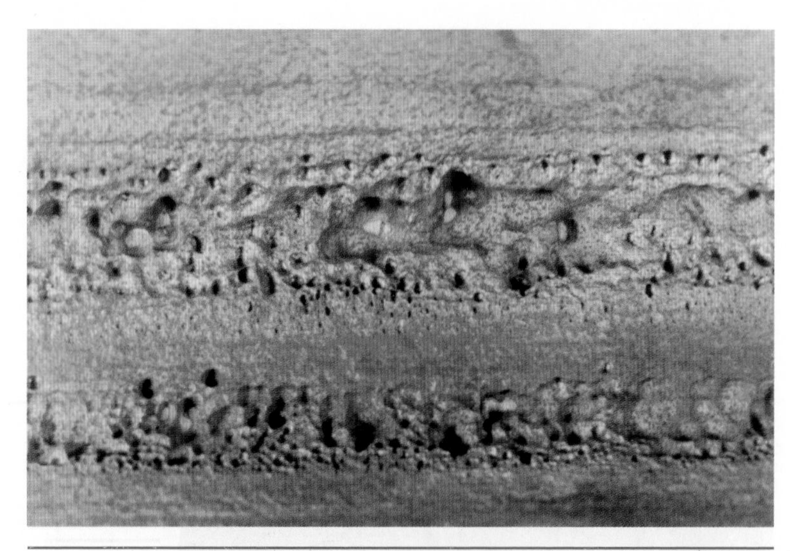

FIGURE 7.3 Mutually intersecting, undercut, rounded pits formed within a narrow zone where phosphate corrosion occurred.

If the waterside surface is not accessible, nondestructive testing techniques such as ultrasonic testing may be used in an effort both to locate areas of internal surface metal loss and to estimate the depth of penetration. The cause of metal loss may not be determined by such testing.

In horizontal or slanted tubes, a pair of parallel or converging, longitudinally oriented trenches or mutually intersecting undercut pits may form (Fig. 7.3). If the tube is nearly full, the parallel trenches will coalesce into a single elongated, longitudinally oriented crater along the crown of the tube. In vertically oriented tubes under rare conditions of stagnant flow, corrosive concentration at a steam/water interface will yield a circumferentially oriented crater.

Once possible phosphate corrosion damage is identified, metallographic analysis must be conducted to confirm that acid phosphate salt corrosion occurred, since acid and caustic corrosion during service may cause similar damage (see Chaps. 4 and 5). Strong supporting evidence for phosphate corrosion includes the presence of maricite and/or iron phosphate corrosion products in layers that cover the corroded surfaces. X-ray diffraction may be used to identify the compounds in corrosion product layers. Metallographic examination of cross sections through the wasted surfaces may reveal dense and possibly stratified iron oxide layers, as shown in Fig. 7.4. In some cases, crystalline material having a needlelike morphology may be identified at high magnification using metallographic and scanning electron microscope (SEM) examinations (Fig. 7.5). Such needlelike features tend to be associated with magnetite needles that are

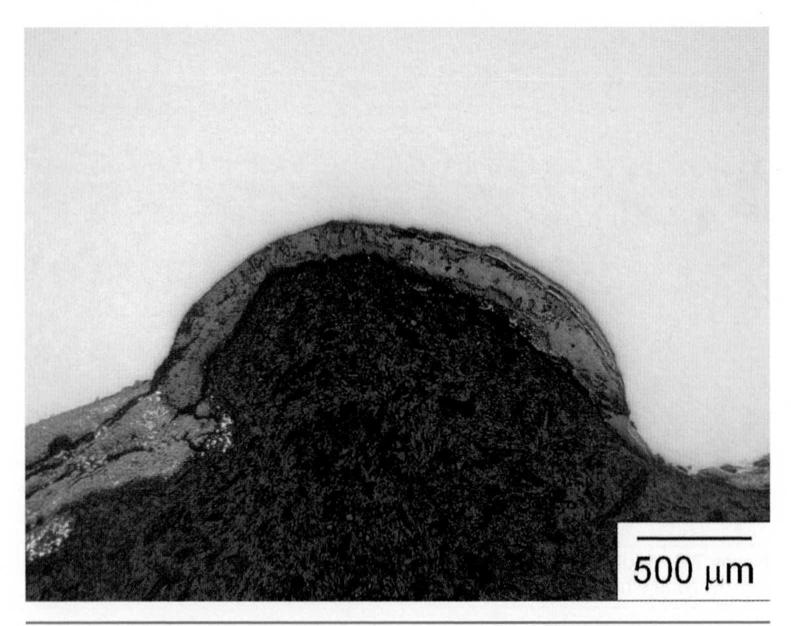

FIGURE 7.4 Metallographic cross section of surface that experienced phosphate corrosion. Dense, stratified corrosion product layers cover the wasted surfaces.

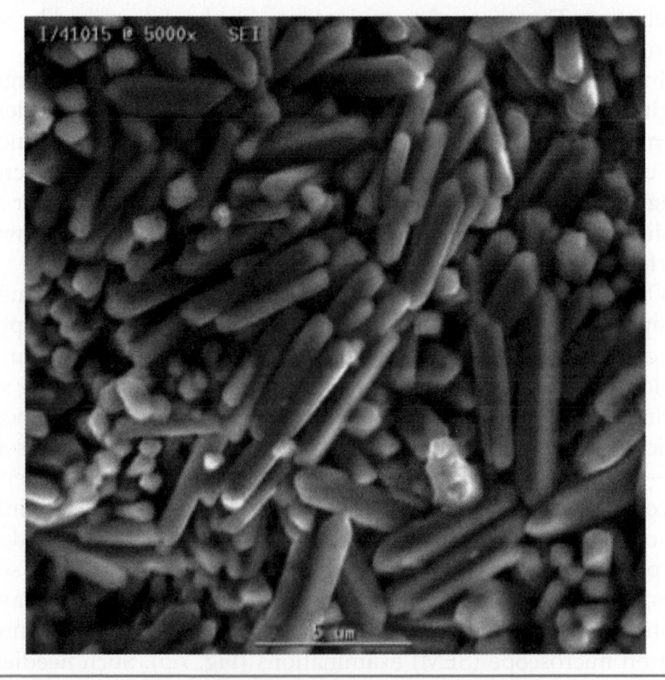

FIGURE 7.5 Needlelike material covering corroded surfaces.

commonly found at sites where caustic corrosion occurs. Therefore, it is necessary to conduct chemical analyses of needlelike corrosion products. Energy-dispersive spectroscopy using the SEM that reveals approximately equivalent concentrations of sodium, iron, and phosphorus will confirm that the crystalline material is indeed maricite. If enough material is present, x-ray diffraction may be conducted to definitely determine if maricite formed. Needle-shaped corrosion products that consist primarily of iron oxide will indicate that they are magnetite needles. Metallic copper particles may also deposit in the crater. This is due to the evolution of hydrogen in the phosphate corrosion reaction, which reduces dissolved copper in the water. The operating conditions, chemical treatment program, and water chemistry historical data should be reviewed to determine if conditions supported acid phosphate corrosion.

Elimination

Corrosion may occur when the boiler water contains dissolved acid phosphate salts and a concentrating mechanism exists simultaneously at a particular location. The following remedies may be used to control or eliminate corrosion by concentrated acid phosphate salts.

- *Reduce the amount of available acid phosphate salts:* If a phosphate treatment program is used, it should be appropriate for the specific conditions of boiler operation. For instance, the congruent phosphate program was designed to prevent caustic corrosion. Below the congruent point, free sodium hydroxide is not produced, thus preventing caustic corrosion. However, under conditions of phosphate hideout, the use of this treatment program and/or a coordinated phosphate program may promote phosphate corrosion. The phosphate program must be designed to control not only phosphate corrosion, but also caustic corrosion and acid corrosion (see Chaps. 4 and 5). Alternate phosphate programs that do not achieve low $Na:PO_4$ molar ratios may be substituted, such as those using only trisodium phosphate and caustic. These programs must be more tightly controlled than the traditional coordinated phosphate programs to avoid caustic corrosion. The most current guidelines for phosphate treatment programs in high-pressure utility boilers are detailed in the phosphate continuum control chart shown in Fig. 7.1. A more in-depth, detailed discussion about phosphate treatment programs that are designed to meet the varying conditions of high-pressure utility boilers can be found in *The Nalco Water Handbook,* third edition.

For programs using low-phosphate concentrations it is imperative that boiler water contamination by chlorides and sulfates be strictly controlled, to prevent hydrogen damage. Because of the powerful concentrating mechanisms that may operate in a boiler, the addition of only a few parts per million of contaminants to the boiler water may be sufficient to initiate or accelerate hydrogen damage.

Prevention of localized concentration is the most effective means of avoiding phosphate corrosion. However, it is also the most difficult to achieve. The methods for preventing localized concentration include these:

- *Prevent DNB:* This usually requires the elimination of hot spots, which is achieved by controlling the boiler's operating parameters. Hot spots are caused by excessive overfiring, misdirected burners, change of fuel, gas channeling, etc. Conditions that reduce flow below appropriate levels, such as reduced firing rates or excessive blowdown, should be avoided. In heat recovery steam generators (HRSGs), excessive heat input due to the use of duct burners and flue gas laning may promote hideout on congruent phosphate treatment and phosphate corrosion.
- *Prevent excessive waterside deposition:* Periodic tube sampling may be performed to measure the relative amount of deposit buildup on tubes. This sampling is referred to as a deposit weight determination, deposit loading, or gram loading. Deposit weight determination practice is described in Chap. 1. Susceptibility to under-deposit corrosion increases as a function of increasing deposit weight and boiler pressure. The boiler manufacturer should be consulted for chemical cleaning recommendations. Aside from the boiler manufacturer's guidelines, associations such as the National Association of Corrosion Engineers (NACE) and the Electric Power Research Institute (EPRI) provide cleaning guidelines. Proper monitoring will ensure that corrosion is controlled in the preboiler system, to reduce the amount of corrosion products introduced to the boiler.
- *Prevent steam/water stratification in tubes:* Proper boiler operating parameters must be followed to ensure proper steam production rates and boiler water circulation. If the design and operating conditions cannot be altered practically to prevent poor circulation, then consistent periodic inspection and cleaning practices may be required for certain tubes.
- *Ensure proper component design:* Intruding welds and other components within the fluid flow path on the water side should be avoided, as they can promote DNB, cause localized deposition, and entrap steam, which can cause acid phosphate salts to concentrate.

Cautions

Because corrosion products may fill the depressions caused by phosphate corrosion, the extent and depth of metal loss in the affected area—and even the existence of a corrosion site—may be overlooked during an internal inspection.

Distinguishing localized attack caused by phosphate corrosion from localized attack by acid or caustic corrosion during service simply by visual examination is difficult. A formal metallographic examination is required. Monitoring the phosphate treatment history and evaluating the boiler water for contaminants will aid in the determination. The presence of crystalline, needle-shaped material at corroded sites may be misleading. Chemical analysis of such material should be conducted. If the composition of the material is consistent with maricite, phosphate corrosion is indicated. If the material is composed of iron oxide, caustic corrosion is probable.

Related Problems

See also Chap. 1, Water-Formed and Steam-Formed Deposits; Chap. 5, Low-pH Corrosion during Service; and Chap. 4, Caustic Corrosion.

Case History 7.1

Industry:	Utility
Specimen location:	Waterwall
Orientation of specimen:	Slanted (nose arch)
Years in service:	10
Water treatment program:	Congruent phosphate
Drum pressure:	2700 psi (18.6 MPa)
Tube specifications:	3-in. (7.6-cm) outer diameter

Figure 7.6 illustrates one of many tubes that had sustained similar severe corrosion. Grooves were located along the top (crown) of each tube. The convergence of the groove marked the position where the slanted tube assumed a vertical orientation. Metal loss was not observed downstream of this point. Microstructural analyses revealed that the grooved portion of the tube wall had sustained mild overheating.

Corroded areas were covered with black, powdery deposits. These deposits covered a second layer of light-colored crystalline material that was directly on the metal. The light-colored material was present only in corroded regions. X-ray diffraction revealed that the light-colored crystalline material was composed of iron phosphate and sodium iron phosphate (maricite). On either side of the groove, a hard coating of protective magnetite covered the smooth internal surface. The contour of the corroded surface following removal of the deposits and the light-colored material are shown in Fig. 7.3.

Gravitational stratification of water and steam established relatively stable steam channels along the tops of the slanted tubes. Evaporative concentration of dissolved solids within this channel furnished the corrodent. The presence of maricite and iron phosphate revealed that the metal loss was caused by phosphate corrosion.

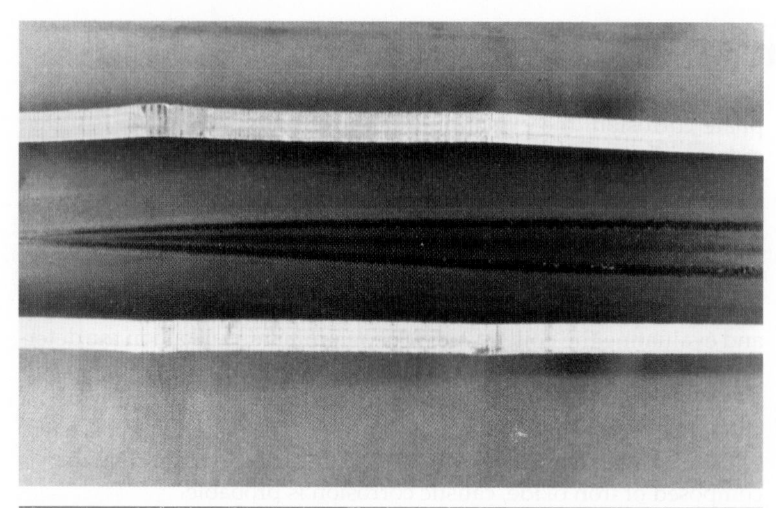

FIGURE 7.6 Convergence of groove.

Case History 7.2

Industry:	Utility
Specimen location:	In bed coil tube inside a fluidized-bed boiler
Orientation of specimen:	Horizontal
Years in service:	8
Water treatment program:	Congruent phosphate
Drum pressure:	1550 psi (10.7 MPa)
Tube specifications:	3.5-in. (8.9-cm) outer diameter
Environment:	
Internal	Base loaded during the week, low loads on weekends. Congruent phosphate treatment program for 7 years (pH 9.0 to 9.8, phosphate at 4 to 10 ppm), changed the program to a phosphate treatment program (pH 9.2 to 9.5, phosphate at 4 to 5 ppm) for several months prior to failure
External	Fluidized-bed boiler using limestone bed and coke as a fuel source, 1500°F (815°C)

The fluidized-bed boiler used U-bend tubes that extended into the limestone and coke bed. Strips of sheet steel were bent into a helical shape and were inserted into the bores of the straight portions of the tubes. These strips are commonly named *turbulators*, as their purpose is to guide the flow of water through the tubes in a helical pattern, promoting turbulent flow and limiting steam/water stratification. Figure 7.7 shows how the turbulators were tack-welded to the internal surface. The edges of the turbulators directly contacted the internal surface of the tube in some locations, while narrow gaps were present between the edge and tube surface in other locations. Undercut pits and grooves on the internal surface of the tube were only located at the contact points of the turbulator (Fig. 7.8). Metal loss in some locations was quite deep (Fig. 7.9). Metallographic examination revealed undercut pits and depressions that were lined with stratified corrosion product layers (Fig. 7.10). Scanning

FIGURE 7.7 Turbulator strip tack-welded to the internal surface.

FIGURE 7.8 Turbulator strip bent away from position to reveal metal loss at contact point with tube internal surface. (*© NACE International, 2005.*)

electron microscope examination of the corrosion products revealed that they contained high concentrations of sodium, iron, and phosphorus, matching the composition of maricite.

Due to the concentration of aggressive salts within crevices between the internal surface and turbulator, metal loss occurred only at locations where the

FIGURE 7.9 Very deep internal surface metal loss at prior tube to turbulator strip contact location. (*© NACE International, 2005.*)

FIGURE 7.10 Metallographic cross section of deep undercut pit that is partially covered with dense, stratified corrosion product layers. (*© NACE International, 2005.*)

turbulator contacted the internal surface. Boiler water could not flow through the tight crevices to rinse aggressive salt deposits from the surfaces. In addition, heat transfer was impeded at the turbulator-to-tube contact points, promoting deposition and aggressive salt concentration. The shapes of the pits and depressions, and the presence of maricite in the corrosion products, indicate that phosphate corrosion occurred at those locations. The boiler was susceptible to phosphate corrosion under the congruent phosphate treatment program. Corrosion may have resulted from periodic phosphate hideout conditions and the resultant feed of acid phosphate.

No matter what phosphate program was used, aggressive salts, including caustic, chloride, sulfate, and acid phosphate, may readily concentrate beneath the turbulator strips to corrosive levels. Therefore, a change in design or in the treatment program was recommended. Other designs, such as internal rifling, could have been considered to promote turbulent flow. Changing to an all volatile treatment program would prevent corrosion by concentrated, aggressive agents, such as caustic and acid phosphate salts. However, the use of such a program would not prevent corrosion by aggressive salts that may contaminate the boiler water, such as chloride. Therefore, boiler water contamination would have to be very strictly controlled.

Case History 7.3

Industry:	Utility
Specimen location:	Waterwall
Orientation of specimen:	Vertical
Years in service:	27
Water treatment program:	Congruent phosphate
Drum pressure:	2500 psi (17.2 MPa)
Tube specifications:	1.5-in. (3.8-cm) outer diameter

The congruent phosphate program used in the boiler was controlled to a sodium-to-phosphate ratio of 2.5:1. In addition, the boiler experienced frequent load swings, causing the boiler load to vary from 200 to 400 MW within a very short time, sometimes more than once per day.

An internally rifled rear waterwall tube failed during hydrotesting at 2850 psi (19.7 MPa). A small water leak was detected during testing, and a pick was used to find the exact source of the leaking water. The perforation occurred at the base of a deep internal surface crater within one of the grooves, between the rifling ridges (Fig. 7.11).

Moderately thick, brown deposit and corrosion product layers overlaid powdery, white deposits and corrosion products within the crater. Significant quantities of metallic copper and needle-shaped particles were embedded within these layers (Fig. 7.12). Chemical analysis using x-ray diffraction indicated that the light-colored material consisted almost entirely of sodium iron phosphate (maricite, $NaFePO_4$), which was formed during phosphate corrosion reactions.

Concentration of phosphate species was caused by wick boiling beneath porous deposits. Wick boiling can occur in high heat flux regions of the boiler and/or in areas where there is poor circulation.

Changing loads promoted phosphate corrosion by producing conditions that were favorable for phosphate hideout. Cleaning was required to remove excessive waterside deposition. The low sodium-to-phosphate ratio congruent phosphate program promoted phosphate corrosion under hideout conditions. The congruent phosphate treatment program was changed to an equilibrium phosphate program to control phosphate corrosion.

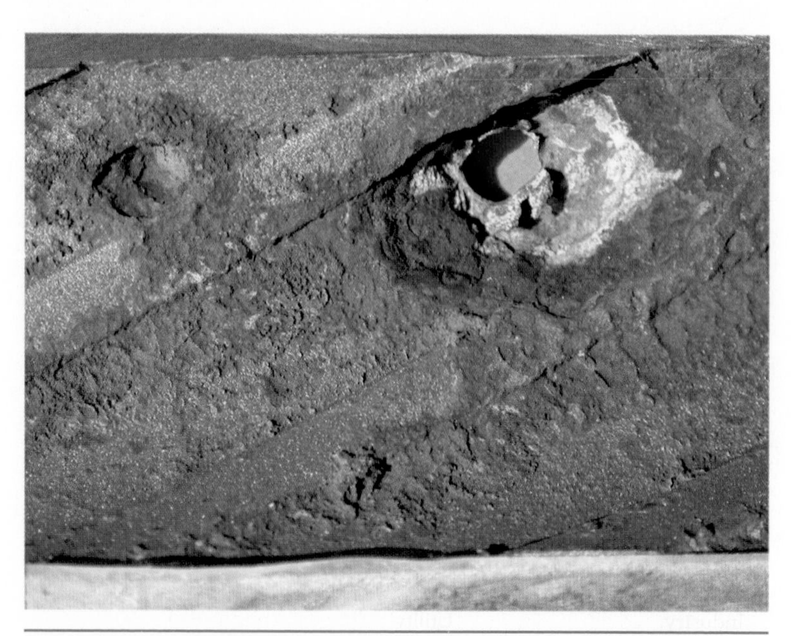

FIGURE 7.11 Perforation at base of deep crater that formed within groove between rifling ridges. (*© NACE International, 2005.*)

FIGURE 7.12 Needle-shaped particles within corrosion products that consist of maricite.

FIGURE 7.13 Deep external surface (waterside) metal loss and failure at hot gas inlet end of tube from a waste heat boiler.

Case History 7.4

Industry:	Refining
Specimen location:	Waste heat boiler
Orientation of specimen:	Horizontal
Years in service:	6
Water treatment program:	Coordinated phosphate
Drum pressure:	1800 psi (12.4 MPa)
Tube specifications:	¾-in. (1.9-cm) outer diameter

Process gas flowed through the tubes at 806°F (430°C), and boiler water flowed across the shell sides. A tube failure occurred due to deep external surface metal loss at the bottom of the waste heat boiler near the process gas inlet end (Fig. 7.13). The region of metal loss extends for approximately 6 in. (15.2 cm). The wasted surface was covered with powdery white deposits that overlaid thin, brown and black iron oxide layers.

In some locations the deposit and corrosion product layers were moderately thick and contained fine platelike particles and magnetite needles. The platelike particles and surrounding deposits and corrosion products contained high levels of iron, sodium, and phosphorus that had approximately a 1:1:1 atomic ratio, consistent with maricite ($NaFePO_4$).

Phosphate corrosion was caused by steam blanketing that resulted from excessive heat flux relative to the water flow rate. Boiler water circulation may be poor on the shell sides of tubes in horizontally oriented waste heat boilers. In addition, heat flux through the tube walls may be excessive due to damage to refractory and ferrules on the gas side. Changes to the phosphate program will not be reliable for controlling corrosion under steam blanketing conditions.

CHAPTER 8

Corrosion by Chelating and Complexing Agents

General Description

Chelating or complexing agents, also known as chelants or sequestrants, such as ethylenediaminetetraacetic acid (EDTA) and nitrilotriacetic acid (NTA) may be used in certain chemical treatment programs to complex calcium, magnesium, and metal ions that enter the boiler water. These ions are strongly bound to the chelating agents and cannot undergo any other reactions, allowing them to be removed through blowdown. Certain polymers that are used in chemical treatment programs will react as weak complexing agents. An in-depth description of chelation and complexation is provided in *The Nalco Water Handbook*, third edition.

Concentrated chelating or complexing agents may cause corrosion by reacting with the protective iron oxide layer (magnetite). Once this layer is removed, more rapid metal loss will occur due to direct reaction with the exposed tube steel. Corrosion will not occur at normal use concentrations of the polymer or chelating agent. However, corrosion rates increase as a function of increasing concentration of chelating or complexing agents above certain levels. Aside from concentration, a strong contributing factor for corrosion by chelating or complexing agents is velocity and turbulence. Corrosion rate increases as a function of increasing flow rate. Other accelerating factors include low pH and exposure to dissolved oxygen.

Currently, chelating agent corrosion failures occur less frequently than in the past, when such programs were originally introduced. This is due to a better understanding of the limits of such programs and to improved monitoring and chemical feed. In addition, traditional chelating agent–based programs have been increasingly replaced by polymer-based programs that are not as aggressive as chelating agents under concentrating mechanisms.

Locations

The steam drum is the area of the boiler most frequently affected by chelating agent corrosion. Steam separation equipment is particularly susceptible to this type of corrosion, due to combined concentration and flow velocity effects. Chelating and complexing agent corrosion may also occur in feedwater distribution lines, in the economizer, on the ends of downcomer tubes, and in regions of high heat flux in water-cooled tubes. Copper alloy impellers of feedwater pumps are highly susceptible to corrosion if exposed to chelating or complexing agents.

Identification

When access is very limited, nondestructive testing (NDT) techniques, such as ultrasonic testing, may be used to detect wall thinning. Chelating and complexing agent corrosion can be visually identified easily if affected equipment is accessible. The corrosion produces general metal loss. Attacked surfaces are typically very smooth and featureless. In situations where there is sufficient fluid velocity, the surfaces may have a smoothly rolling contour marked by "comet tails" and horseshoe-shaped depressions (Fig. 8.1). These features are aligned with the direction of flow. The metal surface will be uniformly covered with a submicroscopic film, of either dull or glossy black iron oxide. A surface under active attack will be free of deposits and corrosion products, since they will already have been removed due to chelation or complexation. Corrosion by chelating or

FIGURE 8.1 Comet tail and horseshoe-shaped depressions on corroded surface that is covered with a thin, glossy, black iron oxide layer.

FIGURE 8.2 Corrosion caused by concentrated chelating agents that was altered by excessive dissolved oxygen. Rough-surfaced, undercut pits formed.

complexing agents does not require overheating. Metallographic analysis will reveal very thin or even unresolvable iron oxide layers covering corroded surfaces. The wasted surface will have an undulating profile if the attack was active shortly before section removal for analysis. Wasted surfaces covered by a microscopically visible iron oxide layer would indicate that corrosion was active in the past.

When excessively high levels of dissolved oxygen are present in the boiler water along with concentrated chelating or complexing agents, the appearance of metal loss is altered. The surface under attack retains sharply defined islands of intact metal surrounded by a smooth plain of metal loss (Fig. 8.2). The attacked surface has a jagged roughness, similar to that resulting from attack by a strong acid.

Steam drum internals are particularly susceptible to chelating or complexing agent corrosion. Severe cases of corrosion have reduced cyclone separator cans to lacelike remnants.

Critical Factors

Chelating and complexing agent corrosion can occur under a variety of circumstances. Corrosion increases as a function of increasing flow rate and/or turbulence and increasing concentration of the chelating or complexing agent. It is possible for one factor to predominate. Locations of high velocity and turbulence are most susceptible to attack. With regard to concentration, overfeed of chelating or complexing agents may be a suspected cause of corrosion. However, a substantial overfeed would be required to cause corrosion, which would be unusual under a properly controlled program. Rather, as with other corrosion mechanisms described in this book, corrosion is possible in regions where a concentration mechanism operates. The two principal concentration mechanisms are described below.

1. *Departure from nucleate boiling (DNB)*: The term *nucleate boiling* refers to a condition in which discrete bubbles of steam nucleate at points on a metal surface. Normally, as these steam bubbles form, minute amounts of boiler water solids will deposit at the metal surface, usually at the interface of the bubble and the water. As the bubble separates from the metal surface, the water will redissolve or rinse soluble material (Fig. 1.1).

   At the onset of DNB, the rate of bubble formation exceeds the rinsing rate of the soluble material. Under these conditions, chelating or complexing agents, as well as other less soluble dissolved solids or suspended solids, will begin to deposit. The presence of sufficiently concentrated chelating or complexing agents will both compromise the thin film of protective magnetite and cause metal loss (Fig. 8.3).

   Under the conditions of fully developed DNB, a stable layer of steam will form. Corrosives concentrate at the steam layer to cause corrosion.

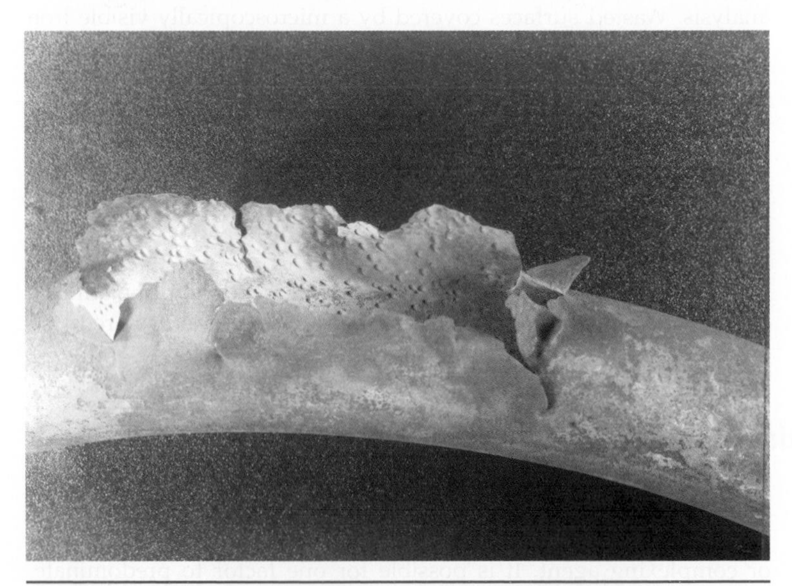

Figure 8.3 Tube rupture caused by severe internal surface metal loss due to corrosion by concentrated chelating agents. (*Courtesy of National Association of Corrosion Engineers.*)

2. *Evaporation along a steam/water interface:* Horizontal or slanted tubes are more susceptible to steam/water stratification, since the critical heat flux for DNB is significantly lower than in vertically oriented tubes. Under unfavorable conditions of heat flux relative to boiler water flow rate, steam/water stratification may occur. In addition, boiler operation at excessively low water levels, or excessive blowdown rates, may create waterlines. Waterlines may also be created by excessive load reduction when pressure remains constant. In this situation, water velocity in the boiler tubes is reduced to a fraction of its full-load value. If velocity becomes low enough in vertical, horizontal, or slanted tubing, steam/water stratification occurs, creating stable or metastable waterlines.

   Carryunder of steam may occur in downcomer tubes to produce steam/water stratification downstream of the steam drum. During boiler operation a steam/water mixture rises from the generating or riser tubes. Ordinarily, the steam rises through the primary steam separators. However, under certain circumstances, some of the steam bubbles will be drawn down toward the bottom of the steam drum. This phenomenon is referred to as *carryunder*. In some cases, steam bubbles find their way into the downcomers at locations where they cannot escape, such as at the tops of tubes at the bends or slanted portions immediately downstream of the steam drum. The steam bubbles will become entrapped and collect within these portions, causing steam/water stratification to occur. At the interface of the entrapped bubbles, aggressive agents and chemicals will concentrate to corrosive levels. Carryunder can also transport steam bubbles all the way to the mud drum, where they can be trapped at the top of the drum.

   There are a few possible causes for carryunder to occur. In a properly operating boiler, steam bubbles that flow out of the riser tubes toward the bottom of the steam drum will pass over the area of the feedwater line that is injecting subcooled water into the drum. The subcooled feedwater collapses the steam bubbles before they reach the downcomers, and carryunder is avoided. However, if modifications were made to the boiler or feedwater system that increased the feedwater temperature, the amount of subcooling available in the steam drum could be reduced enough to allow carryunder. It is necessary to determine if a new economizer or feedwater heater has been installed, or the feedwater temperature has increased. High steaming rates may also produce carryunder. Proper shielding should be used in the steam drum to prevent steam bubbles from being carried under. If such shields are not in use, they should be considered; or if they are installed but were damaged, they should be repaired.

Damaged or missing refractory layers on floor tubes may allow higher fireside heat input to cause steam/water stratification. Blowdown lines may also be susceptible to stratification under certain operating conditions. Intruding welds or welds employing backing rings can create localized pressure drops immediately downstream. The pressure drop may promote localized DNB. At locations of steam/water stratification, corrosives may concentrate by evaporation, resulting in corrosion.

Chelating and complexing agent corrosion of steam separation equipment, which does not contain heat-transfer surfaces, is influenced primarily by high fluid velocities and turbulence. This type of corrosion is especially sensitive to the erosive effects of high-velocity fluids, and it is sometimes encountered where these exist, even in the absence of heat transfer. Excessive concentrations of dissolved oxygen in the boiler water synergistically increase corrosion by concentrated chelating or complexing agents.

Dosage levels for chelating agent or polymer programs depend upon hardness concentration in the feedwater. Corrosion will tend to occur more readily under certain conditions, such as when the boiler water contains very low concentrations of dissolved hardness salts, particularly when demineralized makeup water is used. The lack of hardness may cause the chelating or complexing agents to react with protective iron oxide and exposed tube metal. Treatment programs utilizing combined chelating agents and polymers that are weak complexing agents should be followed with caution, since the combination may be more aggressive than either the chelating agent or the polymer alone.

Elimination

Close control of the chelating agent, complexing agent, and dissolved-oxygen concentrations in the boiler water is imperative. To eliminate chelating and complexing agent corrosion, certain actions can be taken.

First, special care must be exercised both under conditions of inconsistent feedwater quality and when chelating and complexing agents are used in a dirty boiler. Attack from chelating or complexing agents may be detected through use of iron(II) testing of boiler feedwater and blowdown.[1] This testing may be used for monitoring and to optimize dosage levels. In cases where chelating agents or certain polymers are used, less corrosive polymers may be substituted.

Second, elimination of DNB conditions will prevent corrosion by concentrated chelating and complexing agents in water-cooled tubes. DNB may occur at hot spots in the firebox, which are frequently caused by improper boiler operation, improper boiler maintenance, and design deficiencies. Intruding welds and other components within the fluid flow path on the water side should avoid entrapment of steam or DNB.

Proper boiler operating parameters must be followed to ensure proper steam production rates and boiler water circulation. Such parameters may include excessive overfiring, misdirected burners, change of fuel, gas channeling, excessive blowdown, and dislodged refractory, all of which have been linked to DNB.

In some cases, corrosion by chelating or complexing agents can be reduced in severity or eliminated altogether by reducing fluid velocity and eliminating turbulent flow. Primary locations where flow-related damage may predominate include the feedwater system and steam separators.

Cautions

Visual and borescope inspections of tube internal surfaces may be misleading. Chelating and complexing agent corrosion tends to cause uniform general metal loss, which produces a smooth contour on metal surfaces. This feature, coupled with the fact that the corroded surfaces typically have a black passive appearance, can mask the fact that corrosion occurred. In these cases, thickness measurements of suspect areas using appropriate nondestructive testing techniques may reveal a problem.

Flow-accelerated corrosion and simple erosion by high-velocity steam or water may also yield an appearance that closely resembles that of corrosion by chelating and complexing agents. Proper diagnosis should involve a thorough study of the chemical treatment program and the boiler water chemistry data.

Every effort should be made to avoid exposing copper-based alloys to chelating or complexing agents.

Related Problems

See also Chap. 19, Flow-Accelerated Corrosion; Chap. 10, Corrosion during Cleaning; and Chap. 18, Erosion.

Case History 8.1

Industry:	Steel
Specimen location:	Wall tube at bottom waterwall header
Orientation of specimen:	Vertical
Years in service:	14
Water treatment program:	EDTA
Drum pressure:	900 psi (6.2 MPa)
Tube specifications:	2½-in. (6.4-cm) outer diameter
Fuel:	No. 6 fuel oil and various waste gases

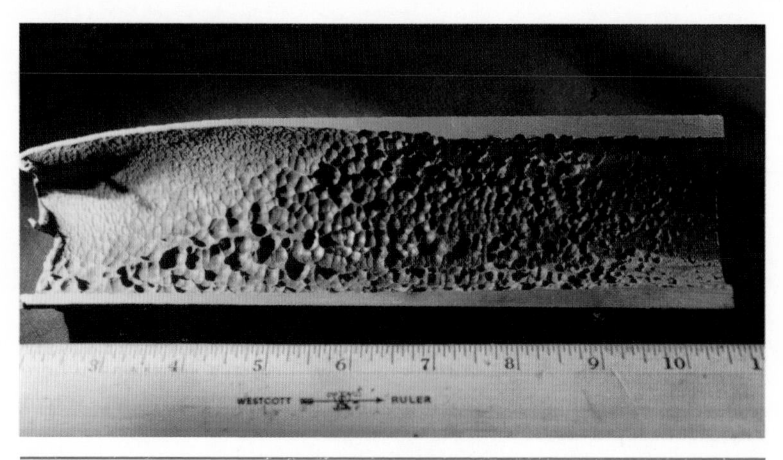

FIGURE 8.4 Corrosion caused by chelating agents on internal surface of wall tube at inlet end.

Severe internal wastage was confined to a 6- to 9-in. (15.2- to 22.9-cm) length of tube at the inlets of each of four adjacent tubes (Fig. 8.4). A tube failure occurred during hydrotesting. Metal loss within the affected region varied. In areas of severe metal loss, the appearance was smooth and wavelike. Discrete isolated pits were evident in areas of moderate metal loss.

The highly localized attack on this section illustrates the importance of velocity and turbulence in an erosion-corrosion process involving chelating agents. Metal loss was most severe in regions of highest turbulence at the inlet end of the tube from the header. The attack moderated with distance along the tube, until it ceased altogether where laminar flow was established. The EDTA concentration alone was not sufficiently aggressive to cause corrosion. Rather, metal loss resulted from the synergistic interaction of the chelating agent with localized turbulence. Modifications to alter the flow in the affected region may help control the corrosion. Changes to the chemical treatment program should also be considered.

Case History 8.2

Industry:	Steel
Specimen location:	Wall tube
Orientation of specimen:	Vertical bend
Years in service:	25
Water treatment program:	EDTA
Drum pressure:	800 psi (5.5 MPa)
Tube specifications:	3-in. (7.6-cm) outer diameter
Fuel:	Blast furnace gas

The massive rupture of a wall tube (Fig. 8.3) was the first to occur in the boiler. The edges of the rupture were very thin, and a population of horseshoe-shaped depressions oriented in the direction of flow formed along the generally wasted internal surface. Microstructural examinations revealed that overheating had not occurred. No external surface metal loss was found.

The rupture occurred on the cold side of the tube. The insulation installed to protect the cold side of this tube apparently had been dislodged by steam impingement from a leaking superheater tube. Hot furnace gases that reached the unprotected back side of the tube caused DNB and resulted in concentration of chelating agents from the boiler water.

The combination of concentrated chelating agents and fluid velocity caused the thinning of the internal surface apparent in Fig. 8.3. Thinning of the tube wall in this manner continued until stresses imposed by normal internal pressure exceeded the tensile strength of the thinned tube wall, causing the massive rupture.

Case History 8.3

Industry:	Chemical process
Specimen location:	Feedwater pump impeller
Orientation of specimen:	Vertical
Years in service:	3
Water treatment program:	EDTA
Tube specifications:	8-in. (20.3-cm) outer diameter
Tube material:	Bronze

This impeller, illustrated in Fig. 8.5, was removed from the fifth stage of a feedwater pump. Broad areas of general wastage were apparent. Areas of metal loss were confined principally to regions of high turbulence such as the vane edges and discharge throat. Horseshoe-shaped depressions were visible in some regions.

Close examination of wasted surfaces under a low-power stereoscopic microscope revealed dendrites and other microstructural features of the casting. Chelating agents are particularly aggressive toward copper and copper-based alloys and should be fed well downstream of copper-alloy impellers.

FIGURE 8.5 Corrosion by chelating agents on bronze feedwater pump impeller.

Case History 8.4

Industry:	Brewing
Specimen location:	Steam drum end of downcomer tubes
Orientation of specimen:	10° to 15° from vertical
Years in service:	36
Water treatment program:	EDTA
Drum pressure:	485 psi (3.3 MPa)
Tube specifications:	2½-in. (6.4-cm) outer diameter

Figure 8.6 illustrates the appearance of the internal surfaces of 12 adjacent downcomer tubes that suffered severe internal corrosion. Although the same water treatment program had been utilized in this boiler for 17 years, this corrosion occurred within a period of 7 months.

The affected tubes were located at one end of the boiler, within 9 in. (22.9 cm) of the feedwater line in the steam drum. It is surmised that feedwater was short-circuiting through the affected area.

Oxygen pitting was discovered along the waterline throughout the steam drum and on the steam separation canisters. This is evidence that excessive levels of dissolved oxygen were present in the feedwater. The excessive oxygen levels coupled with the presence of the chelating agents resulted in the corrosion observed. The metal loss produced smooth, rolling, wavelike surface contours and islands of intact metal in affected areas.

The entire internal surface was lightly covered with deposits and iron oxides. The surface in the corroded areas was covered with a shiny black film under these deposits and iron oxides. The layer of iron oxides and deposits covering corroded surfaces reveals that corrosion was not active shortly before the

FIGURE 8.6 Internal surface of downcomer tube, showing metal loss caused by chelant corrosion that was assisted by dissolved oxygen.

section was removed from service. Therefore, corrosion was not continuous, but intermittent.

This type of damage can be prevented by gaining control of dissolved oxygen in the feedwater. Preventing short-circuiting of the feedwater is also a solution to this type of corrosion.

Case History 8.5

Industry:	Chemical process
Specimen location:	Waste heat boiler (shellside steam generation)
Orientation of specimen:	Horizontally oriented U-Tube
Years in service:	15
Water treatment program:	Polymer
Pressure:	250 psi (1.7 MPa)
Tube specifications:	7/8-in. (2.2-cm) outer diameter

The boiler water in the waste heat boiler flowed across the shell sides of the tubes. Deep general metal loss occurred on the shell sides within the curved portion of the tubes. Some of the tube walls thinned to the point of failure where metal loss was deepest (Fig. 8.7). The surfaces within the wasted areas were very smooth, exhibiting flow-oriented depressions and grooves, some of which were horseshoe- or comet-shaped (Fig. 8.8). The wasted surfaces were covered with thin, lustrous, black iron oxide layers. The metal loss was caused by the concentration of polymer in the boiler water. Corrosion was promoted by the rapid or turbulent flow of fluids across the surfaces. It was reported that water levels in the exchanger dropped to expose the tubes within the topmost portion of the exchanger where the damage occurred. Boiler water evaporated within the steam space, concentrating the polymer to corrosive levels. In addition, turbulent flow conditions were produced within the steam space. Monitoring was instituted to ensure that proper water levels were maintained in the boiler.

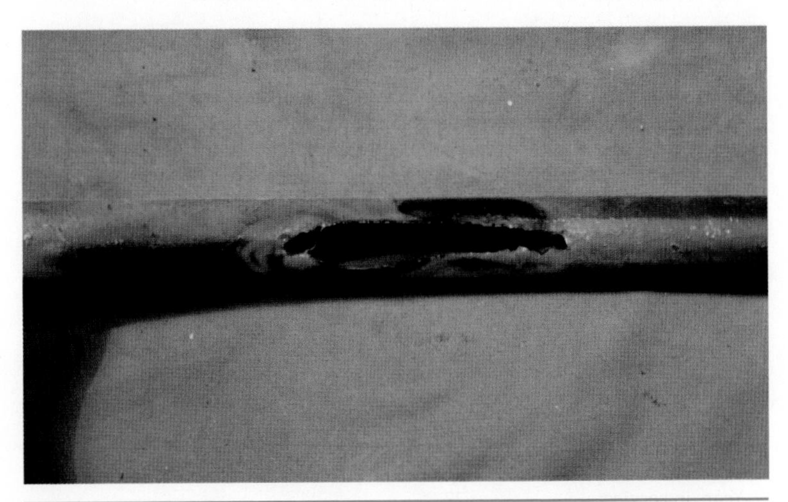

FIGURE 8.7 Thin-edged rupture at location of deep external surface metal loss.

FIGURE 8.8 Flow-oriented, comet-shaped depressions on wasted external surface.

Case History 8.6

Industry:	Chemical process industry
Specimen location:	Boiler feedwater piping
Orientation of specimen:	Horizontal
Years in service:	44
Water treatment program:	NTA/polymer
Pressure:	850 psi (5.9 MPa)
Tube specifications:	4-in. (10.2-cm) outer diameter

The boiler was treated successfully for 10 years using an NTA program. This was followed for another 10 years by a polymer program. Then a combined NTA and polymer program was used for about 3 years. During the 3 years on the combined NTA/polymer program, several leaks occurred within the feedwater lines of six different boilers. A failed section of feedwater piping revealed deep depressions and wide areas of deep general metal loss on the internal surface, where iron oxide and deposit layers were removed (Figs. 8.9 and 8.10). The change from a stand-alone chelating agent or polymer program to a combined chelating agent and polymer program created a more aggressive solution at the use concentration in the boiler water relative to the hardness concentration. This initiated the removal of iron oxide layers from the internal surfaces and rapid corrosion of the exposed metal surfaces. The use of a combined program was abandoned in favor of a polymer program.

Case History 8.7

Industry:	Steel
Specimen location:	Downcomer tube
Orientation of specimen:	Slanted near steam drum connection
Years in service:	27
Water treatment program:	Polymer
Pressure:	950 psi (6.6 MPa)
Tube specifications:	2½-in. (6.4-cm) outer diameter

FIGURE 8.9 Deep depressions and wide areas of deep, general metal loss on internal surface of boiler feedwater line.

FIGURE 8.10 Wide areas of deep, general metal loss on internal surface of boiler feedwater line.

FIGURE 8.11 Deep, smooth, general metal loss on internal surface of downcomer tube.

Several downcomer tube failures occurred within a 3- to 5-year period in an industrial boiler. The failures were in the form of thin-lipped, wide ruptures. Examination of a failed section revealed that the rupture formed where deep localized internal surface metal loss occurred. This was within the slanted portion of the downcomer tube near the steam drum connection (Fig. 8.11). The wasted surfaces were smooth, exhibited flow-oriented grooves and depressions, and were covered with thin layers of lustrous, dark-colored iron oxide. Carryunder of steam occurred during service, producing steam/water stratification. This caused the polymer in the boiler water to concentrate to corrosive levels within the steam space. Modifications were made to the steam drum internals and to operating practice to prevent steam carryunder.

Reference

1. T. Y. Chen and M. R. Godfrey, "Monitoring Corrosion in Boiler Systems with Colorimetric Tests for Ferrous and Total Iron," *Corrosion*, 1995, vol. 51, no. 10, pp. 797–804.

CHAPTER 9

Oxygen Corrosion

General Description

Since the oxides of iron are iron's natural stable state, steels will spontaneously revert to this form if conditions are thermodynamically favorable. During boiler operation, proper boiler water and steam chemistries will spontaneously promote the formation of a protective layer of magnetite on bare steel surfaces due to the well-known Schikorr reactions [see Eqs. (9.1–9.3)]

$$Fe + 2H_2O \rightarrow Fe^{2+} + 2(OH^-) + H_2 \tag{9.1}$$

$$Fe^{2+} + 2(OH^-) \rightarrow Fe(OH)_2 \tag{9.2}$$

$$3Fe(OH)_2 \rightarrow Fe_3O_4 + 2H_2O + H_2 \tag{9.3}$$

This protective layer will form even when there is no dissolved oxygen present in the water due to direct reaction with water or steam. However, the protective magnetite layer may be compromised either mechanically or chemically during service or during idle periods, making the exposed steel susceptible to corrosion by other mechanisms. One of the most frequently encountered corrosion problems in boilers results from exposure of steel to waters containing dissolved oxygen. This is referred to as *oxygen corrosion*. The following reaction occurs:

$$2Fe + H_2O + O_2 \rightarrow Fe_2O_3 + 2H \uparrow \tag{9.4}$$

This reaction is the basis for the intensive mechanical deaeration and chemical oxygen scavenging practices that are typical of a sound water treatment program in most boiler systems. Although these practices are generally successful, corrosion may occur during boiler operation under certain conditions. However, oxygen corrosion occurs more frequently during idle periods than during service.

In addition to tube-wall perforation, oxygen corrosion is troublesome from another perspective. Oxygen pits can act as stress concentration sites, thereby fostering the development of corrosion fatigue cracks (see Chap. 12, Corrosion Fatigue Cracking). Dissolved oxygen may also promote stress-assisted corrosion (see Chap. 11, Stress-Assisted Corrosion).

Locations

Oxygen corrosion most frequently occurs in boilers during idle periods, particularly if wet or dry layup practice is poor. The entire boiler system is susceptible, but perhaps the most common attack sites are in superheater and reheater tubes. During idle periods, steam condenses on the walls of an idle superheater tube and dissolves atmospheric oxygen from in-leaked air. Fractures in the protective magnetite are caused by contraction stresses as the superheater is cooled to ambient temperatures. The fracture sites furnish anodic regions where oxygenated water can react with bare, unprotected steel, resulting in the formation of deep, distinct, almost hemispherical pits (Fig. 9.1). Pitting will occur at the bottom of U-shaped superheater pendants where pools of water can accumulate (Fig. 9.2).

Less commonly, oxygen corrosion may occur during boiler operation in the portions of the boiler that contact only liquid water. The most susceptible locations are in the preboiler, particularly the deaerator storage section, feedwater supply lines, feedwater heaters, and the economizer. Oxygen corrosion may also occur in condensate systems (see Chap. 16, Steam and Condensate System Damage). In a drum-type boiler, once the feedwater reaches the steam drum, any remaining dissolved oxygen will come out of solution and will be

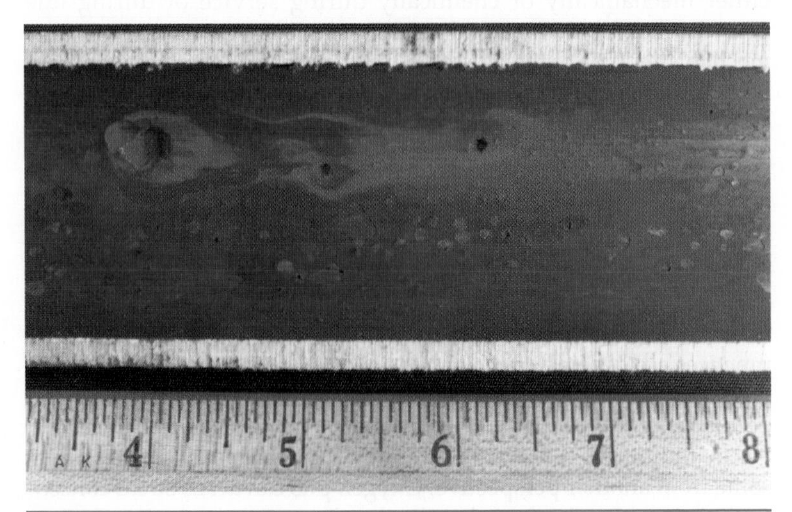

FIGURE 9.1 Oxygen pits in section of superheater tube. (*Courtesy of Electric Power Research Institute.*)

FIGURE 9.2 Oxygen pits along bottom of superheater pendant. (*Courtesy of National Association of Corrosion Engineers.*)

carried out with the steam. In cases of severe oxygen contamination, metal surfaces along the waterline in the steam drum and in the steam separation equipment may experience corrosion (Fig. 9.3). In general, considerable damage can occur even if the period of oxygen contamination is short.

FIGURE 9.3 Metal loss and red iron oxide layer formation above waterline in steam drum.

However, since dissolved oxygen is driven out of solution in the drum, it is extremely rare for oxygen corrosion to occur during service within the waterwalls or generating bank.

Oxygen corrosion may occur during service within the steam-generating portions of low-pressure, low-temperature firetube boilers or with other boiler designs where dissolved oxygen is not allowed to vent from the feedwater before it contacts the tubes and other components. Attack is generally most severe near the boiler feedwater inlet. Oxygen corrosion cannot occur during boiler operation within the portions of the boiler that only contact steam, since such corrosion requires exposure to oxygenated water.

Critical Factors

The critical factors governing the onset and progress of oxygen corrosion include the presence of water containing dissolved oxygen and an unprotected metal surface. Aggressiveness generally increases with increasing dissolved-oxygen concentration. Under normal operating practice, boiler water pH is alkaline, normally within the pH range of 9 to 12. If pH decreases below the proper control ranges for the system and chemical treatment program, oxygen corrosion will occur more readily.

There is a theory that at temperatures exceeding about 400°F (204°C), oxygen corrosion does not occur, since the formation of protective magnetite predominates. Oxygen corrosion occurs more readily at metal temperatures below about 400°F (204°C), since the reaction to produce the protective magnetite film on the surface occurs more slowly.

An unprotected metal surface can be caused by three conditions:

1. The metal surface is bare, e.g., following an acid cleaning.
2. The metal surface is covered with a marginally protective or nonprotective iron oxide, such as hematite, Fe_2O_3. Hematite typically is red, brown, or orange (see Fig. 9.4).
3. The metal surface is covered with a protective iron oxide, such as magnetite, Fe_3O_4, which tends to be black, but cracks exist in the coating.

Breakdown or cracking of the magnetite is due largely to mechanical and thermal stresses induced during normal boiler operation. These stresses are increased and therefore are more damaging during boiler start-up, boiler shutdown, and rapid load swings. During normal boiler operation, the environment favors rapid repair of breaches in the magnetite layer. However, if excessive levels of dissolved oxygen are present during either operation or outages, the cracks in the magnetite may not be adequately repaired and corrosion can occur.

FIGURE 9.4 Caps of iron oxide covering pit sites.

Identification

Simple visual examination may reveal characteristics of oxygen corrosion if affected surfaces are accessible. The presence of red, brown, and orange iron oxides, and distinct iron oxide tubercles, is strong physical evidence of oxygen corrosion (Fig. 9.4). Elongation of tubercles in the flow direction indicates that oxygen corrosion was active while water flowed past surfaces, probably during boiler operation (Fig. 9.5). X-ray diffraction analysis or other suitable techniques may be used to confirm that the corrosion products are iron oxide. X-ray diffraction will identify the type of iron oxide formed. Hematite tends to predominate with oxygen corrosion. However, magnetite and iron oxide hydroxide are also commonly formed.

In many cases, particularly at the beginning stages of oxygen corrosion, metal loss is localized in the form of pits or depressions that are relatively narrow in width. Over time, or under more severe conditions of oxygen corrosion, the metal loss will spread across a wider area. The interiors of the pits or depressions will be irregular to rough and will be filled or lined with iron oxide corrosion products.

Nondestructive testing (NDT), such as ultrasonic testing, may be required if affected surfaces are not accessible.

Metallographic examination can be used to definitively confirm oxygen corrosion. It is common for pits and depressions that form due to oxygen corrosion to be filled with dense iron oxide that billows into the surface similar to that of a cloud shape (Fig. 9.6). In some cases, the iron oxide contains remnants of pearlite colonies or carbides from the microstructure that were present when the oxygen

FIGURE 9.5 Elongated iron oxide tubercles.

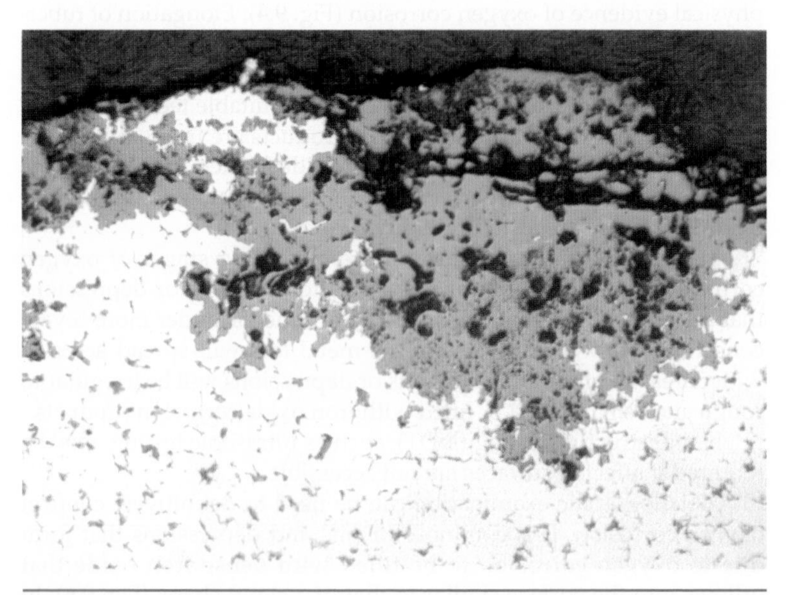

FIGURE 9.6 Metallographic cross section of billowing, dense iron oxide filling depression.

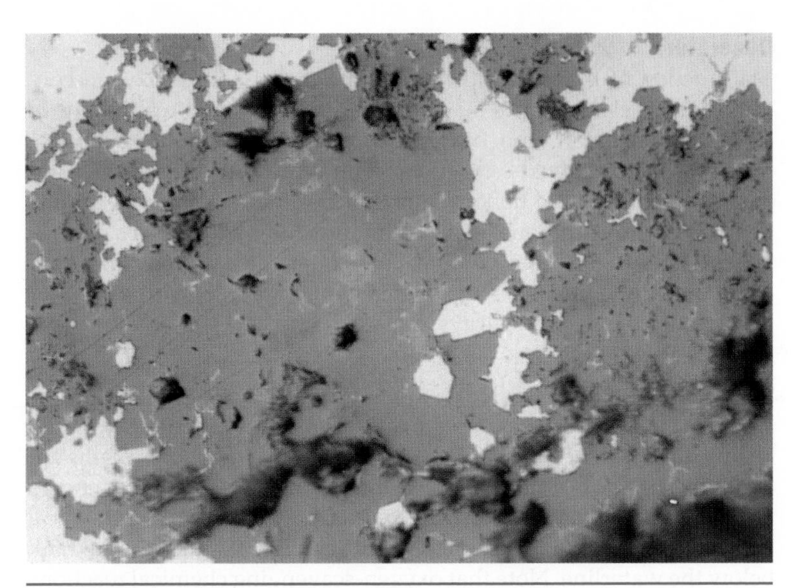

FIGURE 9.7 Remnants of pearlite colonies in iron oxide.

corrosion occurred (Fig. 9.7). Such distinct features in the iron oxide corrosion products that form in situ can easily be distinguished from porous, agglomerated iron oxide deposit layers.

Elimination

Conditions vary between boiler operation and even during different layup practices. In addition, chemical cleaning of the boiler may produce bare surfaces that will readily promote oxygen corrosion unless proper passivating procedures are conducted immediately following the cleaning. Methods used to control oxygen corrosion under each operating condition are discussed separately below.

Operating Boiler

The protective magnetite coating on waterside surfaces exists in a state of continuous breakdown and repair. At any given time, cracks in the magnetite will be present, although the percentage of the entire internal surface they represent will be very small. Therefore, since corrosion sites are present, mitigation of oxygen corrosion is achieved by sufficiently diminishing dissolved-oxygen levels.

Possible causes of excessive levels of dissolved oxygen in boiler water are a malfunctioning deaerator, improper oxygen scavenging, and air in-leakage. Discussions of these contributing factors as well as monitoring practices are detailed below.

Deaerator

Malfunction of the deaerator may allow the dissolved-oxygen content in the boiler water to be excessively high. Due to restrictions in capacity or reaction time, oxygen scavengers may not be capable of reducing the dissolved-oxygen concentration to proper levels prior to introduction to the boiler. Causes of deaerator malfunction must be addressed. Common causes of malfunctions in pressure deaerators include inadequate venting, inadequate steam feed, improper feedwater flow, improper feedwater temperature, and damaged or malfunctioning sprays or trays.

Oxygen Scavengers

Detailed descriptions of the many different oxygen-scavenging chemical treatment programs that may be used in boiler systems are given in *The Nalco Water Handbook*, third edition. Oxygen scavengers must have adequate reaction time to be effective. Ideally, the chemical should be fed to the drop leg of the deaerator or in the storage tank below the waterline. Note that oxygen-scavenging chemical reactions may be deactivated, at least partially, due to exposure and reaction with certain compounds in the water. These compounds may be contaminants or even normal components of the chemical treatment program. If such compounds are introduced to the boiler water prior to the complete reaction of the scavenger with the dissolved oxygen, scavenging may be incomplete.

Air Leaks

There are several possible sources of air in-leakage upstream of the deaerator. Under normal operating conditions, the dissolved oxygen contributed by these sources will be removed by the deaerator to suitable levels. With proper oxygen scavenging, the remaining dissolved oxygen will be reduced to low levels suitable for use in the boiler. However, air leaks may occur downstream of the deaerator and oxygen scavenger feed points. The most common source of air in-leakage following the deaerator is air in-leakage at feedwater heater valves. Low-pressure feedwater heaters are often under negative pressure during low-load operation, allowing air to be drawn into the feedwater through leaky valves, pumps, and flanges.

Monitoring

Monitoring of dissolved-oxygen levels at the economizer inlet, especially during start-up and low-load operation, is recommended. The use of techniques such as at-temperature oxidation reduction potential (ORP) measurements has been found to be very effective in identifying when and how dissolved-oxygen levels become excessive. This allows appropriate corrective actions to be taken.

Certain low-pressure boilers do not employ deaerators. As such, other less efficient methods, such as steam sparging, may be used to reduce dissolved-oxygen concentration. In many cases, the boiler may be exposed to waters containing much higher concentrations of dissolved oxygen than is desired. Very large amounts of oxygen scavenger, primarily sulfite, are fed to reduce the dissolved oxygen to proper levels. In some cases, the reaction of sulfite with the oxygen will produce substantial amounts of sulfate. Sulfate may promote acidic conditions beneath iron oxide corrosion products and deposits, accelerating corrosion. This should not discourage the use of sulfite for such boilers, since its fast reaction time is best suited for the application.

In some low-pressure systems, particularly closed hot water systems used for heating, the dissolved oxygen is not removed. Rather, suitable corrosion inhibitors that are more commonly used in cooling water applications are used to control the oxygen corrosion. These systems typically employ nitrite and/or molybdate corrosion inhibitors. Such inhibitors would not be practical for higher-pressure systems.

Idle Boiler—Wet Layup

An idle boiler during wet layup is subject to conditions similar to those in an operating boiler. Therefore, the preventive method is the same: reduction of the dissolved-oxygen content to a very low level and continuous control that prevents the level from rising. In general, this procedure requires complete filling of the boiler, use of sufficiently high levels of oxygen-scavenging chemicals, maintenance of properly adjusted pH levels, as well as periodic water circulation. Oxygen scavenger replenishment is required. The proper application of nitrogen blanketing may also be useful.

Idle Boiler—Dry Layup

Successful protection of an idle boiler during dry layup depends upon consistent elimination of moisture and/or oxygen. A procedure for boiler protection by dry layup can involve the use of desiccants and nitrogen blankets or the continuous circulation of dry, dehumidified air (< 30% relative humidity).

Boiler after Chemical Cleaning

Limiting the attack of metal surfaces in a boiler following acid cleaning is achieved by developing a protective iron oxide layer on the metal surface. This is usually accomplished by a thorough rinsing followed with a "postboilout." A sodium carbonate solution or other alkaline substance can be used in the postboilout passivation step.

Cautions

The knoblike mounds (tubercles) of corrosion products that frequently cover pit sites are sometimes misidentified as simple deposits. Correct identification of these mounds can be achieved through a consideration of the following:

- Tubercles will overlie corrosion sites. Under the influence of sufficient fluid velocity the tubercles may be elongated in the flow direction.
- A tubercle is highly structured and usually consists of a hard, brittle outer shell of reddish corrosion products that encapsulates an inner core of soft, voluminous, dark corrosion products.

Related Problems

See also Chap. 11, Stress-Assisted Corrosion, and Chap. 12, Corrosion Fatigue Cracking.

Case History 9.1

Industry:	Pulp and paper
Specimen location:	Superheater
Orientation of specimen:	U bend, horizontal to vertical
Years in service:	10
Water treatment program:	Dry layup using nitrogen blanket
Tube specifications:	2-in. (5.1-cm) outer diameter

The oxygen pits illustrated in Figs. 9.1 and 9.8 were discovered during a hydrotest of a recovery boiler that had been in wet layup over the summer. The superheater section had been in dry layup under a nitrogen blanket.

Gravity-induced drainage of corrosion products from sites on vertical sections is apparent in Fig. 9.1. Elliptical rings surrounding the pit on the horizontal portion of the pendant section indicate that a pool of condensate had been present (Fig. 9.8). Close examination under a low-power stereoscopic microscope revealed contraction cracks in the protective magnetite at the pit sites. It is apparent that despite the precautions of dry layup, the tube had been exposed to condensed moisture and atmospheric oxygen.

Case History 9.2

Industry:	Pulp and paper
Specimen location:	Convection bank
Orientation of specimen:	Various
Years in service:	15
Water treatment program:	Wet layup, hydrazine, morpholine
Drum pressure:	700 psi (4.8 MPa)
Tube specifications:	2-in. (5.1-cm) outer diameter

Figures 9.9 through 9.11 illustrate typical oxygen pitting resulting from an improperly executed wet layup. Inadequate feed and replenishment of

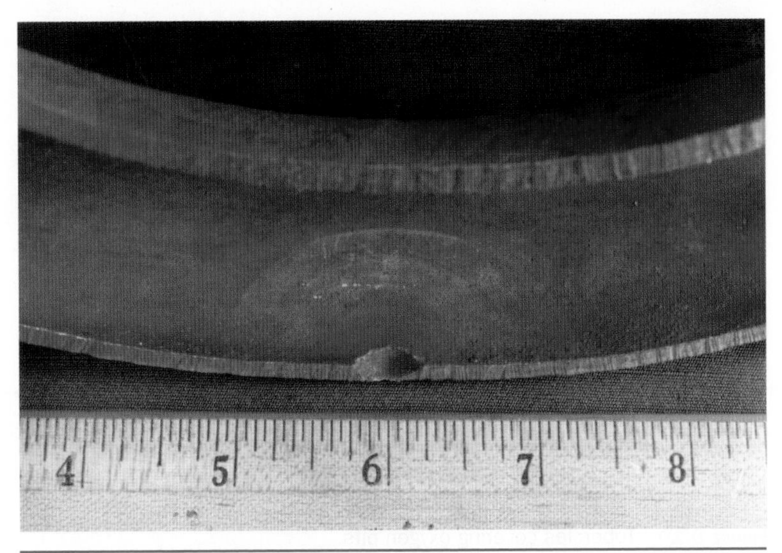

Figure 9.8 Elliptical water rings surrounding oxygen pit in a U bend. (*Courtesy of Electric Power Research Institute.*)

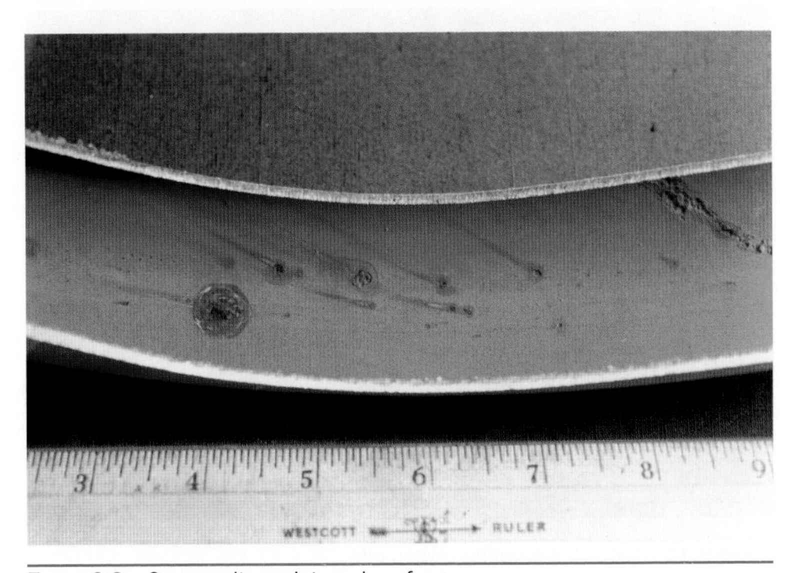

Figure 9.9 Oxygen pits on internal surface.

oxygen-scavenging chemicals and/or failure to maintain proper water circulation may have been responsible.

Gravitationally induced drainage of corrosion products is apparent in Fig. 9.9. This is evidence of stagnant or very low flow of water during layup. Significant water flow would have reoriented the drainage lines in the direction of water flow. Typical knoblike mounds of iron oxide corrosion products (tubercles) are shown in Figs. 9.10 and 9.11.

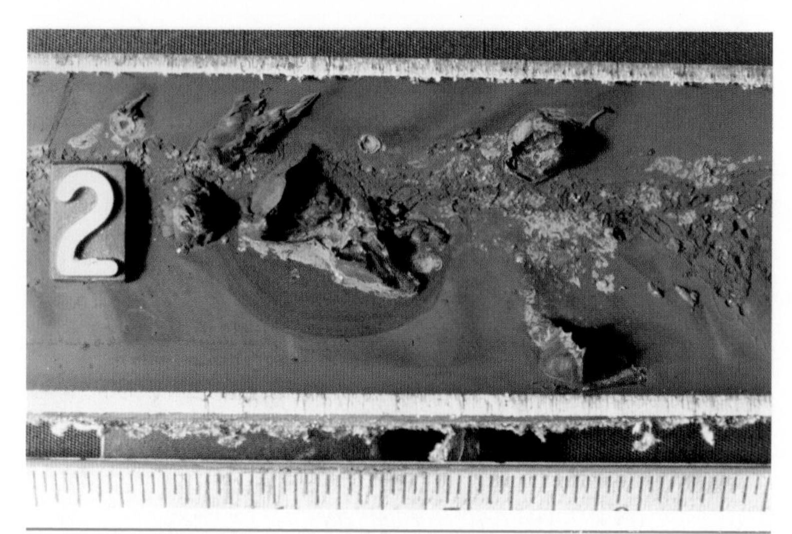

Figure 9.10 Tubercles covering oxygen pits.

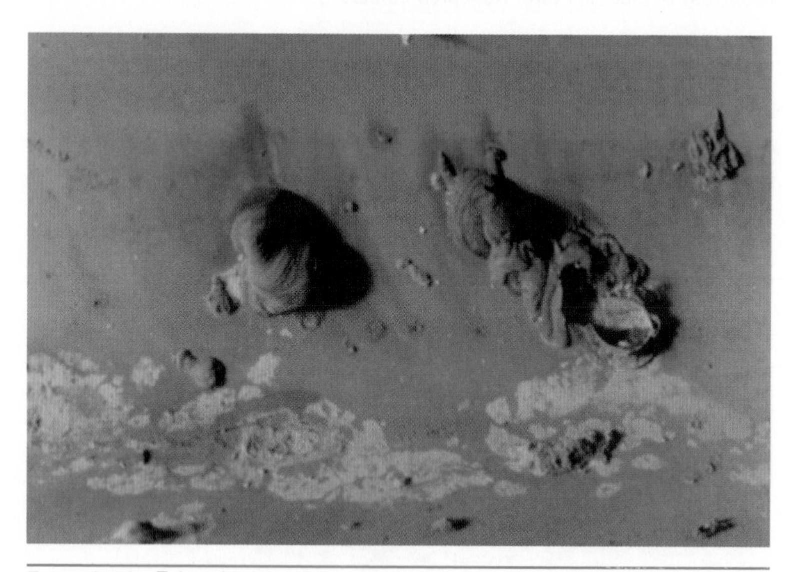

Figure 9.11 Tubercles covering oxygen pits.

Case History 9.3

Industry:	Steam heating
Specimen location:	Firetube boiler
Orientation of specimen:	Horizontal
Years in service:	4
Water treatment program:	Polymer and sodium sulfite
Tube specifications:	3-in. (7.6-cm) outer diameter

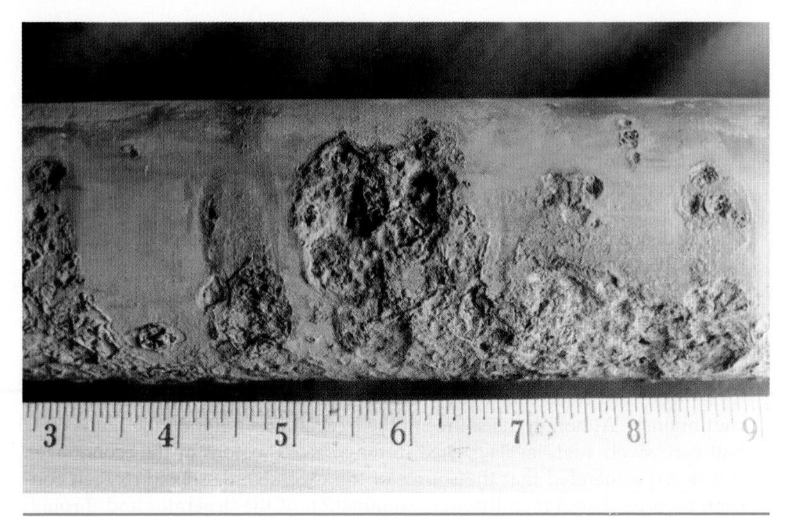

FIGURE 9.12 Oxygen pitting on the external surface of fire tube.

Figure 9.12 illustrates the external surface of a firetube boiler that has sustained severe oxygen corrosion. Interruptions of oxygen scavenger feed lasting as long as one week had occurred. It is probable that the pitting initiated and progressed during these interruptions. Deposits that can also induce pitting (under-deposit corrosion) were not found to be present on the tube.

Case History 9.4

Industry:	Pulp and paper recovery boiler
Specimen location:	Economizer
Orientation of specimen:	Curved, horizontal to vertical
Years in service:	14
Water treatment program:	Coordinated phosphate, organic oxygen scavenger
Drum pressure:	1900 psi (13.1 MPa)
Tube specifications:	2-in. (5.1-cm) outer diameter

The internal surface of the economizer tube section was generally covered with iron oxide tubercles that were oriented longitudinally (Fig. 9.5). These tubercles formed due to oxygen corrosion resulting from exposure to waters containing excessive concentrations of dissolved oxygen. Oxygen corrosion was active during service, causing the tubercles to elongate in the direction of water flow. Metallographic examination revealed dense burrowing iron oxides, within the pits that contained remnants of pearlite colonies (Figs. 9.6 and 9.7), confirming that oxygen corrosion occurred. It was reported that the boiler was always started using cold demineralized water that was not pH-adjusted or treated with oxygen scavengers. Proper deaeration and scavenging practice was followed during normal operation following start-up. The start-up procedure was required to be changed to supply properly deaerated and treated water to prevent oxygen corrosion in the economizer.

Case History 9.5

Industry:	Chemical process
Specimen location:	Economizer
Orientation of specimen:	Horizontal
Years in service:	7
Water treatment program:	Polymer and oxygen scavenger
Tube specifications:	2½-in. (6.4-cm) outer diameter

The reddish color of the iron oxides and the presence of tubercles capping iron oxide–filled pits (Fig. 9.4) are typical of exposure of economizer steel to water containing excessively high levels of dissolved oxygen. Pitting and perforation of economizer tubes were a recurrent problem at this plant. Failures were occurring every 3 or 4 months.

Oxygen content of the water was measured at 5 to 9 ppb. Excursions to higher levels were suspected but could not be documented. The boiler was operated continuously. Although the source of the oxygen was not identified, it is clear that excessively high levels existed in the affected regions of the economizer. It was recommended that the source of the excessive dissolved-oxygen concentration be determined through examination of the deaerator and through monitoring of dissolved oxygen at suitable locations. The use of at-temperature ORP was also recommended. Once the source was found, the proper mechanical, operational, or chemical solutions to the problem were implemented.

CHAPTER 10

Corrosion during Cleaning

General Description

Deposits and corrosion products accumulate on the water sides of tubes throughout a boiler to some degree over time. High-pressure boilers that use high-purity feedwater typically experience low deposition rates. However, even with low deposition rates, after a certain period of service it is typically necessary to clean the boiler, primarily to prevent under-deposit corrosion mechanisms. Another reason for cleaning is to control overheating that can be promoted by the loss of heat transfer through the tube wall, due to insulation provided by the deposits. Deposits in high-pressure boilers are normally composed primarily of transported corrosion products, typically oxides of metals used within the preboiler. However, corrosion products formed in place and components of treatment chemicals, such as phosphates, may result in material on the internal surfaces of boiler tubes as well.

For low-pressure boilers, particularly those that produce steam through convective heat transfer, under-deposit corrosion and overheating are uncommon. As with high-pressure boilers, material on the internal surfaces of the tubes in low-pressure boilers may contain corrosion products from the preboiler and boiler. However, lower-pressure boilers may use lower-purity feedwater, which can produce deposits consisting of compounds produced by the reaction of hardness ions such as calcium and magnesium with species intrinsically present in the boiler water, as well as reaction products of hardness ions with water treatment chemicals, such as phosphate. For such boilers, cleaning may be conducted to increase efficiency to control fuel costs.

The frequency of cleaning is determined by several factors. Information obtained from deposit weight determinations may reveal the need to clean, which is based upon guidelines specific to the boiler type and pressure. Exposure to upset conditions will also influence cleaning decisions. As energy costs increase, cleaning frequency may increase as well.

Cleaning may be conducted mechanically. However, mechanical cleaning tends to be insufficient in many cases, since tenacious material cannot be adequately removed. Appropriate chemical cleaning is much more effective than mechanical cleaning. However, chemical cleaning agents that are most effective for removing boiler deposits within a reasonable time frame are corrosive to steel boiler tubing and components. Hydrochloric acid, HCl, in concentrations ranging from 5 to 10%, is perhaps the most common chemical cleaning agent. Corrosion inhibitors are included in the cleaning solution to limit metal loss. Hydrochloric acid is very effective at removing calcium- and iron-based deposits. Hydrofluoric acid, HF, which can remove silicon-based deposits, can be added to HCl in the acid form. Alternatively, HF may be produced in situ through the mixing of ammonium fluoride with HCl. If HF is used alone, it tends to be less corrosive to boiler steel than HCl. Copper deposits may be removed by chelating agents such as ethylenediaminetetracetic acid (EDTA), which is also effective for removal of hardness deposits, but not for iron oxides. Thiourea is another copper complexing agent that may be used with HCl to keep copper in solution during chemical cleaning. In some cases, two-stage cleaning programs are used for deposits that contain substantial amounts of separate components such as copper and iron oxide. Organic acids, such as a combination of hydroxyacetic and formic acids, may be used for magnetite removal in high-pressure boilers and in superheaters. This acid mixture is useful due to its volatility, which allows it to be removed from nondrainable superheaters. Alkaline cleaning agents, such as caustic, that are used to remove organic material will not cause corrosion or cracking of steel components at the low temperatures used during cleaning.

Corrosion of carbon and low-alloy steels by acids is a natural consequence of steel's thermodynamic instability in these environments. Steel will corrode spontaneously in most acids. During the corrosion reaction, iron displaces hydrogen from solution. That is, the iron is oxidized and iron ions go into solution, as described in Eq. (10.1). Hydrogen ions are reduced and form hydrogen gas bubbles at the metal surface.

$$Fe + 2H^+ + Cl^- \rightarrow H_2\uparrow + Fe^{2+} + Cl^- \qquad (10.1)$$

To stifle this corrosion process, inhibitors are added to acid cleaning solutions used in boilers.

Locations

Corrosion of the internal surfaces of a boiler that results from low-pH exposure may occur during acid cleaning if proper procedures are not followed. In general, any surface exposed to acid is susceptible. Some of the first areas to be affected are the tube ends inside the mud and steam drums. Hand-hole covers, drum manholes, and welds may also be affected. Weldments and locations where applied or residual stresses are high (Figs. 10.1 and 10.2) may experience more

FIGURE 10.1 Acid corrosion on the internal surface of a wall tube. (*Courtesy of Electric Power Research Institute.*)

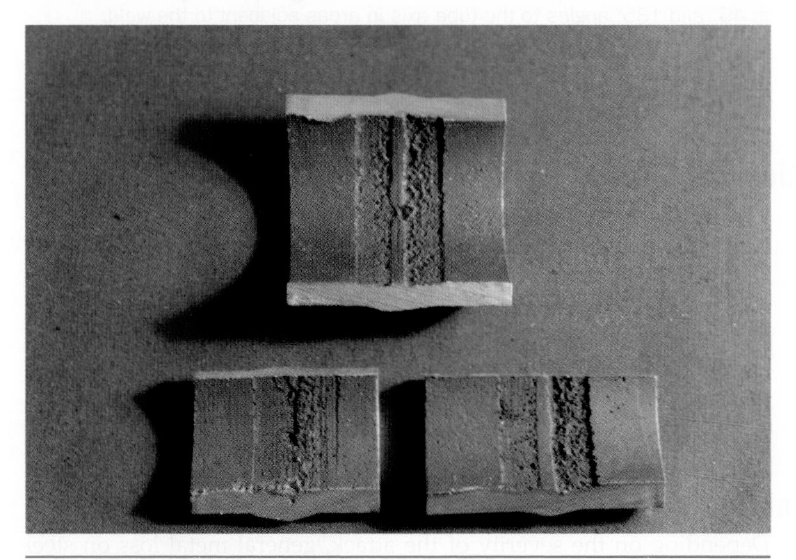

FIGURE 10.2 Acid corrosion at weld.

vigorous attack (see Chap. 11, Stress-Assisted Corrosion). Shielded regions within crevices, behind backing rings (Fig. 10.3), and under remaining deposits may promote localized concentration of aggressive anions and may prevent proper neutralization of the cleaning acid. This results in vigorous localized attack of the metal once the boiler is returned to service.

FIGURE 10.3 Acid corrosion beneath weld backing ring. Note grooves oriented at 45° and 135° angles to the tube axis in areas adjacent to the weld.

Critical Factors

Uncontrolled acid corrosion of a boiler during cleaning generally results from an unanticipated and unintentional deviation from standard conditions or standard practice. Many deviations are possible and may include events such as thermally induced breakdown of the inhibitor, inappropriate selection of cleaning agent or cleaning strength, excessive exposure times, excessive exposure temperatures, and failure to neutralize completely.

Identification

Depending on the severity of the attack, general metal loss on steel surfaces within a boiler by strong acid produces surfaces with a rough or jagged appearance (Figs. 10.1 and 10.4). Metallographic examination of pit cross sections will reveal discrete pits, which are frequently undercut (Fig. 10.5). The pit surfaces may be bare if attack was active shortly before the specimen was examined. If attack occurred sometime in the past, the pits may be lined with thin iron oxide and deposit layers. Attack of metal surfaces by organic acids generally produces significantly different wastage than that produced by strong mineral acids.

FIGURE 10.4 Jaggedness associated with severe acid corrosion. (*Courtesy of Electric Power Research Institute.*)

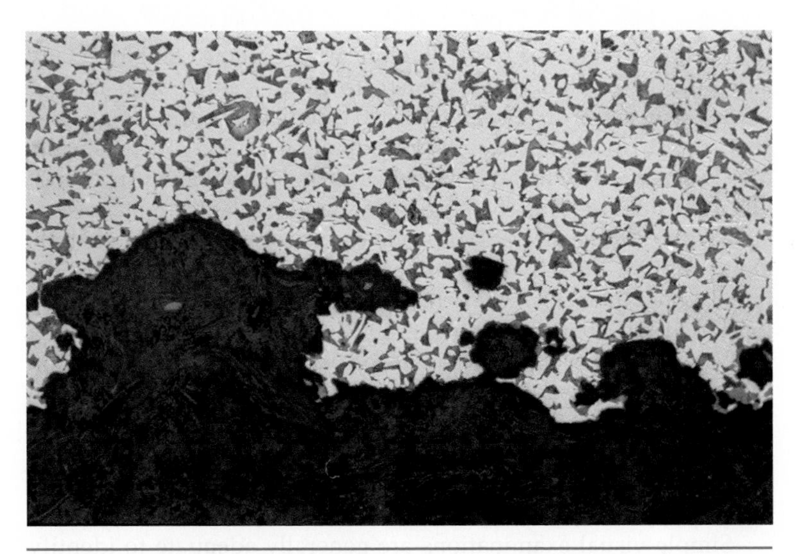

FIGURE 10.5 Metallographic cross section of discrete, undercut pits that are bare-surfaced or lined with very thin iron oxide and deposit layers.

Given equivalent exposure times, organic acids generally cause less metal loss. In addition, surface features of metals subject to excessive exposure to organic acids are typically less jagged and less undercut than those of metals subject to excessive exposure to strong mineral acids, such as hydrochloric acid.

FIGURE 10.6 Metallographic cross section revealing preferential corrosion along inclusion stringers.

On boiler tubes, the pits will frequently be aligned longitudinally along the tube wall, following microstructural features such as weld seams, inclusion stringers, and banded colonies of pearlite (Figs. 10.1, 10.5, and 10.6). In addition, strong mineral acid characteristically attacks areas where metal is at higher stress levels than in surrounding areas; this is because of the greater energy content of metal associated with residual stresses. In some cases, corrosion may penetrate along straight lines that are oriented at 45° and 135°angles to the longitudinal tube axis, where residual stresses are imparted by tube formation and/or installation (Fig. 10.3). A more detailed explanation of the contribution of stress to the corrosion process is described in Chap. 11, Stress-Assisted Corrosion. Deep metal loss may occur in shielded locations, such as beneath weld backing rings, where the acid may concentrate and/or become entrapped following chemical cleaning. In such locations, acids are not easily neutralized and rinsed away.

Simple visual examinations are generally adequate to identify acid corrosion. Attack can first be observed at tube ends in mud and steam drums, at ends of sheet or plate steel, and at ends of bolts. This type of attack is generally associated with corrosion along microstructural features in the metal, especially inclusions. Not all areas of the boiler may be affected to the same degree. Stressed metal, welded joints, crevices, and other shielded regions may suffer more intense damage. Damage assessment in visually inaccessible areas may require either nondestructive testing techniques, such as ultrasonic testing (which may or may not be sensitive enough for accurate measurements), or removal of a specimen for destructive testing.

Elimination

Control of low-pH corrosion of boiler equipment during acid cleaning requires close monitoring of the entire cleaning procedure. The following are only a few examples of the various parameters that must be evaluated prior to cleaning and then monitored and evaluated during the procedure.

Deposit weight determinations: Appropriate selection of several tubes for deposit weight measurements will aid in determination of proper acid strength, exposure time, and total quantity of acid required to adequately clean the boiler.

Deposit analyses: Deposit analyses will help to determine appropriate cleaning agents and the sequence in which the agents should be used.

Temperature of cleaning solution: Both the solution temperature and the metal temperature should be safely below the thermal breakdown point of the inhibitor.

Monitoring: Iron content, copper content, and cleaning solution strength should be monitored at periodic intervals during the boiler cleaning. Chemistry of the neutralizer should be monitored following the boiler's exposure to the acid.

Visual inspection: Inspection of tubes, mud drums, and steam drums should follow the cleaning.

Cautions

Severe oxygen pitting and cavitation damage have been mistaken for attack by strong acid. One of the major distinguishing features of strong acid attack is that it tends to affect all exposed surfaces. Oxygen corrosion tends to occur in specific areas, such as the economizer, return bends in the superheater, or perhaps along the waterline in the steam drum. Cavitation damage also tends to be location-specific, and it most commonly affects pump impellers, valves, and other components in localized areas where pressure drops occur in liquid water. Certain forms of chelant corrosion may resemble acid corrosion, but will tend to occur in localized areas where concentration mechanisms, such as departure from nucleate boiling, occur.

Related Problems

See also Chap. 8, Corrosion by Chelating and Complexing Agents; Chap. 9, Oxygen Corrosion; Chap. 11, Stress-Assisted Corrosion; and Chap. 17, Cavitation.

Case History 10.1

Industry:	Utility
Specimen location:	Right side waterwall in cyclone fired boiler
Orientation of specimen:	Vertical
Years in service:	12
Water treatment program:	Coordinated phosphate
Drum pressure:	2250 psi (15.5 MPa)
Tube specifications:	2½-in. (6.4-cm) outer diameter
Cleaning solution:	Inhibited hydrochloric acid

The boiler was chemically cleaned every 18 months. Scattered, undercut pits formed throughout the internal surface of a section that was removed for examination (Fig. 10.7). Deeper metal loss occurred beneath weld backing rings Fig. 10.3. Crosshatched grooves oriented at 45° and 135° angles to the tube axis were also found in areas adjacent to the weld. Metal loss at all locations was caused by acid corrosion resulting from chemical cleaning. Despite the high cleaning frequency, metal loss was only moderately deep in most areas. Deeper metal loss occurred within the crevices between the weld backing ring and the internal surface due to an ion concentration cell corrosion mechanism,[1] as well as the entrapment of residual acid within the crevices during the neutralization and rinsing steps. The use of backing rings for welding within the boiler was discontinued.

FIGURE 10.7 Undercut pits interrupting internal surface of wall tube.

FIGURE 10.8 Appearance of attack by citric acid.

Case History 10.2

Industry:	Utility
Specimen location:	Wall tube
Orientation of specimen:	Vertical
Years in service:	15
Water treatment program:	Congruent phosphate
Drum pressure:	1960 psi (13.5 MPa)
Tube specifications:	2½-in. (6.4-cm) outer diameter
Cleaning solution:	Inhibited citric acid

The internal surface had an etched, bright, metallic appearance (Fig. 10.8). Close examination revealed irregular islands of unattacked metal, which in this case stood 0.005 in. (0.13 mm) above the surrounding corroded surface. Essentially the entire internal surface, both hot and cold sides, was similarly affected. This condition is consistent with corrosion by weak organic acid during cleaning.

Case History 10.3

Industry:	Utility
Specimen location:	Wall tube
Orientation of specimen:	Vertical
Years in service:	30
Water treatment program:	Congruent phosphate
Drum pressure:	1500 psi (10.3 MPa)
Tube specifications:	3-in. (7.6-cm) outer diameter
Cleaning solution:	Inhibited hydrochloric acid

Figures 10.2 and 10.9 illustrate attack on the internal surfaces of wall tubes at circumferential welds. Cross-sectional profiles of the internal surface in the attacked zones revealed undercutting and jaggedness. Residual stresses remaining from the weld made these sites subject to preferential corrosion when exposed to strong acid.

FIGURE 10.9 Acid corrosion at weld.

Case History 10.4

Industry:	Utility
Specimen location:	Side-wall tube
Orientation of specimen:	Vertical
Years in service:	30
Water treatment program:	Congruent phosphate
Drum pressure:	1500 psi (10.3 MPa)
Tube specifications:	3-in. (7.6-cm) outer diameter
Cleaning solution:	Inhibited hydrochloric acid

A number of fairly deep, overlapping pits were observed in a circumferential zone along the internal surface, corresponding to circumferential welds (Fig. 10.10). Other areas of attack were noted in regions where welds were not present (Fig. 10.11). Attack in these areas was more pronounced along the tube seam.

Deposits overlying the attack sites revealed that the corrosion occurred in the past, probably during the last acid cleaning of the boiler.

Preferential attack of welds may be due to residual stresses associated with the weld or to pores or crevices existing in the weld zone. Preferential attack of a tube seam may be due to segregated impurities in the seam or to a crevice resulting from incomplete fusion of the seam.

Case History 10.5

Industry:	Pulp and paper
Specimen location:	Belly plate, steam drum
Orientation of specimen:	Horizontal
Years in service:	6
Water treatment program:	Polymer
Drum pressure:	600 psi (4.1 MPa)
Cleaning solution:	Inhibited hydrochloric acid

FIGURE 10.10 Acid corrosion at weld.

FIGURE 10.11 Acid corrosion along internal surface.

Visual examinations of belly-plate surfaces revealed shallow patches of irregular metal loss on all top and bottom surfaces. The sites of metal loss tended to be aligned in parallel rows. Figure 10.12 illustrates the appearance of the metal loss, although a thin coating of iron oxide tends to round the normally sharp edges.

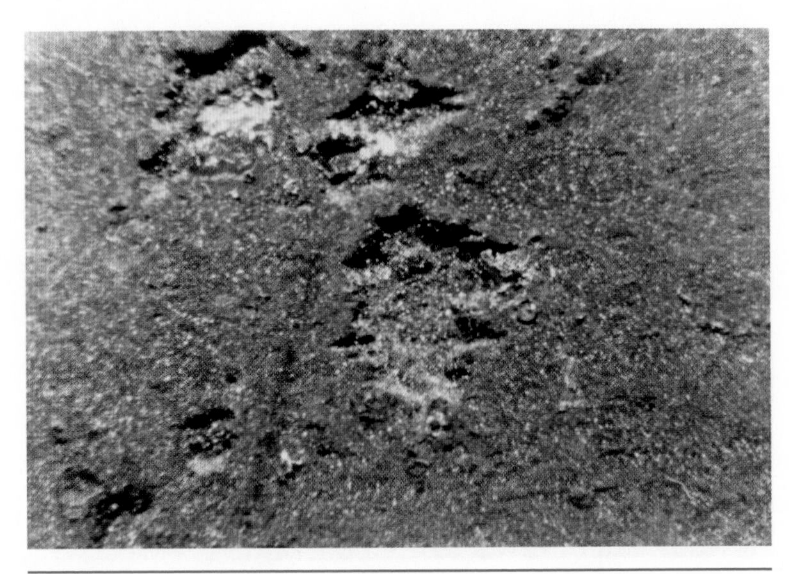

FIGURE 10.12 Metal loss from strong acid.

This case illustrates the importance of supporting observations in establishing low-pH corrosion during cleaning as the cause of metal loss. Figure 10.12 lacks the classic, jagged, undercut surface profile normally associated with attack by a strong acid. However, the following additional observations strongly support the diagnosis:

- Alignment of attack sites in parallel rows oriented in the rolling direction of the steel from which the plate was fabricated
- Preferential attack of welds and tube ends in the steam drum
- Jagged, undercut, cross-sectional profiles of belly-plate surfaces apparent in metallographic examinations

Case History 10.6

Industry:	Refining
Specimen location:	Mud drum cap
Years in service:	15
Water treatment program:	Polymer
Drum pressure:	900 psi (6.2 MPa)
Component specifications:	1¾-in. (4.4-cm) outer diameter
Cleaning solution:	Inhibited hydrochloric acid

The boiler was cleaned approximately every 3 years. Internal inspection of the boiler revealed numerous, deep, undercut pits formed on the surfaces of a mud drum cap (Fig. 10.13). Metallographic examination revealed that the undercut pits penetrated along inclusion stringers in the steel (Fig. 10.6). The cap was made of a free-machining grade of steel that contained a large inclusion population. The large inclusion population allowed the cap to be machined more easily at the expense of resistance to acid corrosion. Inclusions provide preferential sites for the initiation and propagation of pits. The use of free-machining steels for mud drum caps and other components was discontinued.

FIGURE 10.13 Undercut pits on a mud drum cap.

Case History 10.7

Industry:	Textile
Specimen location:	Fire tube
Orientation of specimen:	Horizontal
Water treatment program:	Residual phosphate
Drum pressure:	100 psi (0.69 MPa)
Tube specifications:	2½-in. (6.4-cm) outer diameter
Cleaning solution:	Inhibited hydrochloric acid

Hydrotesting following acid cleaning of the boiler revealed several leaking tubes. Examination of one of the leaking tubes revealed profuse pitting (Figs. 10.14 and 10.15). External surfaces were covered with a film of brown iron oxides.

FIGURE 10.14 Acid corrosion on the external surface of a fire tube.

FIGURE 10.15 Close-up of external surface shown in Fig. 10.14.

Although some pitting may have occurred as a result of excessively high oxygen levels in the boiler water during idle times, the primary mode of attack was insufficiently controlled exposure to strong mineral acid during acid cleaning. Note the large population of pit sites, the general distribution, and the appearance of fine longitudinal grooves, characteristic of attack by a strong acid.

Reference

1. Herro, H. M. and Port, R. D., Nalco Company, *The Nalco Guide to Cooling Water Systems Failure Analysis*, McGraw-Hill, New York, 1993, pp. 13–17.

CHAPTER 11

Stress-Assisted Corrosion

General Description

Stress-assisted corrosion (SAC) is the preferential attack of metal due to elevated static tensile stress levels. SAC is sometimes also referred to as stress-enhanced or stress-influenced corrosion, and it has been known to affect water-contacting surfaces of boilers for many years.[1] SAC can occur where high, localized residual or applied tensile stresses develop in the metal. These stressed areas are susceptible to localized corrosive damage.

Areas of elevated stress are often localized to specific regions on a component, such as at or near a weld. Numerous factors associated with areas of elevated stress can make them more susceptible to attack. The highly stressed regions are surrounded by metal that is at lower stress levels. The difference in stress levels produces a galvanic cell on the surface of the metal, with the regions of higher stress being anodic to surrounding metal. Furthermore, stored energy in the metal is increased at stressed locations, such as an accumulation of crystallographic defects resulting from plastic deformation. The increase in stored energy in the metal makes these locations more chemically active. Exposure to a uniform electrolyte that is aggressive to the metal results in accelerated corrosion rates in the region of higher stresses, as illustrated in Fig. 11.1.

Locations

Stress-assisted corrosion can occur in any type of boiler system, as long as the combination of static tensile stress and an aggressive aqueous environment is present. It has been identified in many types of boiler tubing, including economizer tubes, waterwall tubes, generating bank tubes, screen tubes, and superheater tubes. Some areas identified as susceptible sites for stress-assisted corrosion along waterwall tubes include locations at the floor seal, windbox attachments, and buckstay attachments.[2]

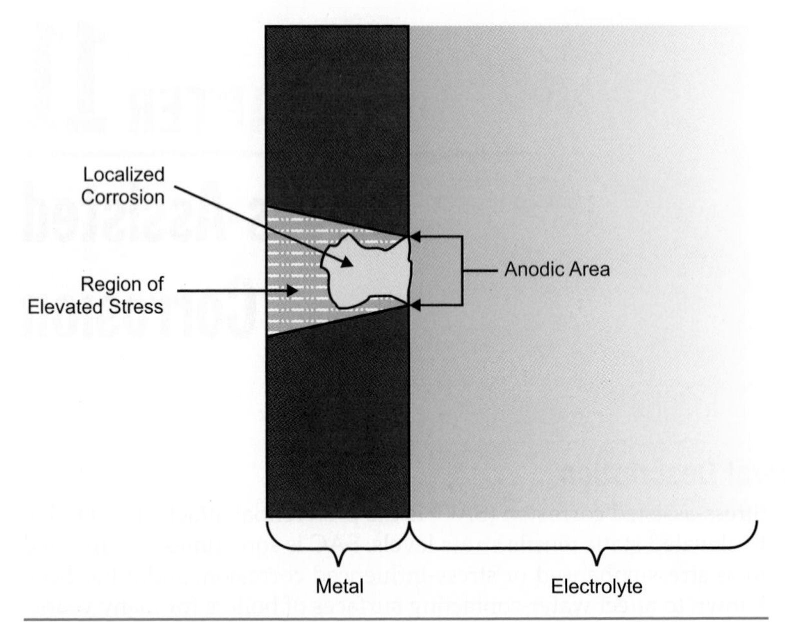

FIGURE 11.1 Schematic of the stress-assisted corrosion mechanism. (© *NACE International, 2004.*)

Locations on water-contacting surfaces at or near welds are generally the most likely areas for SAC. Some welded regions that are susceptible to SAC include attachment welds (Fig. 11.2), membrane welds (Fig. 11.3), and seam welds (Fig. 10.1). In general, stress-assisted corrosion is more common at attachment welds than at membrane welds. This may be related to higher stresses per unit area at attachment welds, or possibly differences between the welding procedures used at the fabrication shop and in the field. Welds that join membranes or other attachments to the external surface typically introduce residual and applied stresses on the internal surface opposite the external welds. Preferential attack can occur at these locations.

Components that have been plastically deformed such as cold formed fasteners, rolled tube ends, or tube bends are also susceptible to SAC. Tube rolling expands the tube by plastic deformation into a tube bore in the drum or a tubesheet. This process can result in considerable residual stresses.

Critical Factors

Two critical factors required for stress-assisted corrosion to occur in a boiler are the presence of elevated tensile stresses in the metal and an environment conducive to corrosion of the metal. Both of these factors can be influenced by multiple conditions, which are discussed in detail below.

FIGURE 11.2 Preferential attack occurring at a location on the internal surface opposite an attachment weld on the cold-side external surface of a waterwall tube.

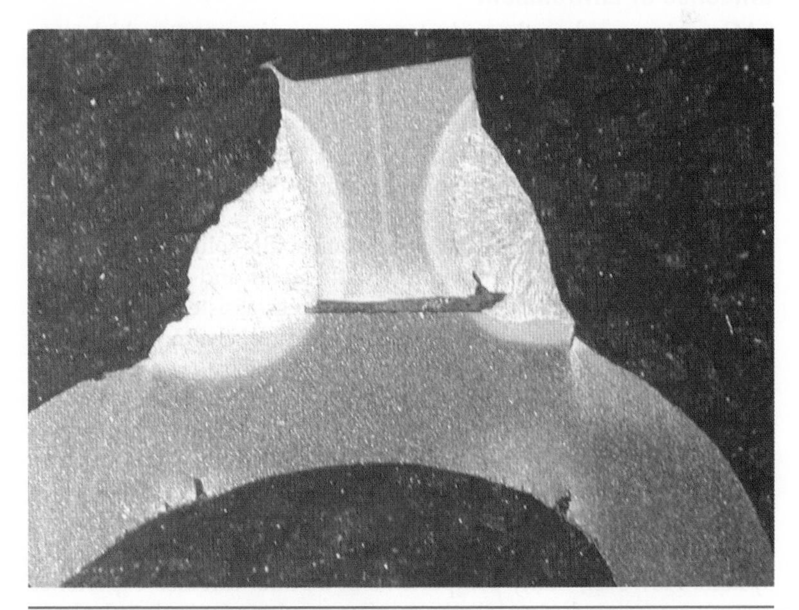

FIGURE 11.3 Moderately deep penetrations primarily caused by acid attack during cleaning, located on the internal surface opposite the membrane welds of a utility boiler waterwall tube (Nital etch).

Influence of Stress

The stresses responsible for SAC may be residual or applied. Residual stresses can be imparted from fabrication processes such as welding and cold working. Residual stresses introduced by welding operations can approach the yield stress of the metal.[3] Figure 10.2 shows preferential wastage due to acid attack along distinct circumferential zones on the internal surface of boiler tubes adjacent to a butt weld. Cold working refers to plastically deforming the metal below its recrystallization temperature. Some typical forming operations that result in boiler components with appreciable residual stresses from cold working are bending, rolling, forging, and swaging.[4] Residual stresses from forming operations can be significant. Residual stresses as high as 40 ksi (276 MPa) have been measured in cold bent superheater tubing.[5]

Applied stresses are caused by mechanical loading of a component, such as stresses that develop from support of its own weight or due to thermal expansion and contraction. Tubes are often subjected to applied bending stresses. Applied stresses are generally intensified at points of attachment or rigid constraint. It is important to note that applied stresses are rarely purely static, and they will vary to some degree during service. If the stresses on the component are more cyclic, then corrosion fatigue cracking may have a greater influence on the failure mechanism (see Chap. 12, Corrosion Fatigue Cracking).

Influence of Environment

SAC occurs in a less than ideal aqueous environment, which may include water with high levels of dissolved oxygen, aggressive chemical cleaning solutions, or possibly low-pH excursions in the boiler water during operation. Both acidic and oxygenated environments have been shown to cause increased attack in stressed areas on boiler components in most industrial and commercial boilers.[6]

Under acidic conditions, the corrosion rate of steel generally increases with decreasing pH. Exposure to acidic solutions can result in dissolution of the protective iron oxide layers and the underlying metal. Acidic environments may cause stress-assisted corrosion in areas with elevated stress levels during chemical cleaning and upset conditions during service. On-line water chemistry excursions during operation can result in depressed pH of the boiler water. Details of these mechanisms are discussed in Chap. 10, Corrosion during Cleaning, and Chap. 5, Low-pH Corrosion during Service.

In most boilers, the corrosion rate of steel also increases with increasing levels of dissolved oxygen. Oxygen corrosion in an operating boiler is normally confined to boiler components in the preboiler system. However, many components of a boiler can also experience oxygen corrosion during idle times, often resulting from air in-leakage and poor layup practices. Chapter 9, Oxygen Corrosion, discusses oxygen corrosion in greater detail.

Identification

The most important distinguishing feature of stress-assisted corrosion is that the attack preferentially occurs in regions of high stresses in the metal. SAC should be suspected when attack is localized to areas generally associated with elevated stress levels. Attack due to SAC will often produce distinct grooves or penetrations that highlight the stress state on the component,[7] as shown in Fig. 11.4. Grooves associated with membrane welds and tube seam welds are longitudinally oriented. Grooves oriented at 45° and 135° angles to the longitudinal tube axis are associated with residual stresses that develop from tube fabrication and/or installation (Fig. 11.5). Comparison of hardness measurements between attacked and unattacked areas, especially microhardness profiles, can be used to identify locations that contain elevated levels of residual stress that are more prone to corrosion.

Although stress-assisted corrosion specifically refers to corrosion from static stresses, applied stresses on a component are seldom purely static. Cyclic tensile stresses may promote corrosion fatigue cracking, which can resemble SAC in some cases (i.e., when the corrosion component of corrosion fatigue is significant). Metallographic examination is required to distinguish SAC from corrosion fatigue cracking. Corrosion fatigue cracking will generally produce a wedge-shaped fissure containing a centerline crack, rather than a penetration

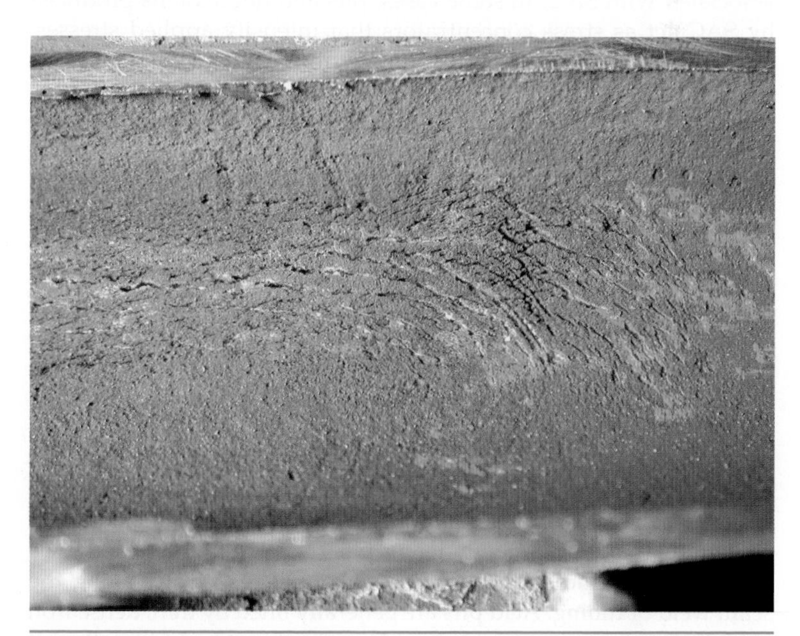

FIGURE 11.4 Close-up of parallel, curved, elongated grooves on the internal surface opposite brackets welded to the external surface.

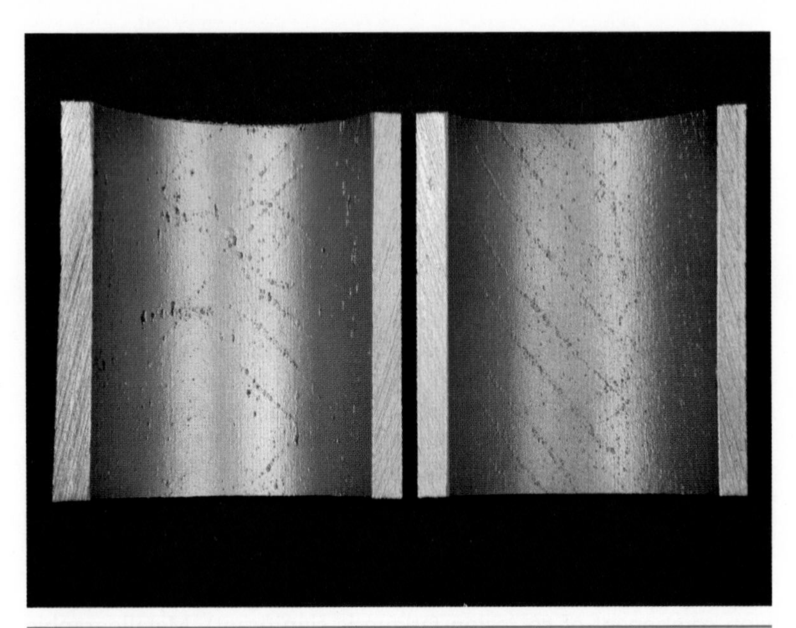

FIGURE 11.5 Internal surface after cleaning, showing shallow pits aligned along 45° and 135° angles to the tube axis. (*© NACE International, 2004.*)

associated with SAC. In some cases, pits and depressions produced by SAC act as stress concentrators that intensify applied stresses. Corrosion fatigue fissures often initiate from penetrations caused by SAC after the penetrations become sufficiently deep (Fig. 11.6).

Metallographic examination is also used to confirm whether preferential attack due to SAC was caused by water containing dissolved oxygen or an acidic environment. Distinguishing characteristics of the wastage can be used to identify whether damage was due to oxygen corrosion or acid attack.

Oxygen corrosion will tend to produce hemispherical depressions, as shown in Fig. 11.7, which are sometimes capped with mounds of iron oxide corrosion products (tubercles). A generally corroded, irregular surface contour may occur if the depressions overlap in areas of severe SAC. Microscopic examination will often reveal iron oxide corrosion products that fill the depressions, which may contain remnant iron carbides. See Chap. 9, Oxygen Corrosion, for more details.

Acid attack from a strong mineral acid, which is a typical component of chemical cleaning solutions, will have an irregular contour containing many discrete, undercut pits (Fig. 11.8). The pits may be elongated longitudinally along the tube axis, especially along the seam weld of tubing. Acid pits are generally lined by thin, dense iron oxide if attack is not recent layers. Deposited material filling the pit interiors may indicate that attack occurred in the distant past.

FIGURE 11.6 Narrow wedge-shaped corrosion fatigue fissure originating from the base of an oxide-filled depression caused by SAC (unetched). (*© NACE International, 2007.*)

FIGURE 11.7 Rounded hemispherical, oxide-lined depressions caused by oxygen corrosion in a region of elevated stress opposite an attachment plate (unetched). (*© NACE International, 2004.*)

A stressed boiler component may be subjected to multiple periods of SAC over its service life. Multiple, distinct events of active corrosion can result in bulbous morphologies with characteristic lobes, as shown in Fig. 11.9. Stratified corrosion product layers can indicate many periods of oxygen corrosion. The characteristics of the

FIGURE 11.8 Moderately deep, bulbous penetrations on the internal surface, opposite an external surface weld overlay. (Unetched).

FIGURE 11.9 Deep internal surface penetration that propagated by the combination of SAC and corrosion fatigue, showing many lobes that are partially filled with oxide. (Unetched).

damage can often be correlated with background information regarding chemical cleaning history and periods where oxygen corrosion was possible, such as during extended idle times.

Elimination

If possible, attack due to stress-assisted corrosion may be reduced by limiting excessive stress levels that can lead to preferential attack. In most cases, avoiding corrosive waterside environments is a more practical approach. Preventing oxygen corrosion, during both service and idle times, is discussed in detail in Chap. 9, Oxygen Corrosion. Essentially, the dissolved oxygen concentration in the boiler feedwater should be below recommended levels during service, and proper layup procedures should be followed during idle times. Corrosion related to acid attack during chemical cleaning can be reduced or eliminated by following standard chemical cleaning procedures. Parameters associated with uncontrolled chemical cleanings are presented in Chap. 10, Corrosion during Cleaning.

Reducing the magnitude of applied and/or residual stresses on components can also decrease the extent of attack due to stress-assisted corrosion. Detailed stress analysis of the system, such as by using finite element analysis, may be useful in predicting stress levels and how design alterations may reduce stresses. Operational changes, such as slower start-ups and shutdowns, may help reduce stresses caused by thermal expansion and contraction. Residual stresses from cold working may be alleviated by stress-relieving the component, if possible. Postweld heat treatments can decrease residual stresses that develop during solidification and cooling of a weldment.

Cautions

Penetrations caused by stress-assisted corrosion can sometimes be mistaken for fissures caused by corrosion fatigue cracking. However, the physical characteristics and contributing factors associated with SAC and corrosion fatigue cracking are distinguishable. The main difference between these two mechanisms is that stress-assisted corrosion is a static process, whereas corrosion fatigue is a dynamic process requiring cyclic stresses. Evidence from metallographic examination can indicate the predominate mechanism. In some cases, evidence of both SAC and corrosion fatigue cracking may be present.

SAC of tubing is difficult to assess prior to failure because attack can be very localized. Damage can be identified in some cases by nondestructive testing methods. X-ray examination has been used with some success to check for SAC in locations known to be susceptible. Other methods such as borescope inspection, eddy current, and ultrasonic testing are generally not as reliable in identifying damage related to stress-assisted corrosion.[8] Because SAC occurs in localized areas of elevated stress, such as near attachment welds, severe attack may tend to occur on the cold sides of waterwall tubes. In this case, failures can be particularly dangerous as they can release steam into areas occupied by personnel. SAC in recovery boilers is a significant concern, as tube leaks can have drastic consequences due to severe water and smelt bed reactions.

Related Problems

See Chap. 9, Oxygen Corrosion; Chap. 10, Corrosion during Cleaning; and Chap. 12, Corrosion Fatigue Cracking.

Case History 11.1

Industry:	Building
Specimen location:	Rolled end of tube in a firetube boiler
Orientation of specimen:	Horizontal
Years in service:	2½
Water treatment program:	Sulfite-based oxygen scavenger, phosphate/polymer
Pressure:	150 psi (1.0 MPa)
Tube specifications:	Carbon steel

This unit was one of two firetube boilers operated in lead/lag mode, cycling between the lead modes on a weekly basis. Leaks developed in numerous tubes of one boiler, primarily in the top tubes at locations adjacent to the tubesheet. Examination of a failed tube showed the leaks were caused by a wide groove adjacent to the tubesheet that penetrated the tube wall along one-half of the tube's circumference (Fig. 11.10). A wide, circumferentially oriented band adjacent to the tubesheet on the external surface was generally corroded (Fig. 11.11). The external surface in this band was covered by layers of red and brown iron oxide that overlaid some deep depressions. The band of corrosion on the external surface was localized to the region of expanded diameter from tube rolling. A microhardness profile revealed significantly higher hardness values in the rolled region than at locations further away (Fig. 11.12). A metallographic cross section through the groove at the failure showed a deep, wide penetration having a rounded profile. The penetration and deep depressions in the region were covered by stratified iron oxide corrosion product layers.

The failure of the tube was primarily caused by stress-assisted corrosion, which resulted in deep localized attack that formed the groove adjacent to the tubesheet. The deep groove that formed on one side of the tube was primarily caused by applied bending stresses that concentrated at the tubesheet. Preferential attack due to SAC also occurred in the rolled region of the tube. Corrosion localized to this band was due to residual stresses generated from the rolling process, which also contributed to the stresses in the metal at the deep groove. Expansion of the tube during the rolling process resulted in cold working of the metal. Residual stresses from this process produced a measurable hardness increase in the expanded region.

The localized attack at areas with elevated stress levels was primarily caused by oxygen corrosion, although some areas at the bases of the deepest depressions

FIGURE 11.10 Rolled end of a firetube, with a narrow groove immediately adjacent to the location of the tubesheet. (*© NACE International, 2004.*)

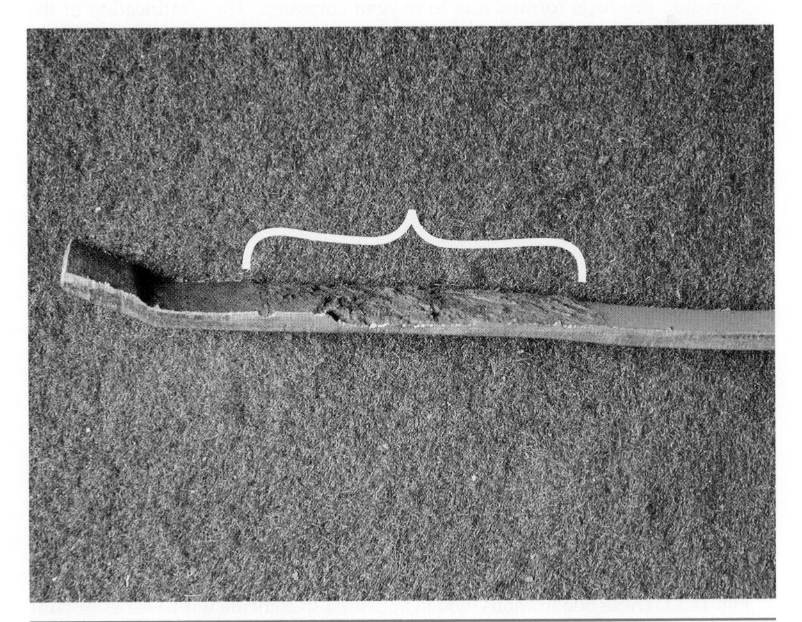

FIGURE 11.11 Profile showing that the corroded region (bracket) on the external surface corresponds to the area expanded by tube rolling. (*© NACE International, 2004.*)

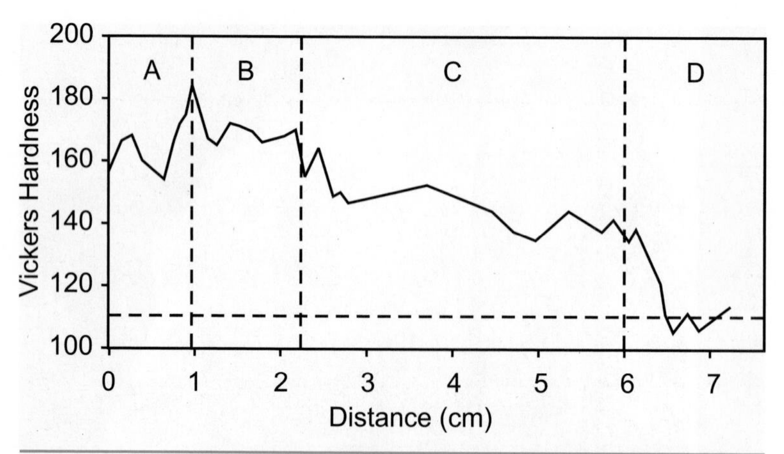

FIGURE 11.12 Hardness profile along the rolled end of the tube. Region A corresponds to the flared end, region B corresponds to the tubesheet, region C corresponds to the expanded area, and region D corresponds to the unexpanded area. The horizontal dashed line indicates the hardness for annealed 1015 steel. (*© NACE International, 2004.*)

exhibited some evidence of acid attack. Concentration and hydrolysis of sulfates present in the boiler water resulted in low-pH conditions beneath the corrosion products formed due to oxygen corrosion. The stratification of the corrosion product layers covering the groove and deep depressions indicated that attack occurred over many separate events, probably during idle times due to air in-leakage. The other boiler reportedly did not show any evidence of attack. Identifying and minimizing locations of air in-leakage during idle times would help reduce oxygen corrosion in this boiler, as well as following proper layup procedures. Excessive rolling of the tubes should also be avoided to limit residual stresses.

Case History 11.2

Industry:	Utility
Specimen location:	Steam-generating tubes
Orientation of specimen:	Vertical
Years in service:	30
Water treatment program:	Phosphate and low-excess hydroxide
Drum pressure:	875 psi (6.0 MPa)
Tube specifications:	Carbon steel

Following an acid cleaning of the boiler with a mineral acid (hydrochloric acid–based solution), localized damage was observed in some tubes. Close visual examinations of these tubes revealed very fine, discontinuous, fissurelike penetrations running longitudinally down the internal surfaces, as shown in Fig. 11.13. These penetrations, which appeared to coincide with a series of faint mandrel marks, were deeper at the end that had been rolled into the drum. The tube illustrated in Fig. 11.14 also shows localized attack where it had been rolled into the drum, in this case as transversely oriented narrow grooves. Examination

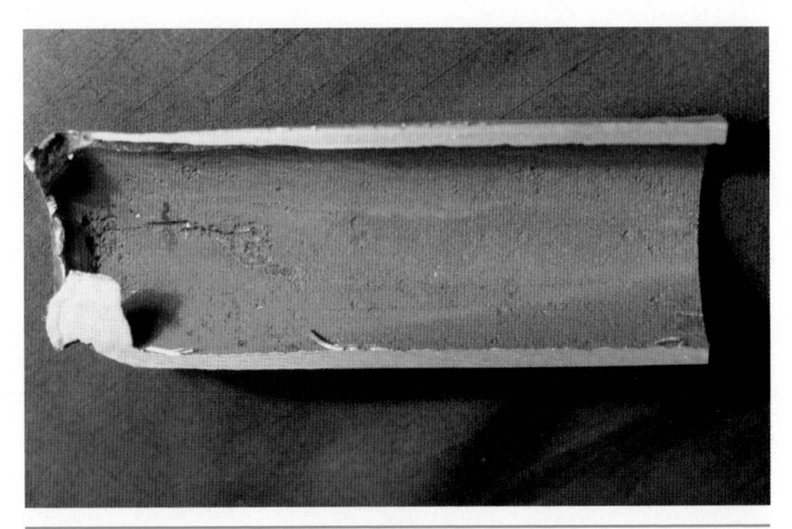

FIGURE 11.13 Narrow, longitudinally oriented grooves near the rolled end of the tube.

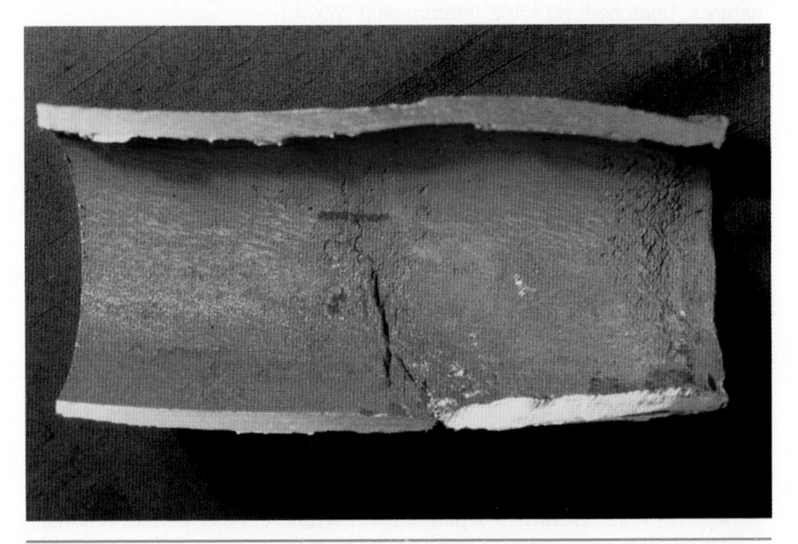

FIGURE 11.14 Narrow, transversely oriented grooves near the rolled end of the tube.

of cross-sectional profiles revealed that the penetrations were bulbous with irregular interiors having undercut features (Fig. 11.15).

Attack by strong mineral acid in locations of stressed metal was responsible for the delineation and deepening of mandrel marks running down the bore of the tube, as well as for the development of deep fissures near the ends of the tubes that had been rolled into the drum. The delineation and deepening of the mandrel marks were associated with residual stresses in the metal at sites impacted

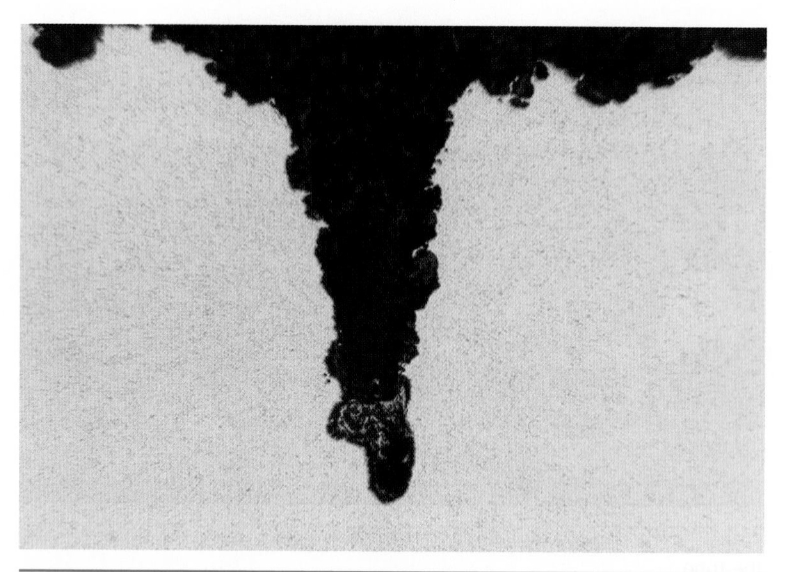

FIGURE 11.15 Deep, narrow, bulbous penetration with some undercut features. Unetched. (*© NACE International, 2004.*)

by the tube fabrication process. Deep penetrations had various orientations at the tube ends that were rolled into the drums. The residual stresses resulted from the tube-rolling process. Bending stresses may have contributed to the formation of grooves with transverse orientations. Strong mineral acid characteristically attacks stressed metal areas more aggressively than unstressed areas, due to the greater energy content of metal associated with residual stresses.

Case History 11.3

Industry:	Utility, coal-fired boiler
Specimen location:	Waterwall tube in lower furnace
Orientation of specimen:	Vertical
Years in service:	50
Water treatment program:	All volatile treatment
Drum pressure:	2500 psi (17.2 MPa)
Tube specifications:	Carbon steel

Ruptures occurred in two separate tubes within a 4-day period, following start-up after a maintenance outage that lasted for several weeks. One failure was at a wall tube in the lower furnace, which experienced a thick-walled rupture along one of the membranes on the hot side (Fig. 11.16). A weld repair patch was present on the cold side in the region of the rupture. Reportedly, this was previously a location of an attached bracket. The section exhibited narrow, longitudinally oriented grooves in locations on both the hot- and cold-side internal surfaces along forming marks. The grooves were much deeper at locations opposite welds on the external surface (Fig.11.17). Visual and metallographic examinations revealed that the internal surface grooves were generally lined with thin iron oxide layers (Fig. 11.18). The groove edges were rounded with distinct, undercut lobes present in places. The tips of some grooves were filled by oxide layers.

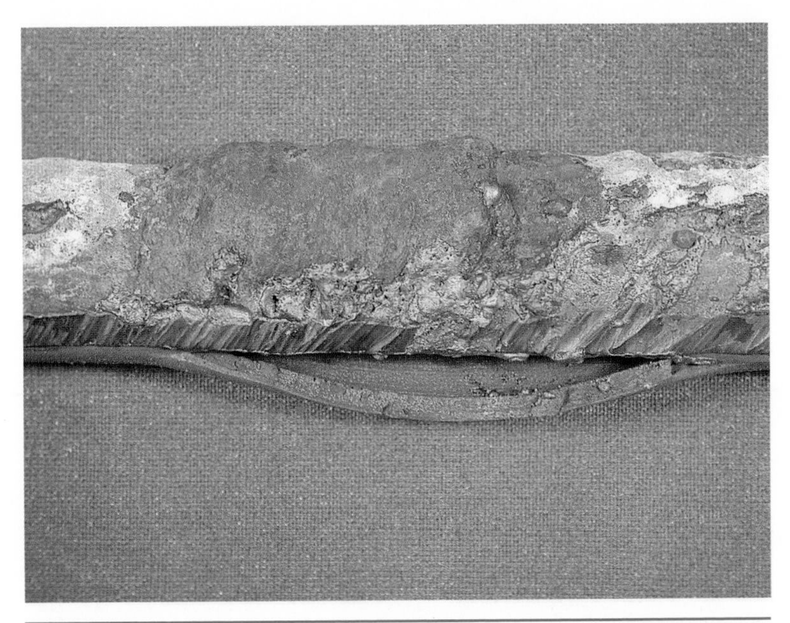

Figure 11.16 Weld repair on the cold side and a thick-walled rupture along a membrane weld on the hot side. (*© NACE International, 2007.*)

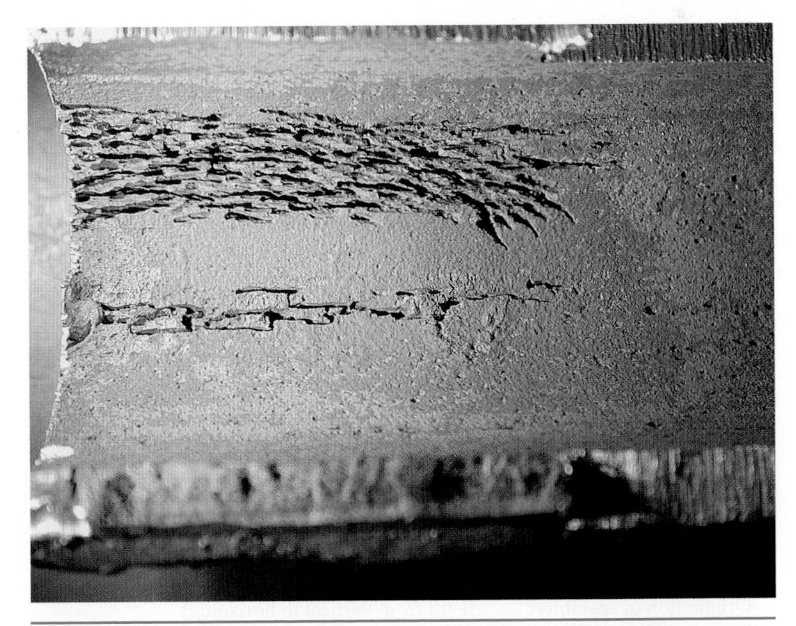

Figure 11.17 Numerous deep, irregularly shaped, discontinuous grooves on the cold-side internal surface of a wall tube at a prior bracket location, where a repair weld patch was applied to the external surface.(*© NACE International, 2007.*)

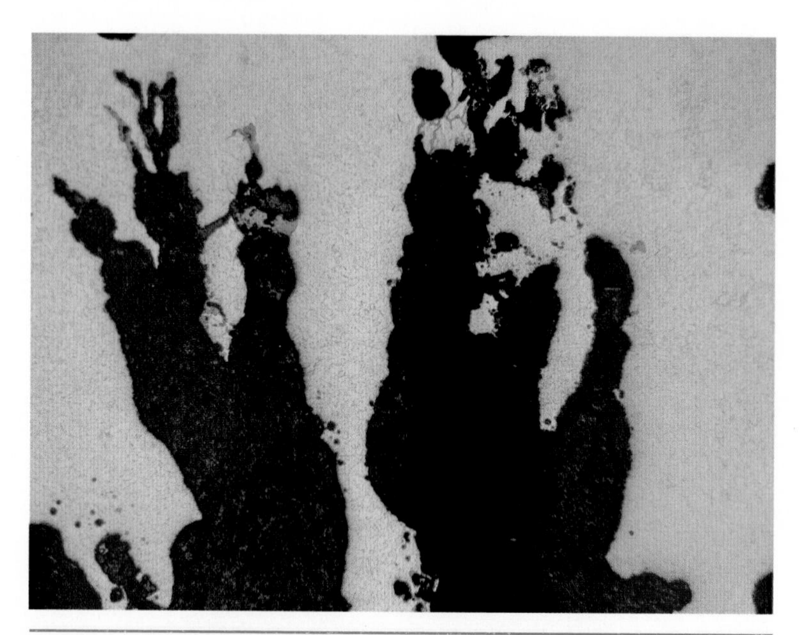

Figure 11.18 Metallographic cross section through multiple, deep grooves on the cold-side internal surface of a wall tube section. Grooves are generally lined with a thin iron oxide layer (unetched). (*© NACE International, 2007.*)

Metal loss at the grooves was caused by stress-assisted corrosion. The deepest attack in the boiler was localized to areas that had considerable residual stresses related to external surface welds. These deep grooves provided preferred sites for initiation and propagation of corrosion fatigue cracks, which extended from the tips of the grooves in places. Failures resulted when the weakened tube wall could no longer support the stresses generated from the internal pressure of the boiler.

The grooves predominately formed due to acid corrosion, which occurred during periodic chemical cleaning in areas with elevated stress levels. The internal surfaces of the boiler were reportedly chemically cleaned every 6 to 7 years during its 50-year service life. The last cleaning was conducted about 5 years prior to the failures. Some oxygen corrosion occurred during the last maintenance outage prior to the start-ups that preceded the failures.

Inspection of the boiler was recommended to determine the extent of damage and to replace tubes exhibiting excessive penetration, in order to limit continued failures from occurring at increased frequency. Chemical cleaning procedures were reviewed and revised to reduce aggressive conditions that could lead to SAC.

Case History 11.4

Industry:	Pulp and paper
Specimen location:	Bottom bend of a waterwall in a recovery boiler
Orientation of specimen:	Bend from vertical to horizontal
Years in service:	22
Water treatment program:	Congruent phosphate, carbohydrazide, neutralizing amine blend
Pressure:	1250 psi (8.6 MPa)
Tube specifications:	Carbon steel

FIGURE 11.19 Side of a waterwall tube, showing areas with deep, longitudinally oriented penetrations located opposite the welded membrane on the external surface.

Localized, moderately deep, internal surface metal loss was found in areas on tubes at the bottom bend region of a recovery boiler waterwall during careful nondestructive inspection. The metal loss was in the form of longitudinally oriented penetrations (Fig. 11.19). The penetrations were located opposite membranes welded to the external surfaces. Cross sections through the penetrations revealed large, undercut lobes (Fig. 11.20). Some small depressions filled with iron oxide corrosion products were present in places.

The penetrations were caused by stress-assisted corrosion. The location and orientation of the penetrations indicated that preferential attack was primarily due to residual and possibly applied stresses associated with the welded membrane. Stresses from the bending operation of the tube also contributed, as appreciable attack was not noted away from the bend.

Microscopic examination of the penetrations indicated features associated with acid attack and oxygen corrosion. Review of the boiler history indicated it was in a wet layup condition for a long period at the start of its initial service life; some idle-time oxygen corrosion likely occurred during this period. The history also indicated the waterwalls were cleaned twice during the boiler's service life. The undercut lobes of the penetration indicated some appreciable deepening that occurred during at least one of the acid cleaning events. Some minor idle-time oxygen corrosion occurred after the most recent acid cleaning.

Case History 11.5

Industry:	Building system, natural gas and oil-fired package boiler
Specimen location:	180° bend of an economizer tube
Orientation of specimen:	Varies
Years in service:	2 after replacement of a failed bend
Water treatment program:	Polymer and sulfite
Pressure:	600 psi (4.1 MPa)
Tube specifications:	Carbon steel

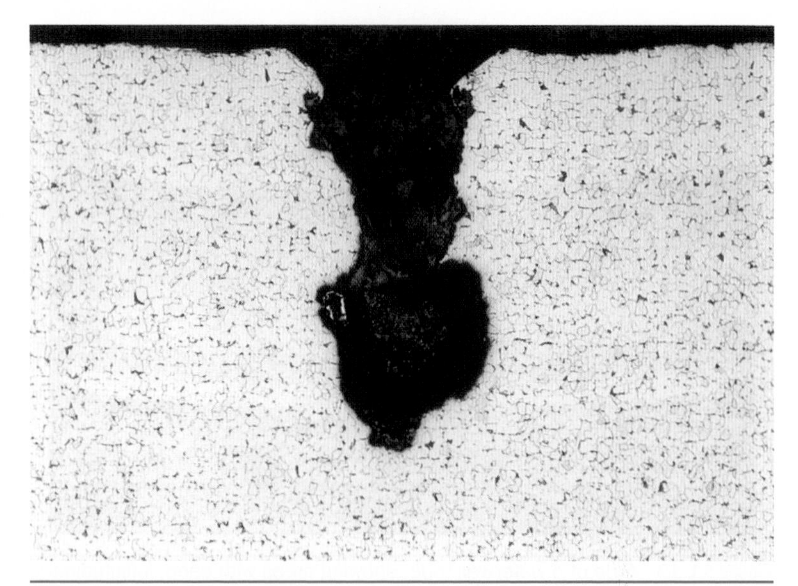

FIGURE 11.20 Metallographic cross section through a penetration, showing an undercut lobe. Etchant: Nital.

Many leaks developed in three boilers over their service life of 10 years. All failures occurred as pinhole leaks or cracks in the elbow bends of the economizer, and the bends required frequent replacement. Cross sections through a failure in the bend region revealed a longitudinally oriented penetration centered along the thin-walled outer bend (Fig. 11.21). Wall thickness measurements indicated

FIGURE 11.21 Tube ring section through the bend region, showing a penetration at the thinned wall along the outer bend of the tube.

FIGURE 11.22 Metallographic cross section, showing deep penetrations filled by billowing oxide at the failure location (unetched).

that the outer bend was less than one-half the thickness measured along the inner bend, caused primarily by thickening and thinning due to deformation during tube bending. The internal surface contained many depressions covered by corrosion product mounds that were elongated along the longitudinal axis of the tube. A metallographic cross section through the failure showed deep oxide-filled penetrations on the internal surface (Fig. 11.22). The oxide billowed into the metal and contained skeletal remnants of pearlite colonies.

The failures were caused by a combination of stress-assisted corrosion and corrosion fatigue cracking. The morphology of the penetration indicated that SAC predominated in this case. The stresses responsible for accelerated attack were primarily hoop stresses that were appreciably elevated in the thin-walled region of the outer bend. The boilers reportedly rotated into and out of service every few months. Cyclic stresses developed during these periods and deepened the penetration due to corrosion fatigue cracking.

Although evidence of moderately deep oxygen corrosion was found away from the failures, it was severe only in the thin-walled region of the section. The boilers were reportedly in wet layup when not in service. However, the elongated characteristics of the corrosion product mounds on the internal surface indicated that attack primarily occurred during service. In addition, oxygen corrosion was actively occurring shortly before removal of the section from the boiler. Eliminating excessive levels of dissolved oxygen in the boiler feedwater was required to reduce oxygen corrosion. In addition, layup practices were reviewed to limit the risk of attack during idle times. Although preventing further oxygen corrosion would eliminate the deepening of the penetrations due to SAC, concentration of cyclic stresses at the deep penetrations could allow for deepening caused by corrosion fatigue cracking.

Case History 11.6

Industry:	Pulp and paper
Specimen location:	Bend region of a superheater tube
Orientation of specimen:	Varies
Years in service:	20+
Water treatment program:	Polymer, sulfite, neutralizing amine blend
Pressure:	475 psi (3.3 MPa)
Tube specifications:	Carbon steel

The boiler was typically idle for over 9 months per year, generally in a wet layup conditions. Condensation of moisture in the superheater, combined with air in-leakage, resulted in deep depressions caused by oxygen corrosion. Attack was generally most severe at the bottom of U bends, where the condensate collected. Weld overlay patches were used to repair leaks. One U-bend superheater section developed a leak in the form of a short, narrow opening along the side of the tube slightly away from a weld overlay patch (Fig. 11.23). Cross sections through the tube indicated the tube was out of round, with a distinct transition along the side of the tube where the failure occurred. Multiple deep, longitudinally oriented grooves were present on the internal surface in the bend region along the side, as shown in Fig. 11.24. Scattered round depressions, highlighted in places by red oxide, were located away from the bend. Metallographic examination of the grooves showed rounded oxide-lined depressions, typical of old oxygen corrosion. In addition, depressions having irregular interiors filled with billowing oxide were superimposed along the edges of the rounded depressions in places (Fig. 11.25).

Stress-assisted corrosion in this case was caused by idle-time oxygen corrosion that occurred over multiple periods. Attack was accelerated in areas of stress along the sides of the tube, predominately due to residual bending stresses. In addition, residual stresses that developed along the edges of the weld overlay patch contributed to the overall stress level in the areas of preferential attack.

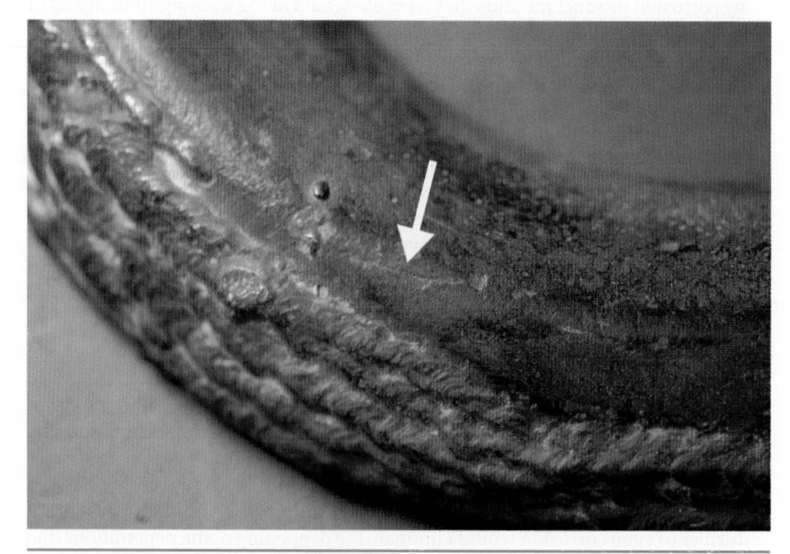

FIGURE 11.23 Tight, narrow opening located along the side of the tube (arrow) near the weld overlay patch on the external surface.

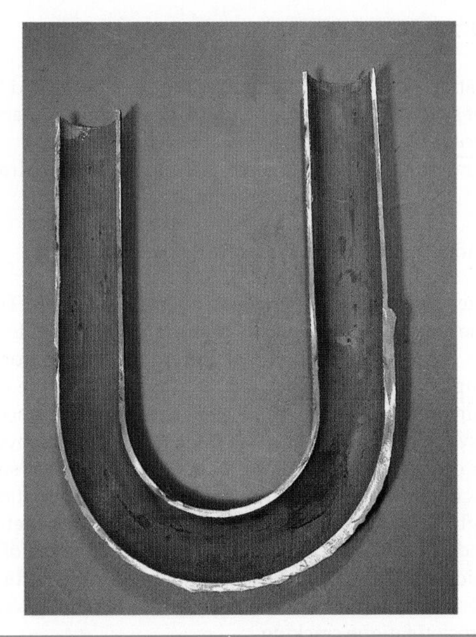

Figure 11.24 Internal surface of the U-bend section, showing deep, longitudinally oriented grooves along the side of the tube in the bend region and a few scattered depressions elsewhere.

Figure 11.25 Edge of a moderately deep internal surface depression, with areas showing attack that occurred during different periods. The area of older attack has a smooth, rounded profile covered by a thin, dense oxide layer, while areas of more recent attack have an irregular profile with billowing oxide (unetched).

Case History 11.7

Industry:	Pulp and paper
Specimen location:	Economizer tube in a bark-fired boiler
Orientation of specimen:	At a bend, vertical segment attached to a header
Years in service:	21
Water treatment program:	Polymer/phosphate, carbohydrazide, neutralizing amine blend
Pressure:	1100 psi (7.6 MPa)
Tube specifications:	Carbon steel (SA-210)

The economizer of the boiler experienced numerous failures. The failures were localized to the areas around the welds connecting the tubing to the inlet header and along the outer bends of tubing slightly downstream from the header. A repair was implemented using a weld overlay patch on the external surface near the circumferential weld to the header. The internal surface contained two areas of transversely oriented grooves (Fig. 11.26). The grooves were significantly deeper opposite the thick weld patch on the external surface. A metallographic specimen through this area revealed deep penetrations, and a cross section through one penetration is shown in Fig. 11.9. The penetration consisted of many rounded lobes and portions of tight fissures. Some of the lobes were filled with oxide, and others exhibited pronounced undercutting. At the tip of the penetration, the fissure was filled with oxide containing a centerline crack, typical of corrosion fatigue cracking.

The transverse orientation of the grooves indicated that longitudinally oriented stresses were responsibly for SAC and corrosion fatigue cracking. These areas had appreciable residual stresses from the weld overlay, which is consistent with the localized attack on the section. In addition, the section was likely

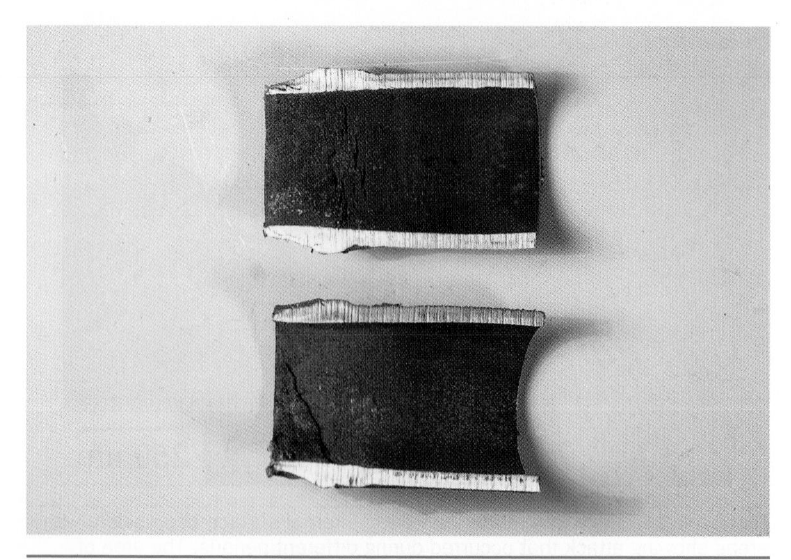

FIGURE 11.26 Internal surfaces of an economizer tube, showing transverse grooves on one side of the tubing, with the deepest grooves located in an area opposite the weld patch on the external surface.

subjected to cyclic bending stresses that promoted deepening of the penetration by corrosion fatigue cracking.

The characteristics of the multiple lobes along the penetration indicated distinct periods of corrosive attack along with some crack propagation. Features of the lobes indicated periods of both oxygen corrosion and acid attack during aggressive chemical cleanings. Numerous acid cleanings were performed on the boiler to remove buildup of internal surface deposition. Some oxygen corrosion was more recent than periods of acid attack. Reportedly, the boiler ran for an extended period without steam going to the deaerator, a condition that allowed for increased levels of dissolved oxygen in the boiler feedwater. Replacement of compromised tubing was recommended, along with ensuring that the boiler feedwater maintained sufficiently low levels of dissolved oxygen. Limiting residual stresses from welding, possibly using a postweld heat treatment, was needed to reduce the attack caused by SAC. In addition, limiting the propagation of corrosion fatigue cracks required identifying and reducing the cyclic stresses. Based on the tight morphology, the cyclic stresses responsible for cracking may have been related to tube vibrations during service, possibly caused by channeling of flue gas or insufficient tube support.

Case History 11.8

Industry:	Chemical processing
Specimen location:	Generating tube
Orientation of specimen:	Vertical
Years in service:	23
Water treatment program:	Phosphate, hydrazine, neutralizing amine
Pressure:	1525 psi (10.5 MPa)
Tube specifications:	Carbon steel

A rupture in the generating section of the boiler resulted in numerous tubes that were removed as part of the repair. One tube section examined from the vicinity of the failure had a baffle plate welded to the tube at a 45° angle (Fig. 11.27). The baffle plate and weld were reportedly austenitic stainless steel alloys. Visual and microscopic inspection of the weld revealed many defects that indicated poor weld quality. Examination of the internal surface revealed a localized area

FIGURE 11.27 Plain carbon steel generating tube with an attached stainless steel baffle plate.

FIGURE 11.28 Band containing clusters of short, narrow penetrations on the internal surface opposite the attachment weld of a baffle plate.

containing clusters of short, narrow penetrations. The penetrations were present in a band that corresponded to the attachment weld of the plate (Fig. 11.28). Microscopic examination revealed many scattered, shallow, undercut pits, with deeper penetrations in the band opposite the attachment weld. The characteristics of the pits indicated they formed due to acid corrosion, likely during chemical cleaning of the boiler.

The deeper penetration of the attack in the band opposite the attachment weld was caused by stress-assisted corrosion, primarily due to residual stresses that developed during solidification and cooling of the weld metal. Applied stresses related to thermal expansion and contraction differences between the different metals of the weld and plate and the tube may have contributed to the overall stress level to some degree. Joining metals with substantially different thermal expansion coefficients should be avoided, or joint design should incorporate filler metals that are intermediate to the plate and tube metals.

References

1. W. Schoch and H. Spahn, "On the Role of Stress Induced Corrosion and Corrosion Fatigue in the Formation of Cracks in Water Wetted Boiler Components," *Corrosion Fatigue: Chemistry, Mechanisms, and Microstructures*, National Association of Corrosion Engineers, Houston, Tex., 1972, p. 56.
2. Babcock & Wilcox, Plant Service Bulletin 29A, "Stress Assisted Corrosion: Boiler Waterside" Barberton, Ohio, 2003.
3. M. G. Fontana and N. D. Greene, *Corrosion Engineering*, 2d ed., McGraw-Hill, New York, 1978, p. 97.
4. *Steam—Its Generation and Use*, 41st ed., Babcock & Wilcox, Barberton, Ohio, 2005, pp. 7–9.

5. L. M. Wyatt and N. G. Germmill, "Experience with Power Generating Steam Plant and Its Bearing on Future Developments," *Proceedings of the Joint International Conference on Creep*, New York, NY, 1963, p. 7–5.
6. P. B. Desch, J. J. Dillon, and S. H. M.Vrijhoeven, "Case Histories of Stress-Assisted Corrosion in Boilers," *Power Plant Chemistry*, pages 82–93, Vol. 7, No. 2, February 2005.
7. H. Thielsch, *Defects and Failures in Pressure Vessel and Piping*, Reinhold Publishing Corporation, New York, 1965, pp. 406–409.
8. S. J. Pawel, A. W. Willoughby, H. F. Longmire, and P. M. Singh, "An Experience with Detection and Assessment of SAC in a Recovery Boiler," Paper #04517, National Association of Corrosion Engineers, New Orleans, LA, 2004.

SECTION 1.4

Cracking

CHAPTER 12
Corrosion Fatigue Cracking

General Description

Corrosion fatigue typically refers to a cracking mechanism in which cracks initiate and propagate due to the combination of cyclic tensile stresses and an environment that is corrosive to the metal. Under normal conditions on the water side, cracking is primarily caused by cyclic stresses that fracture the normally protective iron oxide layer to expose bare metal. The relative contribution of corrosion to the cracking is typically small due to the boiler water's very low dissolved-oxygen concentration and controlled alkaline pH during normal service. Under such conditions, bare steel reacts spontaneously with water and steam to form magnetite in a multistep process called the *Schikorr reaction* (see Chap. 9, Oxygen Corrosion). Unlike stress corrosion cracking, corrosion fatigue does not require exposure to a specific corrosive environment (see Chap. 13, Stress Corrosion Cracking).

Cracks develop according to the following sequence:

- During the first cycle of sufficient tensile stress, the brittle iron oxide layer fractures, opening a microscopic crack through to the metal surface.
- The exposed metal at the root of the crack oxidizes, forming a microscopic notch in the metal surface (Fig. 12.1).
- During the ensuing cycles of tensile stress, the oxide tends to preferentially fracture at the notch, causing the crack to deepen.
- As this cyclic process continues, a wedge-shaped crack propagates through the metal (Fig. 12.2).

Corrosion fatigue cracks always propagate in a direction perpendicular to the direction of the principal stress. Hence, in the case of a boiler tube, if the principal cyclic stress is produced by fluctuations

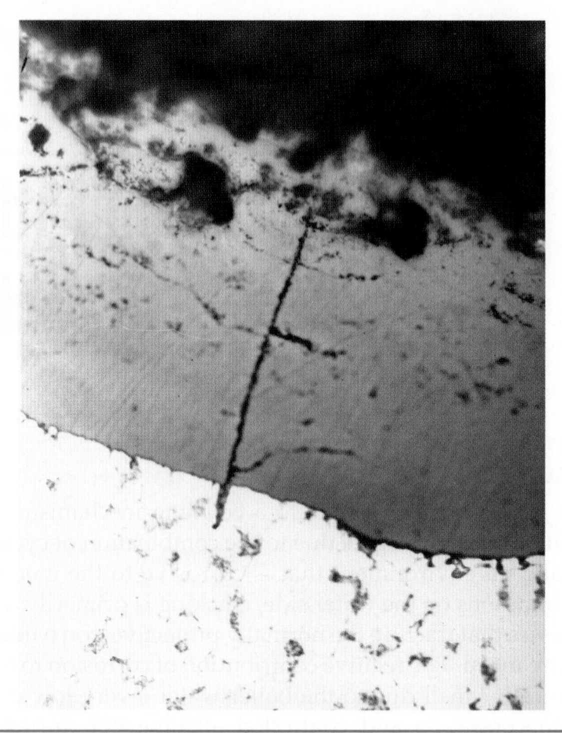

FIGURE 12.1 Incipient corrosion fatigue crack formed at base of cracked layer of iron oxide. Etchant: Picral. (*Courtesy of National Association of Corrosion Engineers.*)

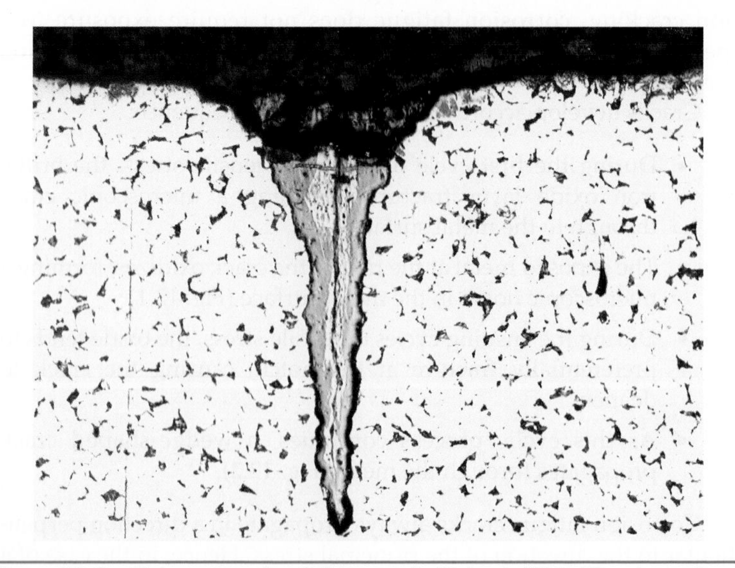

FIGURE 12.2 Mature corrosion fatigue crack. Etchant: Nital. (*Courtesy of National Association of Corrosion Engineers.*)

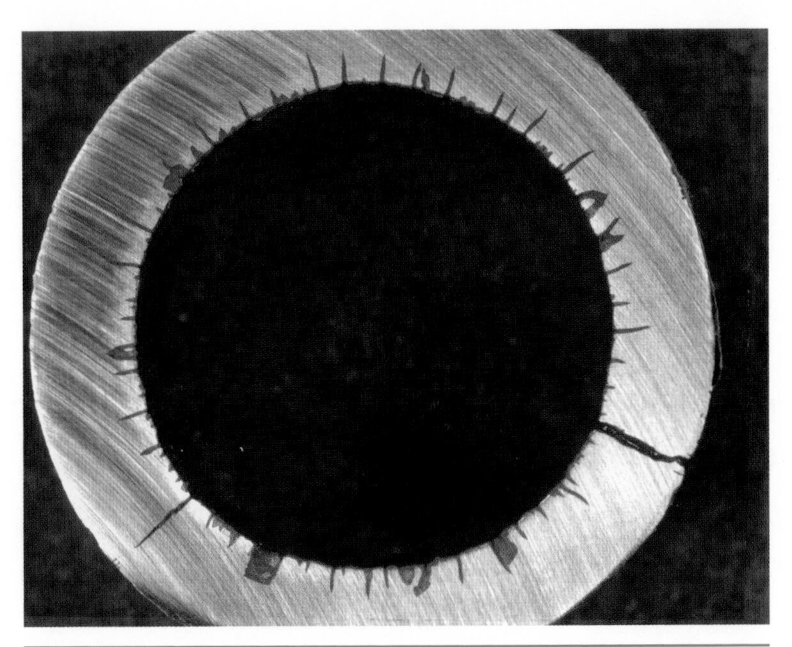

FIGURE 12.3 A family of longitudinal cracks resulting from fluctuation in internal pressure. (*Courtesy of National Association of Corrosion Engineers.*)

in internal pressure (varying hoop stress on the tube), longitudinal cracks are produced (Fig. 12.3). If the principal cyclic stresses are supplied by a bending load or are produced by thermal expansion and contraction of the tube, cracks will be transversely oriented (Fig. 12.4). Corrosion fatigue cracking commonly occurs adjacent to physical restraints, such as brackets, that serve as sites of stress concentration. Cracks may originate on the fire side, the water side, or both simultaneously. Cracks may also originate at pits and other surface features or irregularities. Such features serve as sites of stress concentration, making them preferred sites for initiation of corrosion fatigue cracks.

In air, carbon steels generally exhibit a fatigue limit, or a stress below which failure by fatigue does not occur (Fig. 12.5). A corrosive environment may reduce or eliminate the fatigue limit, and failure at a given stress may occur at fewer cycles. In general, corrosion reduces the magnitude of stress, time, and/or number of stress cycles needed to cause a cracking failure.

Thermal fatigue is a specific type of corrosion fatigue cracking that occurs due to rapid, cyclic cooling of a hot surface. Rapid cooling causes the surface metal to contract at a different rate than the bulk metal beneath it, producing multiaxial stresses that cause cracks to form in a crosshatched or crazed pattern[1] (see Case Histories 12.1 and 12.9).

FIGURE 12.4 Transverse cracks originating on the internal surface.

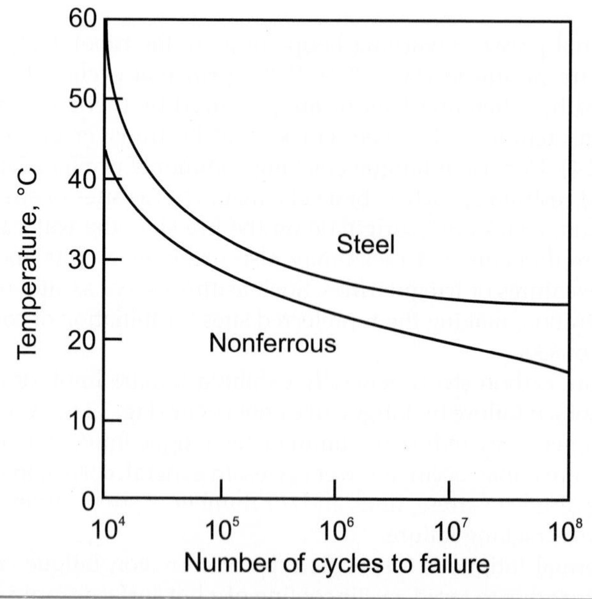

FIGURE 12.5 Schematic plot of stress amplitude vs. number of cycles to failure, showing that the curve for carbon steel asymptotically approaches a fatigue limit, or a stress below which fatigue failure does not occur. Most other materials do not exhibit a fatigue limit.[2] (*Reprinted with permission of ASM International. All rights reserved. www.asminternational.org.*)

Locations

Corrosion fatigue can occur at any location where cyclic stresses of sufficient magnitude are applied. Common locations of corrosion fatigue cracks include wall tubes, reheater tubes, superheater tubes, economizer tubes, deaerators, and at membranes of waterwall tubing. In some cases of long-term overheating, corrosion fatigue fissures may form in bulged areas as a result of intermittent bulge growth (Fig. 12.6). The fissures formed in these situations should not be confused with creep fissures or thermal fatigue fissures (in cases where a crosshatched pattern occurs, as shown in Fig. 12.6). In addition, corrosion fatigue is common at points of attachment or rigid constraint, such as connections to inlet or outlet headers, tie bars, and buckstays.

Corrosion fatigue cracks have also been observed at grooves formed by corrosion along the internal surfaces of boiler tubes that have been only partly full of water, at points of intermittent steam blanketing within generating tubes, at oxygen pits in water-lines, in soot blower lines, and in blowdown lines.

Pits, depressions, craters, gouges, and other surface irregularities may act as stress concentrators and promote the formation of corrosion fatigue fissures (Fig. 12.7), since fissures tend to nucleate where tensile stresses are greatest (Fig. 12.8).

FIGURE 12.6 Crosshatched corrosion fatigue fissures associated with intermittent bulge growth, which should not be confused with thermal fatigue fissures.

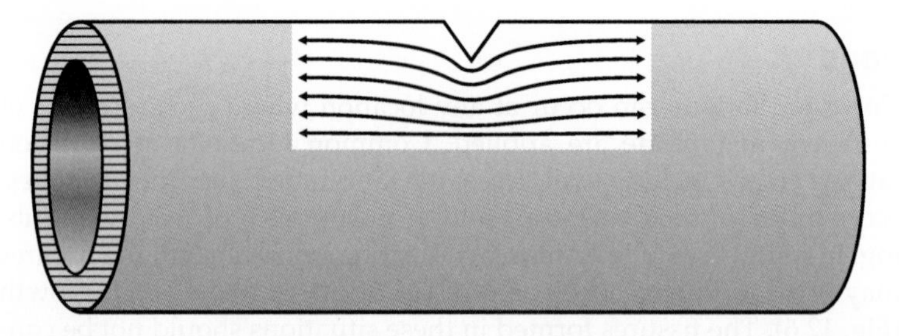

FIGURE 12.7 Schematic diagram of a piece of material containing a surface irregularity. The material is under tensile stress, resulting in tightly spaced imaginary force lines near the tip of the irregularity, indicating the location of maximum stress concentration.

FIGURE 12.8 Corrosion fatigue fissure within a mechanical gouge on the external surface of a recovery boiler tube.

Residual tensile stresses can also promote the formation of corrosion fatigue cracks. The residual tensile stresses combine with the cyclic tensile stresses, facilitating crack formation and propagation. Welding operations are common sources of residual tensile stresses. Postweld annealing, when possible, can relieve such residual tensile stresses.

Other welding issues can also contribute to corrosion fatigue cracking. Welding defects, such as undercutting, slag pockets, or points of incomplete fusion, provide preferred sites for corrosion fatigue fissure initiation. When dissimilar metals (e.g., austenitic stainless steel and carbon steel) are joined by welds, corrosion fatigue can occur due to stresses imparted by differences in their coefficients of thermal expansion (see Case History 12.4).

Critical Factors

Cyclic tensile stresses and an environment that will cause corrosion or spontaneous oxidation of a bare metal surface are two critical factors that govern susceptibility to corrosion fatigue. Sources of cyclic applied tensile stresses include cyclically fluctuating internal pressure, constrained thermal expansion and contraction, and vibrations. Corrosion fatigue failures most frequently occur in boilers that are in "peaking" service, experience frequent and significant load changes, are used intermittently, or otherwise operated cyclically. Rapid boiler start-up or shutdown can greatly increase the susceptibility to corrosion fatigue (see Case History 12.8). Some serious corrosion fatigue problems have been eliminated merely by sufficiently modifying start-up and shutdown rates.

Other factors that may significantly contribute to crack growth are pH level and dissolved-oxygen content. Operation at low pH levels or with excessively high concentrations of dissolved oxygen may promote pitting. The pits then serve as stress concentrators for the initiation of corrosion fatigue cracks. Oxygen corrosion may also contribute to the growth of fissures themselves. Depending upon the severity, such corrosion may cause the fissures to significantly deepen or widen.

Identification

Corrosion fatigue cracking normally results in thick-edged failures and/or families of parallel cracks (Figs. 12.3 and 12.4). The cracks may be very tight, making them difficult to see without very close visual examination. In some cases, they may appear to be only shallow grooves on the surface. Typically, they do not run long distances along the tube surface. Corrosion fatigue cracks often develop simultaneously within two or more components of similar location, if the components are subjected to the same types and magnitudes of cyclic tensile stress.

Ultimately, corrosion fatigue can only be diagnosed with certainty through metallographic examination. Optical microscopy of metallographic cross sections shows that corrosion fatigue cracks are typically straight, unbranched, and transgranular (Fig. 12.2). Crack propagation occurs perpendicular to the metal surface, normal to the direction of the applied stresses. The cracks can range from needle to wedge shaped. In general, low stress frequency and low strain

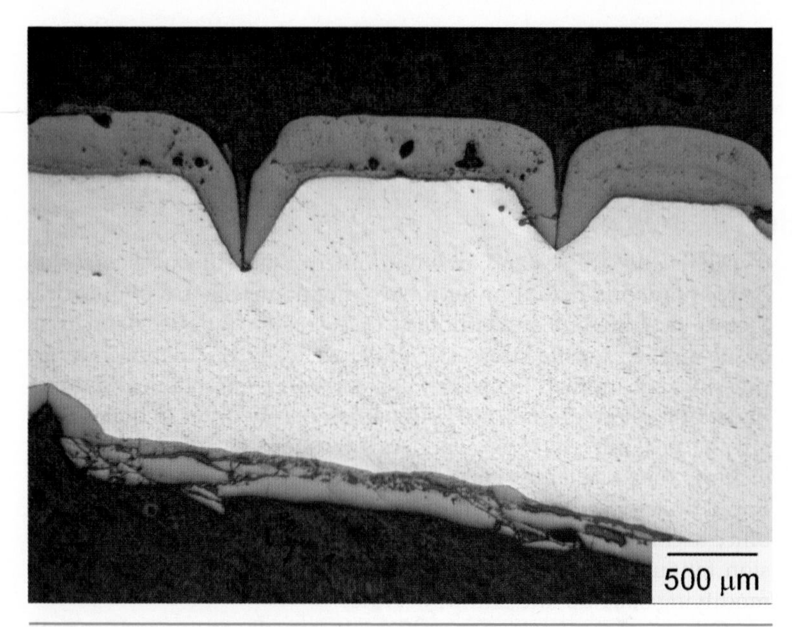

FIGURE 12.9 Wide, wedge-shaped corrosion fatigue fissures that are suggestive of low stress frequency, low strain amplitude, and/or a corrosive environment. Unetched.

amplitude promote the formation of wide, blunt, wedge-shaped fissures (Fig. 12.9).[3] A more corrosive environment also tends to promote more wedge-shaped fissures. High-stress frequency, high-strain amplitude, and a less corrosive environment tend to result in tighter, more needle-shaped fissures (Fig. 12.2). Compared to strain amplitude, stress frequency has a greater effect on the shape of a corrosion fatigue fissure because it directly affects the amount of time allowed for corrosion to occur. At high frequencies, little time is afforded for corrosion to take place, so the environment has a small effect on the corrosion fatigue cracking process. At very high frequencies, the effect of corrosion may be negligible.

Corrosion fatigue cracks are usually lined or filled with iron oxide corrosion products that contain a centerline crack. In cases where oxygen corrosion has contributed significantly to corrosion fatigue, the cracks may exhibit lobes that branch laterally from the edges or at the tips of the fissures (Fig. 12.10). The iron oxide corrosion products may appear to billow into the surrounding metal if oxygen corrosion was actively occurring shortly before the component was removed from service.

Nondestructive methods for crack detection include ultrasonic surveillance, radiographic surveillance, dye penetrant testing, and magnetic particle inspection.

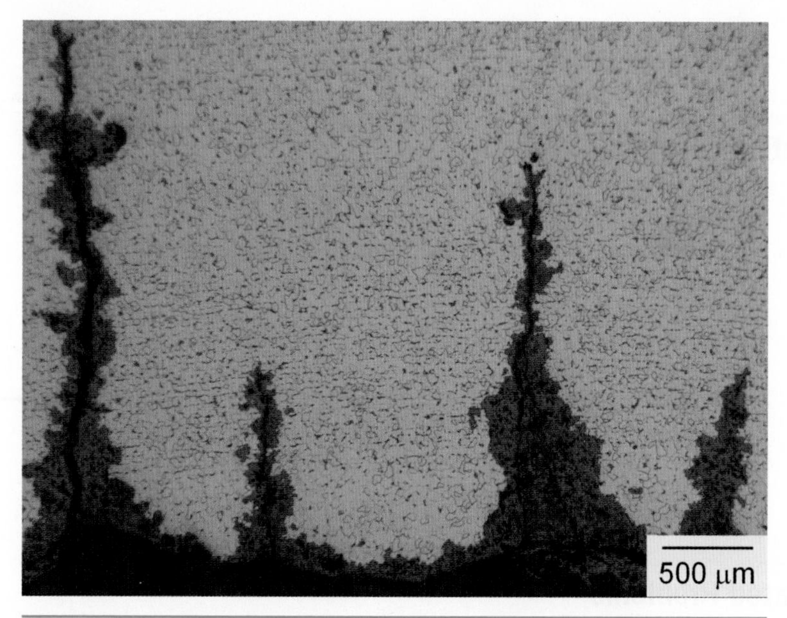

FIGURE 12.10 Micrograph of corrosion fatigue fissures in which oxygen corrosion has played a significant role, showing characteristic laterally branching lobes of billowing iron oxide. Etchant: Picral.

Elimination

Elimination or reduction of corrosion fatigue cracking is realized by controlling cyclic tensile stresses, controlling environmental factors, and boiler redesign. Crack initiation and growth may be controlled by reducing the magnitude and/or frequency of the applied cyclic stresses. Reducing or eliminating cyclic operation of the boiler and extending start-up and shutdown times may also help eliminate or reduce corrosion fatigue cracking.

Elimination of the oxidation process that occurs at a newly exposed crack tip is not feasible. The oxidation will occur spontaneously, even at very low dissolved-oxygen concentrations; however, reducing the general aggressiveness of the environment can decrease crack growth rates. Maintaining the pH and dissolved-oxygen levels within appropriate ranges can limit the formation of pits, which are common points of initiation for corrosion fatigue cracks. However, corrosion fatigue cracks may initiate on a smooth surface without pits or surface discontinuities. Under such circumstances, the magnitude and frequency of the applied cyclic stresses at the crack initiation site predominate.

In persistent cases of corrosion fatigue cracking, measures such as contouring of welds and redesign of tube attachments may be required to eliminate or reduce constraints to thermal expansion and contraction.

Cautions

Complete fractures resulting from corrosion fatigue cracking are typically thick-walled and show very little, if any, ductility. Such fractures might be mistakenly attributed to other failure modes that typically produce thick-walled fractures, such as creep/stress rupture, cracking caused by hydrogen damage, stress corrosion cracking, and some types of severe overheating. Corrosion fatigue cracks are frequently difficult to see since they are often filled with dense iron oxides. In some instances, they may appear as short, shallow grooves in the iron oxide layer covering the surface. For all these reasons, corrosion fatigue cracking failures can only be definitively diagnosed through metallographic examination.

Related Problems

See also Chap. 3, Long-Term Overheating; Chap. 2, Short-Term Overheating; Chap. 6, Hydrogen Damage; Chap. 9, Oxygen Corrosion; and Chap. 13, Stress Corrosion Cracking.

Case History 12.1

Industry:	Utility
Specimen location:	Drain line off reheat header
Tube specifications:	1⅞-in. (4.8-cm) outer diameter, low-alloy steel

Deep, crosshatched cracks and fissures were located on one end of the internal surface of the drain line, in a distinct zone around the entire circumference (Figs. 12.11 and 12.12). Cracks and fissures were not present in areas away from this end.

Microscopic examinations revealed that the fissures formed due to corrosion fatigue. Their crosshatched pattern was consistent with thermal fatigue cracking. Evidence of mild overheating was also observed.

Thermal fatigue can occur when heated metal is rapidly cooled on a repeated basis. Rapid cooling of the metal surface establishes high triaxial stresses that can produce crosshatched cracks. Rapid cooling may have been induced by localized exposure of the line to slugs of water. Elimination of the rapid cooling cycles is necessary to eliminate thermal fatigue cracking.

Case History 12.2

Industry:	Pulp and paper
Specimen location:	Front-wall tube around primary air port, recovery boiler
Specimen orientation:	Slanted
Years in service:	15
Drum pressure:	900 psi (6.2 MPa)
Tube specifications:	3-in. (7.6-cm) outer diameter, studded

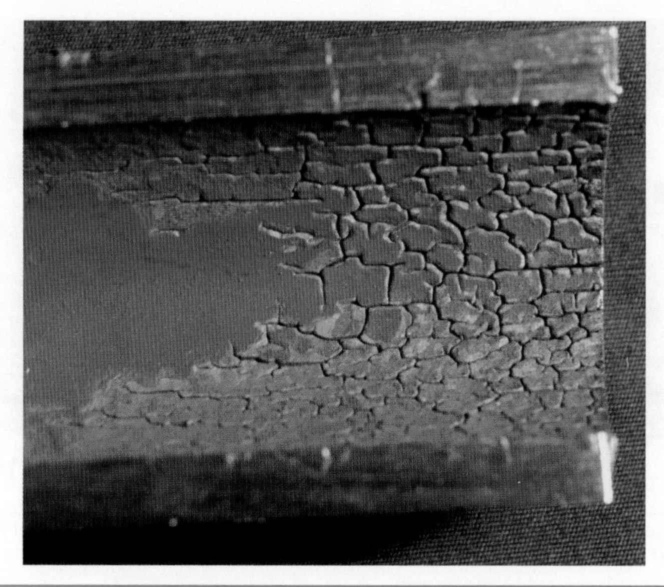

FIGURE 12.11 Thermal fatigue cracks on the internal surface of a drain line off a utility boiler reheat header.

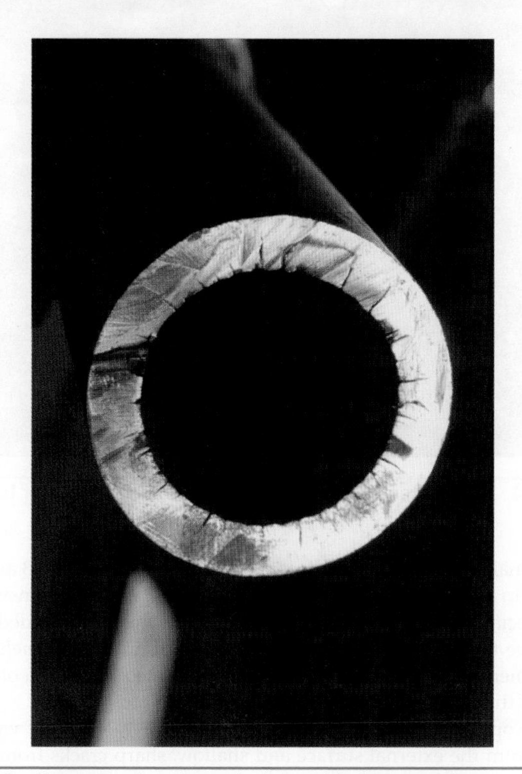

FIGURE 12.12 Cross section through thermal fatigue cracks.

FIGURE 12.13 Transverse fissures at and near stud bases on a recovery boiler tube.

FIGURE 12.14 Transverse fissures near stud on a recovery boiler tube (center).

The external surface (fireside) cracking illustrated in Figs. 12.13 and 12.14 was localized to the area around the primary air port. The short, transverse fissures were especially prominent near the bases of studs that were welded to the external surface for the purpose of retaining layers of protective smelt (Fig. 12.14). Measurements revealed that the fissures penetrated 15 to 20% of the original tube-wall thickness.

Microscopic examinations of surface profiles revealed deep, wedge-shaped fissures from the external surface and shallow, sharp cracks from the internal surface.

The transverse orientation of cracks and fissures reveals that they were produced by cyclic, outward bending of the tubes resulting from thermal expansion and contraction. The prominence of the cracks at the stud bases may have been due to differences in the thermal expansion and contraction characteristics of the studs and the tube wall. The visual appearance of such corrosion fatigue fissures on studded tubes has led to the term *elephant hiding* to describe the phenomenon. Experience suggests that elephant hiding occurs in areas of high heat-transfer rate and may occur on tubes when the smelt layer thins down on an intermittent basis, exposing the studs to the hot furnace environment.

Case History 12.3

Industry:	Chemical process
Boiler type:	Natural gas-fired firetube boiler
Specimen location:	Second pass inlet, near tubesheet
Specimen orientation:	Vertical
Years in service:	18
Tube specifications:	2½-in. (6.3-cm) outer diameter, plain carbon steel

The received section had been rolled and welded to the tubesheet at one end. A tight, transverse fissure was observed near the weld. Reportedly, numerous cracked tubes had been detected during inspections over the previous 3 years. Microscopic examinations of the external surface (water side) near the weld indicated the presence of a family of straight, unbranched, transgranular, transverse fissures that were filled with iron oxide containing centerline cracks (Fig. 12.15). The deepest observed fissure penetrated through about one-third of the tube wall, relative to the nominal measured wall thickness.

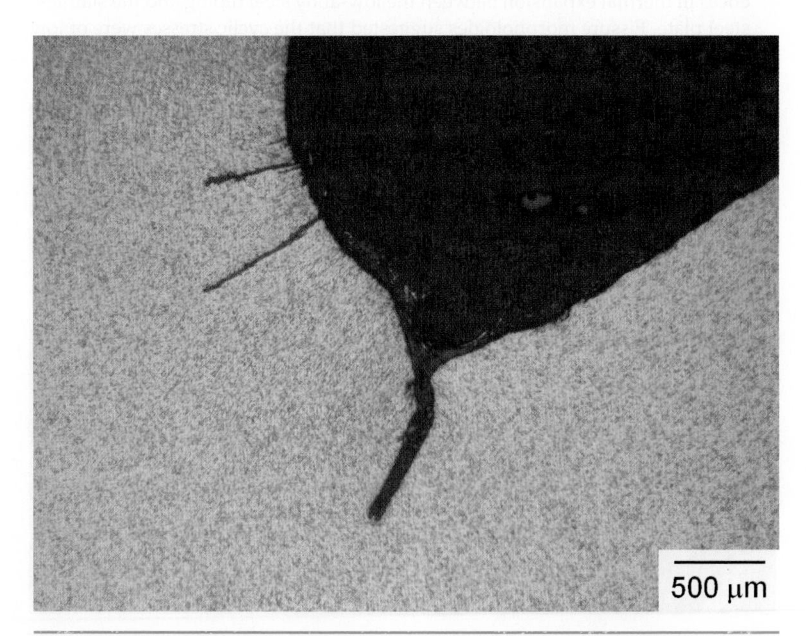

FIGURE 12.15 Micrograph showing numerous straight, transgranular corrosion fatigue fissures near the welded end of a tube. Etchant: Nital.

The narrow nature of the fissures suggested that the cyclic tensile stresses were of high frequency and possibly magnitude. The transverse orientation of the fissures indicated that stresses were likely due to vibrations or tube expansion and contraction caused by rapid temperature changes. The bottom of the second pass, where the section was reportedly located during service, was likely a location of high inlet gas temperatures. Residual stresses from welding likely contributed to the initiation of the fissures. In addition, the tubesheet served as a rigid restraint, providing a natural site for fissure formation.

Case History 12.4

Industry:	Pulp and paper
Specimen location:	Superheater
Years in service:	20
Drum pressure:	900 psi (6.2 MPa)
Tube specifications:	2-in. (5.1-cm) outer diameter, low-alloy steel

A few remnants of welded, longitudinal austenitic stainless steel plate attachments were present on the received section. A through-crack was observed along the toe of a welded attachment that had been ground to the tube surface (Fig. 12.16). Examination of other attachment weld remnants revealed fissures along the weld toes, along one end of the attachments. Microscopic examinations of the through-crack indicated that fissures had initiated from both the external and internal surfaces, meeting in the middle of the tube wall.

The cracks and fissures were localized to regions along attachment welds, indicating that the cyclic tensile stresses responsible for corrosion fatigue cracking were transmitted to the tube at these sites. Cyclic tensile stresses were primarily caused by temperature changes that were intensified due to differences in thermal expansion between the low-alloy steel tubing and the stainless steel plate. Fissure morphologies suggested that the cyclic stresses were of low

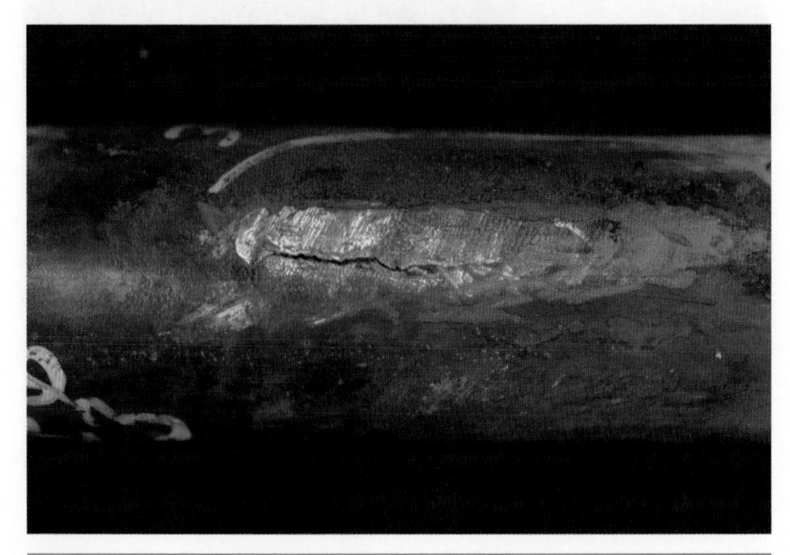

FIGURE 12.16 Longitudinal through-crack at the toe of a weld joining austenitic stainless steel plate to a low-alloy steel tube.

frequency and possibly magnitude, consistent with heating and cooling resulting from intermittent operation of the boiler. To prevent such cracking, welded attachments of metals having disparate thermal expansion coefficients should be avoided.

Case History 12.5

Industry:	Steel
Specimen location:	Basic oxygen furnace (BOF) hood
Years in service:	0.75
Tube specifications:	Mild steel

Several panels of BOF hood tubing were submitted for analysis. Numerous weld repairs were made on the hot side. In many cases, half-sections of tubing were replaced due to the presence of many deep, transverse fissures and some through-wall cracks (Fig. 12.17). Cracks and fissures were observed at and away from repair welds. Many tubercles covered the internal surfaces. Material on the surface consisted primarily of iron, with small concentrations of calcium, magnesium, phosphorus, silicon, aluminum, and sulfur. Moderately deep depressions and transverse fissures interrupted the underlying surfaces.

Corrosion fatigue cracking was caused by thermal expansion and contraction due to intermittent operation of the furnace. Temperatures varied significantly and changed rapidly between service and idle periods. Such dramatic changes in temperature produced substantial stresses in the tube walls. Tubercles on the internal surface were caused by oxygen corrosion.

Since temperature fluctuations in the system could not be controlled, it was recommended that thermally insulating corrosion products and deposits on the internal surfaces be reduced in order to limit the magnitude of metal

FIGURE 12.17 Deep, transverse corrosion fatigue fissures and cracks on the external surface of a BOF hood tubing panel.

temperature fluctuations. Methods of controlling scale formation, as well as oxygen corrosion during both service and idle periods, were to be investigated. It was suggested that thermocouples or infrared sensors could be used to monitor metal temperatures.

Case History 12.6

Industry:	Pulp and paper
Specimen location:	Economizer, power boiler
Specimen orientation:	Horizontal
Years in service:	23
Drum pressure:	995 psig (6.9 MPa)
Tube specifications:	2-in. (5.1-cm) outer diameter, plain carbon steel

Welded attachments were located on opposite sides of the section (Fig. 12.18). A straight, longitudinal through-crack formed between two of the attachments. Numerous longitudinal fissures interrupted the internal surface near the crack. On the opposite side, similar fissures formed between the other two attachments. The internal surface was covered by reddish brown iron oxide and many thin-shelled tubercles. Microscopic examinations showed that the crack and fissures were tight, straight, transgranular, and wedge-shaped. The fissures were filled and lined with iron oxide corrosion products that billowed into the surface, forming lobes in some places (Fig. 12.10). The iron oxide corrosion products contained skeletal remnants of pearlite colonies. The through-crack was located near the weld seam of the tube.

The tightness of the corrosion fatigue fissures suggested that the cyclic stresses were of high frequency and possibly magnitude, potentially related to vibrations or bending stresses localized to the areas between the attachments. Oxygen corrosion, which probably occurred during both service and idle periods, significantly contributed to the corrosion fatigue process. Lobes along some of the fissures suggested points of crack arrest, likely during idle times.

FIGURE 12.18 Longitudinal through-crack between two welded attachments on an economizer tube.

Crack propagation likely occurred over extended periods. Depressions on the internal surface acted as stress concentrators, locally intensifying applied cyclic stresses, which promoted cracking.

Case History 12.7

Industry:	Pulp and paper
Specimen location:	High-pressure economizer, heat recovery steam generator (HRSG)
Specimen orientation:	Curved (elbow)
Years in service:	5
Drum pressure:	310 psig (2.1 MPa)
Tube specifications:	2-in. (5.1-cm) outer diameter, plain carbon steel

Numerous recurrent leaks occurred in the high-pressure economizer, localized to the elbows welded to the economizer header. The leaking sections were removed, and the header and bottom of the tube were plugged. Metallographic analysis of one of the leaking elbow sections revealed two narrow through-cracks located on the extrados, a short distance from the welded end (Fig. 12.19). Examination of the internal surface revealed a circumferentially oriented, thick-walled crack. Numerous shallow fissures filled with billowing oxide containing a centerline crack were present near the fracture edge.

The circumferential crack orientation indicated that the cyclic tensile stresses responsible for corrosion fatigue cracking were due to bending. The stresses were highest where the tube was attached to the header. The cyclic stresses may have been related to boiler design and improper baffling that possibly introduced vibrations. Crack initiation was favored at moderately deep depressions caused by oxygen corrosion, which locally concentrated stresses. Oxygen corrosion was related to poor deaerator performance. In addition, corrosion due to excessive amounts of dissolved oxygen in the feedwater accelerated corrosion fatigue crack growth.

FIGURE 12.19 Transverse through-cracks along the extrados of an economizer elbow, near a weld. (*© NACE International, 2003.*)

Case History 12.8

Industry:	Food and beverage
Boiler type:	Package boiler (vertical tube design, once-through, forced circulation, water on shell side)
Specimen location:	Riser tube
Specimen orientation:	Vertical
Years in service:	2
Pressure:	150 psig (1.0 MPa)
Tube specifications:	Various outer diameters, plain carbon steel

Four package boilers of the same design were employed at a food processing plant, two of which experienced a total of six failures within 2 years of continuous service. During normal service, the loads on the boilers typically varied significantly, depending upon steam demand. The boiler product literature stated that the boiler could be brought on- and off-line quickly and had the ability to be taken from a cold start to full output steam production within 5 min. It was also reported that the boiler was designed in such a way that the tubes were not restrained. It was further claimed that this design feature, combined with forced water circulation, would allow the boiler to operate under severe start-up and shutdown conditions and rapid load variations without any thermal shock effects for as long as 25 years.

A riser tube section containing one of the failures was examined. Several fins were welded to a portion of the hot-side external surface at slight angles to the horizontal orientation. A corrosion fatigue crack originating from the external surface penetrated through the tube wall along one of the fin/tube interfaces (Fig. 12.20). The external surface experienced no significant metal loss or buildup of deposit and corrosion product layers.

FIGURE 12.20 Corrosion fatigue crack along the interface between the tube and a welded fin. (© *NACE International, 2007.*)

The fin weld served as a point of stress concentration for the primary crack to initiate. The cyclic stresses responsible for cracking were supplied by temperature variations, vibration, and possibly bending loads. Residual stresses at the weld also contributed to the cracking. Crack growth rates increase as a function of increasing magnitude and frequency of the applied stresses, both of which are promoted by rapid and frequent start-ups and shutdowns, frequent and/or large load changes, and cold start-ups. It was recommended that start-up and shutdown cycle times be extended, and that cold start-ups be minimized or eliminated. Since process conditions may require rapid load changes, stresses associated with load variations may not be controllable. Although a boiler may be capable of operating under severe conditions, it is prudent to minimize such conditions as much as possible.

Case History 12.9

Industry:	Unknown
Boiler type:	Forced recirculation watertube boiler
Specimen location:	Bottom of combustor coil
Specimen orientation:	Horizontal
Time in service:	2 weeks
Sample specifications:	Tube outer diameter of approximately 2½ in. (6.3 cm), plain carbon steel

The received section was reportedly removed from approximately the 6 o'clock position of the combustor coil tube. The slight concave curvature of the section on the failure side indicated that the failure occurred along the intrados of the coil (top of the tube) during service (Fig. 12.21).

The section contained a short, thick-walled, transverse failure. Numerous tight, short, transverse fissures on the external surface, in line with the rupture, were characteristic of corrosion fatigue cracking. The fissures were most likely

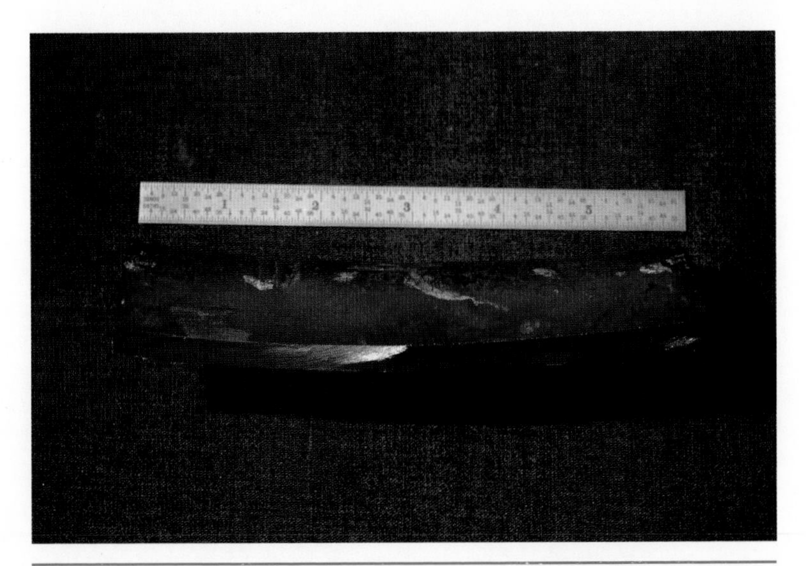

FIGURE 12.21 Transverse corrosion fatigue cracks and fissures along the top of a section of combustor coil tubing.

caused by thermal expansion and contraction due to fluctuations in temperature. Fine crosshatched fissures on the internal surface, in a wide band along the length of the section, were caused by thermal fatigue due to the cyclic formation and collapse of semistable steam blankets along the top of the tube. Fissures on both the internal and external surfaces joined together to cause the failure.

Several other similar corrosion fatigue failures occurred in similar locations of other boilers of the same design. The boilers were designed to produce steam quickly for intermittent operation, resulting in temperature fluctuations. The location where the cracking occurred, on the bottom of the coil along the intrados, was a likely location for steam blanketing to occur due to

- The probable high heat flux from the combustion gases
- The horizontal orientation of the tubing, which promoted steam/water stratification.

Methods to improve water flow through the coil were recommended to control steam blanketing and temperature variations that promoted corrosion fatigue cracking.

References

1. ASM Committee on Failures of Pressure Vessels, Boilers and Pressure Piping, "Failures of Boilers and Related Steam-Power-Plant Equipment," *ASM Metals Handbook*, vol. 10: *Failure Analysis and Prevention*, 8th ed., ASM International, Metals Park, Ohio, 1975, pp. 525–544.
2. D. W. Cameron and D. W. Hoeppner, "Fatigue Properties in Engineering," *ASM Handbook*, vol. 19: *Fatigue and Fracture*, ASM International, Metals Park, Ohio, 1996, pp. 15–26.
3. B. Phull, "Evaluating Corrosion Fatigue," *ASM Handbook*, vol. 13A: *Corrosion: Fundamentals, Testing, and Protection*, ASM International, Metals Park, Ohio, 2003, pp. 625–638.

CHAPTER 13

Stress Corrosion Cracking

General Description

The term *stress corrosion cracking* refers to metal failure resulting from a combination of static tensile stress and a specific corrodent to which the metal is sensitive. The tensile stresses may be either applied, such as those caused by internal pressure or bending, or residual, such as those that develop from welding or other manufacturing processes.

Alloys commonly used in boiler systems that tend to be susceptible to stress corrosion cracking are carbon steels and austenitic stainless steels. Stress corrosion cracking in carbon steel is most often associated with concentrated sodium hydroxide (historically called *caustic embrittlement*). In austenitic stainless steels, the most common corrodents associated with stress corrosion cracking in a boiler environment are concentrated sodium hydroxide and chlorides. Copper-based alloys, which may be exposed to steam in systems such as surface condensers, are most often subject to ammonia stress corrosion cracking (historically called *season cracking*). While numerous other corrodents have been associated with stress corrosion cracking in the above-mentioned alloys, this discussion will focus on those most likely to occur in the boiler water and steam circuit. Stress corrosion cracking may also occur on the fire side if the environment is conducive to attack. A few known causes of fireside stress corrosion cracking will also be discussed.

There are many proposed mechanisms for stress corrosion cracking, although no consensus has been reached. It is likely that a number of different mechanisms can cause stress corrosion cracking, depending on the metallurgy, environment, and stresses. A comprehensive review of proposed cracking mechanisms is beyond the scope of this book, but can be found in the literature.[1] One possible mechanism is based on preferential dissolution of material at the crack tip. This category includes dissolution along grain boundaries or other paths that are more susceptible to corrosion than the bulk metal.

The continuous formation and fracture of a thin, brittle corrosion product layer at the crack tip is also included in this category. Other mechanisms suggest mechanical fracture of the metal at the crack tip, a process that is thought to involve hydrogen. Hydrogen, which may be produced locally due to cathodic electrochemical reactions, may cause embrittlement at the crack tip and result in crack growth. On a microscopic level, hydrogen's role in stress corrosion cracking is mechanistically similar to the type of hydrogen damage discussed in Chap. 6, although with the addition of a significant contribution from tensile stress. The type of hydrogen damage described in Chap. 6 is not necessarily associated with significant tensile stress. Furthermore, such damage is caused by hydrogen evolved during corrosion reactions as a result of localized concentration of boiler water species to aggressive levels.

Locations

General

In principle, stress corrosion cracking can occur wherever a susceptible alloy, a specific corrodent for the alloy, and sufficient static tensile stresses coexist. Typical locations of elevated stresses include bends, welds, locations of change in metal thickness, and supports.

Water Side

Due to improved water treatment programs and improved boiler design, the occurrence of caustic stress corrosion cracking in carbon and low-alloy steel boiler components is much less frequent than it was in the past. However, caustic stress corrosion cracking continues to occasionally affect water tubes, superheater tubes, reheater tubes, steam lines, and turbines (see also Chap. 16, Steam and Condensate System Damage). Caustic stress corrosion cracking may also occur in stressed carbon steel components in the steam drum, such as bolts (Fig. 13.1). Stainless steel boiler components that have experienced caustic stress corrosion cracking include superheaters and reheaters, where the stainless grades are typically used for higher oxidation limits and better corrosion resistance.

Cases of chloride stress corrosion cracking of austenitic stainless steel have occurred in deaerators, superheaters, and, in rare cases, high-pressure utility boiler feedwater heaters. Although the tendency for stress corrosion cracking increases with increasing chloride levels, high chloride concentrations are not necessary to cause stress corrosion cracking. Stress corrosion cracking can occur even where low concentrations exist, provided the alloy is susceptible, liquid water is present, and the tensile stress level and temperature are sufficient (see "Critical Factors").

FIGURE 13.1 Steam drum bolt that failed by caustic stress corrosion cracking. Note the typical brittle character of the fracture. (*Courtesy of National Association of Corrosion Engineers.*)

Stress corrosion cracking of copper-based alloys in the boiler water circuit may occur on the steam side of surface condensers, particularly if a source of ammonia exists. Sulfur-containing compounds, such as sulfur dioxide, have also been associated with stress corrosion cracking of copper-based alloys.

Fire Side

Corrodents that deposit on fireside metal surfaces can result in stress corrosion cracking in the presence of moisture. Since the high temperatures on the fire side of a boiler do not allow for the existence of liquid water during service, fireside stress corrosion cracking generally occurs during idle periods, in locations where condensation can occur, or where liquid water from leaks or external ingress can contact fireside surfaces.

Cooler sections of boilers, such as economizers and low-pressure sections of heat recovery steam generators (HRSGs), are more prone to deposition of potential corrodents such as sulfates and nitrates, so fireside stress corrosion cracking is more likely to occur in such areas than in warmer sections.

Critical Factors

General

Two principal factors govern stress corrosion cracking in the boiler environment. First, the metal in the affected region must be stressed in tension to a sufficiently high level. The stresses may be applied and/or residual. Tensile stresses that cause stress corrosion cracking may approach the yield strength of the metal, although stresses well below the yield strength may also cause cracking if combined with a sufficiently high corrodent concentration. Surface flaws, such as pits, can act as stress concentrators and promote the initiation of cracks, although they are not required. In general, as tensile stress level in the metal increases, the threshold corrodent concentration required to initiate stress corrosion cracking decreases.

Second, a specific corrodent must contact the stressed metal component. Oxygen and, in general, moisture must also be present for stress corrosion cracking to occur. Other environmental factors that can affect the likelihood of stress corrosion cracking are numerous and include temperature, pH, and pressure. Aggressive species can reach elevated concentrations within pits, promoting stress corrosion cracking.

General or localized metal loss is not necessary for stress corrosion cracking to occur and, in fact, does not typically accompany it. High corrosion rates can favor the formation of pits rather than the sharp crack tips that are characteristic of stress corrosion cracking.

Water Side

In boiler systems, stress corrosion cracking of carbon and low-alloy steels is most often associated with sodium hydroxide (caustic). Instances of caustic stress corrosion cracking in carbon and low-alloy steels generally occur at metal temperatures within the range of 390 to 480°F (200 to 250°C)[2] and rarely occur at metal temperatures below 300°F (149°C). Concentrations of sodium hydroxide as low as 5% have been known to cause cracking, but concentrations in the range of 20 to 40% greatly increase susceptibility (Fig. 13.2). As metal temperatures increase, the caustic concentration necessary to produce stress corrosion cracks generally decreases. In boiler components that are normally in contact with liquid water during service, such as steam-generating tubes, a localized concentration mechanism is typically required to achieve a sufficiently high concentration of caustic. See Chap. 4, Caustic Corrosion, for a discussion of concentration mechanisms for boiler water species. Carryover of liquid boiler water in superheaters can produce the high caustic concentration necessary for stress corrosion cracking. It is unusual for cracking to occur while boiler water is actively being carried over due to the rapid boiling of the water to dryness. Cracking tends to occur following an idle period when caustic-containing deposits become hydrated by condensed steam. As the superheater reaches the threshold temperature for stress corrosion cracking, the cracking occurs. Stainless steels are also susceptible to caustic stress corrosion cracking. Such cracking generally begins to occur in the temperature range of 220 to 400°F (105 to 205°C).[2]

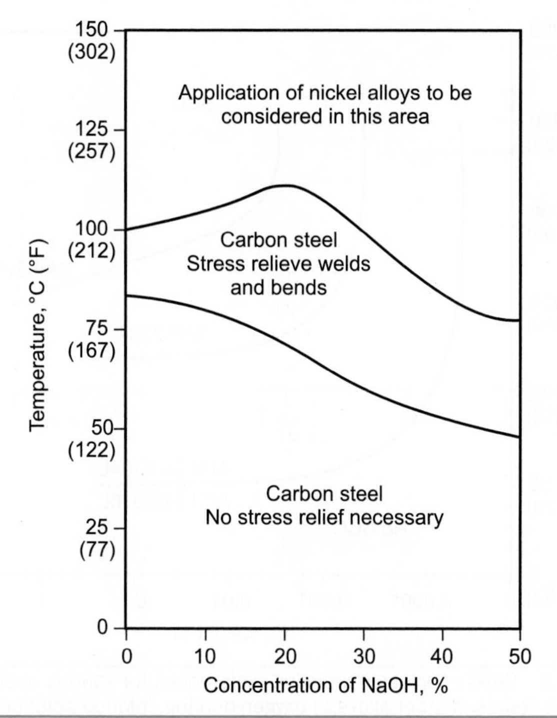

FIGURE 13.2 Temperature and concentration limits for susceptibility of carbon steel to caustic stress corrosion cracking.[3] (*Reprinted with permission of ASM International. All rights reserved. www.asminternational.org.*)

Chloride stress corrosion cracking in austenitic stainless steel does not necessarily require a high chloride concentration (Fig. 13.3). In fact, there is no known threshold concentration below which chloride stress corrosion cracking will not occur. However, there is an apparent minimum temperature; chloride stress corrosion cracking in austenitic stainless steel is rarely observed at temperatures below approximately 120 to 130°F (49 to 54°C).

Sensitization is a microstructural feature that can promote stress corrosion cracking in austenitic stainless steels. Sensitization occurs due to precipitation of chromium-rich carbides at grain boundaries when the metal temperature is within the approximate temperature range of 1020 to 1560°F (550 to 850°C). The metal adjacent to the grain boundaries has a decreased corrosion resistance due to localized chromium depletion, which can promote intergranular attack and cracking in some cases. Low-carbon austenitic stainless steels (e.g., 304L, 316L, etc.) are resistant to sensitization and may be used if appropriate. In some cases, such as in superheaters, in-service sensitization can be expected if a susceptible

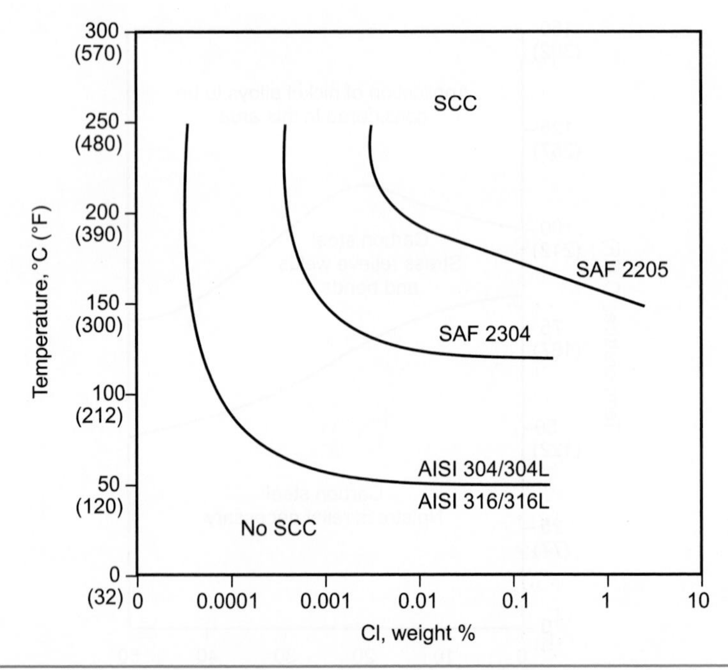

FIGURE 13.3 Stress corrosion cracking resistances for various austenitic and duplex stainless steel alloys in oxygen-bearing chloride solutions of neutral pH.[4] (*Courtesy of AB Sandvik Materials Technology.*)

alloy is used and service temperatures fall within the aforementioned temperature range.

Stress corrosion cracking of some copper-based alloys typically requires a source of ammonia. In boiler systems, ammonia may be present in boiler water due to the use of ammonia itself, breakdown of organic material, ingress of chloramines from some municipal water treatment programs, and chemical breakdown of hydrazine or hydrazine alternatives, some polymers, or amines.

Fire Side

As previously mentioned, numerous chemical species may cause fireside stress corrosion cracking. Corrodents can be incorporated into fireside deposits that form on metal surfaces during service or idle periods. Sources of the corrodents may include combustion products, or even the outside atmosphere (e.g., chlorides in air near the ocean). However, they cannot cause stress corrosion cracking unless moisture is present. Therefore, stress corrosion cracking can only occur if metal temperatures are below the dew point of water, i.e., during idle periods or in areas where the boiler water temperature is below the dew point. Condensation is the most likely source of liquid water on the fire side, though in some cases, nearby leaks or water washing can also provide the necessary liquid water.

Due to the wide range of fuels that may be used in boilers, numerous chemical species may act as corrodents to cause fireside stress corrosion cracking. A few specific examples are discussed below.

In some heat recovery steam generators (HRSGs), ammonia is used in selective catalytic reduction (SCR) systems, which reduce NO_x levels resulting from fossil fuel combustion. If the HRSG utilizes sulfur-bearing fuel (such as fuel oil), even on a part-time basis, unreacted ammonia can react with SO_3 and water vapor in the flue gas to produce ammonium sulfates. Such sulfates can deposit on metal surfaces at temperatures below approximately 300 to 375°F (150 to 190°C). The ammonium sulfates can cause metal loss and stress corrosion cracking in carbon steel when metal temperatures fall below the dew point of water.[5] Nitrates, ammoniacal solutions, and other species have also been implicated in stress corrosion cracking of carbon and low-alloy steels.[6,7] In many cases, lower-pH conditions promote cracking.

Corrodents that can cause fireside stress corrosion cracking in austenitic stainless steel include chlorides, as previously discussed, and sulfur-containing compounds. In units that utilize sulfur-containing oil or gas as fuel, polythionic acids ($H_2S_xO_6$, where x = 3, 4, or 5) can form and deposit on metal surfaces as a result of reactions among sulfides, moisture, and oxygen. When mixed with liquid water, the acids and their decomposition products can cause stress corrosion cracking, particularly in sensitized stainless steels.[8]

Identification

Failures caused by stress corrosion cracking always produce thick-walled fracture faces regardless of the normal degree of metal ductility. Branching is frequently associated with these cracks. Since cracks propagate perpendicularly to the direction of the tensile stresses, the general orientation of the cracks can provide information about the stresses involved. Unless a failure has occurred, stress corrosion cracks may be difficult to see with the unaided eye, since the cracks tend to be very fine and tight. Nondestructive inspection methods, such as dye penetrant, eddy current, magnetic particle, and ultrasonic testing, can aid in identification of flaws such as cracks. However, metallurgical analysis (including deposit characterization and microscopic examination) is required to positively diagnose stress corrosion cracking.

Water Side

In carbon and low-alloy steels, as well as stainless steels, caustic stress corrosion cracks generally penetrate intergranularly (Fig. 13.4). Occasionally, evidence of concentrated sodium hydroxide, often in the form of whitish, highly alkaline deposits, may be observed at the crack site. Chemical analysis of material on or near the crack surfaces, such as by scanning electron microscopy–energy-dispersive x-ray spectroscopy (SEM-EDS), may indicate the presence of sodium. However, due to the high solubility of sodium, such deposits are not always found. Corrosion products containing crystalline magnetite may also be present in some cases where caustic corrosion also occurred.

Chloride stress corrosion cracking in austenitic stainless steels is usually characterized by branched, transgranular cracks (Figs. 13.5 and 13.6). Chemical analysis of material on or near the crack faces by SEM-EDS may indicate the presence of chlorine. However, since a high-chloride concentration is not necessary to cause stress corrosion cracking and chlorides are water-soluble, chlorine is often not detected.

Stress corrosion cracking in copper alloys, such as that caused by ammonia-containing environments, is generally characterized by intergranular penetration. In some environments, however, cracks may propagate transgranularly.[9]

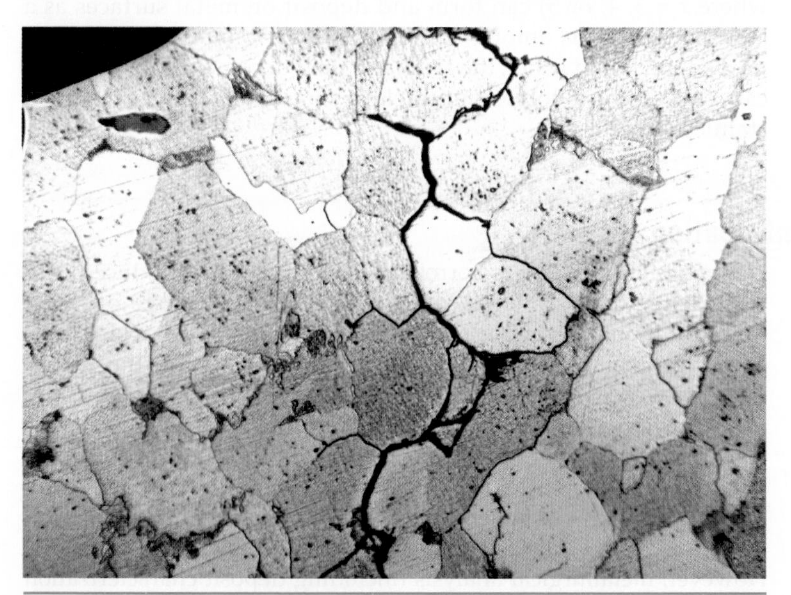

FIGURE 13.4 Continuous intergranular cracks running through a carbon steel tube wall, characteristic of caustic stress corrosion cracking.

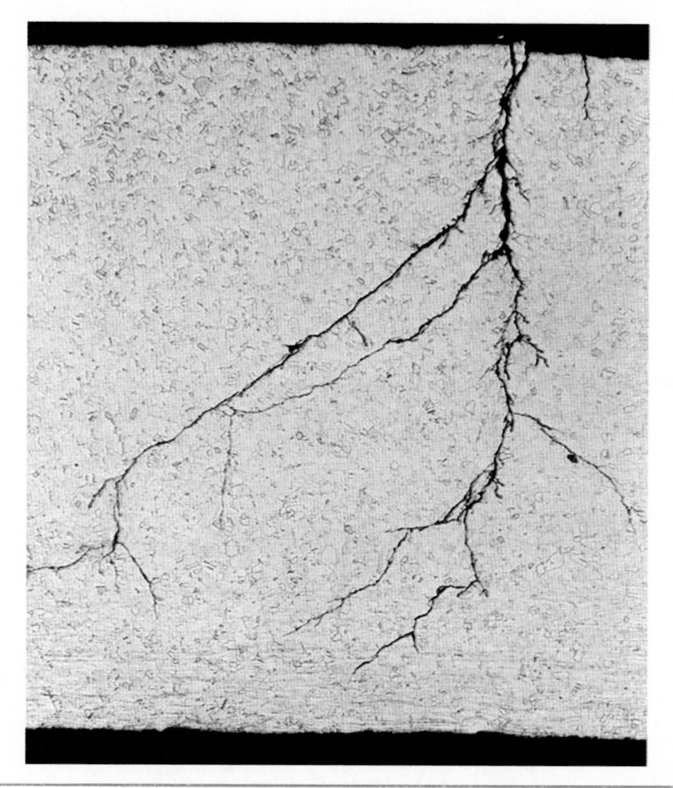

FIGURE 13.5 Tight, branched, transgranular cracks characteristic of chloride stress corrosion cracking in austenitic stainless steel. Etchant: electrolytic oxalic acid.

Fire Side

Cracking may penetrate intergranularly or transgranularly, depending upon the corrodent and situation. Due to the variety of species that may cause fireside stress corrosion cracking, chemical analysis is generally necessary to identify the most likely corrodent responsible for the cracking. Bulk analysis of material on the surface near the crack should include x-ray diffraction (XRD), x-ray fluorescence (XRF) spectroscopy, and measurements of carbon, hydrogen, nitrogen, and sulfur when appropriate. If the material quantity is insufficient for bulk analysis, characterization of material on the component surface and the crack face may be achieved through SEM-EDS. Ion chromatography (IC) of species leached from material on the surface (by mixing the material with water) can also be beneficial for determination of likely corrodents.

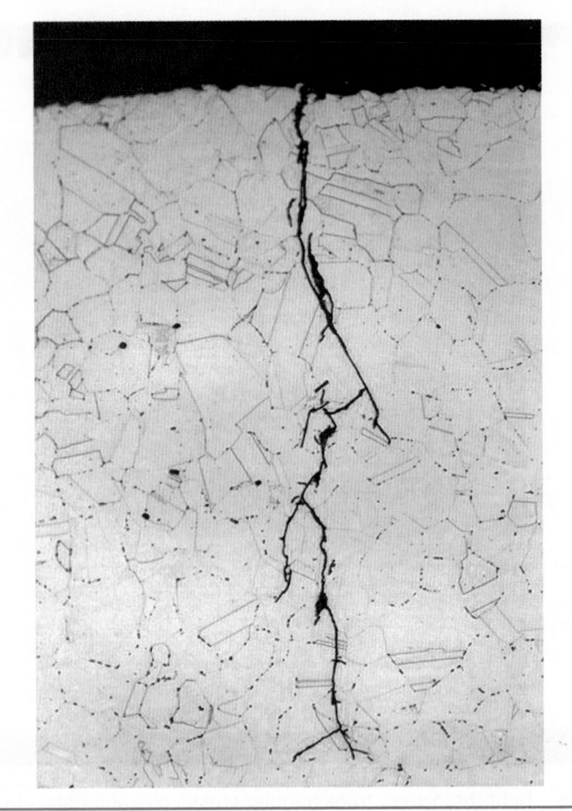

FIGURE 13.6 Chloride stress corrosion cracking in austenitic stainless steel, highlighting transgranular penetration. Etchant: electrolytic oxalic acid.

Elimination

Limiting the likelihood of stress corrosion cracking requires sufficient reduction or elimination of tensile stresses and/or corrodent concentration.

Stresses

Crack orientations and locations provide information on the sources of tensile stress responsible for cracking. Tensile stresses can be either applied or residual. Applied stresses are service-generated stresses, including hoop stresses caused by internal pressure and bending stresses from constrained thermal expansion and contraction. Generally, hoop stresses are subject to minimal control, since the essential function of the boiler tubes and other components is to contain pressurized fluids. Bending stresses, however, may be reduced or eliminated by altering operational parameters or by redesign of the affected components.

The term *residual stress* refers to stresses that are inherent in the metal itself. They are the result of manufacturing or construction processes such as welding or tube bending. In tubes, residual hoop stresses may also remain from the manufacturing process. Residual stresses can be reduced by proper stress relief annealing of the component or by postweld heat treatments.

Waterside Corrodents

Avoiding concentrated corrodents is generally the most successful means of reducing or eliminating caustic stress corrosion cracking. Preventing departure from nucleate boiling (DNB), keeping internal surfaces sufficiently free of deposits, and avoiding formation of steam/water interfaces in components receiving high heat flux are ways to reduce the likelihood of developing concentrated corrodents. Other potential remedies include preventing in-leakage of alkaline-producing salts through condensers, heat exchangers, process streams, and caustically regenerated demineralizer systems; preventing contamination of desuperheating or attemperator water; and preventing boiler water carryover.

The use of inhibitors, such as sodium nitrate or a combination of sodium nitrate and one of many selected organics, has been successful in reducing caustic stress corrosion cracking. Phosphate-based pH control water treatment programs, which are designed to prevent the formation of free sodium hydroxide (caustic), may also be beneficial in limiting the likelihood of caustic stress corrosion cracking as well as caustic corrosion.

In the case of chloride stress corrosion cracking of austenitic stainless steel, preventing contamination of the water or steam with chloride may be necessary to avoid cracking. If chlorides cannot be avoided, other alloys should be considered.

In copper alloys, if the specific corrodent cannot be eliminated through changes in pretreatment or chemical treatment, alternative metallurgies should be considered. Cupronickels are generally resistant to ammonia stress corrosion cracking. In some applications, stainless steels may be suitable alternatives, although it should be ensured that chloride stress corrosion cracking will not pose a problem before making a substitution in higher-temperature applications, such as high-pressure utility boiler feedwater heaters.

Fireside Corrodents

If cracking tends to occur in cooler parts of the boiler, changing the system design or operating parameters to eliminate cold spots may prevent area-specific deposition of potential corrodents. In the case of ammonium sulfate stress corrosion cracking, deposition of the sulfates may be prevented by maintaining gas temperatures above approximately 300 to 375°F (149 to 191°C), using fuel with a lower sulfur content, and/or altering the type or location of the catalyst used in the SCR process.[5]

Sources of liquid water on the fire side, such as leaks, condensation, or ingress of water from the outside environment, should be identified and eliminated. If liquid water on the fire side (such as condensation) is not controllable and deposition of corrosive species cannot be eliminated, washing of the fire side with appropriate solutions may help to control stress corrosion cracking. For example, low-pH conditions often promote stress corrosion cracking and corrosion. If the material on the fire side produces an acidic solution, washing with alkaline neutralizing solutions may substantially decrease the tendency toward cracking and corrosion. Corrosive material should be neutralized as soon as possible after shutdown.

Cautions

Stress corrosion cracks may be difficult to identify through visual inspection. The use of dye penetrants, magnetic particle inspection, and ultrasonic testing in suspect regions may disclose the presence of stress corrosion cracks. Dwell time for dye penetrants may be increased to accommodate the typical tightness of these cracks.

Stress corrosion cracking may be confused with other cracking modes that produce thick-walled fractures, such as hydrogen damage, corrosion fatigue, creep rupture, and some forms of severe overheating. Confirmation of a diagnosis of stress corrosion cracking requires metallographic examination.

Related Problems

See also Chap. 2, Short-Term Overheating; the section entitled "Creep Rupture" in Chap. 3, Long-Term Overheating; Chap. 4, Caustic Corrosion; Chap. 6, Hydrogen Damage; Chap. 12, Corrosion Fatigue Cracking; and Chap. 16, Steam and Condensate System Damage.

Case History 13.1

Industry:	Institutional
Specimen location:	Spray-type deaerator, water box
Years in service:	3
Specimen specifications:	Numerous pieces of austenitic stainless steel plate, 1/8-in. (3.2-mm) thickness

The various pieces of metal sheet exhibited brittle fracture edges as well as cracks and fissures of various orientations (Fig. 13.7). The surfaces were generally covered with a thin layer of reddish brown material. Spot tests on the surfaces of the pieces indicated significant amounts of chlorides and some sulfates. Chemical analysis of material on the surface by SEM-EDS indicated significant concentrations of sodium and chlorine. Metallographic cross sections revealed many tight,

FIGURE 13.7 Austenitic stainless steel metal sheet from a deaerator, showing cracks and fissures of various orientations.

branched, transgranular cracks and fissures of different orientations. Tunnel pits were observed in places along the crack paths (Fig. 13.8).

Reportedly, brine from the softener was introduced into the deaerator, providing a source of chlorides. The areas of the water box that were affected were locations exposed to the water stream from the spray nozzles. The cracks present

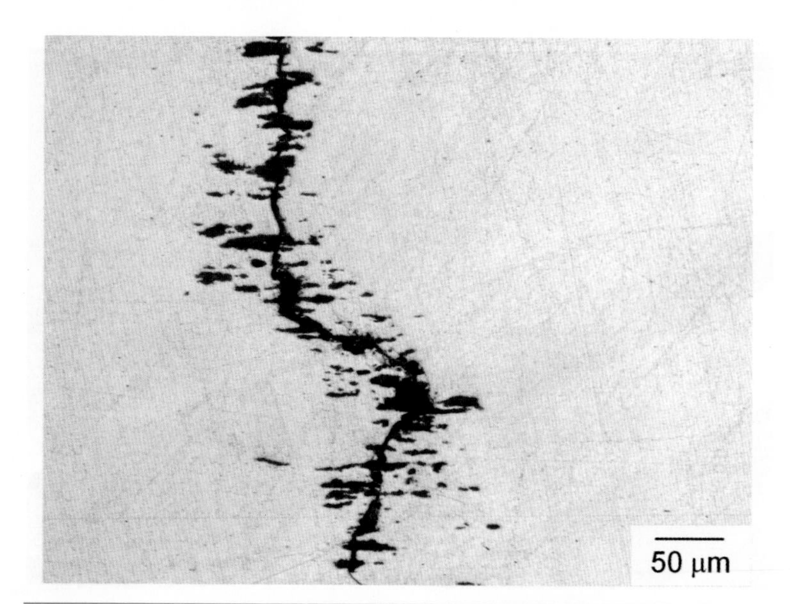

FIGURE 13.8 Tunnel pits along the path of a chloride stress corrosion crack. Unetched.

on the sheet had multiple orientations. Residual stresses associated with sheet manufacture, fabrication, and installation contributed significantly to the stresses responsible for failure.

Case History 13.2

Industry:	Pulp and paper
Specimen location:	Steam supply line to a soot blower, recovery boiler
Specimen orientation:	Vertical
Years in service:	15
Water treatment program:	Oxygen scavenger
Drum pressure:	400 psi (2.8 MPa)
Tube specifications:	2-3/8-in. (6.0-cm) outer diameter, 304 stainless steel

Figure 13.9 shows one of several stainless steel soot blower lines that cracked. Failures of this type can be quite dangerous, since steam can be released into occupied areas.

The line contained several thick-walled, longitudinally oriented cracks that ran as long as 25 in. (63.5 cm) down the tube. Note that the cracks were not tight; rather, they spread apart, an unusual feature for stress corrosion cracking.

Close visual and microstructural examinations revealed that the branched cracks originated on the external surface. Microstructural examinations also revealed severe cold working of the metal.

An uncontaminated segment of the crack face was examined by SEM-EDS. Elemental analysis of material covering the crack face revealed the presence of chlorine.

The line failed by stress corrosion cracking resulting from exposure of the external surface to chlorides from the furnace environment. The longitudinal

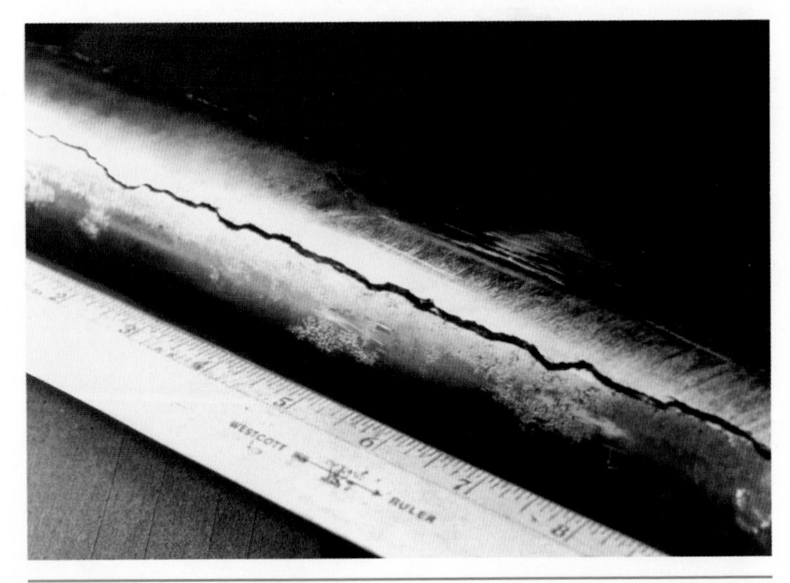

FIGURE 13.9 Extensive longitudinal crack in a stainless steel soot blower line. Note that the crack edges have spread apart.

direction of the cracking revealed that circumferential (hoop) stresses were responsible. It was apparent from the spreading apart of the cracks and the evidence of cold-worked metal in the microstructure that the tube contained high residual hoop stresses resulting from tube-forming processes. The level of residual stresses was calculated to be 29,000 psi (200 MPa). The residual stresses and service stresses generated by internal pressure combined with the chlorides to produce the cracks during idle time.

Case History 13.3

Industry:	Institutional
Specimen location:	Superheater inlet header, heat recovery steam generator (HRSG)
Years in service:	<1
Water treatment program:	Polymer
Drum pressure:	850 psi (5.9 MPa)
Tube specifications:	2-in. (5.1-cm) outer diameter, T22 and T91 low-alloy steels joined by a weld
Fuel:	Natural gas

A thick-walled, circumferential fracture occurred along a circumferential weld joining sections of T22 and T91 low-alloy steel tubing (Fig. 13.10). The fracture edges were generally covered with a thin oxide layer that was overlaid in places by powdery white deposits that produced an alkaline solution when mixed with distilled water. Similar deposits were also observed on the internal surface of the section. SEM-EDS analysis of material on the fracture edge indicated the presence of sodium, silicon, and calcium. Microscopic examinations revealed that tight, highly branched, intergranular cracks originated from the internal surface

FIGURE 13.10 Thick-walled, circumferential fracture along a circumferential weld in a superheater inlet header tube.

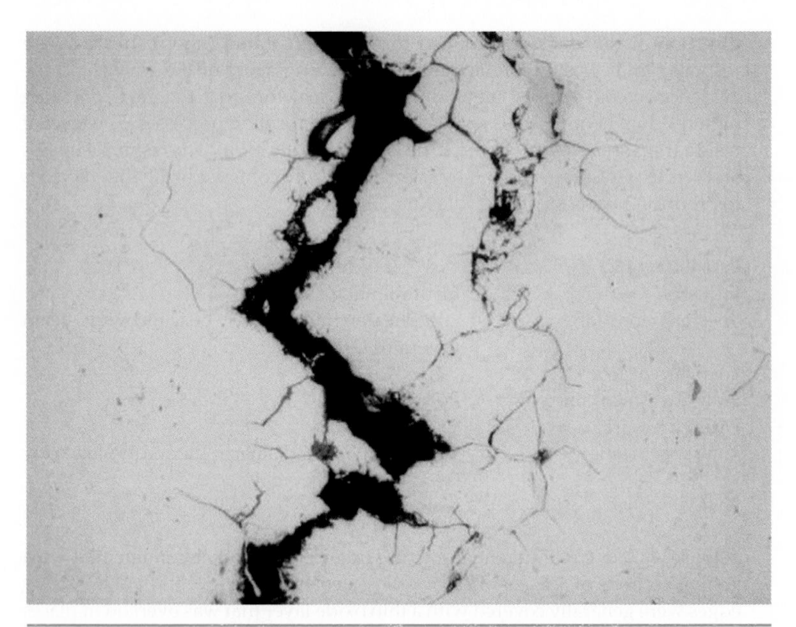

FIGURE 13.11 Tight, highly branched, intergranular fissures emanating from a circumferential through-crack near a circumferential weld. Unetched.

within the fusion zone of the circumferential weld (Fig. 13.11). Metallic copper particles and magnetite needles were embedded within the stratified deposit and corrosion product layers on the internal surface. Microhardness measurements indicated a large increase in hardness within the weld and surrounding heat-affected zone (HAZ), relative to locations away from the weld.

The cracks were caused by caustic stress corrosion cracking due to boiler water carryover. Residual stresses from welding most likely contributed significantly to the formation of stress corrosion cracks. Postweld stress relief annealing was recommended to decrease the likelihood of stress corrosion cracking. Applied bending stresses due to thermal expansion and contraction may have also contributed, particularly since the section was located near a point of rigid constraint. It was emphasized that boiler water carryover needed to be eliminated in order to prevent caustic stress corrosion cracking.

Case History 13.4

Industry:	Petrochemical
Specimen location:	Superheater, first stage
Specimen orientation:	Vertical
Time in service:	3 weeks
Water treatment program:	Phosphate
Drum pressure:	600 psi (4.1 MPa)
Tube specifications:	1½-in. (3.8-cm) outer diameter, 304 stainless steel
Fuel:	Waste gas

The tube illustrated in Fig. 13.12 is one of numerous tubes that failed in this boiler. The tubes had been moderately cold-bent during installation, and were

FIGURE 13.12 Transverse crack resulting from caustic stress corrosion in a bent stainless steel superheater tube. Note small "window" that has been blown out of the wall.

not stress relief annealed. The steam drum lacked adequate devices for separation of steam and water, and load swings were frequent, probably causing carryover of boiler water.

Microstructural analysis revealed plastically deformed grains from the cold bending. Highly branched, intergranular cracks originated from the internal surface.

The primary source of stress in this case was residual stresses from the bending operation. This is apparent from the circumferential orientation of the cracks, the fact that the cracks significantly spread apart, and the proximity of the cracks to the bent zones of the tubes. The corrodent was sodium hydroxide from boiler water carryover.

This case is an excellent example of the value of understanding failure mechanisms and performance of materials. Stainless steel superheater tubes in a boiler of this pressure are unusual. The original tubes were carbon steel that cracked after 9 months of service. Two steps were taken to improve service life. First, stainless steel tubes were specified to replace the carbon steel. Second, moderate bends were put in the tubes, apparently to relieve the thermal expansion and contraction stresses that had caused cracking in the carbon steel tubes. Unfortunately, despite the greater general corrosion resistance of stainless steels, they are still susceptible to caustic stress corrosion cracking. In addition, the bends placed in the tubes to relieve thermally induced bending stresses provided high residual stresses instead. The stainless steel tubes failed after 3 weeks in service.

Case History 13.5

Industry:	Institutional
Specimen type:	U-bend tubing
Specimen orientation:	Horizontal
Years in service:	3
Water treatment program:	None
Tube specifications:	¾-in. (1.9-cm) outer diameter, 316 stainless steel
Fuel:	Unfired

Plant steam at 130 psig (0.9 MPa) was supplied to the steam generator to produce 6 psi (41.3 kPa) clean steam for building humidification. The generator contained a U-tube bundle, with plant steam on the tube side and untreated deionized water on the shell side. Inspection revealed severe damage of the stainless steel U-bend tubing, floats, and humidifier wands.

The tube sections were riddled with cracks and deep pits that penetrated the tube walls in multiple locations (Fig. 13.13). Metallographic analysis determined the attack was caused by chloride-induced pitting (Fig. 13.14) and chloride stress corrosion cracking, as evidenced by the highly branched, transgranular cracks (Fig. 13.15). Spot tests indicated the presence of chlorides in some cracks and pits. Cross sections revealed that cracking initiated on the outer diameter of the tubing, which was exposed to deionized water.

The crack orientations highlighted multiple contributions to the stresses on the tubing. Significant residual tensile stresses from bending operations contributed to transverse cracking along the bend regions of the tubing. Residual tensile stresses also contributed to longitudinal cracking near the longitudinal weld seam of the tubing. Transverse cracks localized to one side of the tube, at locations adjacent to tube baffles or supports, indicated that applied bending

FIGURE 13.13 Severe pitting and cracking at a U bend on the water side of an austenitic stainless steel steam generator tube. (*© NACE International, 2007.*)

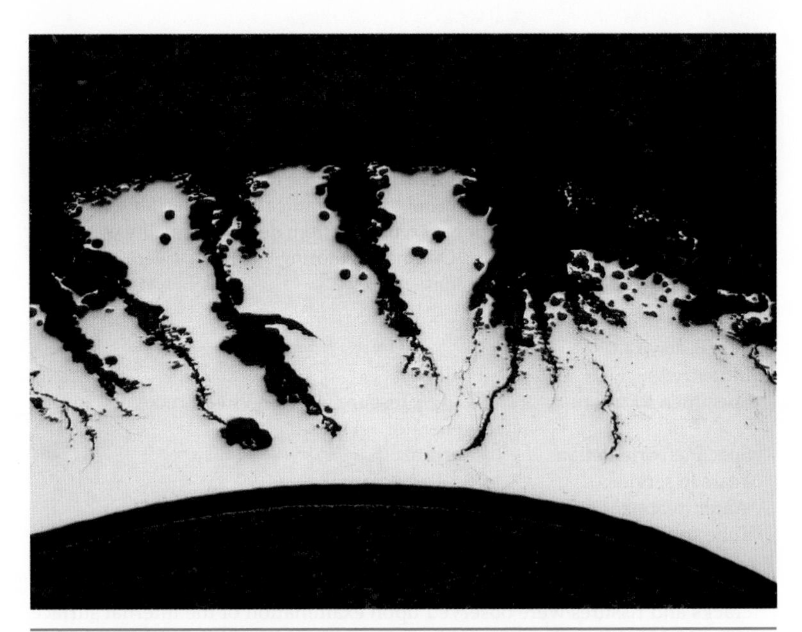

FIGURE 13.14 U bend of a stainless steel tube, showing undercut pits and stress corrosion cracks originating from the external surface, caused by chlorides. Unetched. (© *NACE International, 2007.*)

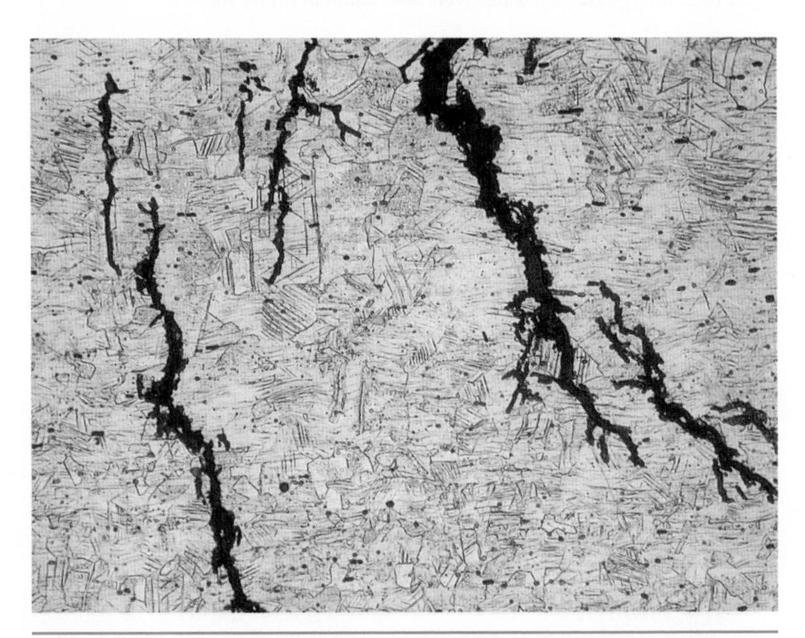

FIGURE 13.15 Branched, transgranular chloride stress corrosion cracks. The many mechanical deformation twins within the grains (parallel lines) are indicative of plastic deformation caused by bending operations. Etchant: electrolytic oxalic acid. (© *NACE International, 2007.*)

stresses were responsible for some of the cracking. Thus, stress relief annealing likely would have reduced the likelihood of cracking, but may not have completely prevented it.

The leaks in the steam generator tubing were caused by exposure to water that unexpectedly contained chlorides. Reportedly, multiple excursions from deionized quality water occurred shortly before the failure was noted. Chemical analysis of deposits on the external surface indicated appreciable levels of magnesium and silicon, confirming excursions from deionized quality water that probably resulted in deposition. On-line monitoring techniques were suggested to alert for water quality excursions.

Case History 13.6

Industry:	Utility
Specimen location:	Low-pressure section, heat recovery steam generator
Specimen orientation:	Vertical
Years in service:	3
Water treatment program:	Phosphate residual
Tube specifications:	2¾-in. (7.0-cm) outer diameter, low-alloy steel

Many of the external surface fins were severely degraded. Short, transverse cracks and fissures were observed upon examination of the internal surface (Fig. 13.16). Microscopic examinations revealed cracks and fissures that were fairly straight, but that exhibited some branching and intergranular penetration (Fig. 13.17). The cracks and fissures originated from the external surface. Intergranular fissures were also found at fin welds.

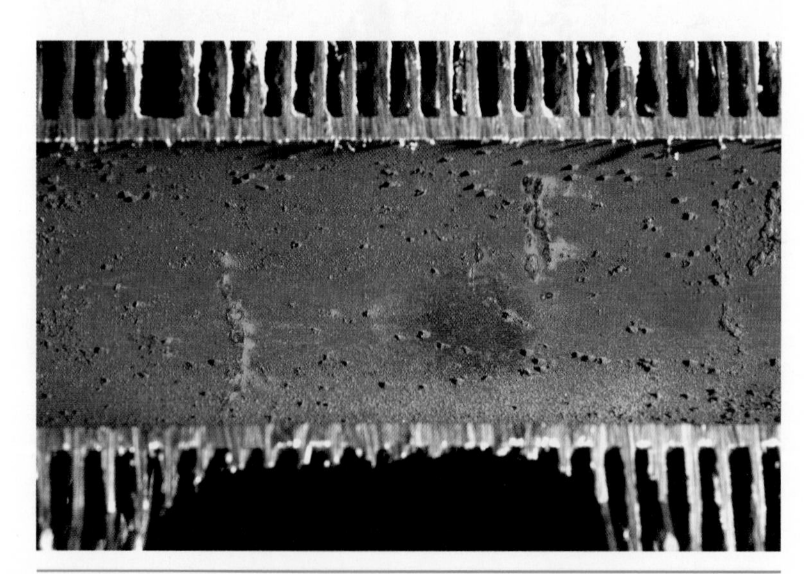

FIGURE 13.16 Internal surface of an HRSG tube, showing short, transverse through-cracks that originated from the external surface. The cracks were caused by both stress corrosion cracking and corrosion fatigue cracking.

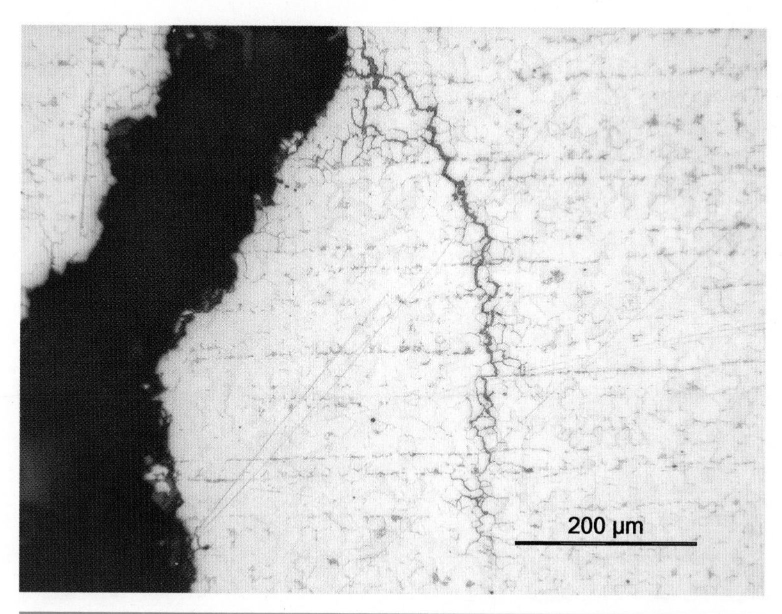

FIGURE 13.17 Branched, intergranular stress corrosion cracks most likely caused by sulfate deposits on the external surface of a low-alloy steel HRSG tube. Etchant: Picral.

The external surfaces of the received sections were covered with layers of deposit material that produced an acidic aqueous solution. Spot tests indicated the presence of sulfates in the material. Chemical analysis of the material indicated that it contained 46 wt % volatile sulfur. X-ray diffraction analysis indicated the presence of iron sulfate hydrate, ammonium iron sulfate, and possibly iron sulfide.

Although corrosion fatigue cracking contributed to the failures, the presence of branching and intergranular penetration indicated that stress corrosion cracking occurred due to the presence of ammonium sulfates on the external surface. The transverse orientation of the cracks suggested that bending, thermal expansion and contraction, and/or possible vibrations contributed to the stresses on the tube. Residual tensile stresses from fin welding likely contributed to the formation of stress corrosion cracks.

Case History 13.7

Industry:	Chemical process
Specimen location:	Inlet end of economizer tube, waste heat boiler
Specimen orientation:	Horizontal to vertical
Years in service:	1 (system in service 30 years)
Boiler pressure:	650 psi (4.5 MPa)
Tube specifications:	1½-in. (3.8-cm), plain carbon steel, L-bend section

Cracking problems reportedly occurred along the extrados of bent economizer inlet tube ends on a recurring basis, approximately every 9 to 14 months. The tubes connected the inlet water distribution header to the economizer in the

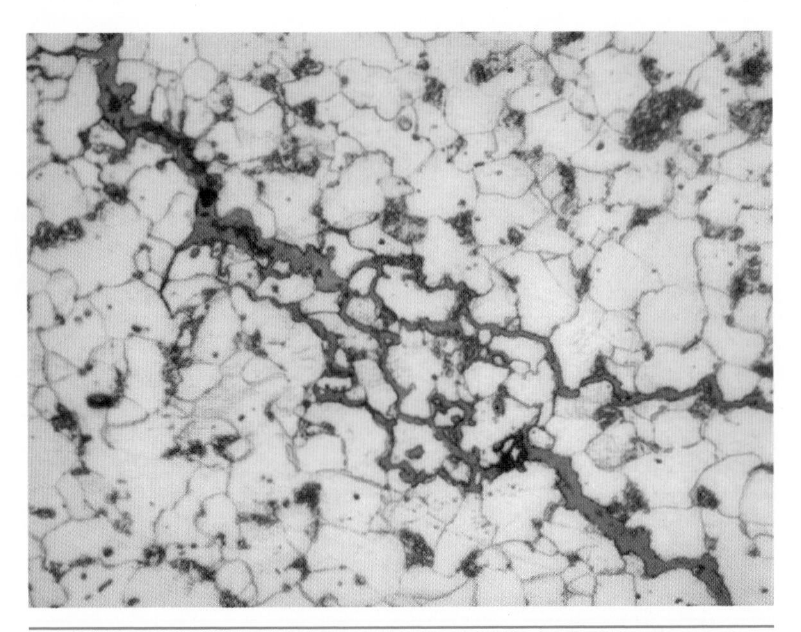

FIGURE 13.18 Intergranular cracks along the extrados of an economizer tube section. The cracks were most likely caused by ammonium nitrate deposits on the external surface. Etchant: Nital. (*© NACE International, 2007.*)

waste heat boiler. On the external surface, waste gases from the catalytic combustion of ammonia were used to heat demineralized water on the tube side. The waste gases included a high concentration of nitrogen gas, as well as nitrogen oxides, water, and oxygen. The cracked tube bends were shielded from the main hot gas flow by a bottom division plate. Periodic washing of the combustion side of the boiler with hot water was done every 8 to 9 months, sometimes using either acetic acid or trisodium phosphate.

Metallographic analysis revealed longitudinally oriented, intergranular cracks along the extrados of the tube bend (Fig. 13.18). The cracks originated from the external surface. Hardness measurements indicated elevated hardness along the extrados and intrados of the bend, highlighting residual stresses that developed due to deformation during cold bending of the tube.

The external surface of the tube was cleaned prior to analysis, preventing the direct determination of the corrodent responsible for stress corrosion cracking. However, based on the constituents in the flue gas, nitrate compounds were suspected. Although other nitrate compounds were considered, ammonium nitrate was the most likely deposit to form. The relatively low decomposition temperature of ammonium nitrate dictated that the deposits only developed in low-temperature areas of the boiler, explaining why only tubes in a shielded location were affected. Pits were observed on the external surface, suggesting the formation of acidic solutions that most likely promoted nitrate stress corrosion cracking.

Cracking only occurred along the bend extrados because this area possessed the highest tensile stress state. Such forming operations imparted considerable residual stresses at the bend, with the highest tensile stresses occurring at the bend extrados.

Recommended actions included stress relief annealing of tube bends or hot bending to avoid excessive residual tensile stresses. Water washing with alkaline solutions, rather than acetic acid, was also recommended. Low-alloy steels were recommended due to their increased resistance to stress corrosion cracking in hot ammonium nitrate solutions. Changing the system design to prevent the accumulation of deposits in colder areas was also suggested, e.g., by exposing occluded areas to the hot combustion gases.

Case History 13.8

Industry:	Utility
Specimen location:	Surface condenser, steam inlet side
Specimen orientation:	Horizontal
Years in service:	18
Water treatment program:	Neutralizing amine, phosphate residual, proprietary nonvolatile oxygen scavenger
Tube specifications:	7/8-in. (2.2-cm) outer diameter, admiralty brass

The external surface of the condenser tubing was exposed to saturated steam. Eddy current inspection indicated external surface flaws in the tubing. Microscopic examinations indicated that a few tight, transverse, branched fissures originated from the external surface in one location. One of the fissures penetrated through at least 30% of the tube-wall thickness (Fig. 13.19).

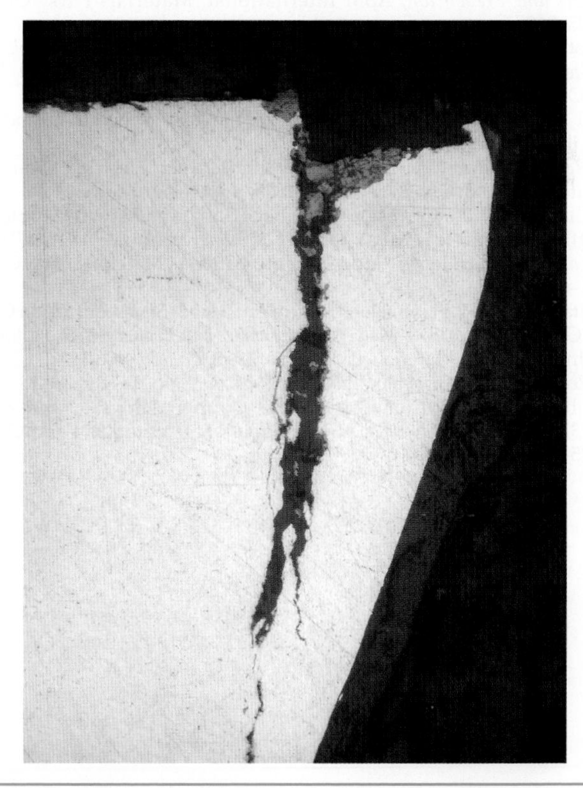

FIGURE 13.19 Slightly branched, primarily transgranular fissure originating from a depression on the external surface of an admiralty brass surface condenser tube. Unetched.

Some of the fissures originated from shallow, corrosion product–filled depressions on the surface. The fissures penetrated intergranularly in places, but were primarily transgranular in nature (mixed mode). No fissures originated from the internal surface.

The cracking was most likely caused by ammonia stress corrosion cracking from the steam side. Ammonia and oxygen dissolved in moisture from condensing steam were necessary for the cracking to occur. The transversely oriented cracks were only found on one side of the tube, suggesting that bending stresses provided the tensile stresses that caused cracking. External surface depressions likely served as stress concentration sites, promoting crack initiation.

Cupronickel was suggested as a material more resistant to ammonia stress corrosion cracking than admiralty brass. It was also suggested that boiler treatment chemicals be reviewed, as the decomposition of some chemicals can contribute to increased ammonia levels in the steam.

References

1. R. H. Jones and R. E. Ricker, "Mechanisms of Stress-Corrosion Cracking," *Stress-Corrosion Cracking: Materials Performance and Evaluation*, R. H. Jones (Ed.), ASM International, Materials Park, Ohio, 1992, pp. 1–40.
2. W. R. Warke, "Stress-Corrosion Cracking," *ASM Handbook*, vol. 11: *Failure Analysis and Prevention*, ASM International, Materials Park, Ohio, 2002, pp. 823–860.
3. P. Elliott, "Materials Selection for Corrosion Control," *ASM Handbook*, vol. 13A: *Corrosion: Fundamentals, Testing, and Protection*, ASM International, Materials Park, Ohio, 2003, pp. 910–928.
4. *Duplex Stainless Steels—Fighting Corrosion Worldwide* (brochure), Sandvik Steel, Sandviken Sweden.
5. *Heat Recovery Steam Generator Tube Failure Manual*, Electric Power Research Institute, Palo Alto, CA, 2002, TR-1004503. pp. 14–1 to 14–9.
6. S. W. Ciaraldi, "Stress-Corrosion Cracking of Carbon and Low-Alloy Steels (Yield Strengths Less than 1241 MPa)," *Stress Corrosion Cracking—Material Performance and Evaluation*, R. H. Jones (Ed.), ASM International, Materials Park, Ohio, 1992, pp. 41–61.
7. S. W. Stafford and W. H. Mueller, "Failure Analysis of Stress-Corrosion Cracking," *Stress Corrosion Cracking—Material Performance and Evaluation*, R. H. Jones (Ed.), ASM International, Materials Park, Ohio, 1992, pp. 417–436.
8. *Heat Recovery Steam Generator Tube Failure Manual*, pp. 16–1 to 16–18.
9. A. Cohen, "Corrosion of Copper and Copper Alloys," *ASM Handbook*, vol. 13B: *Corrosion: Materials*, ASM International, Materials Park, Ohio, 2005, pp. 125–163.

SECTION 1.5

Pre- and Postboiler Damage

CHAPTER 14

Graphitic Corrosion

General Description

Graphitic corrosion is possible only in structures composed of cast irons containing graphite particles. Gray, nodular, and malleable cast irons, which contain graphite particles of different morphologies, are susceptible to graphitic corrosion. Although frequently considered immune to graphitic corrosion, nodular cast iron and malleable iron (which contain rounded forms of graphite) are often attacked. Gray cast iron, which contains flakes of graphite, is more widely used and has more dramatic and recognizable corrosion characteristics than other cast irons.

Although graphitic corrosion is often considered to be a form of dealloying or selective leaching, it actually has more in common with galvanic corrosion. Microscopic galvanic cells form between the metal matrix and graphite particles embedded in the casting when the same mildly aggressive water contacts both materials (Fig. 14.1). The graphite is cathodic to the adjacent metal, causing the anodic metal portion of the casting to corrode. Note the large distance between graphite and iron in the galvanic series (Table 14.1). Although the series is for a seawater environment, iron is also anodic relative to graphite in a freshwater environment. Graphitic corrosion converts the casting to mechanically weak corrosion products containing graphite particles. Graphitic corrosion usually progresses slowly, taking many months or even years to produce significant attack. Corrosion is accelerated if waters are mildly acidic, soft, of high conductivity, and/or contain high concentrations of aggressive anions such as sulfate and chloride.

Locations

Feedwater pump impellers, water supply lines, valves, and other components made of graphitic cast irons may experience graphitic corrosion. Because cast irons are used mainly in preboiler regions, attack occurs primarily in water pretreatment and transport equipment. However, graphitic cast iron components in other parts of the boiler water and steam circuit, such as steam trap bodies, may also experience graphitic corrosion.

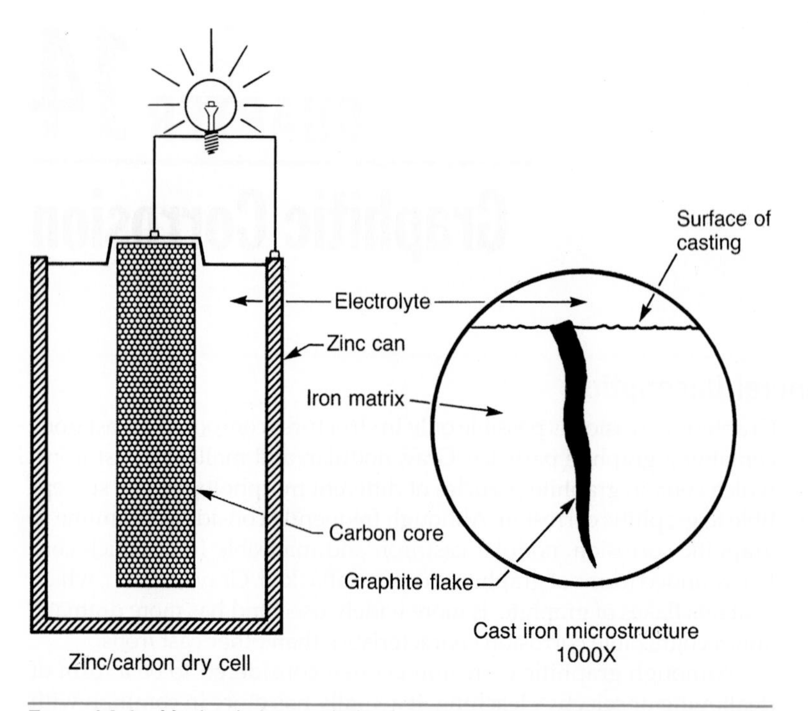

FIGURE 14.1 Mechanistic correlation between a dry cell battery and graphitic corrosion. (*Courtesy of McGraw-Hill, The Nalco Guide to Cooling Water Systems Failure Analysis, New York, 1993.*)

Critical Factors

Graphitic corrosion requires a susceptible graphite-containing cast iron microstructure and a common electrolyte that contacts both the graphite and iron phases.

Environmental factors that promote graphitic corrosion include soft waters, stagnant conditions, slightly acidic pH, high conductivity, and the presence of chlorides, sulfates, or low concentrations of hydrogen sulfide. Graphitic corrosion tends to occur in relatively mild environments. In more aggressive environments, corrosion may be more general, rather than graphitic in nature.

Attack often occurs predominantly during idle periods under stagnant conditions. If turbulence is pronounced (e.g., in pumps), the mechanically weak corrosion products may be dislodged, accelerating wastage.

The flake morphology of the graphite in gray cast iron provides a larger cathode-to-anode surface area ratio compared to nodular graphite. Consequently, gray cast iron is more susceptible to attack than nodular and malleable cast irons.

Cathodic (inert)	Platinum
	Gold
	Graphite
	Titanium
	Silver
	316 Stainless steel (passive)
	304 Stainless steel (passive)
	Inconel (passive)
	Nickel (passive)
	Monel
	Copper-nickel alloys
	Bronzes
	Copper
	Brasses
	Inconel (active)
	Nickel (active)
	Tin
	Lead
	316 Stainless (active)
	304 Stainless (active)
	Cast iron
	Iron and steel
	Aluminum alloys
	Cadmium
	Zinc
Anodic (active)	Magnesium and magnesium alloys

TABLE 14.1 A Galvanic Series, Showing Relative Reactivities of Assorted Pure Metals and Commercial Alloys in a Seawater Environment

Identification

Cast iron is converted to a soft mixture of iron oxide corrosion products and graphite (Figs. 14.2 and 14.3). Pieces of corrosion product tend to smudge hands and can be used to mark paper, just as if the corroded material were lead in a pencil (Fig. 14.4). Attack is often

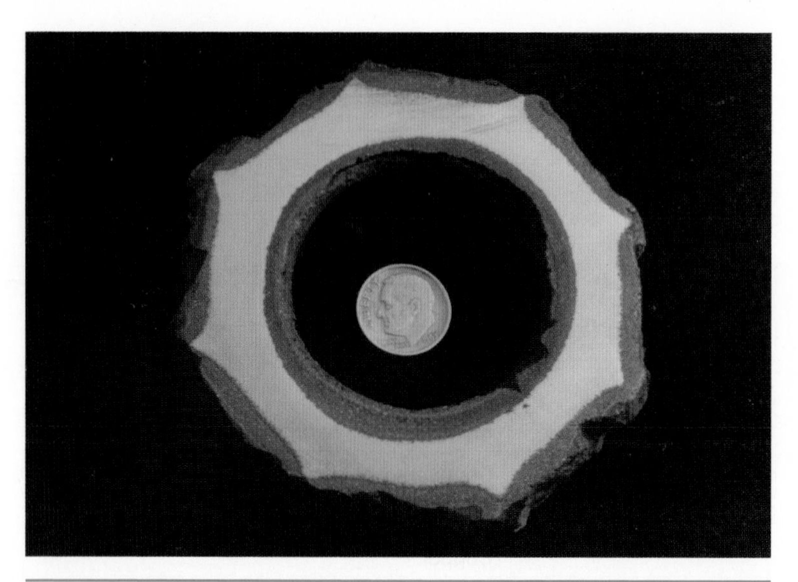

FIGURE 14.2 Pump impeller severely attacked by graphitic corrosion. Material of the impeller is gray cast iron. Note the dark gray areas on both internal and external surfaces where the metal has been converted to iron oxide–based corrosion products and graphite.

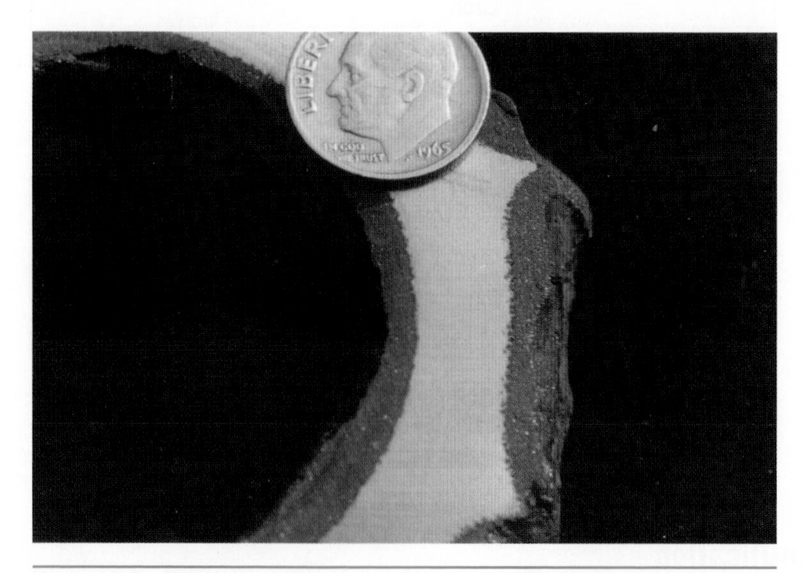

FIGURE 14.3 Note how uniformly metal is converted to corrosion product. Impeller vanes were lost in service because the graphitically corroded metal is brittle and weak.

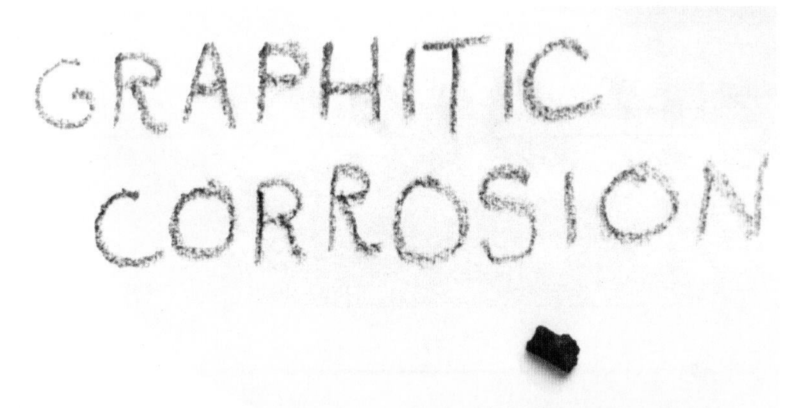

FIGURE 14.4 Graphitic corrosion products can be used to write on paper, similar to lead in a pencil. (*Courtesy of McGraw-Hill, The Nalco Guide to Cooling Water Systems Failure Analysis, New York, 1993.*)

FIGURE 14.5 Graphitically corroded valve butterfly. Original surface contours are preserved. Edges, which were completely converted to brittle corrosion product, have broken.

uniform, with all exposed surfaces corroded to roughly the same depth (Figs. 14.5 and 14.6). Fractures that result from graphitic corrosion are thick-walled due to the brittleness of the corrosion products and the alloy. If localized deposits are present, especially those containing sulfate, chloride, or other acid-producing species, corrosion

FIGURE 14.6 Severe graphitic corrosion of a gray cast iron feedwater pump impeller. Patches of corrosion product have cracked and spalled, revealing uniform general wastage. Note the fracture in the impeller vane.

may be confined to distinct regions beneath such deposits. When attack is severe or prolonged, the entire component is converted to corrosion product. Surface contour and appearance are often preserved, although the graphitically corroded material may appear black, rather than the original metallic gray. Attack is usually not apparent until surfaces are probed or stressed. Craters can be dug in the soft, black corrosion products to reveal the depth of penetration. Corroded areas can be broken with bare hands or by gentle taps with a hard implement.

Metallographic examination will reveal graphite particles from the microstructure (flakes or nodules, depending on the type of cast iron) surrounded by corrosion products and embedded in corrosion product layers (Figs. 14.7 and 14.8).

Elimination

Attack is reduced by alloy substitution, chemical treatment, and/or operational changes. Alloy substitution can entirely eliminate graphitic corrosion if the replacement alloy does not contain graphite. The substitute alloy choice is dictated by requirements unique to each environment. Alloys that may be suitable substitutes include

FIGURE 14.7 Nodular cast iron that sustained graphitic corrosion, showing graphite nodules embedded in the corrosion product layer. Etchant: Picral.

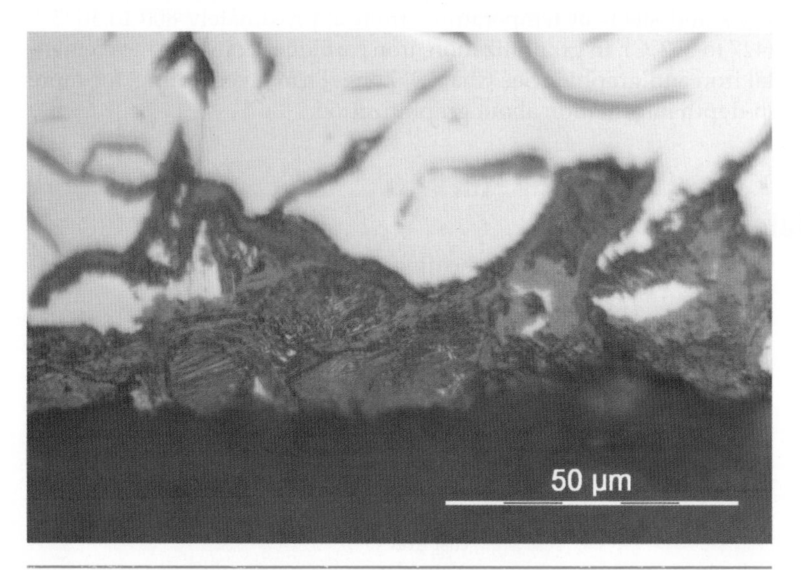

FIGURE 14.8 Gray cast iron, showing graphite flakes surrounded by iron oxide–based corrosion products. Etchant: Picral.

austenitic cast iron; corrosion-resistant cast irons containing chromium, nickel, or silicon; and cast steels because they are essentially immune to graphitic corrosion. When changing alloys, care should be taken to avoid the formation of galvanic couples between components in electrical contact with each other.

Raising water pH to neutral or slightly alkaline levels decreases attack, especially if relatively high concentrations of aggressive anions such as chloride and sulfate are present. When water flow is slight, such as during prolonged shutdowns and lengthy idle periods, attack increases. Stagnant conditions promote graphitic corrosion and should be avoided.

Cautions

Only cast irons containing graphite can corrode graphitically. Due to microstructural differences in graphite particle size, morphology, and distribution, as well as other differences in alloy composition, attack is usually worst in gray cast irons.

Although pipes and other components may be severely corroded, they may not fail. The corrosion product has some mechanical strength, but is brittle. If corroded components are stressed, failure may occur catastrophically.

Graphitic corrosion should not be confused with graphitization, which is a microstructural change that can occur in carbon-containing irons and steels at temperatures from approximately 800 to 1025°F (427 to 552°C). In graphitization, iron carbide is converted to elemental iron and graphite. See Chap. 3, Long-Term Overheating, for more in-depth information about graphitization.

Related Topics

Chapter 3, Long-Term Overheating, and Chap. 15, Dealloying.

Case History 14.1

Industry:	Food processing
Specimen location:	Feedwater pump
Specimen orientation:	Horizontal
Years in service:	5
Water treatment program:	Phosphate
Drum pressure:	600 psi (4.1 MPa)
Tube specifications:	8-in. (20.3-cm) outer diameter, five vanes of gray cast iron
Fuel:	Natural gas

A small feedwater pump impeller was removed during a scheduled maintenance outage. The entire impeller surface was converted into soft, black corrosion products. In some areas, no trace of the original impeller alloy was left. Vane tips were totally converted to corrosion product. Vanes were cracked, and pieces of the corrosion product were dislodged, producing irregular contours (Fig. 14.6).

The impeller had been used intermittently, with greater than 50% idle time in the previous 2 years.

Case History 14.2

Industry: Food processing
Specimen location: Boiler feedwater pump
Years in service: 1.5 or less

Gray cast iron boiler feedwater pump components from various different pumps were analyzed, including a large casing and two suction head rings. Reportedly, the feedwater source was changed from softened lake water to second-effect condensate from an evaporator process approximately 3 years before the components were removed from service. It was reported that pump service life had decreased dramatically after the change in feedwater source.

Reportedly, the pump internals had changed in color from red to gray. When the gray surfaces were wiped, a slippery, powdery gray material rubbed off. Flow-oriented, smooth-surfaced grooves interrupted the surfaces of the suction head flanges of the large casing. In microscopic examinations, the surfaces that visually appeared to have experienced metal loss were bare or covered with a very thin layer of corrosion products. In comparison, the apparently unattacked surfaces were covered with iron oxide corrosion products containing graphite flake particles and skeletal remnants of pearlite colonies.

The grooves were primarily caused by erosion; however, evidence of graphitic corrosion away from the grooves suggested that mechanically weak graphitic corrosion products somewhat promoted erosive metal loss. The condensate was most likely lower in pH than the lake water that was previously used as the feedwater source. Graphitic corrosion is promoted by slightly acidic conditions. Intermittent contamination of the condensate that was used as feedwater probably also contributed to graphitic corrosion.

Case History 14.2

Industry: Food processing
Specimen location: Boiler feedwater pump
Years in service: 1.5 or less

Gray cast iron boiler feedwater pump components from various different pumps were analyzed, including a large casing and two suction head rings. Reportedly, the feedwater source was changed from softened lake water to second-effect condensate from an evaporator process approximately 2 years before the components were removed from service. It was reported that pump service life had decreased dramatically after the change in feedwater source.

Reportedly, the pump internals had changed in color from red to gray. When the gray surfaces were wiped, a slippery, powdery gray material rubbed off. Flow-oriented, smooth-surfaced grooves interrupted the surfaces of the suction head flanges of the large casing. In microscopic examinations, the surfaces that visually appeared to have experienced metal loss were bare or covered with a very thin layer of corrosion products. In comparison, the apparently unattacked surfaces were covered with iron oxide corrosion products containing graphite flake particles and skeletal remnants of pearlite colonies.

The grooves were primarily caused by erosion; however, evidence of graphitic corrosion away from the grooves suggested that mechanically weak graphitic corrosion products somewhat promoted erosive metal loss. The condensate was most likely lower in pH than the lake water that was previously used as the feedwater source. Graphitic corrosion is promoted by slightly acidic conditions. Intermittent contamination of the condensate that was used as feedwater probably also contributed to graphitic corrosion.

CHAPTER 15
Dealloying

General Description

An *alloy* has been defined as "a substance having metallic properties and being composed of two or more chemical elements of which at least one is an elemental metal."[1] Dealloying is a corrosion process in which one or more alloy components are removed preferentially. The corroded region usually has a markedly different structure than the original alloy. However, macroscopic dimensions of the corroded part often remain unchanged. The process is also referred to as *selective leaching* or *parting*. Obviously, dealloying can occur only in alloys containing two or more elements. Particularly susceptible alloys that may be used in the boiler water circuit are brasses (in which zinc and aluminum are leached from copper-zinc and copper-aluminum alloys, respectively) and cupronickels (in which nickel is removed). The name given to a particular dealloying process derives from the leached element. For example, in common brasses where zinc is removed, dealloying is referred to as *dezincification*; where nickel is removed from cupronickels, dealloying is referred to as *denickelification*.

A number of mechanisms have been proposed for dealloying of binary alloys, one of which involves dissolving both the constituents and redepositing of the more noble element. Other proposed mechanisms involve dissolving only the less noble element. The details of the possible mechanisms are beyond the scope of this book and are reviewed elsewhere.[2] Different conditions (i.e., electrochemical potential, pH) likely result in different predominant dealloying mechanisms.

In the case of dezincification, brass alloys containing more than approximately 15 wt% zinc are susceptible to attack. Brasses containing up to 32 wt% zinc only contain alpha phase. Alloys containing more than approximately 32 wt% zinc are particularly vulnerable to dezincification due to the presence of the beta phase, which is richer in zinc than the alpha phase. Two-phase alloys (containing, in the case of brass, both alpha and beta phases) are called *duplex alloys*. Since zinc is less noble than copper, galvanic effects promote preferential removal of zinc from the zinc-rich beta phase first, leaving behind the alpha phase and porous, weak, copper-rich corrosion products (Fig. 15.1). Subsequent dezincification of the alpha phase may still occur.

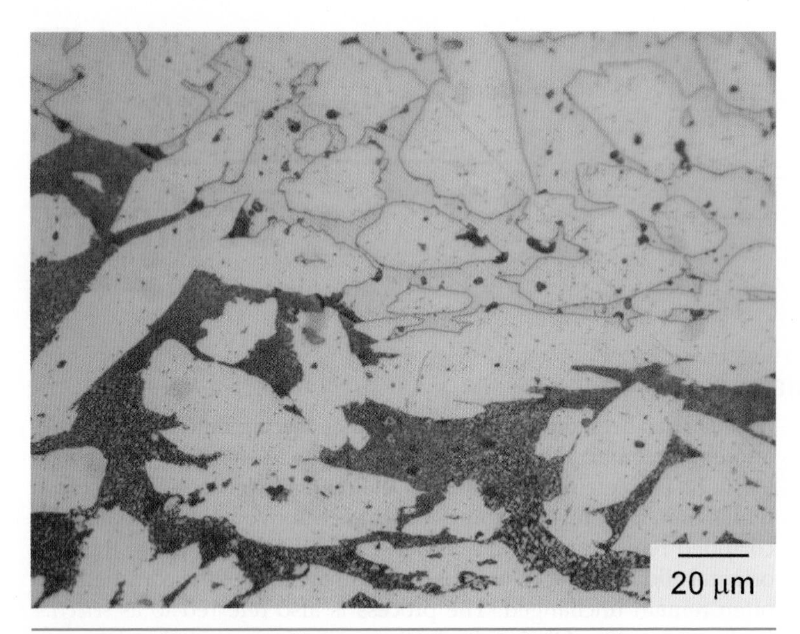

FIGURE 15.1 Micrograph of brass containing both alpha and beta phases, showing copper-rich corrosion products in the former beta phase. Etchant: Potassium dichromate.

There are two commonly recognized forms of dezincification: plug and layer types. In plug-type dezincification, small, localized areas of metal loss in the form of deep, narrow "plugs" occur in the tube wall. Such plugs may be blown out of pressurized tubes (Fig. 15.2). More general attack is called *layer-type dezincification* (Fig. 15.3). It is likely that fundamental mechanisms of plug-type and layer-type attack are similar. However, plug-type attack can produce localized wastage rates of up to several hundred mils per year (mpy), while more uniform, shallower metal loss is common with layer-type dealloying.

Cupronickels are generally more prone to layer-type denickelification than to plug-type deterioration. However, cupronickel wastage is usually slight compared to attack in brasses. Low-flow conditions and high temperatures likely promote denickelification.

Destannification (loss of tin) has been observed in gunmetal, which can be used for steam valves, and tin bronze, which is sometimes used in pump impellers. In boiler systems, steam environments and hot feedwaters are known to promote destannification.

Dealuminification of duplex aluminum bronzes (greater than approximately 9 wt% aluminum) has been reported in waters having both high and low pH.

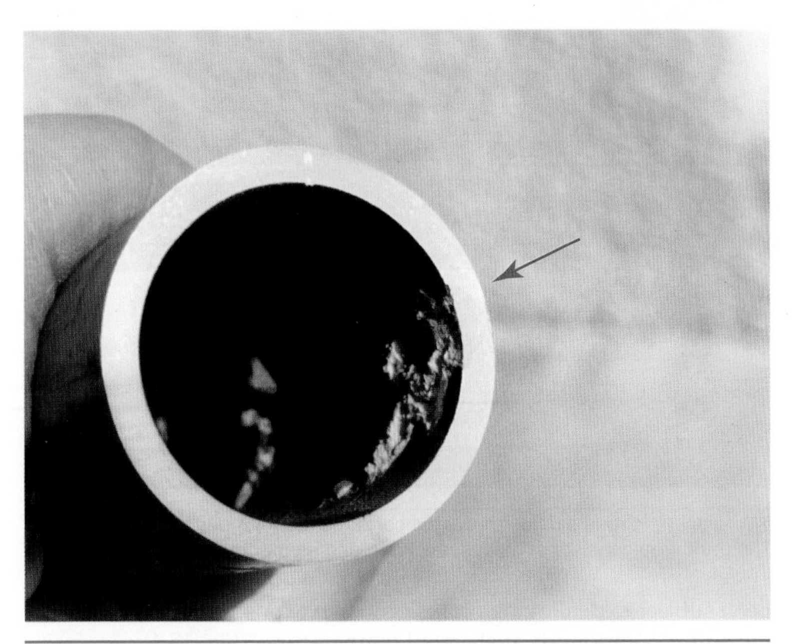

FIGURE 15.2 Plug-type dezincification beneath a deposit layer in an admiralty brass tube.

FIGURE 15.3 Layer-type dezincification of a brass pump component.

Locations

Dealloying most commonly occurs in copper-containing alloys. Corrosion is confined primarily to feedwater systems and afterboiler regions. Components that can suffer attack include bronze pump impellers and boiler peripherals such as pressure-gauge fittings and valves. Monel (a family of nickel-copper alloys) steam strainers have

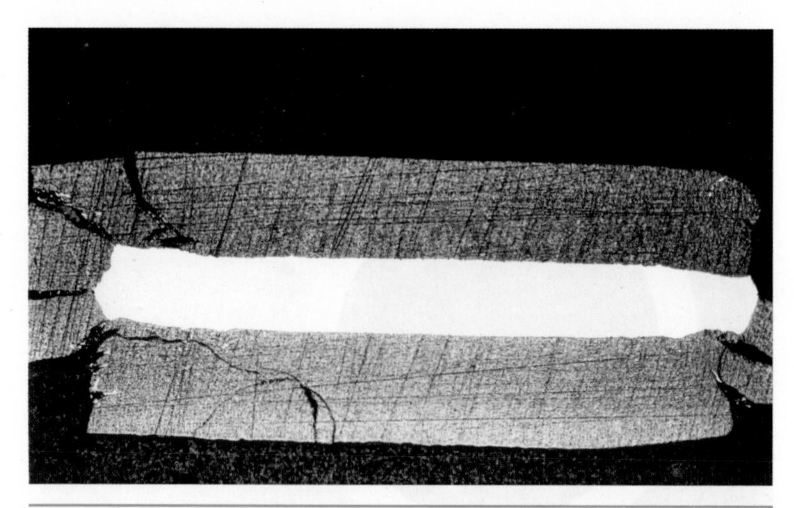

FIGURE 15.4 Cross section through a corroded Monel metal steam strainer. The dark area consists of oxides, sulfides, and elemental copper. Unetched.

been subject to dealloying corrosion when exposed to steam containing sulfur compounds at elevated temperatures (Fig. 15.4).

Condensers and heat exchangers are also frequently affected; however, dealloying generally occurs on the cooling water side (see Case History 15.5). Failures in such systems can result in contamination of the boiler water with cooling water. For more in-depth discussion of dealloying in cooling water environments, see *The Nalco Guide to Cooling Water Systems Failure Analysis*.[3]

Critical Factors

Using a susceptible alloy in an environment conducive to dealloying promotes attack. Deposits, soft waters (especially those containing carbon dioxide), high heat transfer, stagnant conditions, either high- or low-pH waters and high-chloride waters accelerate most forms of dealloying in copper-containing alloys. Addition of small amounts of arsenic, antimony, or phosphorus to some brass alloys, such as admiralty brasses, can reduce the likelihood of dezincification, possibly due to the redeposition of a protective layer on the metal surface. Such alloys are referred to as *inhibited brasses*. However, attack may still occur under extreme conditions.

Identification

In copper-based alloys, copper is almost never selectively removed. Rather, other elements are leached, leaving behind a comparatively soft, porous mass of copper. Dealloying may be discovered through visual examination if damage is severe. However, dealloying is most

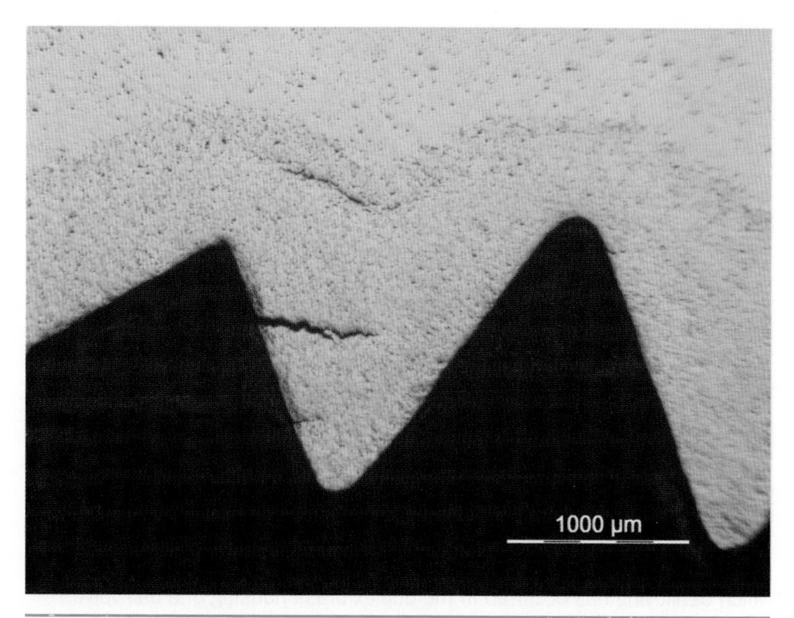

FIGURE 15.5 Micrograph showing dealloying in a threaded region of the housing of a brass valve. Cracks developed in the dealloyed areas due to the weak nature of the corrosion products. Unetched.

effectively diagnosed through metallographic examination. Attacked metal is relatively weak and can be broken easily by stresses from impact or bending. Frequently, surfaces will be riddled with cracks, but will retain original dimensions (Fig. 15.5). Attacked areas will usually change to a deep red or salmon color indicative of elemental copper (Figs. 15.2 and 15.3). In brass alloys that have a two-phase microstructure consisting of alpha and beta brass, the beta brass will be preferentially attacked (Fig. 15.1). Light-colored, zinc-rich corrosion products commonly overlie regions in which dezincification has occurred (Fig. 15.6).

Elimination

Surfaces should be kept free of deposits. In general, outages should be as short as is practical. Air contact should be prevented through the use of steam or nitrogen blanketing. Water and steam quality must be controlled so that concentrations of chloride and sulfur compounds are minimized. Environmental factors that can accelerate dezincification include high temperatures, deposition, and stagnant conditions. Dezincification is also promoted by exposure to soft water, high-pH conditions, and high concentrations of carbon dioxide in the water.

FIGURE 15.6 Light-colored, zinc-rich corrosion products overlying areas in which dealloying has occurred in the duplex brass housing of a ball valve.

Substitution of alternative alloys may be necessary. Use of inhibited grades of admiralty brass (all alpha phase), which contain small concentrations of phosphorus, arsenic, or antimony, should be insisted on when conditions dictate.

Related Problems

Graphitic corrosion (see Chap. 14, Graphitic Corrosion) is sometimes considered to be a form of dealloying, although the mechanisms are actually quite different. In graphitic corrosion, microgalvanic cells are formed between the phases of more-noble graphite and less-noble iron in gray and nodular or malleable cast irons, resulting in corrosion of the iron phase. In true dealloying, one of the alloying elements is chemically removed from an intimate mixture of metals. Formation of microgalvanic cells may contribute to true dealloying, particularly in duplex brasses, although it is not the only driving force for such corrosion.

Exfoliation is a unique form of wastage that may occur in high-pressure utility boiler feedwater heater tubes made of cupronickel. Exfoliation in cupronickel is unrelated to the types of exfoliation that may occur in ferrous boiler reheater and superheater tubes (see Chap. 3, Long-Term Overheating) or in aluminum alloys. In the presence of high-pressure oxygenated steam, distinct layers of oxide material may form on the cupronickel tube surfaces. Eventually, sheets of oxide

FIGURE 15.7 Peeling layers of oxide on a cupronickel high-pressure feedwater heater tube.

corrosion products can peel (exfoliate) off surfaces (Fig. 15.7). Microscopic examinations may reveal finely stratified corrosion products (Fig. 15.8). The exfoliated layers can then be swept downstream into the boiler, where copper- and nickel-containing deposits may form on tubes within the boiler water circuit as well as on turbines.

FIGURE 15.8 Micrograph showing finely stratified, oxide-containing corrosion products on the external surface (steam side) of a cupronickel utility boiler feedwater heater that experienced exfoliation. Unetched.

Although exfoliation in cupronickel is sometimes associated with dealloying, some studies suggest that such attack is not actually a form of dealloying. Chemical analysis of the oxide layers has indicated that they can contain copper and nickel in nearly the same proportion as that of the original alloy, indicating no preferential removal of either of the main alloying elements.[4] In some cases, evidence has suggested that alternating periods of oxidizing and reducing conditions result in alternating oxide and metallic copper layers (see Case History 15.2).

Feedwater heaters in peaking and/or intermittent service tend to be more susceptible to exfoliation, compared to those in base-loaded (continuous) service. High pH, high temperature, and the presence of ammonia also tend to promote attack. The likelihood of exfoliation increases with increasing nickel content. In terms of common commercial alloys, 70:30 cupronickels are particularly vulnerable to exfoliation. Alloys such as 80:20 cupronickels are susceptible to a lesser extent. Such wastage is rare in 90:10 cupronickels. Additions of iron and manganese to the alloy have been shown to increase resistance to exfoliation. Since exfoliation in high-pressure utility boiler feedwater heaters provides contaminants that can produce deposits on turbines, austenitic stainless steel is an alternative to cupronickel that has become the predominant feedwater heater tube material. However, austenitic stainless steels may be subject to stress corrosion cracking if even small amounts of chloride are present in the steam (see Chap. 13, Stress Corrosion Cracking).[5]

Case History 15.1

Industry:	Utility
Specimen location:	Turbine steam strainer
Specimen orientation:	Vertical
Years in service:	16
Water treatment program:	Phosphate
Drum pressure:	1500 psi (10.3 MPa)
Specifications:	Monel wire, 0.15-in. (0.4-cm) diameter
Fuel:	Coal

A Monel turbine steam strainer was discovered to contain many broken elements. The metal was converted to oxide, sulfide, and elemental copper (Fig. 15.4). The deteriorated metal consisted of numerous elemental copper particles embedded in an oxide-sulfide matrix. Cracks in the wasted metal were lined with elemental copper.

Failure was attributed to carryover of sulfur-containing compounds in the steam. The system had a history of poor steam purity and carryover of boiler water into the superheater.

Case History 15.2

Industry:	Utility
Specimen location:	Inlet first pass of high-pressure feedwater heater
Specimen orientation:	Horizontal
Years in service:	7

Water treatment program:	All volatile treatment
Drum pressure:	400 psi (2.8 MPa)
Tube specifications:	5/8-in. (1.6-cm) outer diameter, 70:30 cupronickel

Feedwater heater tubes were thinned by cyclic oxidation followed by reduction of oxides in service (Fig. 15.7). Wall thickness was reduced by as much as 15%.

During the previous 2 years, the boiler had experienced frequent outages in which air leaked into the heater shell and caused surface oxidation. Conversion of oxidized copper to elemental copper occurred during normal operation, as evidenced by layers of metallic copper-colored particles within the corrosion products.

Case History 15.3

Industry:	Institutional
Specimen location:	Low-pressure boiler feedwater return piping
Specimen type:	Ball valve
Metallurgy:	Brass
Years in service:	2

Chemical analysis of the valve housing alloy indicated that it was an uninhibited brass alloy with a zinc content of approximately 38 wt%. The internal surfaces of the valve housing and ball were covered with mounds of light-colored corrosion products (Fig. 15.6). Chemical analysis of the corrosion product mounds by SEM-EDS indicated zinc contents greater than 45 wt% in many areas. Beneath the corrosion products, areas of the metal were copper-colored. Microscopic examinations indicated a duplex microstructure consisting of alpha and beta brass, typical of brass alloys containing greater than 32 wt% zinc (Fig. 15.1). Along the internal surface, porous copper corrosion products lined the surface in many places, indicating that dezincification occurred. Characteristics of both plug- and layer-type attack were observed. Cracks penetrated along the porous corrosion products in threaded regions (Fig. 15.5). It was recommended that inhibited single-phase brass alloys with less than 32 wt% zinc be considered to limit the likelihood of dezincification. The pH of the water was reportedly above 10, which most likely promoted dezincification.

Case History 15.4

Industry:	Institutional
Specimen location:	Boiler feedwater system
Specimen type:	Pump impeller
Years in service:	5
Specifications:	4-in. (10.2-cm) diameter, die-cast silicon brass

The pump supplied feedwater to a boiler that was used for heating. As such, the system experienced extended idle periods. The impeller exhibited smoothened metal loss in most locations. The surface was generally covered with a thin layer of deposits or corrosion products. In a few scattered locations, white corrosion product mounds containing an average of 38 wt% zinc (as analyzed by SEM-EDS) were observed. Trace amounts of sulfur and chlorine were also detected on the surface by SEM-EDS. In some places, the surface was bare and revealed the typical dendritic structure that results from solidification during the casting process (Fig. 15.9). Chemical analysis of the impeller alloy indicated that it contained approximately 13 wt% zinc and 5% silicon.

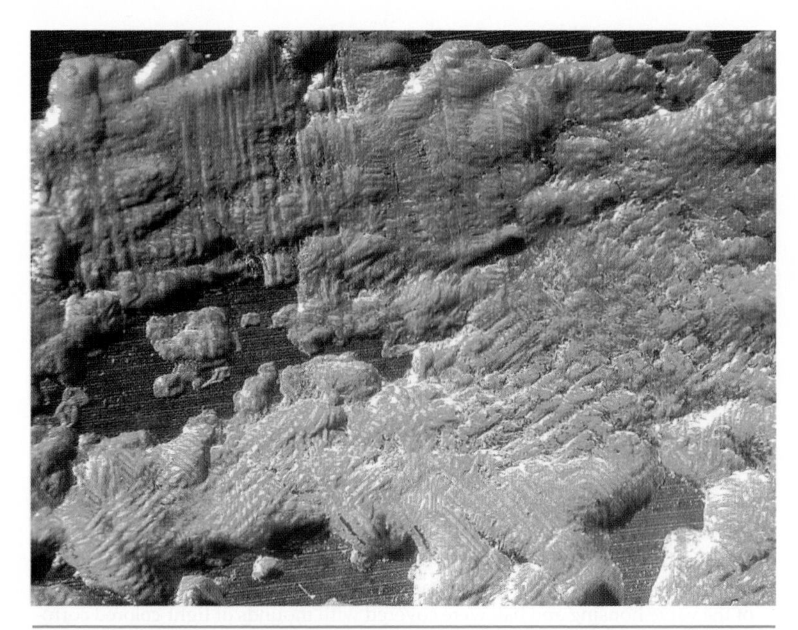

FIGURE 15.9 Bare-surfaced metal loss areas on the surface of a die-cast silicon brass feedwater pump impeller, showing the normal dendritic structure that resulted from the casting process.

Microscopic examinations indicated that the microstructure consisted of dendrites of the alpha solid solution phase, with significant coring (compositional segregation) evident in the dendrites. Porous, copper-rich corrosion products indicative of dezincification were present on all surfaces, although dezincification penetrated preferentially along zinc-rich areas in the cored microstructure (Fig. 15.10). Elongated depressions were superimposed on the dezincified areas.

Dezincification was most likely promoted by stagnant conditions during idle periods and the presence of sulfates, sulfides, and/or chlorides. Erosive forces easily removed the soft, copper-rich corrosion products produced by dezincification. It was recommended that pumps be drained and their components dried prior to extended idle periods. Alternative metallurgies were also suggested, including aluminum bronze, leaded nickel bronze, and cast 400-series stainless steels. Materials containing secondary phases that are particularly susceptible to dealloying (such as beta phase in brass) or that exhibit segregation of alloying elements were not recommended.

Case History 15.5

Industry:	Utility
Specimen location:	Condenser, inlet side
Metallurgy:	Brass
Years in service:	29

The failure in this case history occurred due to dealloying on the cooling water side; however, it is included due to the relatively common nature of dealloying in condensers, which are part of the boiler water circuit. For more in-depth

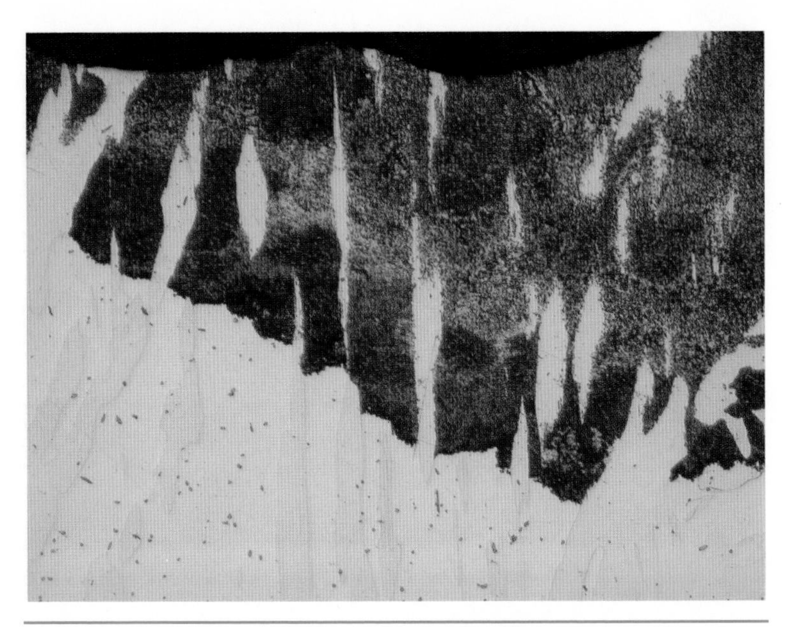

FIGURE 15.10 Preferential dezincification along coring in the all-alpha-phase dendritic microstructure of a silicon brass boiler feedwater pump impeller. Etchant: Potassium dichromate.

information regarding dealloying in cooling water environments, see *The Nalco Guide to Cooling Water Systems Failure Analysis.*[6]

Eddy current testing indicated both general corrosion and pitting on the internal surface of the condenser tubes (cooling water side). A few perforations occurred in deep internal surface depressions (Fig. 15.11). The depressions were generally covered with hard, light-colored deposits and corrosion products. Material close to and away from the depressions was found to contain a high concentration of silicon, in addition to numerous other elements. Chemical analysis of material within a depression by SEM-EDS indicated that it contained an elevated concentration of zinc, relative to material elsewhere on the surface. The deep depressions were filled with porous, copper-based corrosion products, indicative of plug-type dezincification (Fig. 15.12). The patches of deposit material on the surface most likely promoted plug-type attack due to the formation of concentration cells beneath the material.

The internal surface was generally covered with a layer of porous copper-based corrosion products, indicating that layer-type dezincification had occurred in addition to plug-type attack. High-salt content of the water and high temperatures likely promoted layer-type dezincification. Water analyses indicated high conductivity and sulfate concentration in the water. The system had reportedly experienced biological fouling several years before the metallurgical analysis was conducted; occlusive material in general, including biological material, promotes dezincification.

Microscopic examinations indicated that the microstructure was single-phase, consisting of twinned grains of alpha brass. Chemical analysis of the tube alloy indicated that it was consistent with admiralty brass. Small concentrations of arsenic and phosphorus were detected in the metal, but they were below the levels typically used in inhibited admiralty brass alloys. The addition of

FIGURE 15.11 Internal surface (cooling water side) of a brass condenser tube, showing a deep depression caused by plug-type dezincification.

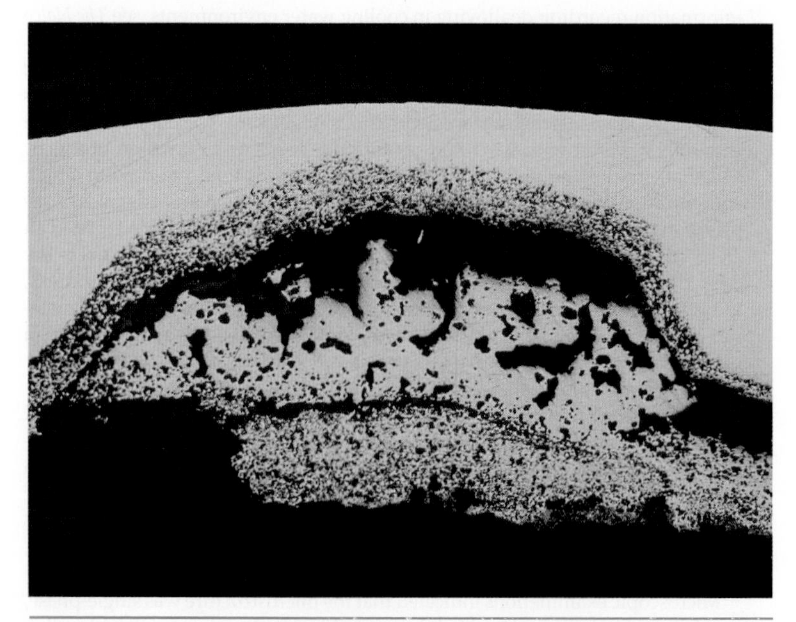

FIGURE 15.12 Micrograph showing deep, plug-type dezincification on the cooling water side of a brass condenser tube. Unetched.

an adequate concentration of a corrosion-inhibiting element may have prevented the plug-type dezincification and limited layer-type dezincification. It was recommended that tube surfaces be kept clean to minimize the possibility of concentration of aggressive species beneath occlusive material. The reported service life of 29 years was considered to be long for uninhibited admiralty brass exposed to water with a high conductivity and sulfate concentration.

References

1. "Definitions Relating to Metals and Metalworking," *ASM Metals Handbook*, vol. 1: *Properties and Selection of Metals*, American Society for Metals, Metals Park, Ohio, 1961, pp. 1–41.
2. S. G. Corcoran,"Effects of Metallurgical Variables on Dealloying Corrosion," *ASM Handbook*, vol. 13A: *Corrosion: Fundamentals, Testing, and Protection*, ASM International, Metals Park, Ohio, 2003, pp. 287–293.
3. H. M. Herro and R. D. Port, *The Nalco Guide to Cooling Water Systems Failure Analysis*, McGraw-Hill, New York, 1993.
4. J. A. Beavers, A.K. Agrawal, and W. E. Berry, *Corrosion-Related Failures in Feedwater Heaters*, Electric Power Research Institute, Palo Alto, CA, 1983, CS-3184, pp. 5–1 to 5–3.
5. T. J. Muldoon, "Stress Corrosion Cracking (SCC) of 304SS Tubes at Outlet of the Desuperheating Zone in Feedwater Heaters,"*Feedwater Heater Technology Symposium*, Electric Power Research Institute, Palo Alto, CA, 2004.
6. Herro and Port, *The Nalco Guide to Cooling Water Systems Failure Analysis*, McGraw-Hill, New York.

CHAPTER 16

Steam and Condensate System Damage

General Description

Steam that is produced in boilers may be used as a source of energy for power generation and to provide sources of heat and energy for a variety of industrial uses. The environments and some materials of construction that are used in steam and condensate systems are very different from those in boiler systems. In particular, temperatures and pressures are often reduced beginning at the point of use of the steam. The steam condenses at lower temperatures and pressures to produce liquid water that is called *condensate*. Consequently, corrosion and failure mechanisms are very different from those in boilers. It is important to control corrosion within steam and condensate systems for two main reasons:

1. To preserve the integrity of the components within these systems
2. To prevent corrosion products produced by corrosion in these systems from contaminating condensate that is returned to the boiler

This can be a concern for systems that do not employ equipment that will remove the corrosion products, such as condensate polishers. In this chapter, the most common corrosion and failure mechanisms in steam and condensate system components will be described.

Acid Corrosion

Under normal operating conditions, steam will be quite pure and will be mixed with gases that are evolved from the boiler water at high temperatures. Depending upon the pressure of the boiler from which the steam is generated and the required feedwater quality requirements, amounts of certain gases such as carbon dioxide and oxygen will be present. Carbon dioxide will be generated by the thermal breakdown of bicarbonate and carbonate alkalinity in the boiler water. Oxygen will be evolved from the water if it is not completely removed prior to introduction to the boiler. Carbon dioxide will dissolve in condensate to produce solutions of carbonic acid, as in Eq. (16.1).

$$CO_2 + H_2O \leftrightarrow H_2CO_3 \leftrightarrow H^+ + HCO_3^- \qquad (16.1)$$

Materials such as carbon steel and copper alloys are susceptible to corrosion by carbonic acid. The carbonic acid corrosion reaction for iron (in steel) is

$$Fe + H^+ + 2HCO_3^- \rightarrow Fe(HCO_3)_2 + H_2 \qquad (16.2)$$

Gaseous sulfur compounds such as hydrogen sulfide and sulfur dioxide may be produced by degradation of sulfite oxygen scavengers, at pressures exceeding approximately 600 psi (4.1 MPa). Volatile organic compounds (VOCs) may be produced by the breakdown of neutralizing amines, used as condensate corrosion inhibitors, and other organic materials such as contaminants in the boiler water. All these gaseous compounds may produce acidic solutions in condensate to cause corrosion. Martensitic stainless steel components used in turbines are highly resistant to carbonic acid. However, aggressive sulfur compounds and organic acids may cause corrosion on turbine components.

Damage by Oxygen

Oxygen will also dissolve in condensate. Dissolved oxygen will accelerate carbonic acid corrosion of carbon and low-alloy steels and copper alloys. Additional oxygen and carbon dioxide other than that carried through the boiler may be introduced to condensate due to in-leakage of air into the steam and condensate system. The main source of oxygen is from air in-leakage, which occurs at locations such as seals, joints, steam traps, vacuum breakers, and vented condensate receivers due to reduction of pressure that occurs as the system cools. Dissolved oxygen in water will cause oxygen corrosion on carbon and low-alloy steel components, as described in Chap. 9. As the dissolved-oxygen concentration increases in the condensate, the amount of corrosion increases by both oxygen and carbonic acid corrosion. Stainless steel components are essentially immune to oxygen and carbonic acid corrosion.

Damage by Ammonia

Ammonia may be produced by degradation of components of the chemical treatment program such as hydrazine, polymers, amines, or other organic compounds in the boiler water. High-temperature breakdown of neutralizing amines used to control carbonic acid corrosion may also produce ammonia. Chloramines used in the feedwater provide another possible ammonia source.

Ammonia in condensate solution will increase its pH, so it will not be corrosive to steel components. Dissolved ammonia may actually reduce corrosion of steel due to neutralization of carbonic and other acids in condensate and to some degree due to the pH increase. However, condensate containing dissolved ammonia and some oxygen at pH exceeding about 9.2 is very corrosive to copper alloy components. The dissolved ammonia complexes the copper to cause corrosion. The cupric ammonia complex is soluble in water and is carried away in water solution. Ammonia corrosion of copper alloys increases as a function of increasing dissolved-oxygen concentration.

In addition, certain copper alloys, particularly copper-zinc alloy brasses, are susceptible to stress corrosion cracking when exposed to waters containing high concentrations of dissolved ammonia.

Boiler Water Carryover

Under conditions that allow boiler water to carry over from the boiler with the steam, to components downstream of the boiler, other damage mechanisms may occur. Damage such as stress corrosion cracking and/or other corrosion mechanisms may occur on carbon, low-alloy, and even stainless steel components, depending upon the metal temperatures and concentration of corrodents in water solution. Stress corrosion cracking is described in Chap. 13.

Stress corrosion cracking of carbon, low-alloy, and stainless steel components may result from exposure to caustic or alkaline salts in condensed steam. Austenitic stainless steel components such as 304 and 316 grades may experience stress corrosion cracking due to exposure to chlorides when metal temperatures exceed about 130°F (54°C).

Pitting on turbine components may result from exposure to concentrated aggressive agents such as chloride and sulfate.

Erosion

Steam and condensate may flow at high rates and impinge across surfaces to produce erosion damage. Erosion damage that is found in steam and condensate systems is described in Chap. 18, Erosion and Erosion-Corrosion (Water and Steam Sides).

Fatigue and Corrosion Fatigue Cracking

Fatigue and corrosion fatigue cracking is caused by the application of cyclic stresses. These stresses may be supplied from forces applied by steam impingement and vibration on turbine components. Temperature variations and vibration may promote fatigue and corrosion fatigue on condenser tubes and to a lesser degree on components in steam and condensate systems. Corrosion fatigue cracking is described in Chap. 12.

Locations

Acid Corrosion

Locations where steam may condense are susceptible to acid corrosion, provided that the steam contains sufficient amounts of carbon dioxide or other gases as well as oxygen that may produce acidic solutions when dissolved in water. Acid corrosion occurs within steam lines, condensate return lines, condensers, steam traps, and other locations where acidic condensate forms and collects. Corrosion by acid condensate tends to be most significant at points of initial condensation. In such locations, thin films of condensed water will tend to dissolve higher concentrations of carbon dioxide and/or other gases to produce more aggressive solutions. Condensate may collect and drip down the faces of baffles or tubesheets (Fig. 16.1). This condensate may cause metal loss on tube surfaces located

FIGURE 16.1 Carbonic acid corrosion on external surface of brass condenser tube.

Figure 16.2 Carbonic acid corrosion on bottom-side internal surface of carbon steel condensate return pipe.

directly adjacent to the baffle or tubesheet contact points. Streams of acidic condensate may collect and flow down the sides of tubes and along the bottoms of steam and condensate lines, causing metal loss along the flow paths (Fig. 16.2). In some cases, acid corrosion damage will form in horizontally oriented condensate piping within the airspace located above the condensate level, while the surfaces beneath the condensate level are not attacked (Fig. 16.3). This may result from dilution of the acidic condensate and/or elevation of condensate pH. A secondary cause of failure, related to acid corrosion of steel and copper alloy components, is obstruction or pluggage of components downstream. Components that contain narrow orifices, such as steam traps, are particularly susceptible. When the condensate containing dissolved copper and iron ions are released by the trap, the drop in pressure will cause it to vaporize. The dissolved iron and copper will then deposit, causing obstructions or pluggage (Fig. 16.4).

Oxygen Corrosion

Oxygen corrosion damage may produce localized metal loss at any location where oxygenated water contacts surfaces of carbon and low-alloy steel components (Fig. 16.5). General metal loss may be produced when corrosion becomes more widespread. Oxygen corrosion is most significant beneath pools of condensate or beneath waterlines in piping.

FIGURE 16.3 Carbonic acid corrosion on top-side internal surface of carbon steel condensate return pipe.

FIGURE 16.4 Iron oxide deposits on steam trap component.

Figure 16.5 Pits formed by oxygen corrosion on internal surface of condensate return pipe.

Ammonia Corrosion and Stress Corrosion Cracking

Corrosion by dissolved ammonia in condensate solution will be found on copper alloy condenser tubes and copper alloy components within steam and condensate systems. Ammonia concentration must be high enough to allow corrosion. Condensate containing dissolved ammonia may collect and drip down the faces of baffles or tubesheets to cause metal loss on closely adjacent tube surfaces (Fig. 16.6).

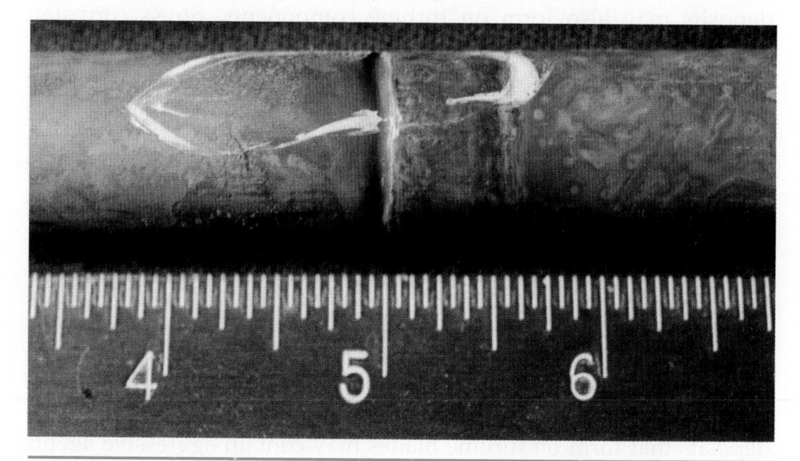

Figure 16.6 Ammonia corrosion on external surface of brass condenser tube.

FIGURE 16.7 Ammonia stress corrosion cracking on external surface of brass condenser tube.

Ammonia stress corrosion cracking may also only occur in locations where ammonia concentration is sufficient, and only on susceptible alloys, particularly zinc-containing brasses (Fig. 16.7). Perhaps the most common brass component that experiences ammonia stress corrosion cracking is tubing used in surface condensers.

Carryover Deposits

Boiler water may carryover in the steam to reach significant distances within steam lines where evaporation will produce deposits. Such deposits may also form on turbine components. Stress corrosion cracking may occur on carbon and stainless steel components at any location where caustic solutions of sufficient concentration are produced in condensate, depending upon the temperature (Fig. 16.8).

Deposits may form on turbine components within various locations, depending upon the solubility of the agents in the fluids. For instance, sodium hydroxide deposits may form even in locations where the steam is dry. Sodium hydroxide deposition may extend from the high-pressure sections through the rest of the turbine if the solubility limit is exceeded. Deposition of other impurities may form in lower-pressure locations where saturated steam first forms. Moisture from the condenser hotwell may migrate to the turbine as well. Therefore, impurities supplied by leaks of cooling water into the condensate may be transported to the turbine. Under such conditions deposits that form on turbine blades may contain aggressive agents such as chlorides and sulfates. These aggressive agents may become

Figure 16.8 Carbon steel steam pipe containing caustic stress corrosion cracks.

highly concentrated in condensate solution to cause corrosion (see Figs. 1.7 and 1.8). Vaporous carryover of silica and copper at high pressure will produce deposits, but not corrosion damage.

Erosion

Turbines may experience metal loss due to erosion at locations where steam impinges, such as on turbine buckets (blades) and other components (Fig. 16.9). Refer to Chap. 18 for more information on erosion. Steam traps and steam and condensate lines experience erosion damage primarily where there are changes in flow direction, such as at bends, T-joints, reducers, and valves. [See Fig. 16.10 or Chap. 18, Erosion and Erosion-Corrosion (Water and Steam Sides)].

Fatigue and Corrosion Fatigue Cracking

Components that experience the application of cyclic stresses such as turbine components and condenser tubes are particularly susceptible to fatigue and corrosion fatigue cracking. The cyclic stresses applied to turbine components are primarily supplied by vibration. Susceptible locations for cracking are adjacent to points of rigid restraint (Fig. 16.11). Condenser tubes may experience cracking at locations of rigid restraint, such as at tubesheets and baffle contact points (Fig. 16.12). The stresses may be supplied by vibration and expansion and contraction. Fatigue and corrosion fatigue cracking is less common in steam and condensate lines where cyclic stresses are lower in magnitude and frequency, and components tend to be supported to minimize the concentration of stresses.

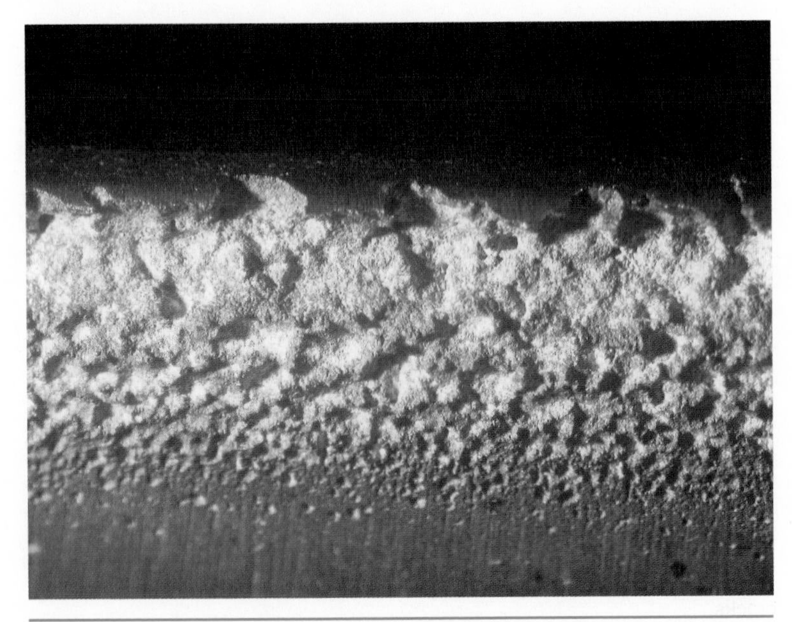

FIGURE 16.9 Liquid impingement erosion on turbine blade.

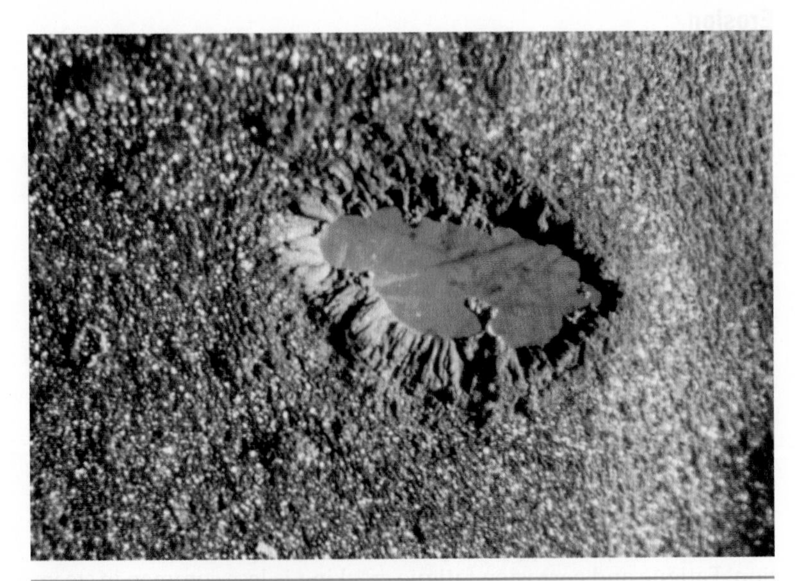

FIGURE 16.10 Erosion damage on component in steam condensate system component.

FIGURE 16.11 Corrosion fatigue cracking on turbine blade housing.

FIGURE 16.12 Corrosion fatigue cracking of condenser tube section.

Critical Factors

Acid Corrosion

Carbonic acid corrosion of carbon and low-alloy steel components is dependent upon the carbon dioxide and oxygen concentration in the steam, the amount of condensate, the pH of the condensate, and the temperature. These factors will define the possible concentration and corrosivity of the carbonic acid solution. The use of condensate corrosion inhibitors to adjust pH to control acid corrosion is described below in the section entitled "Elimination."

Oxygen Corrosion

Oxygen corrosion of carbon steel components requires a source of oxygen that is carried over in the steam from the boiler and/or is supplied by in-leakage of air to the steam and condensate system. Susceptibility to corrosion depends upon the amount of condensate and the ability of the condensate to dissolve it, which is dependent upon the temperature. Dissolved-oxygen concentration exceeding about 50 ppb is typically caused by air in-leakage.

Ammonia Corrosion and Stress Corrosion Cracking

Ammonia corrosion and stress corrosion cracking of copper alloys requires a source of ammonia from the feedwater, or from degradation of treatment chemicals or contaminants in the boiler water. Condensate containing dissolved ammonia and some oxygen at pH exceeding about 9.2 is very corrosive to copper alloy components. The recommended maximum tolerable ammonia concentrations in condensates at various dissolved-oxygen concentrations are shown in Table 16.1. Ammonia stress corrosion cracking will occur on specific copper alloys. High concentrations of dissolved ammonia are required to cause such cracking. In steam and condensate systems, zinc-containing brasses are the most commonly affected alloys. Cupronickel alloys are highly resistant to ammonia stress corrosion cracking.

Dissolved Oxygen (ppb)	Maximum Allowable Ammonia (ppm)
0–20	0.5
20–50	0.3
>50	Any level may cause problems

TABLE 16.1 Relationship of Oxygen and Ammonia

Boiler Water Carryover

Significant corrosion may occur when thin moisture films hydrate aggressive agents in deposit layers that cover martensitic stainless steel turbine components. Low-alloy steel turbine components will experience more severe metal loss. The solution is typically highly concentrated. The deposits are primarily supplied by boiler water that carries over to the steam and condensate system. Boiler water carried over from lower-pressure industrial boilers that have lower feedwater quality requirements will tend to provide more aggressive agents. Carryover may be caused by mechanical, operational, or boiler water chemistry problems. Mechanical problems may include damaged steam separation equipment. Operational problems may include boiler operation at high drum levels. Boiler water chemistry problems may include high dissolved-solids concentration or contamination by organic material that causes foaming.

Stress corrosion cracking of carbon and stainless steel components requires exposure to specific corrodents supplied by boiler water carryover in water solution. Caustic or alkaline salt concentration should be in the percent levels and at temperatures exceeding about 300°F (149°C). Components made of austenitic stainless steels, such as the 304 or 316 grades, may experience chloride stress corrosion cracking at metal temperatures exceeding about 130°F (54°C), at chloride concentrations as low as 20 ppm.

Erosion

Erosion increases as a function of increasing fluid flow rates above a threshold for a particular alloy. Thresholds are described in Chap. 18, Erosion and Erosion-Corrosion (Water and Steam Sides). Entrainment of liquid water or solid material in steam and corrosive condensate will accelerate erosion. Susceptibility to erosion increases as metal hardness decreases.

Fatigue and Corrosion Fatigue Cracking

Fatigue and corrosion fatigue cracking is influenced by design and operating conditions. Proper tube support that is provided by baffles is important, since the contact points serve as locations for stress concentration where the cracks may initiate and grow. Due to the stiffness of certain condenser tube materials such as titanium, it is necessary to install more baffles and tube supports than for other materials such as stainless steel, cupronickels, and brass. Vibrational stresses may cause or contribute to cracking of turbine components. Vibration may increase due to deposition or metal loss on turbine components, which changes the weight balance.

Identification

Acid Corrosion

Carbonic acid corrosion produces smooth, general metal loss and grooves on carbon steel and copper alloy components that follow the path of the condensate stream (see Figs. 16.1 to 16.3). Parallel, narrow striations may be superimposed on the corroded surfaces. Metallographic examination may reveal that striations are produced by corrosion that preferentially penetrates along microstructural features in the metal, particularly inclusions. The wasted surfaces may be lined with thin layers of corrosion products containing carbonates and oxides. Acid spot tests conducted on corrosion products that cause them to effervesce may indicate that carbonate is present. Green corrosion products on surfaces of copper alloy components suggest that that they contain copper carbonate (Fig. 16.13). Blue or purple condensate suggests the presence of dissolved copper-ammonia complex due to corrosion of copper alloys. X-ray diffraction analysis may be conducted on the corrosion products to determine what compounds are present. Metallographic examination will reveal wasted surfaces that are bare or are lined with thin corrosion product and deposit layers that formed following the most recent acid corrosion event (Fig. 16.14). In some cases, undercut pits may be produced.

FIGURE 16.13 Green copper carbonate corrosion products.

FIGURE 16.14 Metallographic cross section of undercut pits lined with thin iron oxide corrosion products.

Oxygen Corrosion

Metal loss due to oxygen corrosion is commonly in the form of iron oxide–filled pits or depressions (Fig. 16.5). Wider areas of metal loss may form as corrosion spreads outward from the original pit sites. Voluminous iron oxide mounds or tubercles may form atop the corroded surfaces. Metallographic examination may reveal burrowing depressions filled with dense iron oxides, as shown in Fig. 16.15.

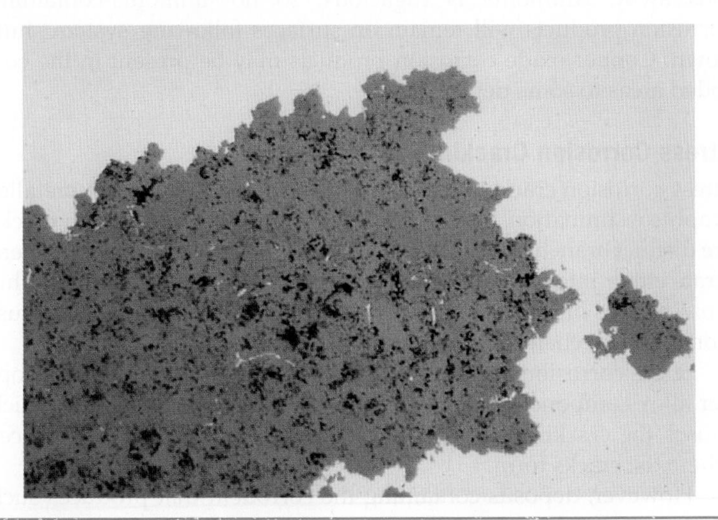

FIGURE 16.15 Metallographic cross section of pit formed by oxygen corrosion.

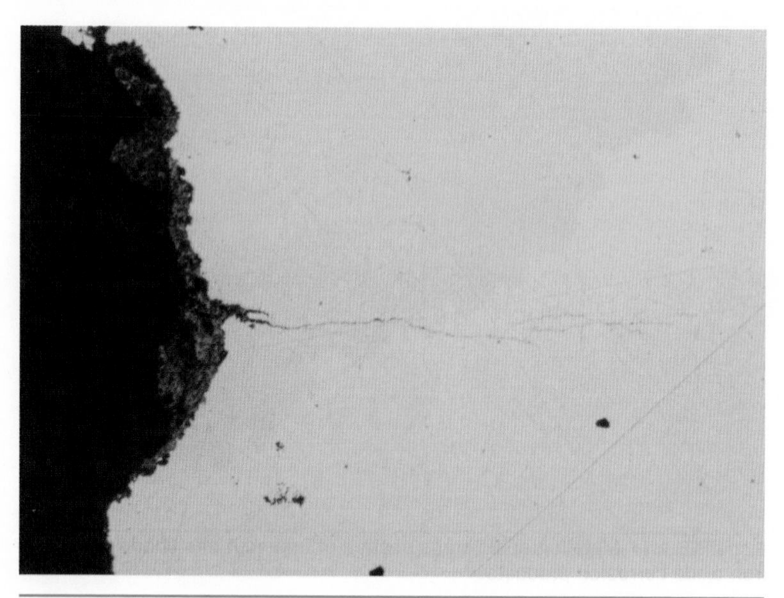

FIGURE 16.16 Metallographic cross section of branched, transgranular stress corrosion crack on brass condenser tube.

Ammonia Corrosion

Ammonia produces metal loss on copper components, which is similar to that produced by carbonic acid corrosion, such as grooves and smooth, general metal loss (Fig. 16.16). In some cases, little or no corrosion products form, since the complexing solution carries the metal ions away. Ammonia is fugacious, so no ammonia-containing corrosion products will remain on surfaces following system shutdown. Copper oxide corrosion products may be present in the corroded areas to some degree.

Stress Corrosion Cracking

Stress corrosion cracks must be properly identified through metallographic examination. The typical identifying features of such cracks are distinct branching (see Figs. 16.16 and 16.17). The cracks may penetrate either intergranularly or transgranularly. More metallographic crosssections of typical stress corrosion cracks and a detailed discussion of the mechanism are presented in Chap. 13.

Stress corrosion cracks may form on surfaces of susceptible copper alloy components, typically zinc brasses. Since ammonia, which causes the cracking, is volatile, no residue will be left on surfaces where the cracks form.

However, deposits containing the corrodent that produces such cracking on carbon, low-alloy, and stainless steel components may remain on the cracked surfaces if the surfaces were not rinsed by

FIGURE 16.17 Metallographic cross section of branched, transgranular stress corrosion crack on carbon steel steam pipe.

water subsequent to the cracking event. In some cases, pH strip tests conducted on deposits that are hydrated with deionized water will indicate an alkaline solution, which is strong evidence for caustic stress corrosion cracking. The material may be identified by SEM or other techniques. The presence of sodium in the deposits is strong evidence for caustic stress corrosion cracking.

Boiler Water Carryover

Corrosion due to boiler water carryover on turbine components tends to be in the form of pitting, or localized corrosion (see Figs. 1.7 and 1.8). Undercut pits may be produced (Fig. 16.18). Corrosion products and deposits may fill or line the pits. However, subsequent operation or cleaning may remove such material. The pH of the deposits in distilled water solution may reveal if acidic conditions may have been produced. Deposits and corrosion products may be identified with x-ray fluorescence and diffraction or SEM analysis.

Corrosion monitoring may be conducted through the use of corrosion coupons. Corrosion may be identified by measuring iron and copper in the condensate. The measurements may be determined by wet chemistry techniques on water samples or by collecting membrane filter samples. Membrane filter samples can be used to estimate the amount of corrosion that occurs within the system based upon the darkness of the color on the filter. On-line monitoring techniques such as conductivity and total organic carbon (TOC) and meters designed

FIGURE 16.18 Undercut pits on turbine blade.

to detect turbidity may be useful for detecting carryover. This will allow the contaminated condensate to be drained from the system.

Fatigue and Corrosion Fatigue Cracking

Fatigue and corrosion fatigue cracks form in directions that are normal to the applied cyclic stress. They may form at weld interfaces or at points of rigid restraint, such as at tubesheets or baffles (Fig. 16.12). As with any cracking failure, metallographic examination is required to positively identify the cause of cracking. Corrosion fatigue cracks are transgranular and tend to be straight or wedge-shaped, and they are filled with dense corrosion product layers that contain centerline cracks (Fig. 16.19). Fatigue cracks tend to be lined with thin corrosion product layers because corrosion is typically not a substantial accelerating factor for crack growth (Fig. 16.20). It is also common for several fissures or cracks to form within localized areas where cyclic stresses are applied.

Erosion

Erosion is characterized by localized metal loss that produces features that are shaped by flow, such as parallel grooves or depressions that travel in a common direction. In some cases, clusters of deep, sharp-edged pits form on turbine components (Fig. 16.9). Since erosion removes material, surfaces in the wasted areas will be devoid of corrosion products and deposits. Metallographic examination may reveal plastic deformation at the eroded surfaces.

FIGURE 16.19 Metallographic cross section of straight, transgranular corrosion fatigue crack on brass condenser tube.

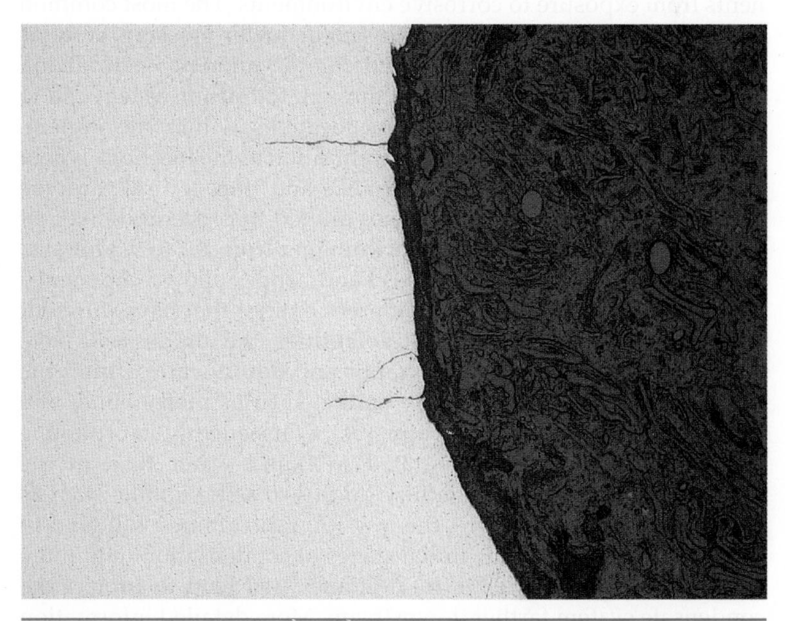

FIGURE 16.20 Metallographic cross section of straight, transgranular corrosion fatigue cracks on turbine blade.

Elimination

Acid Corrosion

Controlling the corrosivity of carbonic acid, dissolved oxygen, and ammonia to condensate system components requires two approaches. It is necessary to both remove these gases as effectively as possible and to treat the condensate properly, since neither method will be successful alone. Methods of oxygen removal in the preboiler include boiler water pretreatment, deaeration, and oxygen scavenging. Pretreatment processes, such as dealkalizing, demineralization and reverse osmosis, that remove bicarbonate and carbonate alkalinity will significantly reduce or remove sources of carbon dioxide that can be carried through to the steam system. Deaerators will also reduce carbon dioxide in the preboiler. In steam systems, venting of carbon dioxide and oxygen may be conducted by components such as air ejectors in surface condensers and steam traps within steam lines.

However, such components cannot be relied upon to remove the gases completely. Therefore, it is necessary to employ suitable condensate corrosion inhibitors, which are fed to the boiler water or steam, as appropriate. These inhibitors are conveyed by the fluids through the steam and condensate system to protect the components from exposure to corrosive environments. The most common condensate corrosion inhibitors used in lower-pressure systems consist of neutralizing amines and filming amines. Neutralizing amines are designed to be carried through the steam system and to dissolve in condensate at various locations within the system. When there is sufficient amine concentration at the locations where the steam condenses, it will neutralize and thus control carbonic acid corrosion. Neutralizing amines are fed to concentrations that will produce condensates with pH ranging from 8.2 to 9.3 for systems that contain both carbon steel and copper alloy components. It is common to use blends of different amines that have different volatilities. Due to the different volatilities, the amines will have different concentrations in the vapor and liquid water within the system. That is, a low-volatility amine, such as morpholine, will have a low vapor-to-liquid ratio V/L. Consequently, morpholine will be more highly concentrated in liquid water than in the steam relative to a high-volatility (high V/L ratio) amine, such as cyclohexylamine. Therefore, the low V/L ratio amines will tend to protect components at the initial stages of condensation within the system, while the high V/L ratio amines will tend to protect the condensate system farther downstream. More detailed information concerning condensate corrosion inhibitors is presented in *The Nalco Water Handbook*, third edition.

Oxygen Corrosion

Filming amines are most applicable for control of oxygen corrosion. Filming amines do not neutralize acids. Rather, they will form a water-repellent film on surfaces where condensate forms. If the water cannot contact the metal surfaces, corrosion cannot occur. Filming amines may also be used when carbonic acid corrosion cannot be adequately controlled through use of neutralizing amines alone. The amine films may be compromised if exposed to condensates with pH exceeding about 9.0, or less than about 6.5, or under erosive conditions. Neutralizing amines are typically not effective enough to control oxygen corrosion. At high pH, some reduction in oxygen corrosion rates may be afforded. However, at such high pH, copper alloy corrosion may increase. Volatile oxygen scavengers may be used in steam and condensate systems to reduce dissolved-oxygen concentration in the condensate, to control oxygen corrosion. Specific condensate corrosion inhibitors and oxygen scavengers that are used in steam and condensate systems are described in detail in *The Nalco Water Handbook*, third edition.

Ammonia

Ammonia may be removed to some degree by air ejectors in condensers, or in deaerators if the pH is suitably high. Venting within the condensate system may also be useful for ammonia removal. If such removal is not applicable or sufficient, then the source of ammonia may be reduced. Ammonia contaminants from process in-leakage into returned condensate or the feedwater should be removed. If the source of ammonia is from degradation of treatment chemicals, the use of alternate programs may be considered.

Stress Corrosion Cracking

For susceptible copper alloys, the ammonia control recommendations described above should be followed. For carbon, low-alloy, and stainless steel components, boiler water carryover and contamination of attemperation water by agents that may cause stress corrosion cracking must be prevented.

Carryover

The sources of impurities in the boiler water that may carry over to the steam system and turbine are minimized to some degree by maintaining proper feedwater and boiler water chemistries. However, such control does not guarantee that deposition will be controlled completely. Carryover may be caused by mechanical, operational, or boiler water chemistry problems. Mechanical problems may include damaged steam separation equipment. Operational problems may include boiler operation at high drum levels. Boiler water chemistry problems may include high dissolved-solids concentration or contamination by organic material that causes foaming.

Erosion

To control erosion, entrainment of water droplets, entrained hard particles such as exfoliated iron oxide from superheater and reheater components, and boiler water, carryover in the steam should be controlled.

Fatigue and Corrosion Fatigue

In general, fatigue and corrosion fatigue may be controlled through reduction of the magnitude and frequency of applied cyclic stresses. Vibration may be reduced on turbine components by controlling deposition by carryover of boiler water, and by vaporous carryover at high pressures. Clean turbine surfaces may be maintained by water washing. Proper design and installation of components, particularly turbine components and condenser tubes, will provide increased resistance to these cracking mechanisms.

Related Problems

See also Chap. 9, Oxygen Corrosion, and Chap. 10, Corrosion during Cleaning.

Case History 16.1

Industry:	Chemical process
Specimen location:	Condensate return line
Orientation of specimen:	Horizontal
Years in service:	9
Steam treatment program:	Neutralizing amine blend
Tube specifications:	2½-in. (6.4-cm) outer diameter

Reportedly, the condensate return system operated intermittently. The piping carried condensate from a reactor jacket to a condensate receiver approximately once per week, for a period of a few hours.

Deep localized metal loss occurred within an area along the reported bottom side, where a shallow pool of condensate collected. Very deep, sharply defined, curved, longitudinal grooves formed within the wasted area. The metal loss area terminated abruptly (Figs. 16.21 and 16.22). Brown and black corrosion products that covered the surfaces of the grooves effervesced strongly when acidified, suggesting the presence of carbonates. This indicated that carbonic acid corrosion occurred.

The neutralizing amine program required adjustments to ensure that proper neutralization was achieved at the location that experienced carbonic acid corrosion.

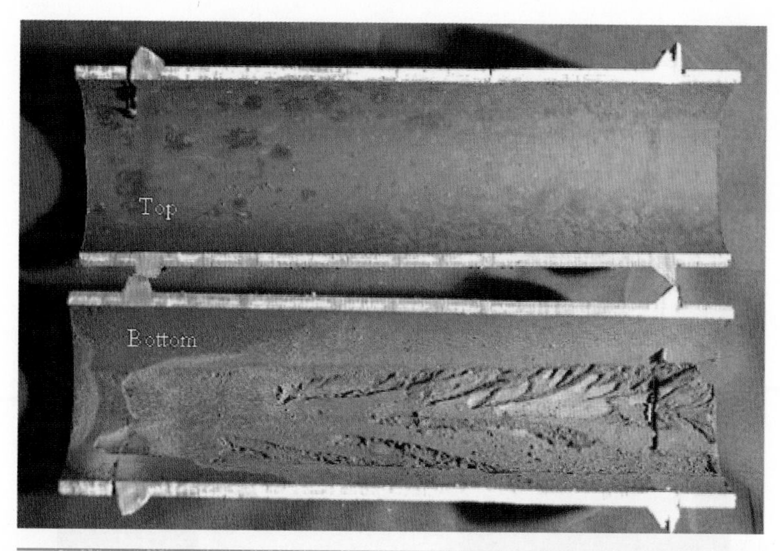

FIGURE 16.21 Grooves formed on bottom side of condensate line due to carbonic acid corrosion.

FIGURE 16.22 Close-up of grooves shown in Fig. 16.21.

Case History 16.2

Industry:	Oil refining
Specimen location:	Main steam discharge line
Orientation of specimen:	Horizontal
Steam treatment program:	Neutralizing amine blend
Tube specifications:	18-in. (45.7-cm) outer diameter

Due to forced shutdowns caused by multiple hurricanes, the boiler that supplied steam to the line was out of service for an extended period. This long-term shutdown was followed by intermittent operation while the refinery was being returned to steady-state operation. A failure in the line during boiler operation prompted an internal inspection. Deep pits, depressions, fissures, and cracks were found on the internal surface of the line within a narrow band along the reported bottom side (Figs. 16.23 and 16.24).

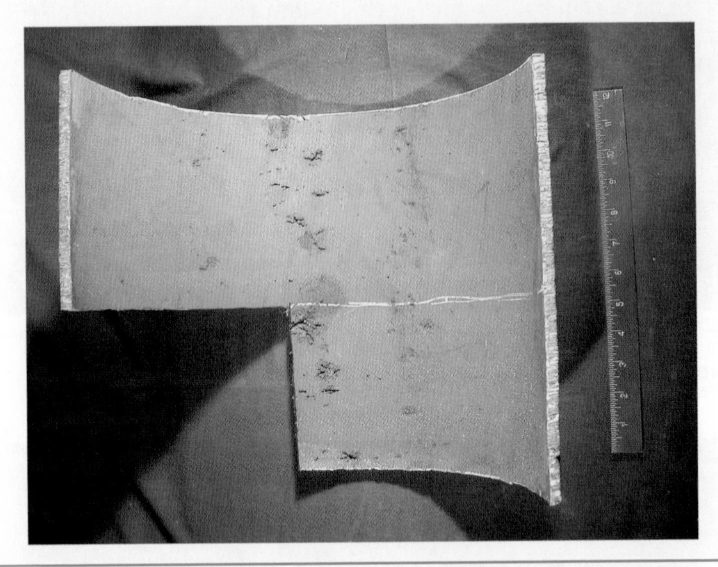

FIGURE 16.23 Oxygen pits and fissures formed within localized area on bottom side of steam pipe.

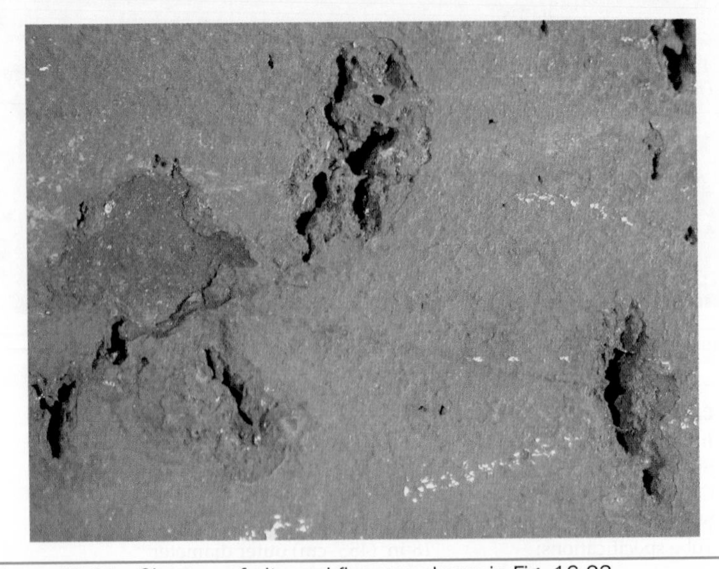

FIGURE 16.24 Close-up of pits and fissures shown in Fig. 16.23.

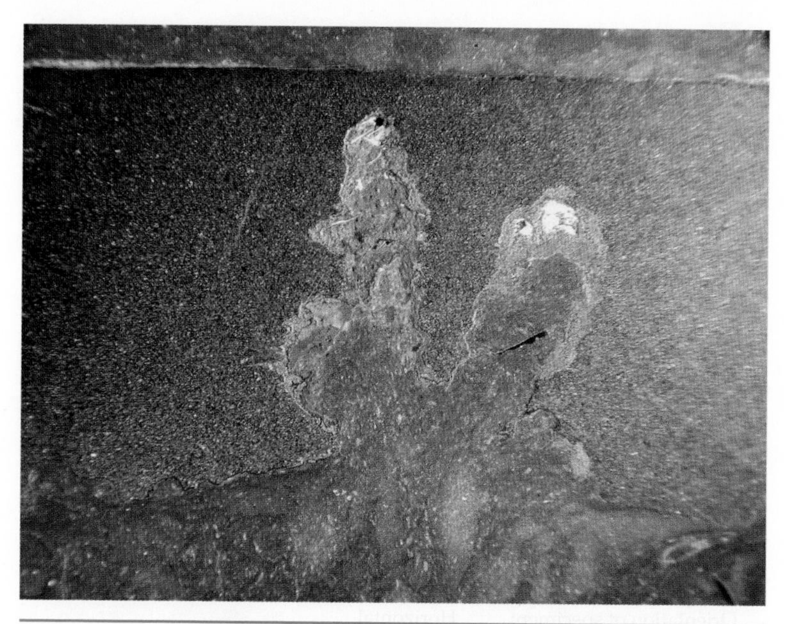

FIGURE 16.25 Metallographic cross section of pits and fissures that are filled with dense iron oxide.

Metallographic analysis revealed that the pits and depressions formed and grew in part due to oxygen corrosion resulting from exposure to condensate containing dissolved oxygen. Oxygen corrosion occurred during idle periods when steam condensed and collected along the bottom of the pipe and became oxygenated by in-leaked air. Corrosion during idle periods may be controlled by using proper layup procedures.

In addition to the corrosion, several deep fissures and cracks formed at the sites of the pits and depressions within a narrow band along the bottom most portion of the line (Fig. 16.25). The fissures and cracks resulted from corrosion fatigue cracking. Corrosion fatigue cracking was caused by the application of cyclic tensile stresses. The pits and depressions that formed due to oxygen corrosion supplied points of stress concentration for initiation and growth of the corrosion fatigue fissures. It is probable that the cyclic stresses responsible for cracking were primarily supplied by temperature variations and from intermittent operation.

Case History 16.3

Industry:	Chemical process industry
Specimen location:	Surface condenser
Material:	Aluminum brass
Orientation of specimen:	Horizontal
Years in service:	9
Water treatment program:	Boiler (congruent phosphate, hydrazine) Steam (neutralizing amine blend)
Tube specifications:	¾-in. (1.9-cm) outer diameter

Deep grooves formed on the external surfaces of several tubes in a surface condenser. The grooves were oriented in the transverse direction and were located closely adjacent to baffle contact points (Fig. 16.6). The grooves formed due to corrosion resulting from exposure to condensate containing significant concentrations of dissolved ammonia and oxygen. In localized areas where thin films of condensate cover surfaces, ammonia concentrations may be much higher than in the bulk condensate.

Corrosion may be controlled by reducing ammonia concentration in the steam. It is probable that the primary source of ammonia was from breakdown of the hydrazine used as an oxygen scavenger for the boiler water. Other sources such as process contaminants were possible. When ammonia cannot be removed adequately, or the amount contributed by hydrazine breakdown or contamination cannot be reduced, the use of alternate materials is recommended. Cupronickel alloys are more resistant to ammonia corrosion than aluminum brass. Stainless steel is immune to corrosion by dissolved ammonia. Ammonia stress corrosion cracking did not occur due to the resistance of the aluminum brass alloy.

Case History 16.4

Industry:	Refining
Specimen location:	Surface condenser
Material:	Admiralty brass
Orientation of specimen:	Horizontal
Years in service:	19
Water treatment program:	Boiler (polymer, organic oxygen scavenger) Steam (neutralizing amine blend)
Tube specifications:	¾-in. (1.9-cm) outer diameter

After many years of successful service, numerous admiralty brass condenser tube sections experienced cracking failures. The failed tubes were located exclusively within the section of the condenser where noncondensable gases collected (Fig. 16.7). The cracks were primarily oriented in the transverse direction, and formed closely adjacent to baffle contact points. Metallographic examination revealed that the cracks that penetrated from the external surface were tight, transgranular, and branched (Fig. 16.16). These features revealed that the cracks resulted from stress corrosion cracking.

The primary contributing factor for the cracking was exposure to condensate containing significant concentrations of dissolved ammonia and oxygen. This was verified by the pH of the condensate, which was periodically as high as 10 in the bulk solution. In localized areas where thin films of condensate cover surfaces, ammonia concentrations may be much higher than in the bulk condensate.

Improved removal of oxygen by the air ejector may help to control ammonia stress corrosion cracking. However, ammonia reduction in the steam would be more beneficial. It is probable that the primary source of ammonia was from breakdown of the polymer used in the internal treatment program for the boiler water. Other sources, such as process contaminants, were possible. If the ammonia could not be removed adequately, or the amount contributed by polymer breakdown or contamination cannot be reduced, the use of alternate materials is recommended. Cupronickel alloys are very resistant to ammonia stress corrosion cracking. Tubes made of these alloys are recommended for condensers that employ admiralty brass within the noncondensable gas section, particularly when tubes in those locations experience ammonia stress corrosion cracking.

Case History 16.5

Industry:	Oil refining
Specimen location:	Steam line to deaerator
Orientation of specimen:	Horizontal
Material:	Carbon steel
Years in service:	14
Tube specifications:	6½-in. (16.5-cm) outer diameter
Treatment program:	Boiler: residual phosphate treatment, organic oxygen scavenger Steam: Neutralizing and filming amine blend

Several cracking failures occurred exclusively at welds within a steam line. The cracks penetrated along irregular, branched paths through both the weld beads and the adjacent pipe walls (Fig. 16.8). The internal surface was fairly clean in most areas. However, light-colored deposits were found in scattered locations. Components of these deposits produced an alkaline solution in distilled water. Metallographic examination revealed that the cracks were tight, intergranular, and highly branched, typical of stress corrosion cracks (Fig. 16.17). These cracks resulted from exposure of the steam line to condensate that contained concentrated dissolved caustic or alkaline salts, while metal temperatures exceeded about 300°F (149°C). The alkaline salts were provided by carryover of boiler water in the steam. The cracks formed preferentially at welds and heat-affected zones where residual stresses were greater.

Caustic stress corrosion cracking may be controlled most effectively by preventing boiler water carryover. If it is known that a steam line experienced carryover, proper water washing is recommended.

Case History 16.6

Industry:	Electric utility
Specimen location:	Steam turbine
Components	Turbine buckets
Material:	Martensitic stainless steel
Water treatment program:	Equilibrium phosphate treatment, hydrazine

A turbine was out of service for an extended period. When the system was being prepared for restarting, the turbine was out of balance, so it was taken out of service for inspection and repairs. Inspections revealed metal loss on about one-third of the buckets within one portion of the turbine. Metal loss on the buckets was in the form of hemispherical shaped pits and depressions, some of which were slightly undercut (Fig. 16.26). Some depressions along the blade root were covered by red and black corrosion products that contained chloride. Chloride was supplied by the carryover of boiler water to the turbine by the steam.

The pits and depressions formed due to corrosion during idle periods, when the chloride in the deposits became hydrated by water supplied by condensed steam and water vapor from in-leaked air. The solution produced a corrosive, acidic environment.

Corrosion on turbine components may be controlled through use of appropriate rinsing practice. This is particularly important prior to long-term idle periods.

FIGURE 16.26 Metallographic cross section of pits on turbine bucket. Pits are filled with corrosion products and deposits.

Case History 16.7

Industry:	Alcohol production
Specimen location:	Turbine for compressor
Components:	Turbine blade housing
Material:	Low-alloy steel
Treatment program:	Boiler: congruent phosphate treatment, organic oxygen scavenger Steam: neutralizing amine blend

Inspection of the boiler revealed severe damage of the steam separation equipment. During service, the turbines required water washing to operate properly. Inspection of the turbine revealed deposits as well as cracking within the turbine blade housings, where they contacted the blade roots (Fig. 16.27). Metallographic examination revealed that the cracks were tight, intergranular, and branched (Fig. 16.28).

These cracks formed due to stress corrosion cracking. Stress corrosion cracking of metals results from exposure to specific corrodents for the particular alloy while being subjected to applied and/or residual stresses. Residual stresses are supplied by original manufacturing operations. Based upon the locations of the cracks, it is probable that the stresses responsible for the cracking were primarily applied.

The specific corrodent that was responsible for stress corrosion cracking was caustic or alkaline salts in water solution. Caustic or alkaline salts were exposed to the turbine due to boiler water carryover and/or contamination of attemperator water. The caustic solutions concentrated to levels that were sufficient to cause cracking within the tight crevices between the blade roots and the slot in the disk.

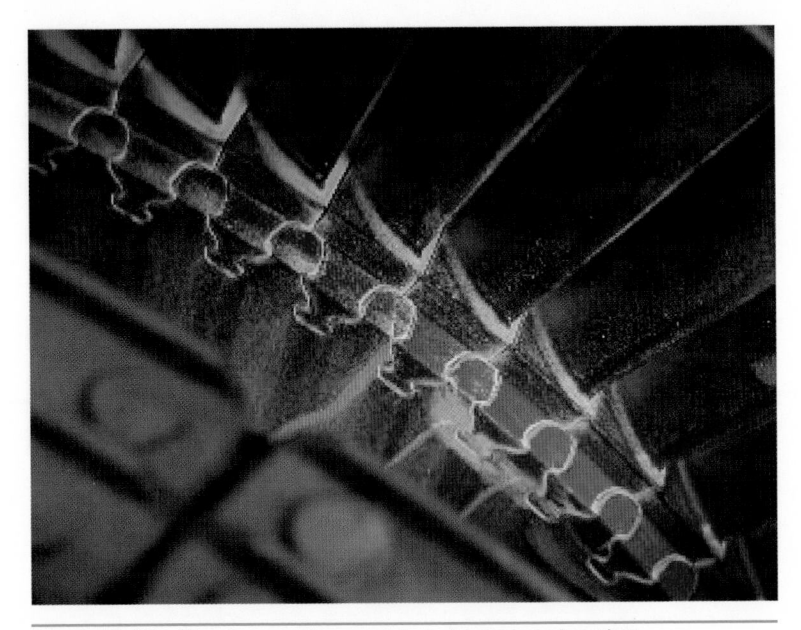

FIGURE 16.27 Stress corrosion crack in turbine blade housing.

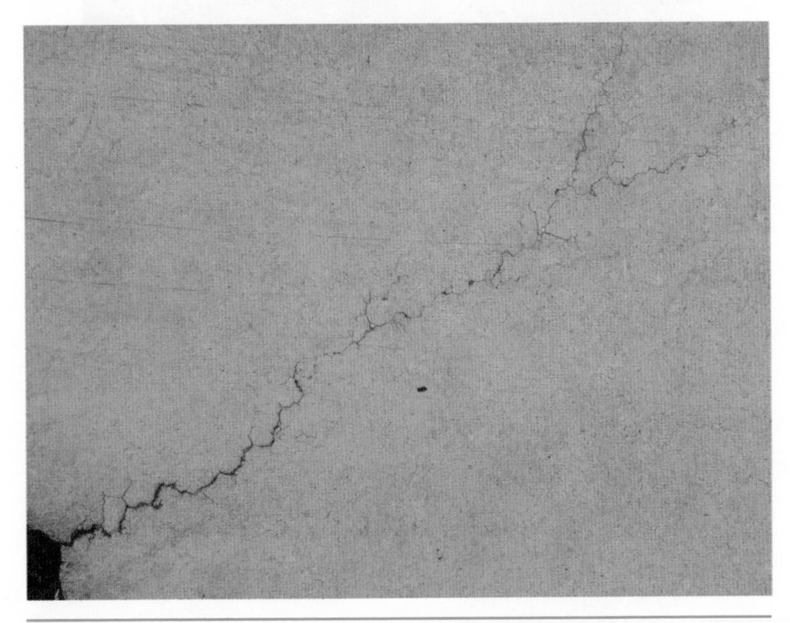

FIGURE 16.28 Metallographic cross section through branched, intergranular stress corrosion crack.

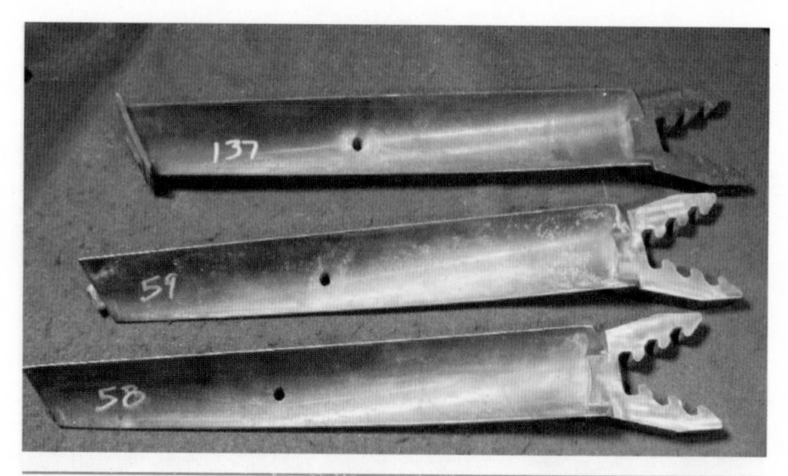

FIGURE 16.29 Failed turbine blades.

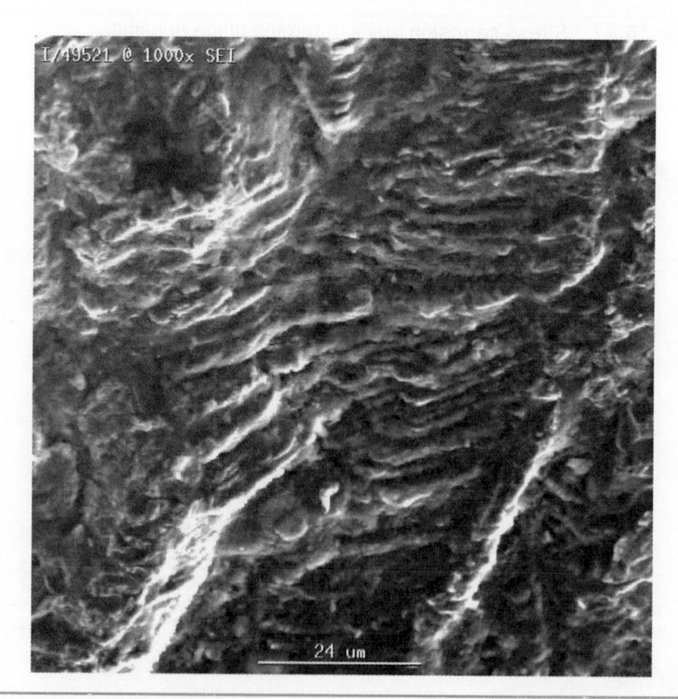

FIGURE 16.30 SEM photo of fatigue striations on fracture surface.

It was necessary to repair the damaged steam separation equipment to control boiler water carryover that was primarily responsible for the cracking.

Case History 16.8

Industry:	Electric utility
Specimen location:	Steam turbine
Components:	Turbine blades
Material:	Martensitic stainless steel
Treatment program:	Boiler: congruent phosphate treatment, organic oxygen scavenger Steam: neutralizing amine blend

Turbine blades fractured at the tenon joints (Figs. 16.29 and 16.11). The fracture surfaces were flat and bare-surfaced, exhibiting crack arrest marks. Metallographic examination revealed families of parallel, tight, straight, transgranular, unbranched fissures (Fig. 16.20). SEM examination revealed striations on the fracture surface (Fig. 16.30). These features are consistent with fatigue cracking. Fatigue cracking is caused by the application of cyclic tensile stresses. The stresses were supplied by vibrational loading at the tenon joint. It is possible that the band that attached the blade to the tenon was loose.

In addition to fatigue cracking, the edges of some blades in the turbine experienced erosion due to impingement of steam containing entrained water droplets (Fig. 16.9).

SECTION 1.6

Waterside Flow-Related Damage

CHAPTER 17
Cavitation

General Description

Cavitation is a process whereby small vapor or gas bubbles rapidly form and collapse in a liquid (Fig. 17.1 depicts a simplified schematic of the process). Figure 17.1A shows a sealed vessel containing a liquid at rest. Rapid, localized pressure changes in the liquid, represented by upward movement of the plunger in Fig. 17.1B, cause vapor or gas bubble formation. The liquid actually boils at the reduced pressure. The bubbles quickly collapse (represented by downward movement of the plunger in Fig. 17.1C), producing microjets of liquid that impinge on metal surfaces (Fig. 17.2). The microjets may have velocities ranging from 330 to 1650 ft/s (~100 to 500 m/s).[1] The microjets may damage only the normally protective oxide layer or may directly attack the underlying metal in severe cases, physically dislodging less resistant phases in some alloys. Bubble formation and collapse may occur in just a few milliseconds. Each bubble collapse produces a relatively small amount of damage; however, the damage accumulates over thousands of cycles.

Energy is required to form a cavitation bubble. Part of that energy is consumed in creating the bubble surface, or vapor/liquid interface. Cavitation bubbles form most readily on existing surfaces, where less vapor/liquid interface area is required for bubble formation. Pressure may be lowest and turbulence highest at or near moving surfaces. Surface discontinuities afford easy bubble nucleation sites. Once surface irregularities are formed due to attack, cavitation will tend to concentrate at the irregularities, eventually producing deep, localized attack. Corrosion and cavitation can act synergistically to result in accelerated metal loss.

Locations

Cavitation is favored anywhere low-pressure regions form in a liquid. Abrupt pressure changes and turbulent flow promote attack. Damage may occur only where liquid contacts surfaces. Pump impellers and other pump parts are the most commonly attacked components in the

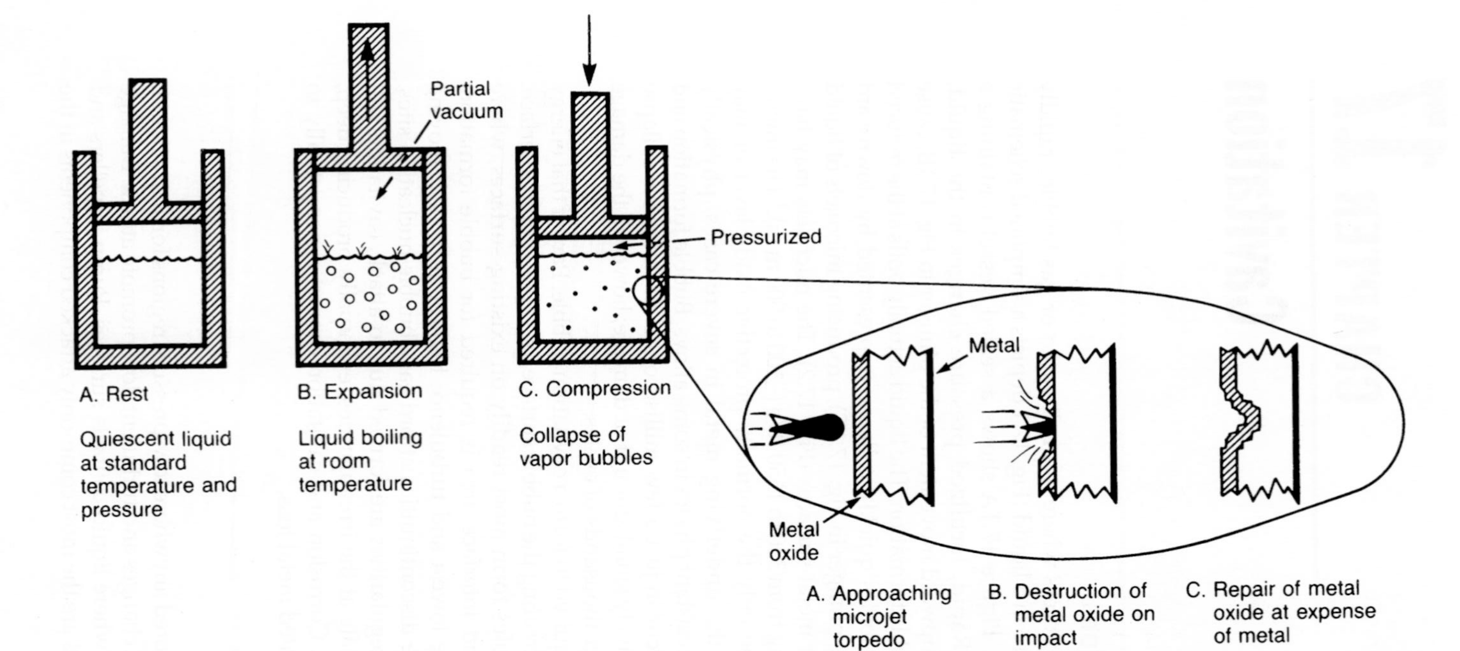

FIGURE 17.1 Simplified schematic representation of the cavitation process, depicting liquid in a sealed vessel with a plunger. (A) Liquid at standard temperature and pressure, plunger stationary; B) plunger is withdrawn, lowering pressure and causing liquid to boil at standard temperature; (C) plunger is advanced, collapsing bubbles. (*Courtesy of McGraw-Hill, The Nalco Guide to Cooling Water Systems Failure Analysis, New York, 1993.*)

FIGURE 17.2 Microjet of water (represented by a torpedo), destroying a layer of protective corrosion products on a metal surface. (*Courtesy of McGraw-Hill, The Nalco Guide to Cooling Water Systems Failure Analysis, New York, 1993.*)

boiler water and steam circuit. Blowdown lines and valves are also frequently affected. Pump impellers are usually damaged on the suction side, and valves typically show wastage on discharge sides. Condensate lines and, less commonly, turbine components may also be attacked.

Critical Factors

Pump cavitation is often caused by an excessive pressure differential between the suction and discharge sides. Insufficient head pressure is usually the precipitating cause. Throttling on the suction side of pumps promotes a large pressure differential. Gas entrainment due to leaking packing, decomposition of water treatment chemicals, and gas effervescence can also promote bubble formation. In a surprising number of cases, incorrectly sized impellers and other pump components cause difficulties.

Blowdown lines are especially susceptible to damage if flow is excessive and the discharge direction changes abruptly. Attack usually occurs during intermittent manual blowdown when flow direction is severely changed in pipe tees and elbows. Attack can be intense if the blowdown rate is high and lines are undersized (Fig. 17.3).

Alloy composition also influences attack. Soft, ductile metals and brittle, low-strength alloys such as gray cast iron are especially susceptible (Fig. 17.4). Alloys such as stainless steels are resistant to attack in many environments.

FIGURE 17.3 Longitudinally split blowdown line with severe localized metal loss caused by cavitation.

FIGURE 17.4 Pump housing section severely attacked by cavitation. Material of the pump housing section is gray cast iron.

Turbulent flow and abrupt pressure changes promote attack. In many cases, cavitation is a threshold phenomenon. Cavitation does not simply gradually decrease in intensity, but ceases altogether below some critical turbulence level.

Identification

Cavitation damage produces localized areas of jagged metal loss. Undercutting is usually pronounced. Close visual inspection reveals surfaces having a spongy, honeycombed texture. Attacked regions may or may not be covered with corrosion products. Corrosion products will be minimal if attack was active shortly before the section was inspected. Although corrosion usually accelerates metal loss, attack can occur in its absence.

A striking feature of cavitation damage is the localized nature of attack. Wastage is generally most severe on pump impeller vanes near the outer periphery where impeller speed and presumably turbulence are highest (Fig. 17.5). Attack occurs preferentially on trailing vane edges, where low-pressure areas are formed (Fig. 17.6). Pump shafts can be attacked locally where surfaces are exposed to turbulent conditions (Fig. 17.7). Adjacent surfaces are usually only lightly polished or apparently free of damage. Edges, corners, and projections all intensify turbulence, provide bubble nucleation sites, and become preferred damage sites.

FIGURE 17.5 A cast steel feedwater pump impeller severely damaged by cavitation. Note how damage is confined to the outer edges of the impeller, where vane speed was at its maximum.

FIGURE 17.6 Vane damage on low-pressure sides of a small, special-purpose, bronze feedwater pump impeller. Some vanes have been penetrated.

FIGURE 17.7 Highly localized cavitation damage on steel feedwater pump shaft. Adjacent regions are free of metal loss.

If components are vibrating, damage may be present on all surfaces in contact with liquid; however, vibratory modes involving movement in a single plane are more common and produce attack on opposite sides of the component.

Cavitating pumps can sometimes be recognized by the sounds they make. Active cavitation can sound like the impact of stones against metal surfaces. However, pump noise and vibration usually mask these sounds.

In many cases, including those occurring in plain carbon steel, microstructural examination of areas damaged by cavitation will often reveal the presence of deformation twins (or Neumann bands) near the surface. Deformation twins appear as well-defined parallel lines that often terminate at grain boundaries (Fig. 17.8). Such twins are indicative of a high rate of plastic deformation, such as that caused by mechanical shock.[2] In other cases, alloys that are damaged by cavitation may exhibit microstructural evidence of cold work.

Elimination

Cavitation damage can be reduced by operational changes, design, alloying, coating, and/or surface finishing. Maintaining sufficient head pressure, preventing packing leaks, and discharge-side throttling all reduce pump damage. Design strategies concentrate on reducing

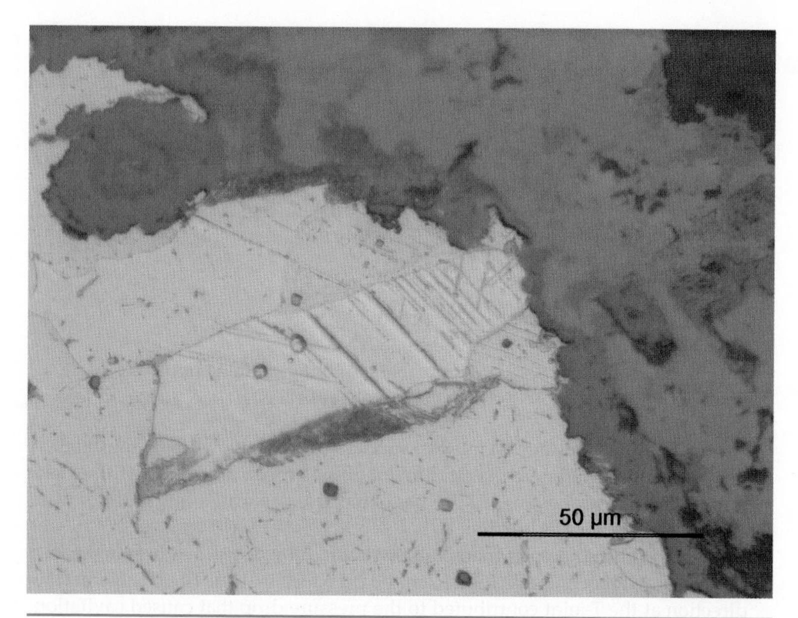

FIGURE 17.8 Micrograph of an area of cavitation damage in carbon steel, showing deformation twins (Neumann bands) near the surface. Billowing iron oxide corrosion products on the surface indicate that oxygen corrosion occurred subsequent to cavitation. Etchant: Nital.

turbulence, vibration, and rapid pressure changes. Smoothing of impeller surfaces and use of elastomeric coatings prevent damage by reducing nucleation sites and by absorbing the implosion energy of bubbles, respectively. In piping, increasing radii of curvature and eliminating abrupt changes in cross-sectional area would aid in preventing the occurrence of cavitation.

Hard, resistant alloys including 18–8 stainless steels are often recommended. Hard-facing and weld overlays, using suitable more resistant materials, may also be effective.

Cautions

Cavitation damage resembles corrosion by a strong acid. However, metal loss caused by a strong acid is generally widespread due to its lack of dependence on flow characteristics, whereas cavitation usually causes damage in localized areas. Although somewhat similar to other impingement and erosion phenomena, cavitation produces distinctly different wastage. Because of the unique characteristics of this damage, microscopic observations often conclusively show that cavitation has occurred.

Related Problems

See also Chap. 10, Corrosion during Cleaning, and Chap. 18, Erosion and Erosion-Corrosion (Water and Steam Sides).

Case History 17.1

Industry:	Chemical manufacturing
Specimen location:	Blowdown line threaded into a tee
Specimen orientation:	Horizontal
Years in service:	6
Water treatment program:	Phosphate
Drum pressure:	1200 psi (8.3 MPa)
Tube specifications:	2½-in. (6.4-cm) outer diameter, mild steel
Fuel:	No. 6 fuel oil, natural gas

Severe, localized wastage on internal surfaces caused a blowdown line to fail. A perforation occurred near the attachment of the pipe to a tee (Fig. 17.3). Attack was confined to spongy, jagged patches of metal loss; surrounding surfaces were free of significant deterioration.

Failure occurred during manual blowdown. Blowdown had been increased due to feedwater contamination from the preboiler system. The change in flow direction at the T-joint contributed to the pressure drop that caused cavitation. The source of contamination was eliminated, thus decreasing the need for blowdown. The T-joint was redesigned to minimize the pressure drop.

Case History 17.2

Industry:	Pulp and paper
Specimen location:	Boiler feedwater pump impeller
Specimen orientation:	Vertical
Years in service:	1
Water treatment program:	Phosphate
Drum pressure:	1200 psi (8.3 MPa)
Specifications:	10½-in. (26.7-cm) diameter, seven vanes, cast steel
Fuel:	Black liquor

Feedwater supply could not meet boiler demand. The feedwater pump impeller was severely wasted, and the vanes were almost completely destroyed. Attack was confined to the periphery of the impeller (Fig. 17.5).

Wastage was apparently caused by cavitation induced by insufficient head pressure. The feedwater supply pressure was increased to eliminate the problem.

Case History 17.3

Industry:	Chemical
Specimen location:	Boiler feedwater line valve
Specimen orientation:	Horizontal
Valve specifications:	Gate valve (Fig. 17.9); bore: 9¾ in. (24.8 cm) long, 1-5/8-in. (4.0-cm) diameter, cast steel

Deep, rough, mutually intersecting pits were present on the inlet and outlet faces of the bore, although the metal loss was much more severe at the outlet (Fig. 17.10). Similar pits were also present on the internal surface of the bore, along the top. The internal surfaces of the pits were covered with a layer of

FIGURE 17.9 Boiler feedwater line gate valve that sustained severe cavitation damage.

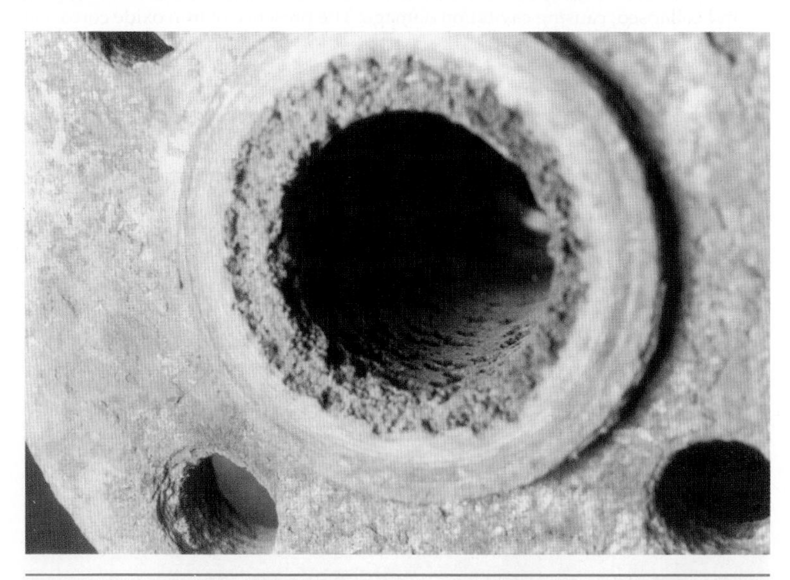

FIGURE 17.10 Outlet side of the valve, showing deep pits on the face of the bore and on the top.

dark-colored iron oxide. The deep pits were surrounded by generally unattacked metal that was overlaid by a layer of dark brown iron oxide.

The jagged nature of the pitting and the unattacked metal near the pits indicated that the wastage was caused by cavitation. The diameter of the bore decreased by approximately ¼ in. (0.6 cm) near the gate and valve seat, which likely created downstream turbulence on the outlet side. High-flow velocities may have also promoted cavitation.

Case History 17.4

Industry:	Institutional, heating system
Specimen location:	Condensate return line
Specimen orientation:	Slanted
Years in service:	40
Condensate treatment program:	Neutralizing amines
Pipe specifications:	2-3/8-in. (6.0-cm) outer diameter, 0.154-in. (0.391-cm) wall thickness, mild steel

The steel piping contained many deep, iron oxide–lined depressions along the reported top (Fig. 17.11). The depressions had an irregular surface contour, and some were undercut. Along the bottom, many shallow depressions were lined and capped with mounds of iron oxide corrosion products. Microscopic examinations revealed the presence of deformation twins near the surfaces of some deep depressions along the top.

Oxygen corrosion caused most of the damage, but cavitation contributed to the metal loss along the reported top of the section. The section was reportedly removed from a location near a steam trap. A failure had reportedly occurred in the vicinity. It is possible that the steam trap malfunctioned and allowed some steam to pass through, bubbles of which floated along the top of the piping and collapsed, causing cavitation damage. The presence of iron oxide corrosion

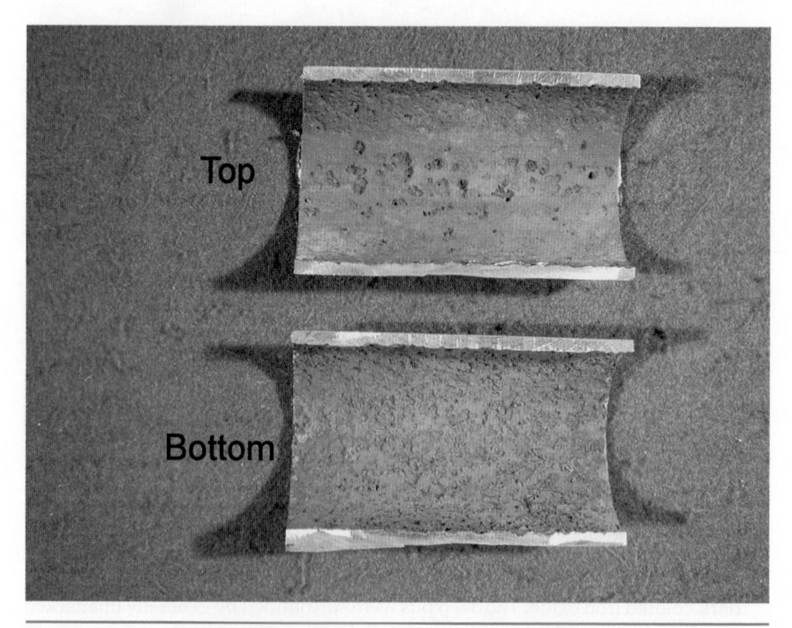

FIGURE 17.11 Condensate return line, showing many deep depressions along the top, where cavitation and oxygen corrosion both contributed to metal loss. The many shallow depressions along the bottom were caused by oxygen corrosion.

products lining the depressions along the top indicated that oxygen corrosion had occurred subsequent to at least some of the cavitation damage.

Case History 17.5

Industry:	Pharmaceutical
Specimen location:	Condensate return line
Specimen orientation:	90° elbow with vertical and horizontal orientations
Years in service:	3
Condensate treatment program:	Proprietary film-forming chemistry
Line pressure:	135 psi (0.9 MPa)
Sample specifications:	Cast iron, inner diameter 1¾ in. (4.4 cm)

A 90° pipe elbow developed a perforation along the extrados (outer bend). The perforation was centered in a localized, triangle-shaped area of deep internal surface metal loss, which had irregular contours along the surrounding edges (Fig. 17.12). Some moderately deep depressions with irregular interiors and slight undercutting were present in places surrounding the larger metal loss area. These depressions were bare-surfaced, indicating that the attack was active shortly before the elbow was removed from service.

Microscopic analysis indicated deformation twins in the grains near the metal loss area (Fig. 17.13). The presence of the deformation twins and the characteristics of the metal loss indicated attack was primarily caused by cavitation damage. The elbow was located between a heat exchanger and a steam trap; the flow of steam was regulated by a thermostatically controlled valve. Pressure variations from operation of the valve were likely responsible for cavitation. The localization of the attack to the outer bend of the elbow indicated that turbulence from abrupt changes in flow direction promoted cavitation. Operating parameters for the system were investigated to limit sources of locally intense pressure changes and turbulent flow to reduce cavitation intensity, as film-forming chemical treatments cannot prevent cavitation.

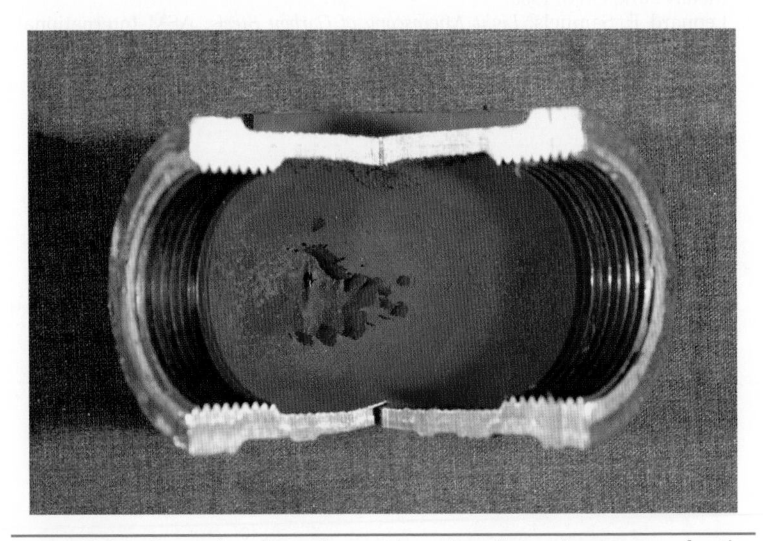

FIGURE 17.12 Internal surface of a cast iron pipe elbow, showing a perforation located in a deep region of metal loss having irregular edges.

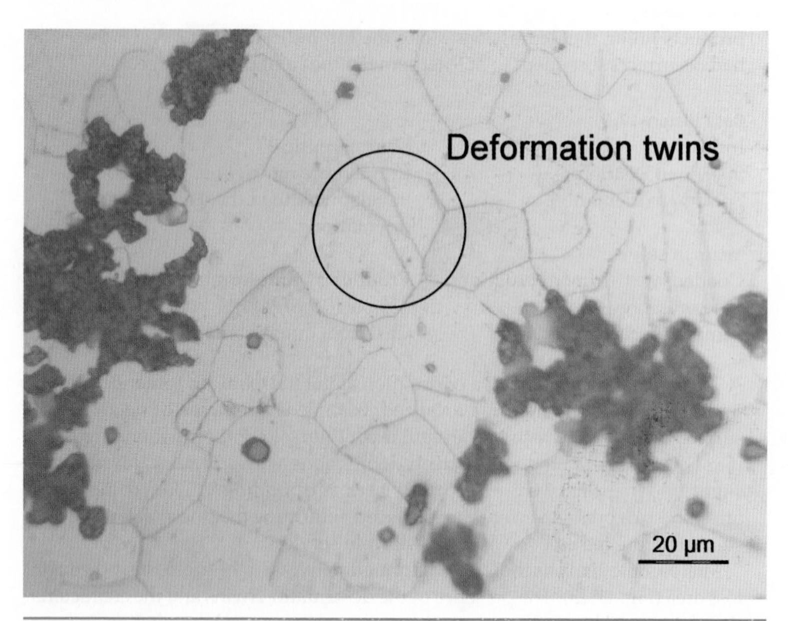

FIGURE 17.13 Deformation twins (Neumann bands) in some ferrite grains in the area of deep metal loss. The black constituents are graphite from the cast iron. Etchant: Nital.

References

1. Frederick G. Hammitt and Frank J. Heymann, "Liquid-Erosion Failures," *ASM Handbook*, vol. 11: *Failure Analysis and Prevention*, 9th ed., ASM International, Metals Park, Ohio, 1986.
2. Leonard E. Samuels, *Light Microscopy of Carbon Steels,* ASM International, Materials Park, Ohio, 1999, pp. 136–140.

CHAPTER 18

Erosion and Erosion-Corrosion (Water and Steam Sides)

General Description

Erosion is metal loss caused by impact of fluids or solids. Attack is promoted by turbulent, high-velocity fluid flow. Rapid pressure changes promote water jetting and turbulence. Abrupt changes in flow direction and the entrainment of hard particulate matter in fluids also contribute to wastage. Although appearing simple, the erosion process is complex. It is thought that plastic deformation occurs in a shallow zone near the metal surface, and that metal loss occurs when the plastically deformed metal is subsequently removed by the fluid. During the early stages of erosion, while the initial plastic deformation is occurring, an incubation period is observed, during which the erosion rate is nonexistent or very low (Fig. 18.1).

Erosion-corrosion occurs when erosion and corrosion act jointly to cause accelerated metal loss. Erosive forces can remove corrosion products that form on the metal surfaces. Corrosion products re-form and are again removed; the continued cycle of corrosion product formation and removal results in increased or accelerated metal loss rates, relative to erosion alone. In severe cases of erosion-corrosion, metal can be removed as dissolved ions. Almost all metals contacting steam or water experience some corrosion, including normal formation of an oxide layer. Therefore, all waterside or steamside erosion is an erosion-corrosion process to some extent. If evidence suggests that corrosivity of the fluid had a very small effect on the metal loss, however, the process may be classified as erosion rather than erosion-corrosion.

Flow-accelerated corrosion (FAC) is a phenomenon that is related to erosion-corrosion. FAC is sometimes considered to be a special case of erosion-corrosion. In FAC, the water primarily dissolves the magnetite coating on the waterside surface or prevents it from forming, whereas

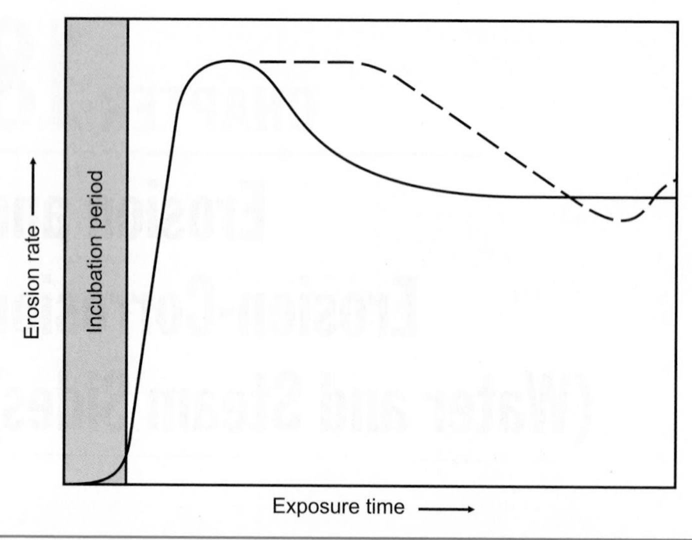

FIGURE 18.1 Schematic of liquid erosion rate as a function of time, showing an incubation period in the initial stages and a rapid rise to a maximum rate. The duration of the maximum wastage rate can vary, as shown by the solid and dashed lines. The erosion rate then decreases to a lower value, which can fluctuate or remain steady.[1] (*Reprinted with permission of ASM International. All rights reserved. www.asminternational.org.*)

mechanical damage of the oxide layer and underlying metal occurs in the case of erosion and erosion-corrosion. FAC and erosion or erosion-corrosion can produce wastage with very similar characteristics and thus may be misdiagnosed. However, the water must meet certain temperature, pH, dissolved-oxygen concentration, and other criteria for FAC to occur. See Chap. 19 for an in-depth discussion of FAC.

Locations

Metal loss due to erosion on waterside surfaces is relatively rare in boilers, since normal fluid velocities are generally too low to cause such damage to typical alloys used in boiler construction. Erosion may occur, however, in locations that are exposed to high-velocity and/or turbulent fluid flow; such cases are often associated with entrained hard particulate matter. Common velocities in various parts of steam-generating systems are listed in Table 18.1. Typical flow velocities in economizer tubes and boiler water circuits are 150 to 300 ft/min (0.8 to 1.5 m/s) and 70 to 700 ft/min (0.4 to 3.6 m/s), respectively. However, modern heat recovery steam generators (HRSGs) may be designed to produce large mass fractions of steam through the use of externally finned tubes, which can result in velocities close to, or in excess of, 1970 ft/min (10 m/s), particularly in the low- or intermediate-pressure sections. Higher heat input in selected

Type of service	Velocity	
	ft/min	m/s
Steam lines		
High pressure	8,000–12,000	40.6–61.0
Low pressure	12,000–15,000	61.0–76.2
Vacuum	20,000–40,000	101.6–203.2
Superheater tubes	2,000–5,000	10.2–25.4
Water		
Boiler circulation	70–700	0.4–3.6
Economizer tubes	150–300	0.8–1.5
Pressurized water reactors		
Fuel assembly channels	400–1,300	2.0–6.6
Reactor coolant piping	2,400–3,600	12.2–18.3
General water lines	500–750	2.5–3.8

TABLE 18.1 Common Fluid Velocities in Steam-Generating Systems[2]

tubes, e.g., along side walls where laning of hot flue gas may occur, can result in even higher steam fractions and higher velocities. Due to typically high flow velocities in superheater tubing, any water that is carried over can cause erosion, in addition to the corrosive effects that result from evaporation to dryness and concentration of boiler water salts. First bends in superheater tubing typically exhibit the most severe wastage in cases of carryover. In general, flow velocities in steam tubing and transport lines increase with decreasing pressure. Any magnetite chips that are loosened and carried with the steam can have a considerable erosive effect downstream, removing additional oxide and creating more mechanical damage in their path.

Turbulence may develop as a result of a change in flow direction, abrupt cross-sectional change, thick buildups of deposits, or the presence of an obstruction that disrupts flow. In the case of water- or steam-cooled tubes, an obstruction or disruption in flow will typically cause other problems, such as overheating, that usually result in failure before erosion can cause severe damage. In some components, such as pumps and valves, turbulence is an intrinsic characteristic of the fluid flow through the component.

Preboiler attack is confined primarily to feedwater systems. Feedwater pump components, including impellers, fittings, valves, and housings, may experience erosion. Less commonly, transfer lines, pipe elbows, and blowdown components are attacked.

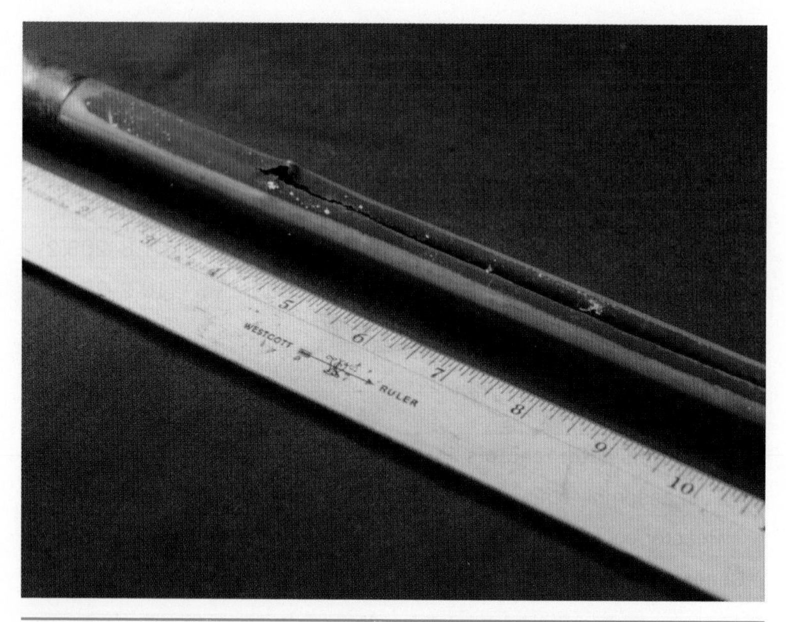

FIGURE 18.2 Erosion-corrosion damage on the surface of a brass tube facing the steam inlet nozzle. (*Courtesy of McGraw-Hill, The Nalco Guide to Cooling Water Systems Failure Analysis, New York, 1993.*)

After boiler erosion often occurs in turbine components because of the high velocities inherent in turbine operation. Turbine blades are wasted both by hard particles and by water droplet impingement. Latter-stage buckets are most frequently affected by water droplet impingement. Any components that are contacted by wet, high-velocity steam (such as condenser tubing near a steam inlet, shown in Fig. 18.2) can sustain damage by erosion or erosion-corrosion. Valve stems, nozzle blocks, diaphragms, and early-stage buckets may suffer hard-particle erosion due to exfoliated iron oxide particles from superheaters, reheaters, main steam leads, and hot reheat piping. Deposits supplied by boiler water carryover can also promote erosion. Blowdown piping, condensate return lines, steam traps, and many other boiler components can be affected by erosion.

Critical Factors

In the case of waterside erosion, a threshold velocity exists below which metal loss is negligible.[3] The threshold velocity is dependent on environmental conditions, alloy type, and other factors. Above the threshold velocity, the rate of metal loss generally increases with the erodent kinetic energy. Erodent velocity has a very significant effect on metal loss; doubling the velocity may increase metal loss by a factor of 4 or more.[4]

If particulate matter is entrained in the fluid, many factors can affect attack. These factors include particle size, shape, density, hardness, size distribution, impact angle, velocity, and fluid viscosity. On the water side, particulate matter may consist of deposits, corrosion products, or thermally formed oxides that are dislodged from surfaces during cleaning or service (such as resulting from the formation of a bulge due to overheating). If gases become entrained in liquids, cavitation (see Chap. 17, Cavitation) is more likely than "simple" erosion. On the steam side, solid-particle erosion is typically caused by exfoliated oxides, primarily from superheaters, reheaters, and main steam piping. Oxides are often dislodged by stresses associated with thermal transients. In turbines, impingement of liquid water droplets on ferrous and copper alloys produces wastage on latter-stage buckets, drain lines, condenser tubing, and piping. If condensed fluids have a low pH, such as can result from dissolved carbon dioxide, erosion-corrosion becomes more significant.

In the absence of significant corrosion, alloys with higher hardness values tend to be more erosion-resistant than softer alloys. Note that this rule of thumb may only be valid when comparing similar alloys; it may not be a good way to predict relative erosion resistances of different alloy types. Microstructural features such as small grain sizes and fine dispersions of hard second-phase particles contribute to the erosion resistance of some alloys, such as Stellites used for hard-facing and tool steels.[5] Copper and copper-containing alloys are generally more susceptible to erosion and erosion-corrosion than carbon steel and other ferrous alloys, with the exception of cast iron. Figure 18.3 shows approximate normalized relative erosion resistances of various

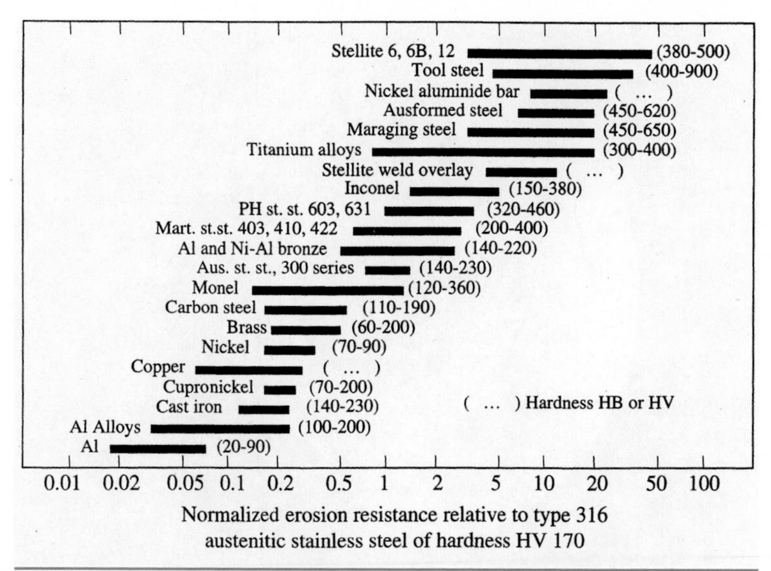

Figure 18.3 Approximate normalized erosion resistances of various metals and alloys, relative to type 316 austenitic stainless steel with a hardness of HV 170.[6] (*Reprinted with permission of ASM International. All rights reserved. www.asminternational.org*)

alloy types. It should be emphasized that such erosion resistance values should only be used as rough estimates.

In the cases where corrosion plays a significant role in causing metal loss, factors affecting the contributing corrosive attack logically have a resulting effect on erosion-corrosion. Fluid pH, dissolved-oxygen concentration, and temperature can have various effects on erosion-corrosion, depending on the alloy, fluid, and operative corrosion mechanism(s).

Identification

Water Side

Erosion caused by turbulent water flow usually produces flow-related features that may include smoothened surfaces, rolling contours, grooves, sand dunelike patterns, horseshoe-shaped depressions, and/or comet tail-shaped depressions. Very high velocities may result in general thinning, although metal loss is more commonly localized to regions of disrupted or turbulent flow (Fig. 18.4). Compared to unattacked adjacent regions, the oxide layer in eroded regions is thinner or nearly nonexistent, can have a different color or sheen, and has a smoother surface texture. A drop of acid placed on freshly eroded carbon steel will reveal bright metal almost immediately. Unattacked surfaces usually require considerable exposure to

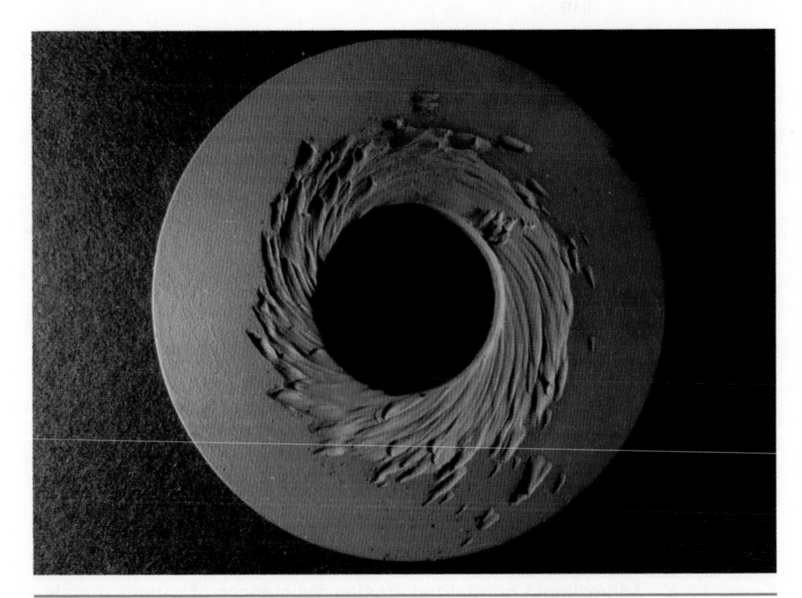

FIGURE 18.4 Feedwater pump spacer eroded by turbulent high-speed water. Flow patterns are obvious.

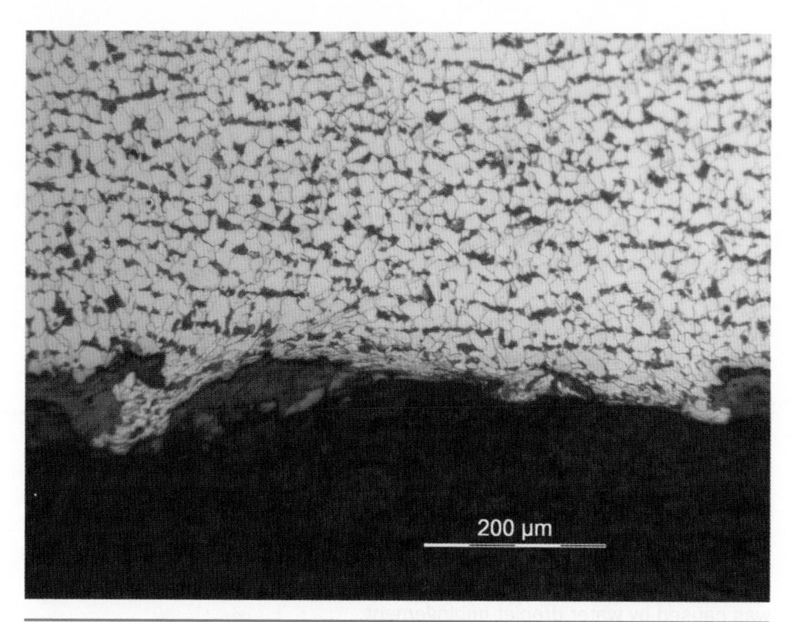

FIGURE 18.5 Micrograph of the internal surface of a boiler tube in an area of localized metal loss caused by erosion-corrosion. A shallow zone of plastic deformation suggests that hard particles contributed to the wastage. Etchant: Nital.

reveal bare metal. The extent of corrosive damage outside of eroded areas may give some indication as to the magnitude of corrosion's contribution to the wastage. If no significant corrosion is observed outside of the eroded areas, erosion most likely played a larger role in causing the damage.

Unless particulate matter is entrained, close visual inspection will reveal no evidence of individual scratches, streaks, or dents. Usually, such damage is apparent only when a larger amount of hard particulate matter is entrained within the carrying fluid. If impingement of hard particulates contributed to the wastage, microscopic examinations will often reveal a shallow zone of plastic deformation near the surface (Fig. 18.5).

Steam Side

Erosion caused by impingement of water droplets in latter turbine stages produces general wastage, which is most noticeable on leading edges of buckets (Fig. 18.6). Blade edges are marked by fine, transverse serrations and grooves (Fig. 18.7). Cone-shaped projections may rise from surfaces (Fig. 18.8). Often the cones will be tilted so that the conical axes are parallel to the impingement direction. Hence, cones are usually present on leading bucket faces and are less numerous or absent on back blade faces. Similar impingement

FIGURE 18.6 Ragged leading edges of final-stage turbine blades. Damage was caused by water droplet impingement.

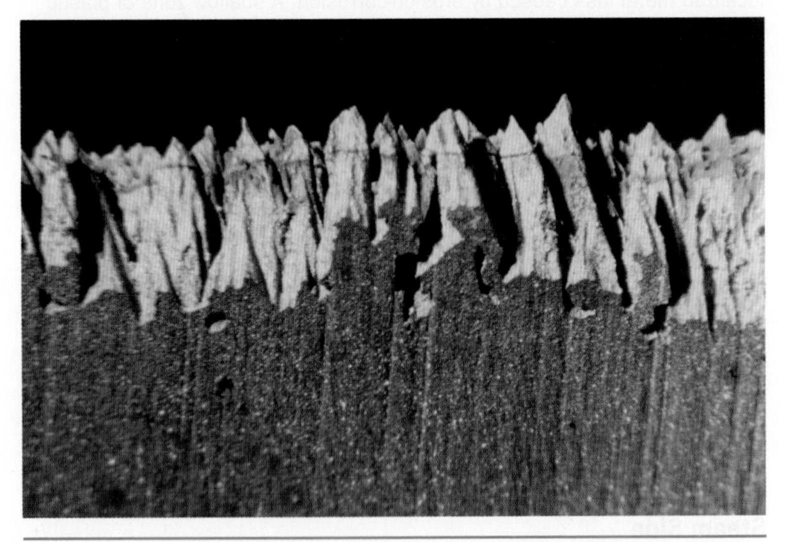

FIGURE 18.7 Fine, striated grooves on front face of final-stage turbine blade caused by water droplet impingement.

cones can occur on mild steel surfaces such as turbine drain lines (Fig. 18.9).

Damage caused by hard-particle erosion causes tearing and microdenting of leading bucket edges. Striations and tilted cones are usually absent. A ragged feather edge usually develops.

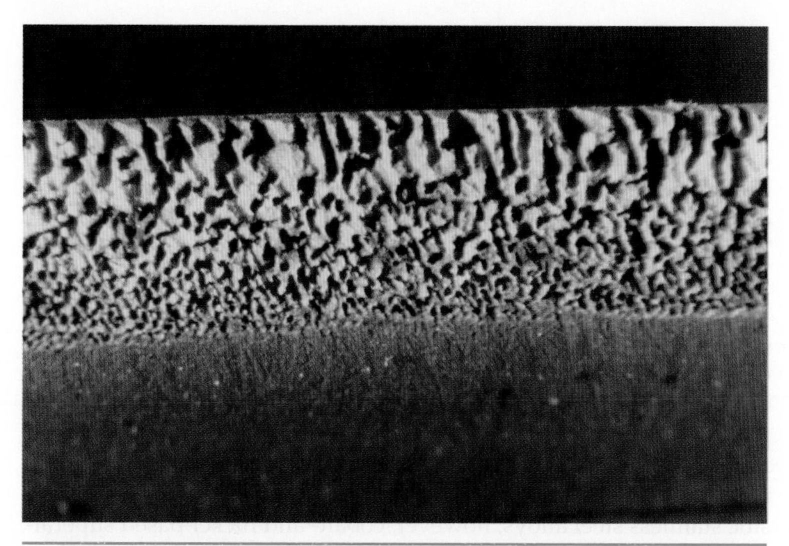

FIGURE 18.8 Dimpling of back edge of bucket surface shown in Fig. 18.7.

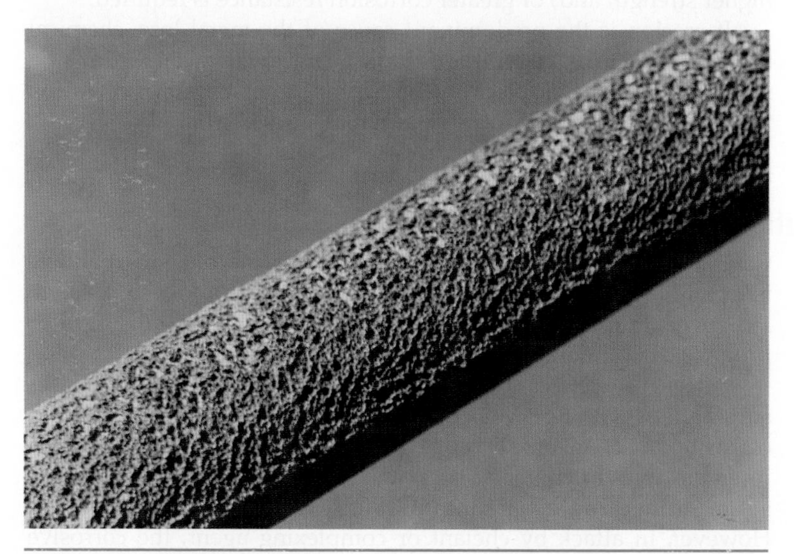

FIGURE 18.9 Erosion on mild steel turbine drain line. Surface is festooned with small, conical projections pointing toward the direction of impingement.

Elimination

Reducing attack from erosion or erosion-corrosion requires eliminating the erodent, decreasing fluid velocity and/or fluid turbulence, shielding surfaces, or substituting alloys that are more erosion-resistant. If pumps, valves, and piping are sized, designed, and operated properly, attack is rare. Turbine erosion problems can be more troublesome.

Magnetite exfoliation in superheaters, reheaters, and steam-transfer lines (see Chap. 1, Water-Formed and Steam-Formed Deposits) is the major cause of solid-particle erosion (SPE) in turbines. The exfoliation process is lessened when thermal stress, tube temperature, and oxide layer thickness are reduced. The internal surfaces of superheater and reheater tubes that were in service for many years may be covered with thick, stratified layers of thermally formed iron oxide. Tube replacement or chemical cleaning (if possible) may be necessary to remove thick oxide layers. Other measures that may prevent exfoliation from steamside tube internal surfaces include the application of diffusion-type coatings (e.g., chromizing) or overlay-type coatings of carbides or hard-facing materials, which may be applied via methods such as plasma spray, thermal spray, vapor deposition, electroplating, or cladding.[7] Turbine screens and shielding devices, which may be used at start-up, should be maintained in good repair. The most common materials used to fabricate turbine buckets are martensitic stainless steel alloys; however, cobalt- and nickel-based superalloys, titanium, and proprietary metals are also used, usually when higher strength and/or greater corrosion resistance is required.

If erosion is the predominant cause of the metal loss, chemical inhibition can usually do little to reduce attack; however, if corrosion is significant, pH modification and/or the judicious use of corrosion inhibitors appropriate for the system may be beneficial.

Cautions

Cavitation is closely related to erosion. In both mechanisms, damage is promoted where high-velocity turbulent fluids are present. However, cavitation does not occur in steam and is not likely to produce smooth, undulating, or grooved surfaces.

Attack by concentrated chelant or complexing agent increases substantially with flow velocity, and it may be considered an erosion-corrosion mechanism. Because of similar features that strongly resemble erosion characteristics, chelant attack in feedwater lines, steam drums, and generating tubes can be confused with erosion. However, in attack by chelant or complexing agent, the corrosive component to the metal loss dominates.

Damage caused by flow-accelerated corrosion, in which oxide dissolution plays a very significant role, may also be diagnosed as erosion; however, FAC is different and is active in only well-defined regimes of water chemistry and other environmental conditions.

Related Problems

See also Chap. 8, Corrosion by Chelating and Complexing Agents; Chap. 17, Cavitation; and Chap. 19, Flow-Accelerated Corrosion.

Case History 18.1

Industry:	Food and beverage
Specimen location:	Boiler feedwater pump impeller
Time in service:	7 weeks
Metallurgy:	Leaded semired brass

Two feedwater pump impellers sustained deep general metal loss and severe localized metal loss along the outer edges, where fluid velocities were expected to be highest (Fig. 18.10). Numerous perforations occurred in the locations of deepest wastage. Many vanes were bent in the direction of flow, as a result of weakening from thinning due to the metal loss. In the wasted areas, many intersecting depressions produced rough or spongy contours. The spongy appearance of the metal loss areas suggested that cavitation may have contributed to the wastage; however, microscopic examinations did not reveal evidence of plastic deformation that is often indicative of cavitation damage.

The surfaces were mostly free of significant deposits or corrosion products and had an etched appearance with a dendritic, as-cast structure evident upon inspection with a stereomicroscope (Fig. 18.11). Thin, brown deposit and/or oxide layers were present in scattered locations along the hub and vane edges where fluid velocity and turbulence were not as great.

The severe metal loss was caused by erosion-corrosion. The presence of general metal loss suggested that corrosion contributed significantly to the severe wastage. Reportedly, the feedwater experienced periods of low pH, particularly during the winter. It was recommended that the pH be raised to the range appropriate for the metallurgies in the system. If a change in pH proved to be ineffective, it was recommended that alternative metallurgies, such as certain

Figure 18.10 Leaded semired brass boiler feedwater pump impeller that sustained deep general metal loss and severe localized metal loss along the outer edges. Many vanes were bent in the direction of flow.

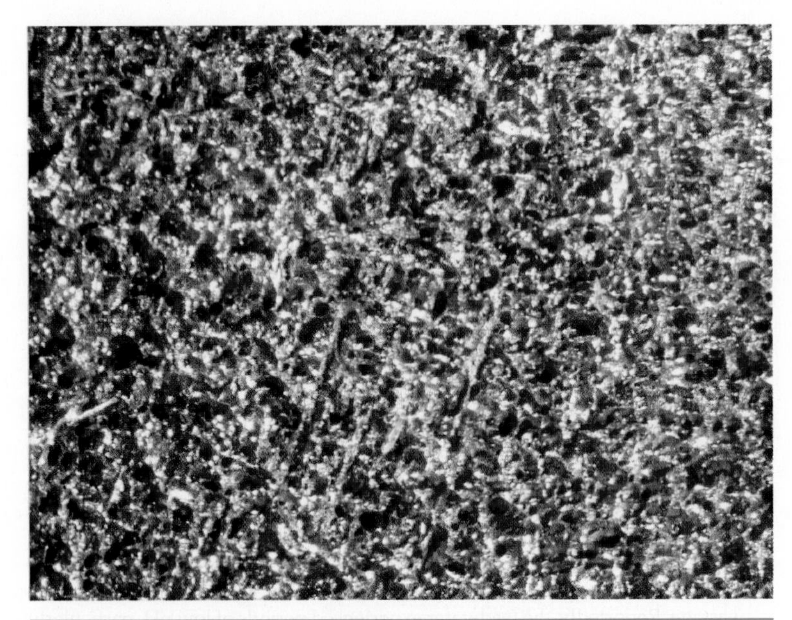

FIGURE 18.11 Surface of a leaded semired brass pump impeller, showing an etched appearance with a dendritic, as-cast microstructure.

cast stainless steels, be considered. Possible galvanic effects and other interactions should always be considered when evaluating alternate metallurgies.

Case History 18.2

Industry:	SAGD oil recovery
Boiler type:	Heat recovery steam generator (HRSG)
Specimen type:	Two elbows and one finned, straight tube section, high-pressure evaporator section
Metallurgy:	Mild steel
Tube specifications:	Approximately 3½-in. (8.9-cm) outer diameter, 0.438-in. (1.11-cm) nominal tube wall thickness, lengths ranging from 10.5 to 21 in. (26.7 to 53.3 cm)

No failures were present in the received sections, but ultrasonic thickness testing indicated appreciable wall loss. The most severe metal loss occurred in the straight tube section, where the tube-wall thickness was decreased by as much as 66%, relative to the nominal wall thickness. The sections experienced internal surface metal loss that was characterized by the presence of horseshoe-shaped depressions and a sand dunelike contour in places (Fig. 18.12). In the elbows, the wastage was most severe along the extradoses, where fluid impingement and turbulence were greatest. The straight tube section was most severely wasted in a location between the reported top and side, most likely in line with, and downstream from, the outer bend of a previously connected elbow. A thin layer of black iron oxide on the surface stripped quickly when exposed to a strong acid. Microscopic examinations indicated the presence of a shallow zone of plastic deformation near the internal surface in some metal loss areas (Fig. 18.5).

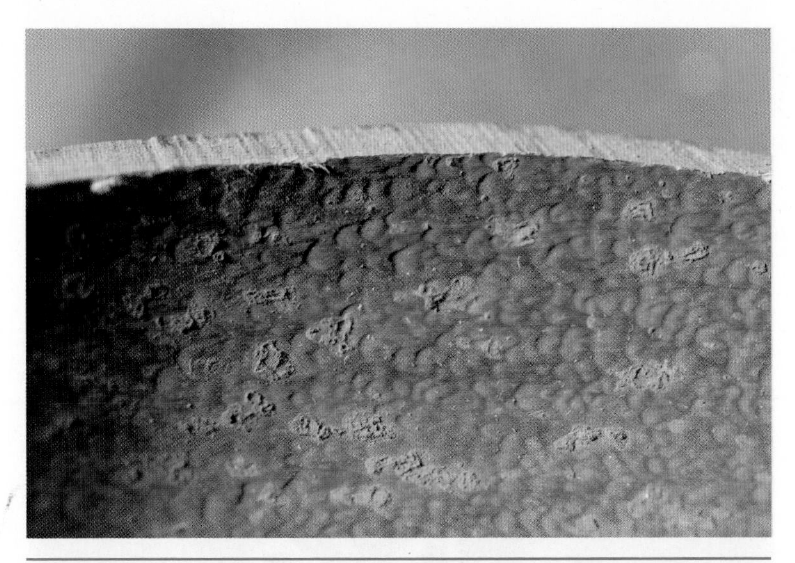

FIGURE 18.12 Overlapping, horseshoe-shaped depressions in an area of internal surface metal loss along the extrados of an elbow section from an HRSG.

The localized nature and flow-related features of the metal loss areas indicated that they were caused by erosion-corrosion. Fluid velocities in the passes increase with increasing steam quality in the once-through unit. The thin nature of the iron oxide layer overlying the metal loss areas suggested that attack was actively occurring shortly before the sections were removed from service. Plastic deformation in the metal loss areas suggested that hard particles entrained in the water contributed to the erosive damage. Deposit and water analyses suggested that some particles likely consisted of silicate scale that was periodically released from the tube surface upstream from the wastage. No significant corrosion was observed away from the areas of erosive metal loss, suggesting that erosion contributed much more significantly to the wastage than corrosion. A more erosion-resistant alloy, such as a low-alloy steel, was suggested to limit the metal loss observed in this case. The small amount of chromium in low-alloy steels can result in an iron oxide layer that is slightly more erosion- and corrosion-resistant than mild steel.

Case History 18.3

Industry:	Food and beverage
Specimen location:	Ball float steam trap, branch steam line
Specimen orientation:	Horizontal
Time in service:	12 to 15 months
Metallurgy:	Ductile cast iron body and cover, stainless steel internals with a graphite gasket

A small perforation was located on the outlet side of the trap, near the interface between the cover and body (Fig. 18.13). The perforation was located in an area of smoothened metal loss in the internal passage that allowed condensate to flow from the interior of the body side to the trap outlet (Fig. 18.14). The metal loss occurred along the outer bend in a location of abrupt change in flow direction in

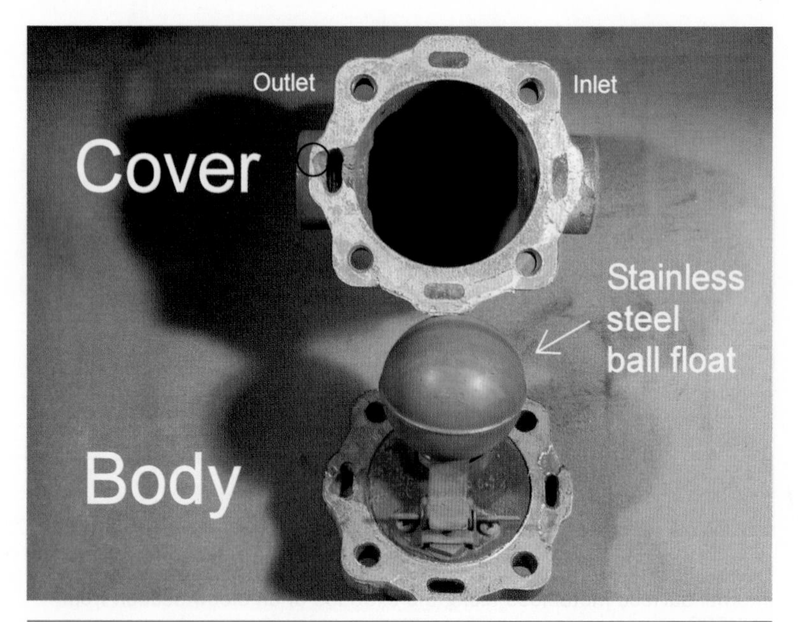

Figure 18.13 Ball float steam trap, showing the location of a small perforation near the interface between the cover and body (circled).

Figure 18.14 Perforation in an area of smoothened metal loss in an internal condensate passage of the steam trap shown in Fig. 18.13.

the passageway, where impinging and turbulent flow would be expected. In this location, the cross-sectional area of the passageway also decreased, resulting in a fluid flow velocity increase. The metal loss area exhibited flow-related features.

Jagged metal loss on the side of the passageway opposite the perforation suggested that an initial leak may have occurred at the body/cover interface, where a leaking mixture of steam and condensate may have impinged on the opposite side of the passageway and caused erosive damage. Possible reasons for an insufficient seal at the body/cover interface include improper tightening, a defect in the reinforced exfoliated graphite gasket, or debris that prevented a complete seal.

Iron oxide deposits were observed on components of the system, including the air vent valve ball and seat. Such hard, abrasive particulate deposits should be limited to reduce the likelihood of steam pass-through and enhanced erosive damage due to entrained particles.

Case History 18.4

Industry:	Pulp and paper
Specimen location:	High-pressure section of condensing turbine, last row, condensing section
Specimen orientation:	Horizontal turbine shaft
Years in service:	6
Water treatment program:	Chelant
Drum pressure:	45-MW turbine, 3600 rpm, and 830°F (443°C) superheated steam

Turbine buckets were damaged severely on leading edges. Edges were striated and grooved (Fig. 18.7). Erosion was caused by high-velocity fluids in which water droplets were entrained. Erosion damage due to water droplet impingement is not uncommon in late stages of condensing turbines.

A 65 psig (0.45 MPag) steam extraction nonreturn valve failed to seat properly during an electrical turbine trip. The 65 psig (0.45 MPag) header emptied through the turbine (vacuum condition). The turbine speed increased to 5000 rpm (design 3600 rpm) before manual shutdown.

Case History 18.5

Industry:	Utilities
Specimen location:	Feedwater system
Specimen type:	Equalizing valve and piping U-bend
Years in service:	20+
Metallurgy:	Mild steel
Pipe specifications:	1-1/16 in. (2.7-cm) outer diameter, 23 in. (58.4 cm) long, and 0.129-in. (0.328-cm) maximum measured wall thickness

The piping U-bend was joined to the valve on the indicated inlet side. Both the valve and piping U-bend sustained internal surface metal loss that exhibited flow-related features, including many overlapping depressions (Fig. 18.15). The metal loss was deepest along the extrados of the U bend and on the outlet side of the valve, where turbulence was expected to be greatest. The short segment of piping that extended into the valve body on the outlet side sustained nearly complete metal loss. The metal loss surfaces were covered with patches of orange flash rust and a thin layer of black iron oxide that stripped quickly when exposed to a strong acid, suggesting that metal loss was actively occurring shortly before

FIGURE 18.15 Internal surface of a boiler feedwater valve, showing severe metal loss due to erosion-corrosion, particularly on the outlet side.

removal from service. The internal surface metal loss was caused by erosion-corrosion. The boiler feedwater's reported temperature was 126°F (52°C), which was significantly lower than the range where flow-accelerated corrosion generally occurs (see Chap. 19).

On the outlet side, a thick-walled fracture occurred where the piping was welded to the valve (Fig. 18.16). The fracture at the valve outlet was caused by

FIGURE 18.16 Thick-walled fracture on the outlet side of the boiler feedwater valve shown in Fig. 18.15, at the weld between the valve and piping.

fatigue. The corner formed by the valve and piping served as a site of concentration for applied bending stresses. At the valve outlet, the severe pipe wall thinning due to internal surface erosion-corrosion promoted the formation of the crack, as the applied cyclic stress increased considerably with decreasing cross-sectional area for a given load.

Case History 18.6

Industry:	Institutional
Specimen location:	Feedwater system
Specimen type:	Various soldered feedwater return piping fittings
Years in service:	2
Metallurgy:	Copper

No failure was found in the received longitudinally divided piping; however, several pinhole leaks had reportedly occurred in the system, especially above pumps and at 90° bends. The piping contained areas of deep internal surface metal loss with flow-related features, such as horseshoe-shaped depressions, on the indicated downstream end (Fig. 18.17). Islands of mostly unattacked metal interrupted the metal loss areas. The metal loss areas were bare-surfaced, suggesting that erosion was actively

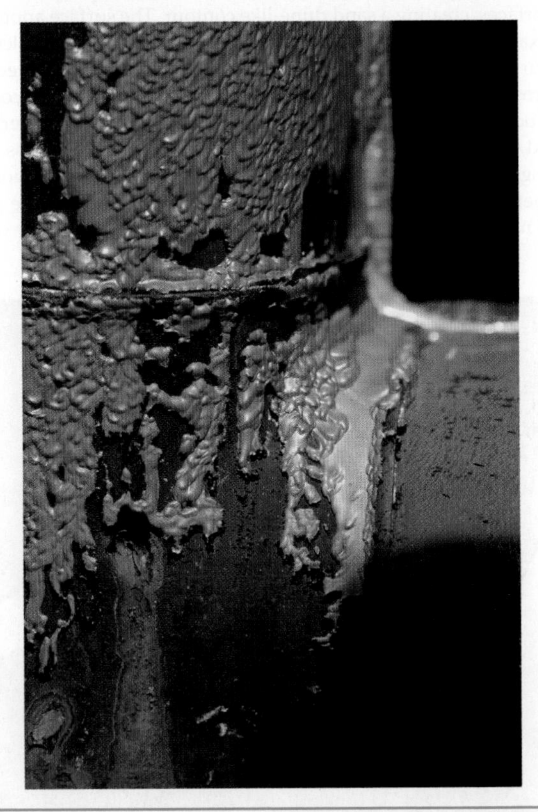

FIGURE 18.17 Copper feedwater return piping, showing a T-joint with severe erosion-corrosion damage, including horseshoe-shaped depressions and islands of apparently unattacked metal.

occurring shortly before the piping was removed from service. In locations away from the metal loss areas, the internal surface was mostly unattacked and was covered with a thin layer of deposits and oxide, suggesting that the water conditions were generally not corrosive. Bends and elbows are areas where turbulence occurs, and they can be especially susceptible to erosion damage. It was recommended that turbulence be decreased through the use of larger-diameter piping or reduced pumping speed, if practical. Alternate metallurgies were recommended if changes in system operation or design were not feasible, as copper is rare for feedwater systems.

Case History 18.7

Industry:	Pulp and paper
Specimen location:	Condensate flash tank
Specimen type:	Pump impeller
Years in service:	5+
Metallurgy:	Nodular cast iron
Impeller specifications:	6½-in. (16.5-cm) diameter, 2 in. (5.1 cm) deep

Severe general metal loss occurred across the entire impeller. The most extreme wastage occurred along the outer edges, where the metal was thinned to a knife-edge (Fig. 18.18). Many overlapping, horseshoe-shaped depressions were present on the surfaces, creating a sand dune-like contour. The surface appeared etched during examination using a low-power stereoscope; graphite nodules embedded in the surface were also visible. The surface was generally free of significant corrosion product or deposit layers, suggesting that erosion-corrosion was actively occurring shortly before removal from service. However, patches of black and/or red-orange corrosion products were scattered across the surface, suggesting that oxygen corrosion contributed to the attack from erosion-corrosion. In the metal loss areas, microscopic examinations indicated the presence of graphite nodules along the surface (Fig. 18.19). If the corrosivity of the water

FIGURE 18.18 Metal thinned to a knife-edge around the outer edges of a cast iron pump impeller, as well as many horseshoe-shaped depressions.

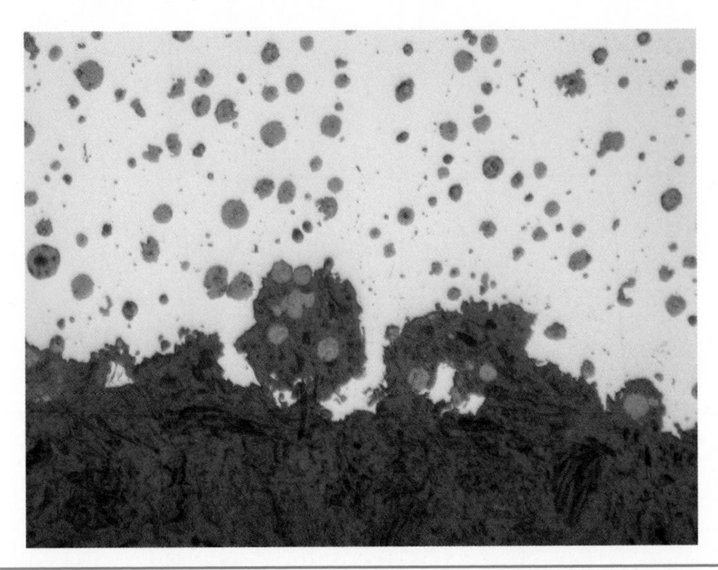

FIGURE 18.19 Micrograph of a nodular cast iron pump impeller in an area of deep metal loss, showing many graphite nodules along the surface of the metal. Unetched.

could not be decreased, it was recommended that alternate metallurgies, such as cast steels, be considered for the application. It was emphasized that potential galvanic effects should always be considered when changing metallurgies.

In one location, a deep, extremely jagged, undercut metal loss area penetrated through one of the pump-out vanes on the back side. Considerable plastic deformation and casting porosity were evident near the jagged metal loss during microscopic examination. The metal loss was caused by cavitation. Due to the repeated nature of the impeller's features, the localization of the cavitation damage to one area suggested that cavitation may have occurred preferentially in that location due to casting porosity.

References

1. Frederick G. Hammitt and Frank J. Heymann, "Liquid-Erosion Failures," *ASM Handbook*, vol. 11: *Failure Analysis and Prevention*, 9th ed., ASM International, Metals Park, Ohio, 1986.
2. John B. Kitto and Steven C. Stultz (Eds.), *Steam: Its Generation and Use*, 41st ed., Babcock & Wilcox Company, Barberton, Ohio, 2005.
3. Mars G. Fontana and Norbert D. Greene, *Corrosion Engineering*, McGraw-Hill, New York, 1978, p. 76.
4. J. Postlethwaite and S. Nesic, "Erosion-Corrosion in Single and Multiphase Flow," in R. Winston Revie (Ed.), *Uhlig's Corrosion Handbook*, 2d ed., John Wiley & Sons, New York, 2000, pp. 254–272.
5. Hammitt and Heymann, "Liquid-Erosion Failures."
6. F. J. Heymann, "Liquid Impingement Erosion," *ASM Handbook*, vol. 18: *Friction, Lubrication and Wear Technology*, 10th ed., ASM International, Materials Park, Ohio, 1992.
7. *Program on Technology Innovation: Oxide Growth and Exfoliation on Alloys Exposed to Steam*, Electric Power Research Institute, Palo Alto, CA, 2007, pp. 7–5 to 7–7.

CHAPTER 19

Flow-Accelerated Corrosion

General Description

Normally, the internal surfaces of steel piping and tubing used in boiler systems are covered by a layer of protective iron oxide (primarily magnetite, Fe_3O_4). Flow-accelerated corrosion (FAC) is a wastage mechanism caused by the dissolution of the magnetite layer in a flowing stream of high-purity water (single-phase FAC) or a steam/water mixture (two-phase FAC).[1] Normally, the metal surface is subjected to steady-state conditions of oxide formation and removal. When the metal dissolution rate exceeds the rate of oxide growth, rapid corrosion can occur. Corrosion rates as high as 120 mpy (3 mm/year) have been reported.[2]

The generally accepted mechanism of FAC takes into account the balance of processes across a porous magnetite layer that forms on a steel surface, as shown schematically in Fig. 19.1. At the metal/oxide interface, the iron oxidizes by reaction with water to form ferrous species and hydrogen. The reaction species diffuse through the porous oxide layer to the oxide/water interface. The oxide layer at the oxide/water interface dissolves by a reductive process aided by the presence of hydrogen, causing ferrous ions to collect in a boundary layer adjacent to the oxide/water interface. The ferrous ions that accumulate in the boundary layer are removed by water flowing past the surface. Decreasing the ferrous ion concentration provides a driving force for continued solubility, thus increasing corrosion rates.

The corrosion rate due to FAC is dependent on numerous factors that influence the oxide layer that forms on the metal surface. Some important variables in the dissolution process include oxide solubility, the presence of a porous magnetite layer, the transfer rate of iron ions to the fluid stream, and the concentration of iron ions in the stream. The factors can be mechanical, such as flow velocity and turbulence; environmental, such as temperature and steam quality; chemical, such as pH and oxidation-reduction potential (ORP); and

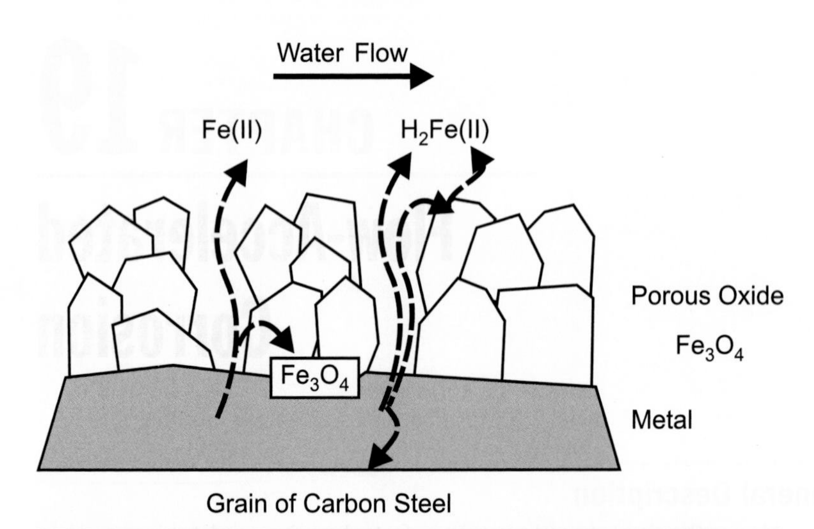

FIGURE 19.1 Schematic of the FAC mechanism, highlighting oxide formation at the metal/oxide interface, diffusion of iron species across the porous oxide layer, dissolution of the oxide layer at the water/oxide interface, and removal of the iron ions from the oxide surface in flowing water.[3] (*Courtesy of Electric Power Research Institute.*)

metallurgical, such as alloy composition of the steel. Some of these factors are discussed in detail in the "Critical Factors" section.

There are other corrosion mechanisms that produce wastage with similar appearances to FAC, such as erosion-corrosion and chelant and complexing agent corrosion. Although considerable flow or turbulence is required for or greatly enhances these mechanisms, there are inherent differences in how the metal is removed from the surface. FAC operates by a dissolution process; erosion-corrosion involves a mechanical removal of the oxide and possibly the underlying metal; chelants or complexing agents can chemically react with the oxide layer and metal. Velocity is a major factor that can be used to conceptually differentiate between FAC and erosion-corrosion mechanisms. In water free of particulate, there is a breakaway velocity for single-phase flow at which oxide removal changes from dissolution to mechanical damage, as illustrated in Fig. 19.2. Above this velocity, higher metal loss rates can occur due to mechanical damage by an erosion-corrosion mechanism. Erosion-corrosion is discussed in detail in Chap. 18.

Locations

Components in nuclear power plants, fossil fuel utility plants, and industrial plants have experienced FAC attack. Because the FAC mechanism produces appreciable metal loss only for well-defined conditions, locations susceptible to attack are generally limited to

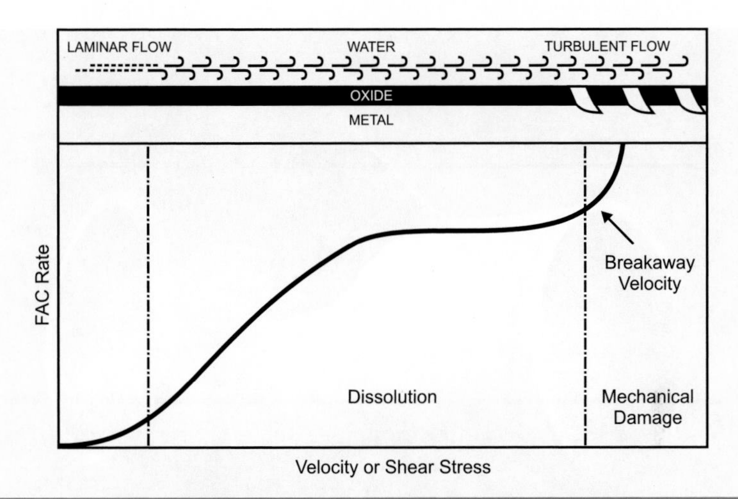

FIGURE 19.2 Effect of flow velocity on metal loss rate, highlighting the region of dissolution associated with FAC and the breakaway velocity where erosive damage begins to occur.[4] (*Courtesy of Electric Power Research Institute.*)

specific areas and components that meet the criteria. The critical factors required for FAC are discussed in the next section. Susceptible components may be exposed to single-phase flow (water) or two-phase flow (steam/water mixtures). Because water must be present for attack, FAC does not occur in components exposed to dry steam under normal operating conditions, such as in superheaters. In addition, FAC typically only affects plain carbon steel components. Steels with certain alloying additions are much less susceptible to wastage (see "Critical Factors" section).

Areas in the boiler water circuit that can be subjected to FAC include feedwater, deaerator, blowdown, and condensate systems.[5] Some components exposed to single-phase flow conditions that have experienced wastage caused by FAC include feedwater piping (Fig. 19.3) high- and low-pressure feedwater heaters, and economizers. Pump impellers and valves (Fig. 19.4) in feedwater systems can be especially susceptible due to turbulence that is intrinsically associated with those components. Regions at or downstream from bends or T-fittings are more likely to be attacked (Fig. 19.5). In feedwater heaters, preferred locations for FAC are the tubesheet (Fig. 19.6), tube inlets, and shellside drain lines, primarily due to increased turbulence at these locations. Similarly, failures are often more frequent at economizer tubing inlets from the headers (Fig. 19.7) or economizer outlet tubing to the headers, depending on water temperature. In general, areas at locations of increased turbulence will have higher rates of metal loss than areas exposed to less-turbulent flow. Components exposed to two-phase flow that are susceptible to FAC include steam separation equipment

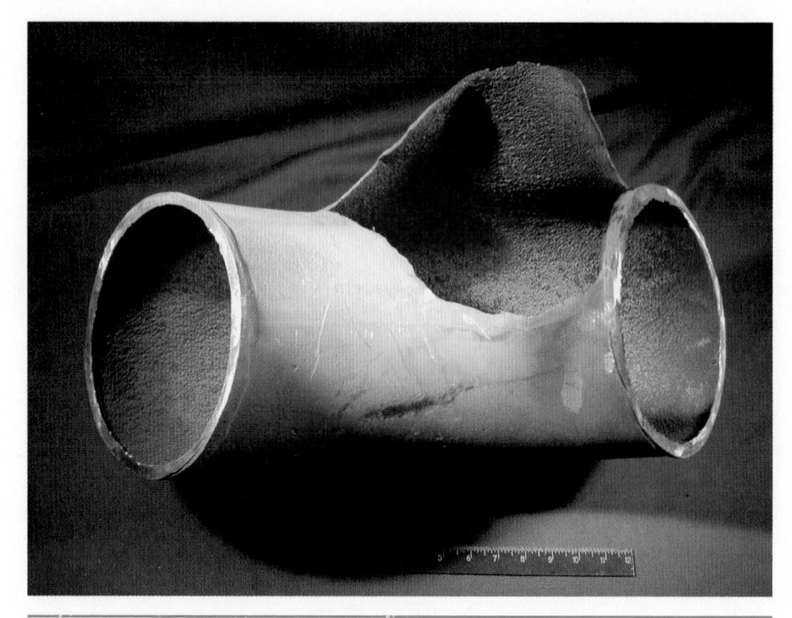

FIGURE 19.3 Ruptured feedwater piping due to severe internal surface metal loss.

FIGURE 19.4 Longitudinally split feedwater valve exhibiting deep metal loss on the downstream side (right) that resulted in a perforation.

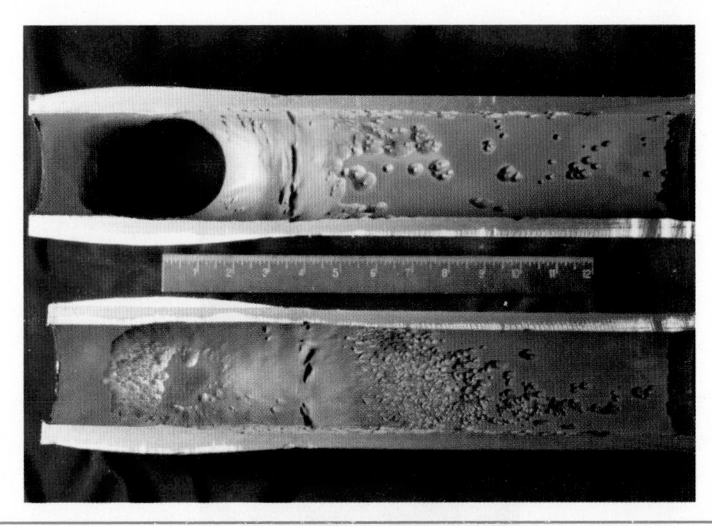

Figure 19.5 Longitudinally split boiler feedwater supply line, showing wastage downstream from a T-fitting.

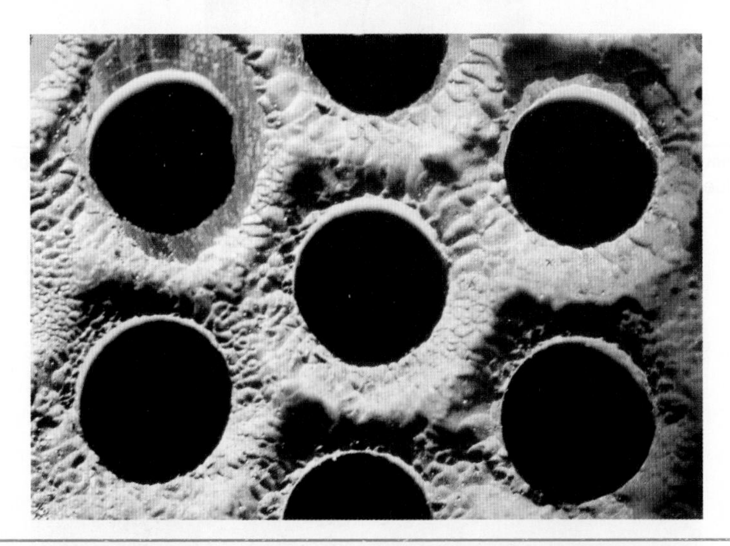

Figure 19.6 General wastage on the inlet tubesheet of a sweet water condenser.

(Fig. 19.8) and condensate piping subjected to steam and water mixtures.

Surveys have indicated that FAC is one of the primary tube failure mechanisms in heat recovery steam generator (HRSG) equipment.[6] Wastage in HRSGs has been generally associated with low-pressure evaporator circuits. These units typically use high-purity water in low-pressure systems, and the temperatures are in the range where

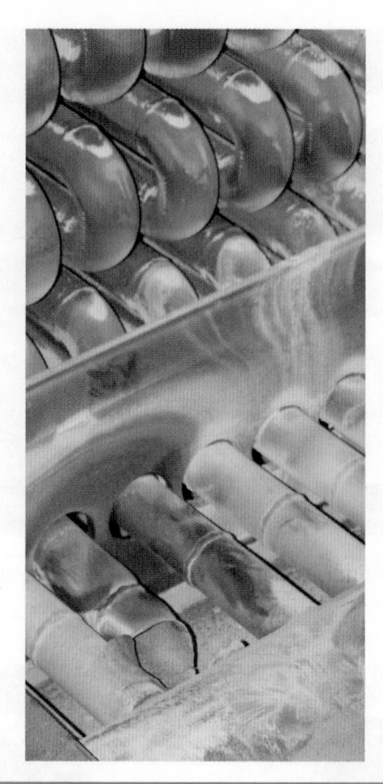

FIGURE 19.7 Ruptured tubing downstream from an economizer inlet header, due to tube wall thinning from FAC.

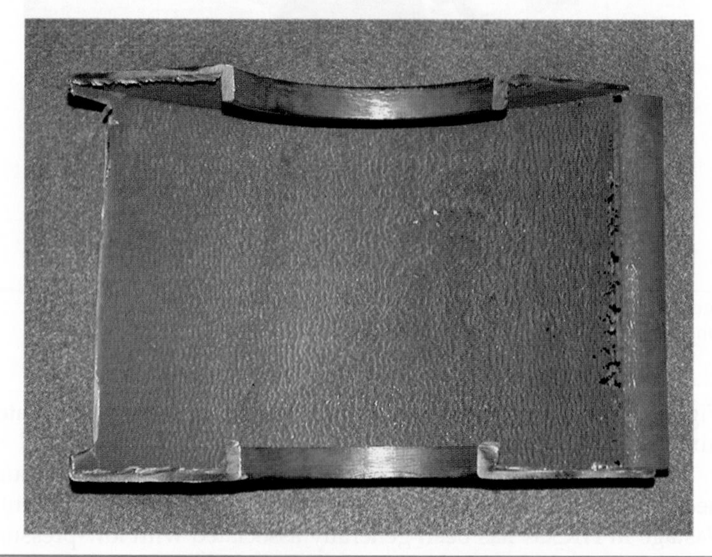

FIGURE 19.8 General metal loss due to FAC along the interior of a cyclone separator that resulted in numerous perforations.

FAC can be appreciable.[7] Two-phase flow regions in evaporator tubing and outlet piping to the steam drum can experience accelerated metal loss, primarily in low-pressure boiler systems and sometimes in intermediate-pressure systems as well. Single-phase FAC in HRSGs predominates in preheater, feedwater, and economizer tubing, similar to other systems.

Critical Factors

A number of major factors have been shown to dramatically influence the FAC mechanism. Many of these factors are interrelated. Simplified approaches that only consider the effect of one factor can be misleading. Models have been developed by the Electric Power Research Institute (EPRI) to address the role and interaction of numerous parameters on FAC rates.[8] This section highlights some of the factors and their influence on corrosion rates. The specific factors discussed include flow velocity, turbulence and piping geometry, temperature, pH, dissolved oxygen and ORP, and metallurgy.

Flow Velocity

FAC rates increase with increasing bulk velocity of the fluid. This is schematically shown in Fig. 19.2. When the flow velocity is sufficiently high, the iron ions produced from the dissolution process are removed from the surface, maintaining the driving force for continued dissolution. Bulk water velocities above 8 to 10 ft/s (2.4 to 3.0 m/s) can reportedly be sufficient to induce FAC; however, there does not appear to be a practical velocity threshold below which FAC completely ceases.[9] In addition, correlations based on bulk velocity alone can be very misleading. Other factors such as local turbulence and geometry can dramatically influence attack. Figure 19.9 illustrates, for single-phase flow conditions, how the corrosion rate increases with increasing flow rate as a function of temperature. At very high-flow velocities, the mechanism changes from dissolution of the oxide to mechanical removal (erosion-corrosion).

Turbulence and Piping Geometry

The transport of soluble iron away from the oxide surface is greatly increased with turbulent conditions. Fittings with geometries that result in dramatic changes in flow direction, such as 180° bends and 90° elbows, will experience greater turbulence. The locations at and downstream from bends will be more prone to FAC. Significant turbulence will also develop at T-fittings, in headers, at tubing or piping inlets and outlets, as well as at pump impellers and valves. These components will be more susceptible to attack due to localized turbulence, and they should be considered as primary locations to examine for possible metal loss if FAC is suspected.

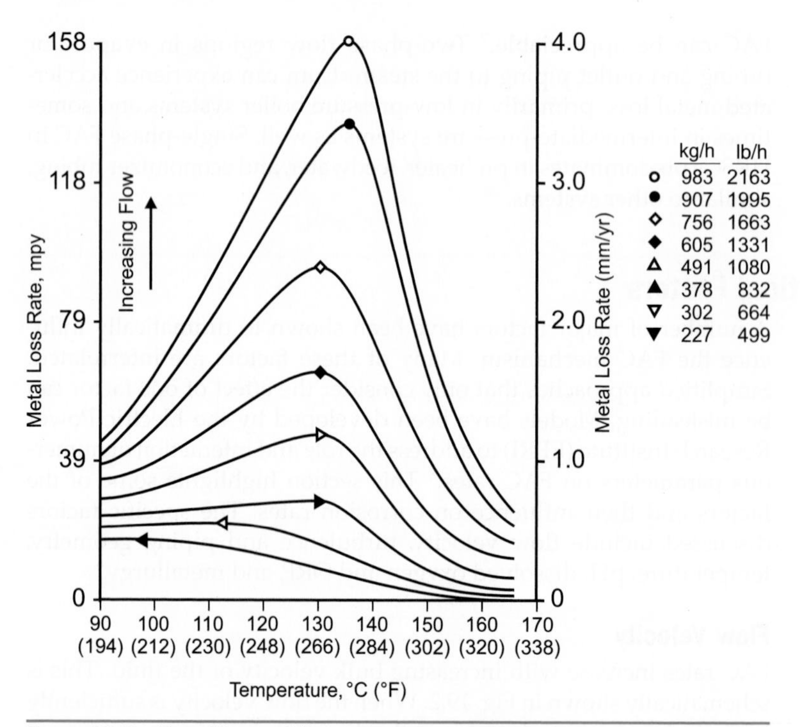

FIGURE 19.9 The effects of flow rate and temperature on the metal loss rate due to FAC with single-phase conditions. Note the increase in corrosion rate with increasing flow velocity. The maximum rate of metal loss occurs between 130 and 140°C (266 and 284°F) in this case (pH was 9.04, measured at 25°C).[10] (*Courtesy of Electric Power Research Institute.*)

Temperature

Temperature can affect numerous factors that influence FAC and the subsequent solubility of magnetite. Corrosion rate curves due to FAC as a function of temperature are typically bell-shaped, both for single-phase flow (Fig. 19.9) and for two-phase flow (Fig. 19.10). FAC has been found to occur at temperatures in the range of 212 to 482°F (100 to 250°C).[11] For single-phase flow, the maximum metal loss rate typically occurs at temperatures from 265 to 300°F (130 to 150°C). The temperature range for maximum corrosion rates for two-phase flow conditions is higher, typically ranging from 300 to 390°F (150 to 200°C).[12]

pH

The pH of the water or water phase is a primary parameter in the water chemistry that can influence the FAC rate. The dramatic effect of decreasing pH on magnetite solubility and the metal loss rate is illustrated in Fig. 19.10 for two-phase flow conditions, and a similar effect occurs for single-phase flow. In two-phase streams, the pH of the liquid phase in the mixture must be considered in the localized

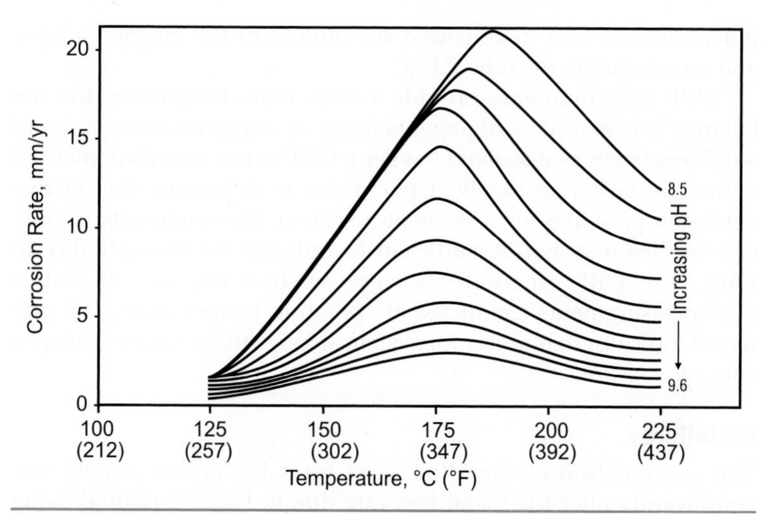

FIGURE 19.10 The effects of pH and temperature on the corrosion rate due to FAC with two-phase conditions having 65% steam quality. Note the increase in corrosion rate with decreasing pH. The maximum rate of metal loss occurs between 175 and 185°C in this case.[13] (*Courtesy of Electric Power Research Institute.*)

areas where water and steam separate, rather than simply the bulk water pH. The pH of the separated water can be considerably different from the bulk water pH in the boiler, due to partitioning of species such as amines into the separate steam and water phases.

Dissolved Oxygen and ORP

Dissolved oxygen has a dramatic effect on corrosion rates due to FAC in high-purity water, as low dissolved-oxygen concentrations are required for attack to occur in single-phase conditions. Corrosion rates due to FAC significantly decrease with increasing dissolved-oxygen content. This relation is used in some treatment programs, such as oxygenated treatment and all volatile treatment without reducing agents, to limit the susceptibility to FAC. However, in two-phase flow conditions, attack may still occur even if the bulk water contains some dissolved oxygen. In this situation, oxygen will preferentially partition to the steam phase, allowing liquid water with a low concentration of dissolved oxygen to promote dissolution of the magnetite layer.

In many treatment programs, oxygen scavengers are used to reduce dissolved-oxygen levels in the feedwater to limit oxygen corrosion. Oxygen scavengers, such as hydrazine utilized as part of an all volatile treatment (AVT) program, are reducing agents. Some studies indicate that the FAC rate increases with increasing hydrazine level in the 0 to 150 ppb range.[14] The increase in hydrazine level results in a decrease in the oxidation-reduction potential. Lower potentials can lead to enhanced dissolution of the magnetite layer and an increase in the rate of FAC.

ORP measurements provide values that characterize the net balance between reducing agents (such as oxygen scavengers) and oxidizers (such as dissolved oxygen). ORP is not a measurement of corrosion, but is an excellent parameter to determine the relative corrosion potential for a feedwater system. Measurements of only one species may not be sufficient to indicate the susceptibility to FAC. Thus ORP, especially the very sensitive response of on-line ORP measurements conducted at operating temperatures, is a preferred indicator to monitor for system stresses that can cause oxygen corrosion or FAC.

Metallurgy

The composition of the alloy used for a boiler component can significantly alter the metal loss rate due to FAC. Certain alloying elements in the steel can increase the stability of the oxide layer that forms on the surface. In the case of plain carbon steel, chromium is one of the most beneficial alloying elements in terms of limiting FAC. Molybdenum and copper also effectively improve resistance to FAC in steel; however, their influence is not as great as that of chromium. Low-alloy steels, such as T11 and T22 alloys, containing at least 1% chromium are expected to have very low to negligible FAC rates for most conditions. This is shown in Fig. 19.11, which highlights that corrosion rates can be significantly reduced with chromium levels above 0.1%.

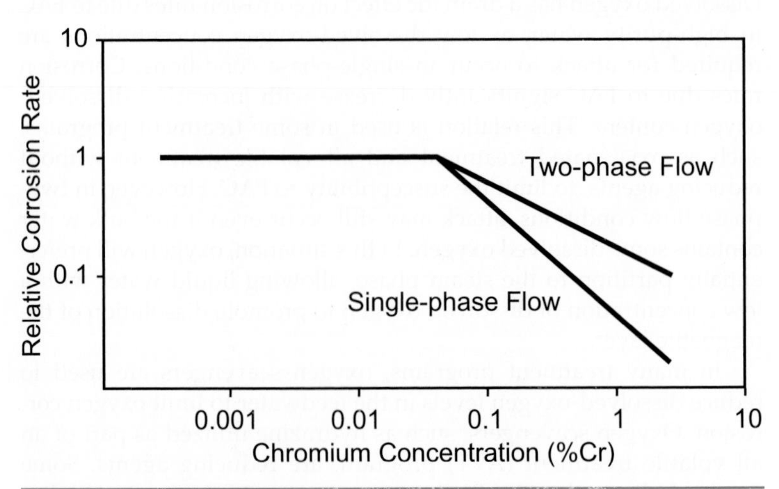

FIGURE 19.11 Relative metal loss rate as a function of chromium content in steel, including trends for single-phase and two-phase flow conditions.[15] (*Courtesy of Electric Power Research Institute.*)

Identification

Typical features associated with attack due to FAC depend to some extent on the location, as attacked surfaces can have differences in appearance depending on the environmental conditions. Although wastage may be localized to regions experiencing high degrees of turbulence, metal loss associated with FAC often occurs in tubing and commonly results in gradual wall thinning within the attacked region. Failures generally are in the form of thin-lipped ruptures (Figs. 19.3 and 19.7) or thin-walled perforations (Figs. 19.4 and 19.8).

The features typically associated with FAC include horseshoe-shaped depressions, as shown in Fig. 19.12. The rounded end of the horseshoe points upstream. The depressions tend to overlap in regions of deeper metal loss, especially in areas of turbulence or change of flow direction (see Fig. 19.13). The overlapping horseshoe-shaped depressions can result in a sand dune type of appearance (see Fig. 19.14), also referred to as a scalloped or orange peel appearance. This appearance usually is associated with locations of severe metal loss. In cases where two-phase flow is present, areas of FAC damage may have multiple distinct regions of attack. Adjacent areas or bands of rapid FAC, and slow or nonexistent FAC, can create a morphology known as tiger stripes. This feature is generally associated with large pipes.

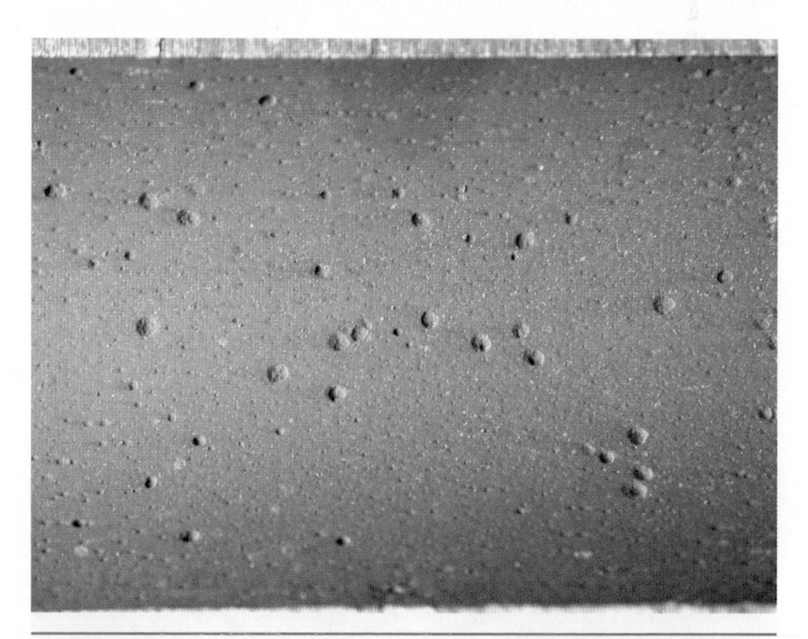

FIGURE 19.12 Scattered horseshoe-shaped depressions on a smoothened surface contour on the internal surface of an economizer tube slightly downstream from the inlet header.

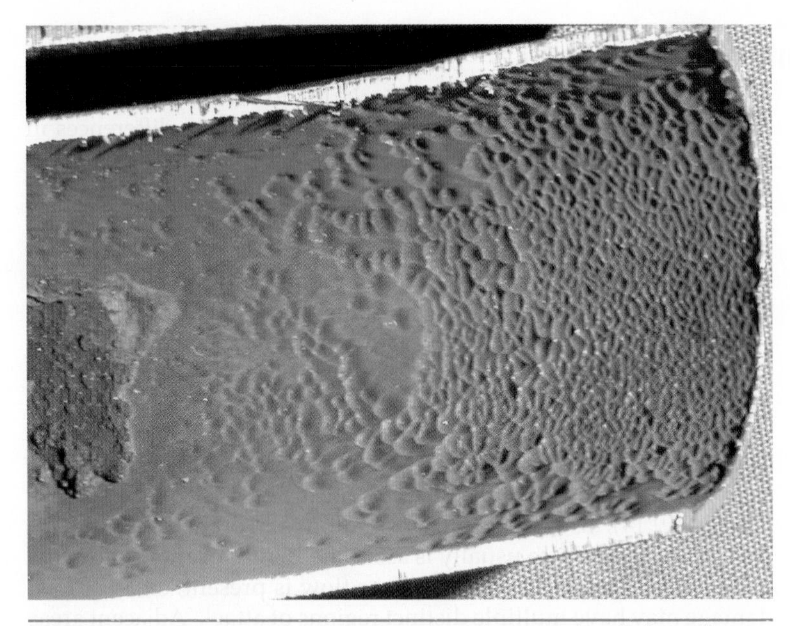

FIGURE 19.13 Deep metal loss region along the outer bend of an elbow downstream from a manifold of the low-pressure section of a heat recovery steam generator (HRSG). Note the gradual change from a smoothened contour with a few horseshoe-shaped depressions to a region containing many overlapping depressions. Flow is from left to right.

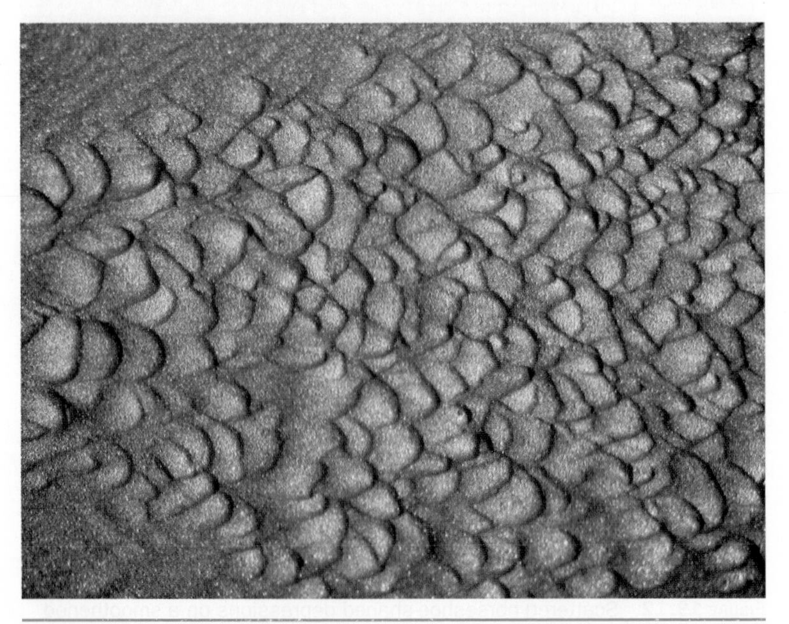

FIGURE 19.14 Many overlapping horseshoe-shaped depressions that create a sand dune appearance.

FIGURE 19.15 Very thin black oxide layer that was quickly stripped when exposed to concentrated acid, resulting in a shiny metallic surface (acid was applied to the area on the left).

Regions that experienced recently active FAC attack will be covered by an extremely thin, nearly transparent black oxide layer. This layer will be quickly stripped (in seconds) from the surface when exposed to concentrated acid, as shown in Fig. 19.15. A thicker oxide layer that is not removed quickly tends to indicate attack that was not recently active, prior to removal of the section from service.

Microscopic examination of cross sections through wasted regions will show a smoothly undulating, wavy profile (Fig. 19.16). The surfaces will be generally free of corrosion product and oxide layers if attack was actively occurring shortly before removal of the section from service. The metal microstructure adjacent to the wasted region should not exhibit evidence of deformation, as this is a feature associated with an erosive mechanism. Unattacked areas are generally covered by a thin, dense oxide layer. In some cases, blistered or blocky magnetite layers have been reported.[16] Blocky, angular oxide layers are typically noted on the surface, away from regions of metal loss, in systems on oxygenated treatment.

If the waterside surfaces of the component are not accessible, then nondestructive testing can be used to check for wall thinning. Some inspection techniques include ultrasonic testing, radiography, pulsed eddy current testing, and borescope inspections.

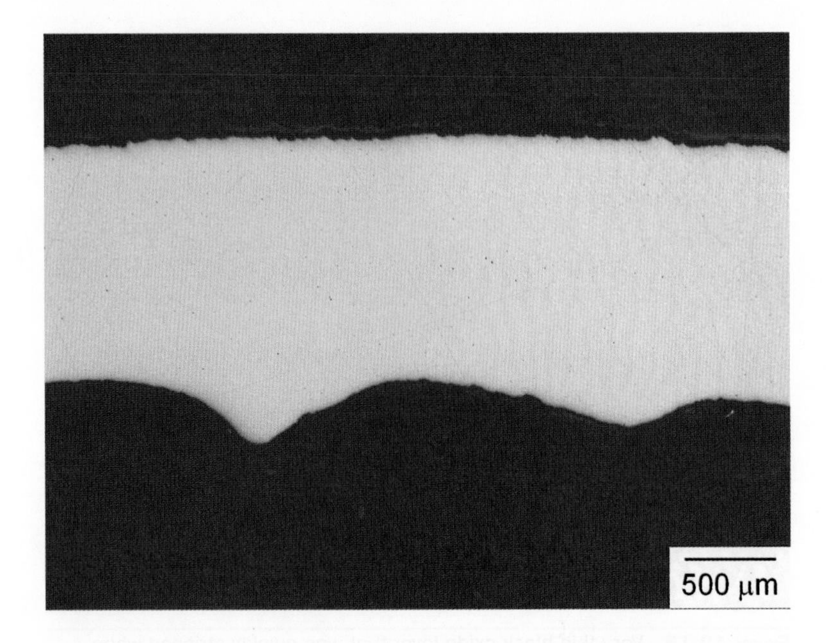

FIGURE 19.16 Comparison of microscopic features associated with FAC damage, showing a smoothly undulating, wavy profile free of deposit and oxide layers of appreciable thickness (top). Note the oxide layer covering the surface at a location away from metal loss (bottom).

Elimination

Reducing wastage caused by FAC requires an understanding of the numerous factors that can contribute to the mechanism, and how they influence metal loss rates. Because of the specific conditions that allow for dissolution of the magnetite layer, identifying the factors that promoted corrosion and changing them, where possible, can significantly reduce corrosion rates.

Determining the influence of the different factors on corrosion requires monitoring and correlation of changes in water chemistry to periods of active corrosion. Active corrosion testing in boiler feedwater systems can be achieved by reduced iron (Fe^{2+}) measurements. In the boiler feedwater system, sample points before and after a component known or suspected to be experiencing FAC are recommended to determine the soluble iron pickup across the component. These measurements can be used to optimize the water chemistry parameters, minimizing the potential for FAC. On-line measurements of at-temperature ORP can also be used as part of a monitoring program to help assess and control variability in reduction potential that can promote FAC.

Some remedial actions regarding changing parameters to limit attack are addressed separately below. However, it is important to understand that the parameters are often interrelated.

Metallurgy

Alloy upgrades from plain carbon steel will dramatically reduce the susceptibility to FAC. Steel containing 1% chromium has greatly reduced oxide solubility relative to plain carbon steel. Additions of copper and molybdenum to steel are also beneficial to limit solubility. Low-alloy steels, such as T11 tubing and P11 piping, should be considered as replacement materials for plain carbon steel components affected by FAC. These alloys are typically specified in locations susceptible to FAC in many HRSGs.

pH

The pH has an appreciable effect on FAC metal loss rates. Increasing the pH to greater than 9.3 should be beneficial to reduce corrosion rates. However, elevating the pH may be restricted for mixed metallurgy systems, as increasing the pH can promote corrosion of yellow metals in some cases.

Dissolved Oxygen and ORP

Increasing the dissolved-oxygen contents in high-purity water systems can reduce FAC rates, apparently by creating a mixed oxide layer that limits solubility and diffusion through the layer. However, this approach, known as oxygenated treatment, or all volatile treatment without reducing agents are not suitable for most industrial applications, as oxygen corrosion can result if treatment is not applied properly. Oxygenated treatment and all volatile treatment without reducing agents can only be used in units with all ferrous metallurgy in the preboiler system, and they require very restrictive water chemistry limits more typically associated with utility boilers than with industrial systems. Details on the requirements for oxygenated treatment and all volatile treatment without reducing agents can be found elsewhere.[17]

Because FAC is promoted by reducing conditions, excessive levels of reducing agents can increase metal loss rates. Oxygen scavengers are reducing agents, and optimum treatment levels can limit potential corrosion issues from FAC and oxygen corrosion. On-line monitoring techniques such as at-temperature ORP can be used to evaluate redox potential trends and control oxygen scavenger feed. Details on at-temperature ORP monitoring are presented elsewhere.[18]

Velocity and Turbulence

Increasing bulk water velocity and localized turbulent conditions will increase FAC rates. Redesign of components to reduce velocity and turbulence, such as increasing pipe diameters and avoiding abrupt changes in flow direction, can reduce metal loss rates where such changes are possible.

Temperature

Although FAC is active only in specific temperature ranges, from 212 to 482°F (100 to 250°C), altering temperature is generally not considered a practical approach to limit attack in most cases.

Cautions

One of the primary issues regarding FAC is that it can affect pressurized systems external to the boiler. Wastage and subsequent failure of a component may result in inadvertent release of steam into areas that can be occupied by plant personnel. In many cases the affected components may not show any signs of degradation until failure. Unless monitoring and mitigation programs for FAC are in place, unexpected failures from severe wall thinning may occur. Numerous fatalities have occurred in multiple incidents related to FAC damage.[19]

Metal loss morphologies of FAC can appear similar to wastage from other mechanisms, such as attack due to erosion-corrosion and chelant and complexing agent corrosion. Proper diagnosis of the cause of failure requires consideration of the environment and parameters of the fluid. Specific conditions are required for FAC to be appreciable.

Related Problems

See Chap. 8, Corrosion by Chelating and Complexing Agents, and Chap. 18, Erosion and Erosion-Corrosion (Water and Steam Sides).

Case History 19.1

Industry:	Pulp and paper
Specimen location:	Economizer inlet
Orientation of specimen:	Horizontal
Years in service:	15
Water treatment program:	Congruent phosphate, carbohydrazide
Drum pressure:	1525 psi (10.5 MPa)
Tube specifications:	2-in. (5.1-cm) outer diameter, type SA-178A

Multiple economizer tubes experienced leaks near the economizer inlet header. External surface metal loss was believed to have occurred in the area. Pad welds were used to repair leaking tubes, and shields were used to limit attack along the external surfaces. Continued failures prompted removal of a tube section for analysis. The inlet end of the tube experienced severe wall thinning along one side (Fig. 19.17). Tube diameter measurements indicated metal loss was almost entirely from the internal surface. Examination of the internal surface revealed a large, rounded area highlighted by orange flash rust on the severely thinned side. The surface contour of this area had a wavy, sand dunelike appearance, as shown in Fig. 19.18. The thin iron oxide layer overlying the surface indicated that attack was active shortly before removal of the section from service. Less severe attack was noted on the opposite side of the tube and at downstream locations.

FIGURE 19.17 Severely thinned tube wall on the inlet end of the section. Note the variation in wall thickness around the circumference. Remnants of a pad weld are present at the top of the tube.

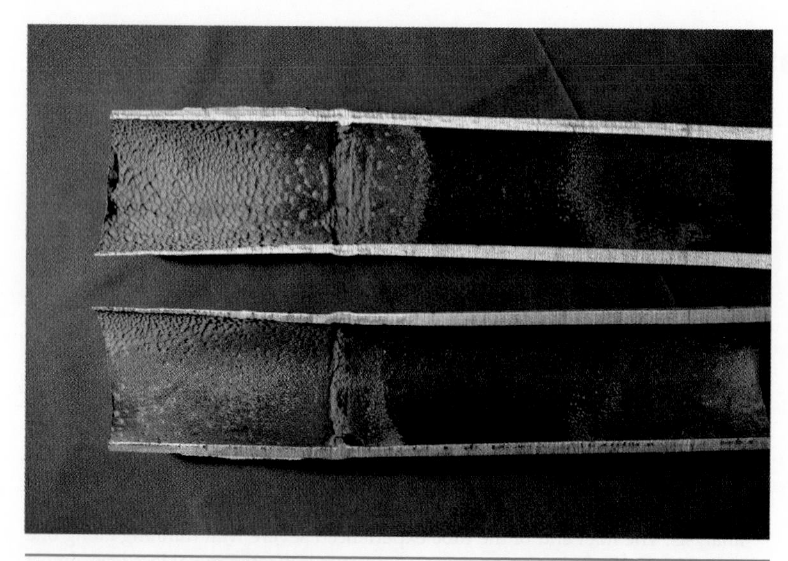

FIGURE 19.18 Longitudinally split section showing a region of deep localized wastage having a sand dune morphology on one side of the inlet end (left) and some shallow metal loss downstream from the circumferential weld.

Slightly deeper metal loss was also noted on the internal surface downstream from the circumferential weld on the section.

The characteristics of the metal loss on the section were typical of FAC that was promoted by localized turbulence at the inlet header and downstream from an intruding weld. Review of background information indicated a consistent feedwater temperature of 268°F (131°C) at the economizer inlet, pH ranges between 8.8 and 9.3, and dissolved-oxygen levels consistently below 1 ppb. The alloy composition was checked, and no measurable quantities of chromium were noted. These factors are consistent with conditions in which FAC can occur. In addition, review of water chemistry data indicated higher levels of residual oxygen scavenger than the targeted range for some periods, which may have slightly promoted attack.

Extensive nondestructive evaluation by ultrasonic testing methods was used to determine the extent of the damage in nearby tubes, the economizer header, and bend regions in the feedwater piping. Replacement of the inlet header and stubs with a low-alloy steel containing chromium was recommended. Monitoring of reduced iron (Fe^{2+}) and at-temperature ORP was suggested to evaluate the impact of any water chemistry changes on FAC.

Case History 19.2

Industry:	Utility
Specimen location:	Steam separation equipment, low-pressure steam drum in an HRSG
Orientation of specimen:	Varies
Years in service:	15
Water treatment program:	Organic amines, carbohydrazide
Drum pressure:	60 psi (0.4 MPa)

Numerous failures occurred in the steam separators located along one side of the low-pressure (LP) steam drum of a heat recovery steam generator, as shown

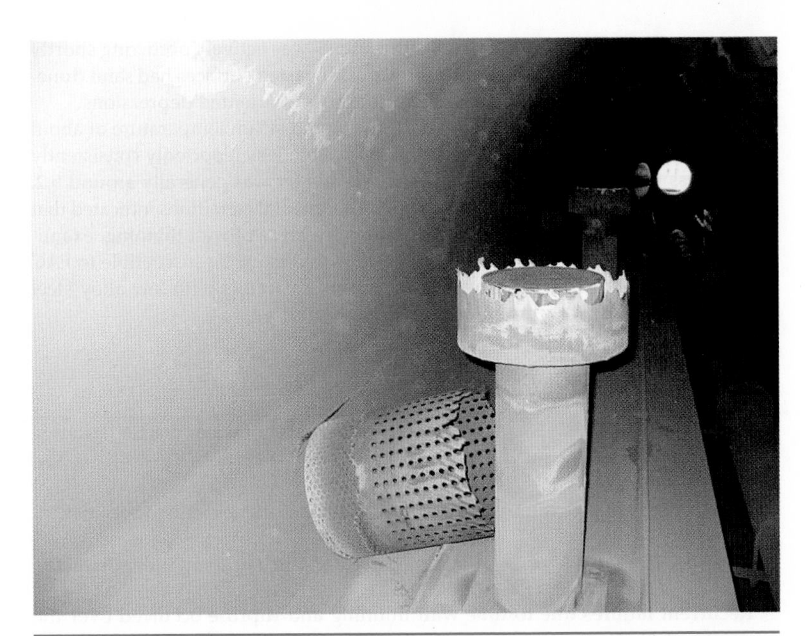

FIGURE 19.19 Failed steam separation equipment in the LP steam drum.

in Fig. 19.19. Deep metal loss occurred in most places along the interior surfaces of the inlet, separator cylinder, and vanes (Fig. 19.20). Severe wastage of the cylinder resulted in numerous perforations and detachment of the top portion from the remainder of the steam separator. The wasted areas were essentially baresurfaced, or covered with only very thin, gray-black oxide layers and

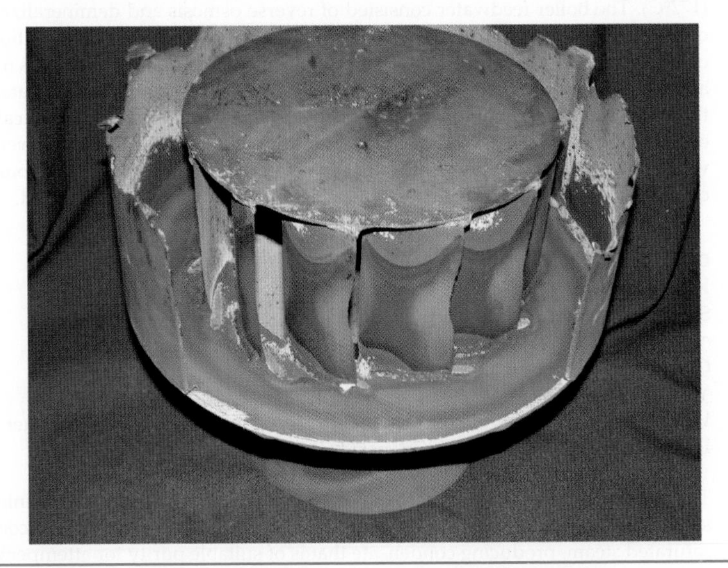

FIGURE 19.20 Severe wastage along the interior surface of the steam separator.

scattered patches of flash rust, indicating attack was actively occurring shortly before removal of the section from service. The wasted surfaces had sand dune-like contours, containing many, overlapping, flow-oriented depressions.

The drum pressure corresponded to a saturated steam temperature of about 290°F (143°C). The dissolved oxygen in the system was reportedly consistently low (less than 5 ppb), and the pH of the feedwater was generally around 9.2. The morphology of the wastage and the environmental conditions indicated that FAC with two-phase flow occurred. To identify areas of wall thinning, examination of other locations in the system that would likely be susceptible to FAC was recommended. Replacement of the steam separators with low alloy steel containing at least 1% chromium was strongly advised.

Case History 19.3

Industry:	Chemical process
Specimen location:	180° bend, piping in a hot water reactor jacket
Orientation of specimen:	Varies
Years in service:	39
Water treatment program:	Polymer, erythorbic acid, neutralizing amines
Pressure:	436 psi (3.0 MPa)
Pipe specifications:	Plain carbon steel, 4-in. (10.2-cm) outer diameter

Recurrent failures due to tube wall thinning and rupture occurred over the years in the hot water reactor system. The U-bend section submitted for analysis contained a thin-lipped rupture along the outer bend on the downstream side, shown in Fig. 19.21. Severe internal surface metal loss occurred along the outer bend. The internal surfaces had a smooth, wavy contour covered by an extremely thin oxide layer (Fig. 19.22). These characteristics are typical of FAC that was actively occurring shortly before removal of the section from service.

The pipe reportedly carried boiler feedwater at temperatures up to 350°F (177°C). The boiler feedwater consisted of reverse osmosis and demineralized quality makeup water, with appreciable amounts of condensate. The addition of large quantities of condensate resulted in feedwater pH values as low as 6.5 at times. Dissolved-oxygen levels were reportedly low. The conditions indicated that metal loss was predominately due to FAC. The polymer treatment chemicals can contribute to wastage under certain conditions, but the treatment levels were very low and likely did not contribute significantly to the wastage. Additional controls to ensure appropriate pH values in the system were recommended.

Case History 19.4

Industry:	Pulp and paper
Specimen location:	Inlet tubesheet of a sweet water condenser for a recovery boiler
Orientation of specimen:	Varies
Years in service:	11
Water treatment program:	Carbohydrazide, neutralizing amines, polymer
Drum pressure:	1250 psi (8.6 MPa)

The sweet water condenser was taken out of service due to failures at the inlet side of the tubesheet. The sweet water condenser used boiler feedwater to cool saturated steam, producing condensate that is of suitable purity for attemperation. Inspection revealed deep metal loss on the inlet tubesheet and little to no wastage on the outlet end. Close examination revealed wavelike surface contours

FIGURE 19.21 Thin-lipped rupture along the extrados on the downstream side of a U bend of feedwater piping.

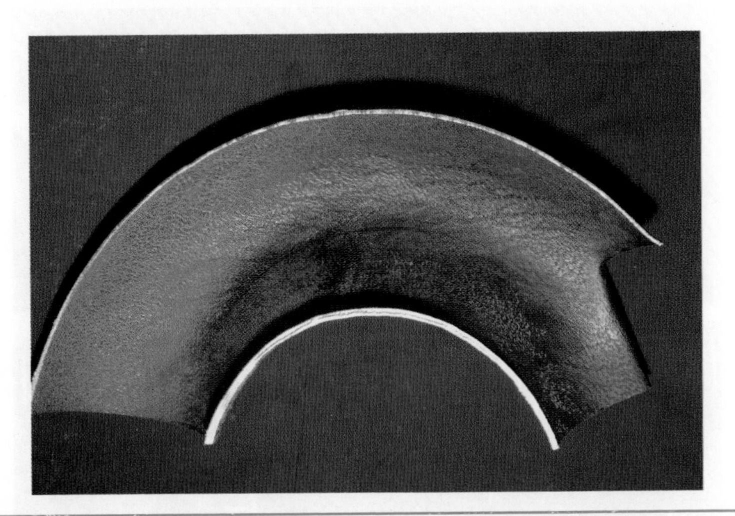

FIGURE 19.22 Longitudinally split section showing severe wall thinning from the internal surface along the outer bend. Note the smoothened, wavy appearance covered by a very thin black oxide layer and areas with orange flash rust.

FIGURE 19.23 Failures of the tubes due to consumption of the tube wall adjacent to the inlet tubesheet.

on the tubesheet surface (Fig. 19.6). The tube walls adjacent to the tubesheet were completely consumed in places, as shown in Fig. 19.23. Examination of the internal surfaces of the tubes showed a smoothened zone of wastage that extended about 1 to 2 in. (2.5 to 5.1 cm) from the inlet end (Fig. 19.24). Deposits covered the surface downstream of the metal loss. Damage was caused by FAC, appreciably promoted by turbulence on the inlet end of the tubes.

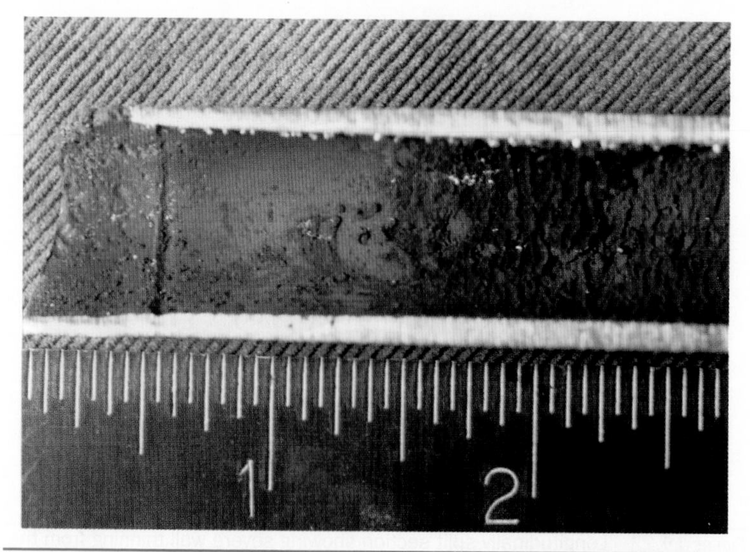

FIGURE 19.24 Zone of deep metal loss with a smoothened contour on the inlet end of the tube (left).

Over the years, the water chemistry, environmental parameters, and the extent of damage varied. Dissolved-oxygen levels were always less than 5 ppb. Mixed-bed polished demineralized water was used for the feedwater. Some metal loss was noted on the inlet end during an inspection 3 years prior to the failure. The feedwater typically was about 430°F (221°C), and the pH ranged from 8.6 to 9.0. After the inspection, the oxygen scavenger was changed from erythorbic acid to carbohydrazide, amines were increased to raise the pH to about 9.1, and a polymer dispersant was added to the program. Inspection of the inlet end of the condenser 20 months later did not indicate any appreciable metal loss to the tubes or tubesheet. After this outage, the location of the condenser was changed, and the inlet feedwater temperature decreased from 430°F (221°C) to about 330°F (166°C). The failures occurred within 1 year of operation at these conditions. This case history illustrates the beneficial effect of increasing pH to limit corrosion rates and the dramatic effect that temperature can have on attack. Possible remedial measures to limit attack included increasing feedwater temperatures, further increasing pH, substituting tubes containing at least 1% Cr, and installing tapered ferrules on the inlet end of the tubes to mitigate turbulence.

Case History 19.5

Industry:	Utility
Specimen location:	Low-pressure evaporator section of a heat recovery steam generator, near the outlet header
Orientation of specimen:	Vertical
Years in service:	8
Water treatment program:	Oxygenated treatment (OT)
Drum pressure:	91 psi (0.6 MPa)
Tube specifications:	Externally finned, 1¼-in. (3.2-cm) outer diameter, type SA-178A

Several localized failures occurred in the unit. The received tube section contained a perforation located on a side that had experienced severe tube-wall thinning on the internal surface, as shown in Fig. 19.25. The perforation was

FIGURE 19.25 Severe tube-wall thinning along one side of the tube.

FIGURE 19.26 Smoothened contour of the internal surface containing a few horseshoe-shaped depressions. Note the perforation of the tube wall at the location of a horseshoe-shaped depression.

located in a deep depression that was covered by an extremely thin, shiny, black oxide film that was quickly stripped when exposed to acid. The surface contour was smoothened and contained some scattered horseshoe-shaped depressions (Fig. 19.26). Microscopic examination revealed a blocky, angular oxide layer covering the surface away from metal loss. Characteristics of the wastage indicate that FAC was responsible for the attack.

The alloy of the tube did not contain measurable levels of chromium or molybdenum. The pH of the low-pressure feedwater reportedly ranged from 9.0 to 9.4. The drum pressure corresponded to a saturated steam temperature of about 321°F (161°C), a temperature at which maximum corrosion rates may occur with two-phase flow. Two-phase flow conditions occurred near the top of evaporator tubes and bends of riser piping from the LP evaporator section. Although oxygenated treatment can limit attack due to FAC in single-phase flow conditions, corrosion in two-phase flow is still possible with oxygenated treatment. Preventing attack in this area of the HRSG required changing the metallurgy of the tubing to a low-alloy steel such as SA-213 T11.

References

1. *Flow-Accelerated Corrosion in Power Plants*, TR-106611, Electric Power Research Institute, Pleasant Hill, Calif., 1996, pp. 1–4.
2. R. B. Dooley and V. K. Chexal, "Flow-Accelerated Corrosion," Paper 347, *NACE Corrosion 99*, NACE International, Houston, 1999.
3. *Flow-Accelerated Corrosion in Power Plants*, pp. 3–19.
4. Adapted from T. Y. Chen, A. A. Moccarri, and D. D. Macdonald, "Development of Controlled Hydrodynamic Techniques for Corrosion Testing," *Corrosion*, 1992, vol. 48, no. 3, p. 239.
5. *Flow-Accelerated Corrosion in Power Plants*, pp. 2–11 to 2–30.

6. *Guidelines for Controlling Flow-Accelerated Corrosion in Fossil and Combined Cycle Plants*, TR-1008082, Electric Power Research Institute, Palo Alto, CA, 2005, pp. 1–4 to 1–5.
7. A. Banweg, "Flow Accelerated Corrosion in Industrial Boiler Feed Water Systems," International Water Conference, Pittsburgh, Pa., October 18–21, 1999.
8. *Flow-Accelerated Corrosion in Power Plants*, pp. 4–1 to 4–119.
9. R. D. Port, "Flow Accelerated Corrosion," Paper 72, *NACE Corrosion 98*, NACE International, San Diego, 1998
10. Adapted from G. J. Bignold, K.Garbett, R.Garnsey, and I. S. Woolsey, "Erosion Corrosion of Mild Steel in Ammoniated Water," *Proceedings of the 8th International Congress on Metallic Corrosion*, Mainz, 1981.
11. *Flow-Accelerated Corrosion in Power Plants*, pp. 4–20.
12. Ibid., p. 4–27.
13. Adapted from M. Bouchacourt, EDF Internal Report, 1982, Ref.: HT-PVD. XXX MAT/T.42
14. Dooley and Chexal, "Flow-Accelerated Corrosion," *NACE Corrosion 99*, Paper 347.
15. Adapted from *Flow-Accelerated Corrosion in Power Plants*, pp. 4–84.
16. *Heat Recovery Steam Generator Tube Failure Manual*, TR-1004503, Electric Power Research Institute, Palo Alto, CA, 2002, pp. 9–4.
17. Dooley and Chexal, "Flow-Accelerated Corrosion."
18. D. J. Flynn (Ed.), *The Nalco Water Handbook*, 3d ed., McGraw-Hill, New York, 2009, p. 10.42.
19. *Guidelines for Controlling Flow-Accelerated Corrosion in Fossil and Combined Cycle Plants*, TR-1008082, Electric Power Research Institute, Palo Alto, CA, 2005, pp. 1–1 to 1–2.

SECTION 2

Fireside Corrosion and Damage

In simple terms, combustion involves the rapid reaction of oxygen with the basic chemical elements in fuels—carbon, hydrogen, and sulfur—with a consequent release of heat and the formation of combustion products (Fig. S2.1). Other material present in fuel forms combustion by-products, typically referred to as ash.

Regardless of the original physical state of the fuel, combustion may convert fuel components to any of or all three states of matter—solid, liquid, or gas. The constituents in the material formed by combustion depend strongly on the type of fuel and the many reactions that can occur. The reactions are dependent on the environment and temperature.

In a combustion device, the flue gas temperature may range from 3000°F (1650°C) in the flame to 250°F (121°C) or less at the exhaust stack. As combustion products cool at various locations throughout the boiler on their way to the exhaust stack, gaseous products may condense to liquids, and liquids may freeze to solids. This cooling may occur rapidly on heat-transfer surfaces that are cold relative to the flue gas, such as boiler tubing cooled by water and/or steam. For instance, condensation of acidic vapors in the flue gas in the low-temperature region of the boiler can result in attack, generally referred to as cold-end corrosion.

In addition, combustion products rarely remain as individual oxides, but generally interact to form new families of compounds

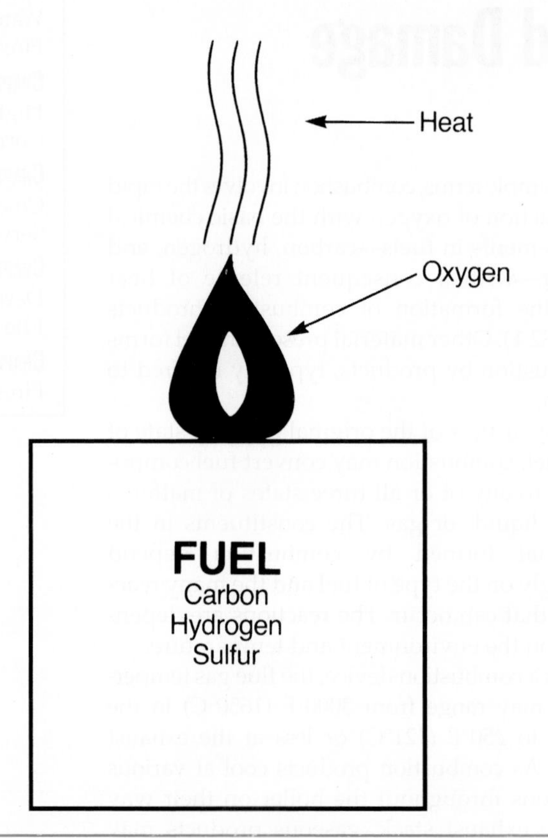

FIGURE S2.1 Combustion products resulting from burning of fuels.

and complexes. At times, these new substances may have melting temperatures that are lower than those of the substances from which they were formed. The presence of certain liquid substances on tube surfaces can result in various fireside corrosion mechanisms that depend on fuel type and temperature.

The type of environment is also of importance in considering causes of fireside attack. In the combustion zone, the environment may be reducing in places. A reducing environment can provide conditions suitable for corrosion mechanisms such as sulfidation. Combustion parameters, such as the level of excess air, will also impact corrosion mechanisms.

Chapters 20 through 23 in this section discuss some forms of fireside corrosion, with respect to conditions in different regions in the boiler. Many different types of fuels can be burned to produce heat in boilers, such as coal, natural gas, oil, refinery waste products, black liquor, biomass, and refuse. Because of the wide variety of materials, the chapters in this section do not cover every mechanism associated with fireside attack, but provide an overview of factors that can cause or promote attack during service.

In addition to corrosive attack that can happen during service, wastage related to aggressive deposits on tube surfaces can occur during idle periods. This topic is addressed in the Chap. 23, Dew Point Corrosion. Damage by various erosive mechanisms, which generally depend on the type of fuel and boiler operation, is discussed in Chap. 24, Fireside Erosion.

CHAPTER 20

Waterwall (Furnace) Fireside Corrosion

General Description

The furnace of most boilers consists of four walls. These walls, called waterwalls, are panels of tubes that are joined together, generally either in a tangential arrangement or by welded membranes between tubes. The resulting construction encloses the furnace combustion zone to contain combustion gases. The surfaces of the waterwalls are exposed to radiant heat from combustion. In addition, the environment that forms in the lower furnace is often reducing due to combustion practices.

The conditions that develop in the lower furnace can be very corrosive for standard plain carbon boiler tubing in many coal-fired boilers, black liquor recovery boilers, and refuse-fired boilers. If these corrosive conditions are not mitigated through design and operation, then severe wastage can result.

Two important corrosion mechanisms that can occur in the lower furnace include attack from the development of molten phases in the deposits and attack due to reactions with reduced sulfur gases (sulfidation). These mechanisms are discussed separately below.

Corrosion from Molten Phases

The molten phases that can form in the slag deposits on the tube surfaces depend on the constituents in the fuel, the metal surface temperature, and the surrounding environmental conditions. The molten phases in the slag layers may flux the protective magnetite on tube surfaces, causing accelerated metal wastage. They can also promote attack by increasing the rate of reaction and transport of aggressive species in the slag layer.

Pyrosulfates can cause corrosion due to the production of molten phases on waterwalls of coal-fired boilers. Incomplete combustion

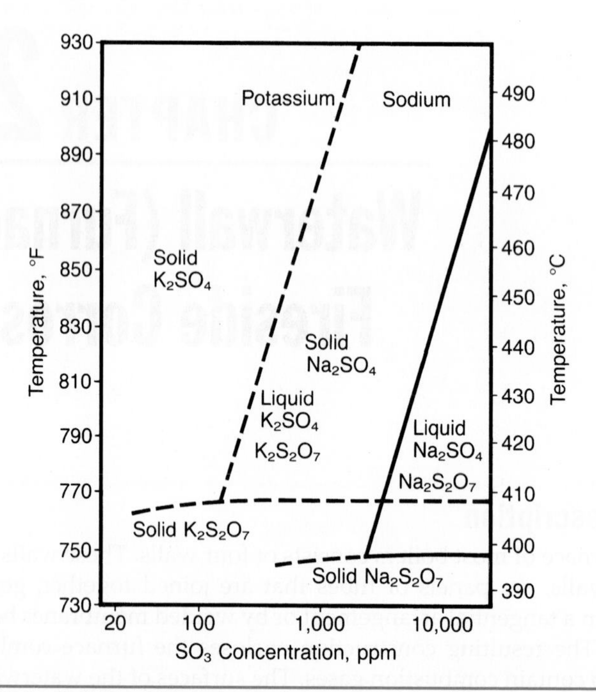

FIGURE 20.1 Relationships between temperature and SO_3 concentration to produce solid and liquid phases in the $Na_2SO_4 \bullet SO_3$ and $K_2SO_4 \bullet SO_3$ systems. (*Reprinted with permission from William T. Reid, External Corrosion and Deposits: Boilers and Gas Turbines, American Elsevier, New York, 1971, p. 106.*)

causes release of volatile sulfur compounds, which may form pyrosulfates. Sodium and potassium pyrosulfates ($Na_2S_2O_7$ and $K_2S_2O_7$) have relatively low melting points and high chemical activity.

The melting point for the pyrosulfates depends on the amount of sulfur trioxide (SO_3) in the salt, as shown in Fig. 20.1. The melting point may be less than 770°F (410°C) for $K_2S_2O_7$ and about 754°F (401°C) for $Na_2S_2O_7$. As temperature increases, the amount of SO_3 required to form a liquid phase also increases significantly. Consequently, neither sodium nor potassium pyrosulfates are likely to be present as liquid except on relatively cool surfaces, such as the waterwalls.

Because of the wide variety of constituents in refuse, waste-to-energy boilers (incinerators) are very susceptible to attack from molten phases. Refuse typically has an appreciable amount of chlorine-containing material, which is the main element responsible for low melting point phases. Refuse also contains metallic elements, such as lead, zinc, tin, and alkali metals, that form low melting point eutectics.[1] Some low melting eutectic combinations are shown in Fig. 20.2.

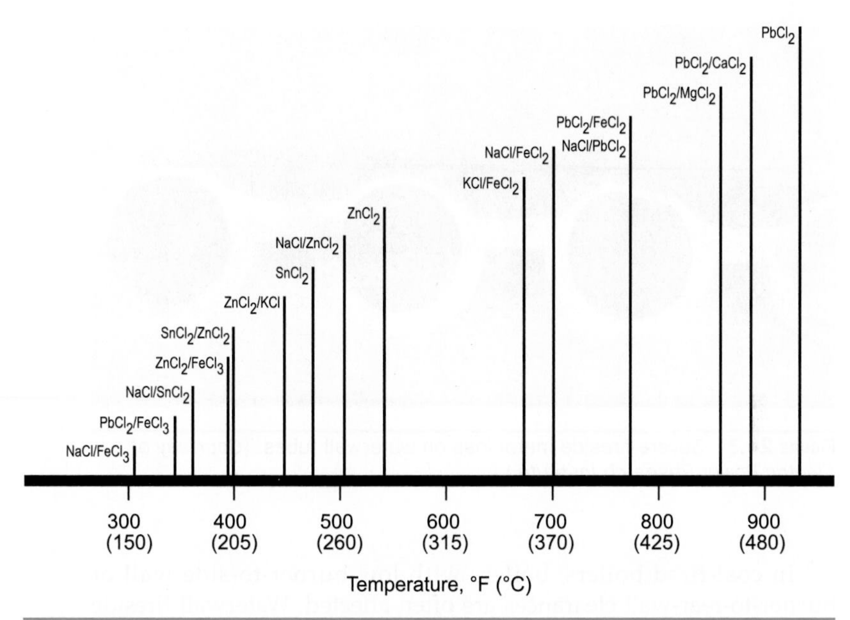

FIGURE 20.2 Melting points of various metal chloride and eutectic combinations that may occur in refuse-fired boilers.

Corrosion from Gaseous Reactions (Sulfidation)

The reducing environment due to incomplete combustion in the lower furnace allows for sulfidation attack. Sulfidation is the main mechanism of thinning for waterwalls of recovery boilers.[2] Reduced sulfur gases are formed during pyrolysis of black liquor. In coal-fired boilers, the slag that forms on the tube surface can contain unburned carbon and unoxidized pyrite.[3] Reduced sulfur gases on the tube surface react to form iron sulfide (FeS) layers. These layers are not as protective as the iron oxide layers that form under oxidizing conditions. Appreciable wastage can also occur due to alternating oxidizing and reducing environments, which develop at higher elevations of the furnace. Corrosion rates will generally exceed 5 mpy (0.13 mm/year) when attack is severe.

Locations

Areas that experience the highest heat flux tend to suffer the most severe corrosion. Waterwall fireside corrosion frequently produces severe metal loss along the crown of the tube due to radiant heat flux (Fig. 20.3). Attack may extend uniformly across several tubes in a particular location.

FIGURE 20.3 Severe fireside metal loss on waterwall tubes. (*Courtesy of Electric Power Research Institute.*)

In coal-fired boilers, boilers with low burner-to-side-wall or burner-to-rear-wall clearances are often affected. Waterwall fireside corrosion is frequently found in the windbox and burner areas.

In recovery boilers, wastage is most severe near the center of the walls due to higher gas temperatures at these locations. Attack can be more severe near the elevation of the black liquor guns, as this location will have increased concentration of reduced sulfur gases at the tube surface.[4]

Critical Factors

Waterwall fireside corrosion depends on the constituents in the fuel, the development of a reducing environment, and sufficient surface temperatures to cause or accelerate attack.

The composition of coal varies considerably between different grades. Coals that are capable of forming an ash with a low fusion temperature are required for attack by molten pyrosulfates.

Severity of attack generally increases with increasing sulfur content of the fuel. Fuels that contain appreciable amounts of chlorine can also be especially aggressive, either by forming molten phases or by disrupting the formation of protective corrosion product layers on metal surfaces.

Because reducing conditions are required for the development of waterwall corrosion, insufficient oxygen in the burner zone is a primary factor. This was a major cause of accelerated wastage on the lower furnace walls of coal-fired units that were retrofitted with low-NO_x burner technology.[5] Staged combustion in these units suppresses NO_x formation. However, this staged combustion creates areas where the fuel is partially oxidized, and the sulfur from the fuel exists primarily as hydrogen sulfide that causes sulfidation. The severity of sulfidation increases with increasing concentration of reduced sulfur gases. In general, poor combustion conditions and steady or intermittent flame contact with the furnace walls combine to produce a hot, fuel-rich corrosive environment. In recovery boilers, the concentration of reduced sulfur gases is affected by liquor sulfidity, smelt bed temperature, and the local gas temperature, in addition to the oxygen partial pressure.[6]

Temperature also has a significant influence on attack. Metal surface temperatures will increase with increasing boiler pressure, as the steam-water saturation temperature is raised (see Table 1.3). In addition, the accumulation of thermally insulating deposits on the internal surface will increase metal temperatures (see Fig. 3.1). For attack from molten slag phases in the ash deposits on tube surfaces, the surface temperature must exceed the melting point of the lowest-melting constituent in the slag. For sulfidation, corrosion rates due to reduced sulfur gases have been shown to increase significantly with increasing metal temperature, as illustrated in Fig. 20.4.

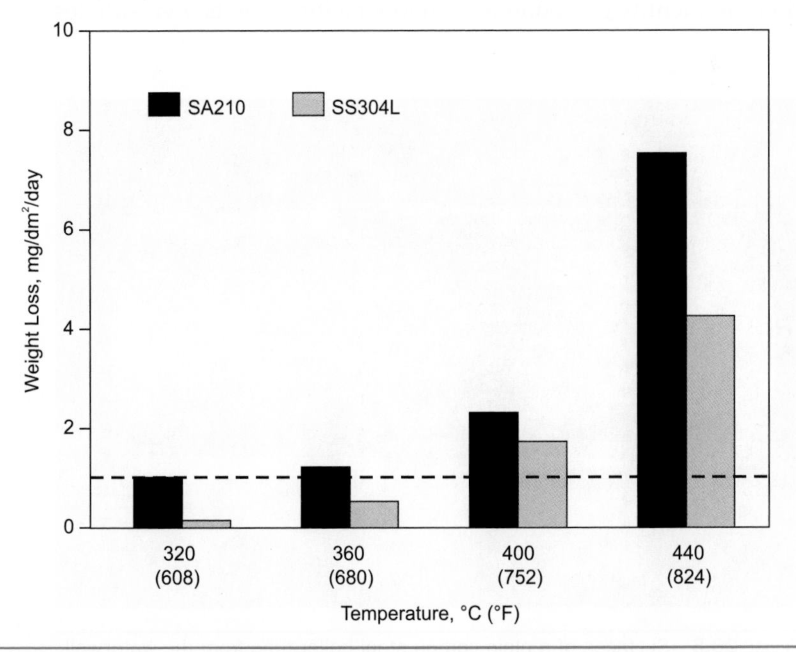

FIGURE 20.4 Relative corrosion rates at various temperatures shown for plain carbon steel (SA210) and stainless steel (SS304L) in a gaseous environment of 1% H_2S and nitrogen.[7] (*Courtesy of the Technical Association of the Pulp and Paper Industry.*)

Identification

Corroded regions due to molten pyrosulfates are covered by abnormally thick layers of iron oxide and iron sulfide corrosion products. The underlying surface contour may exhibit flow patterns from molten phases. The metal surface will typically have an irregular contour with smoothened features in areas of general wastage due to sulfidation, as shown in Fig. 20.5.

Corrosion of waterwalls in coal-fired boilers can be accompanied by deep penetrations of circumferential fissures, as shown in Fig. 20.6. The fissures are the result of corrosion fatigue cracking due to the combination of the corrosive environment on the external surface and cyclic, longitudinally oriented tensile stresses on the tube. Thermocouple measurements on waterwall tubes have shown the stresses to be associated with temperature fluctuations, likely associated with the accumulation and shedding of slag from tube surfaces.[8]

Chemical analysis of both fuel and ash may be required for identification of corrosive substances. The presence of unburned carbon in the deposit layers (especially 3% or more) is evidence of incomplete combustion. X-ray fluorescence can be used to indicate the presence of aggressive elements in the slag layers, and x-ray diffraction can identify compounds with low melting points. Pyrosulfates

FIGURE 20.5 Surfaces of a plain carbon steel boiler tube from the waterwall of a recovery boiler that experienced wastage due to sulfidation attack, shown after removal of overlying corrosion products.

FIGURE 20.6 Circumferentially oriented fissures on the generally wasted surface of a waterwall tube in a coal-fired boiler.

may not be detected due to the small amount required for attack. The presence of volatile sulfur in an analysis can be associated with pyrosulfates in the deposits, as they will liberate SO_3 when heated and form alkali sulfates.

Microscopic examination of attacked surfaces generally reveals layers of sulfides in the overlying corrosion products, as shown in Fig. 20.7. The presence of eutectic phases in the slag layers (see Fig. 20.8) is indicative of attack due to molten phases. Attack due to molten phases may propagate intergranularly through the microstructure. Chemistries of the different constituents in the ash deposit and corrosion product layers on the external surface of corroded tubing can be determined on cross sections through the layers. Spot analysis or element mapping using a scanning electron microscope (SEM) equipped for energy-dispersive (x-ray) spectroscopy (EDS) is a very useful technique for examination of the chemistry of the constituents on a microscopic level. An element map is shown in Fig. 20.9, identifying elevated sulfur and iron in eutectic corrosion products covering a corroded region (see Case History 20.6 for details).

In suspect areas, the use of ultrasonic thickness surveys may determine the extent and rate of any corrosion that has occurred.

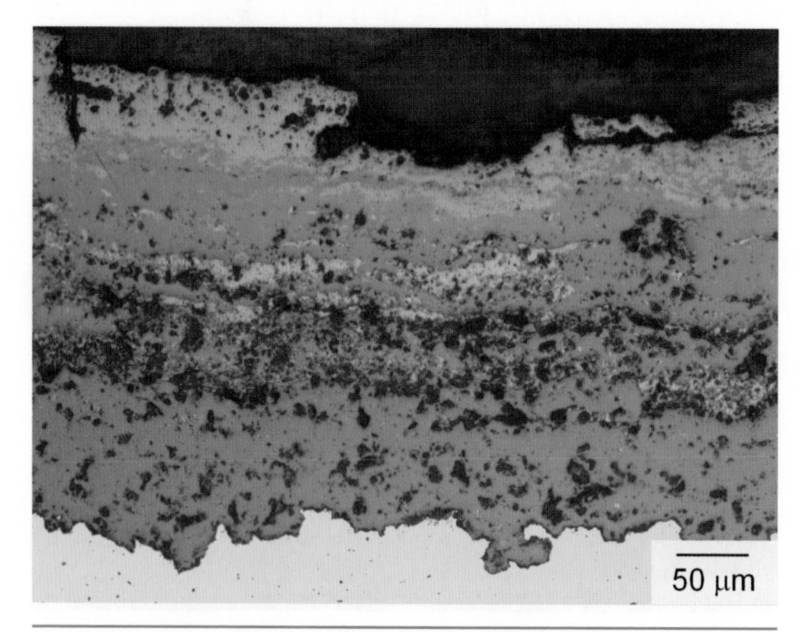

FIGURE 20.7 Distinct layers of sulfide in the corrosion products covering a corroded external surface of a coal-fired waterwall tube. Unetched.

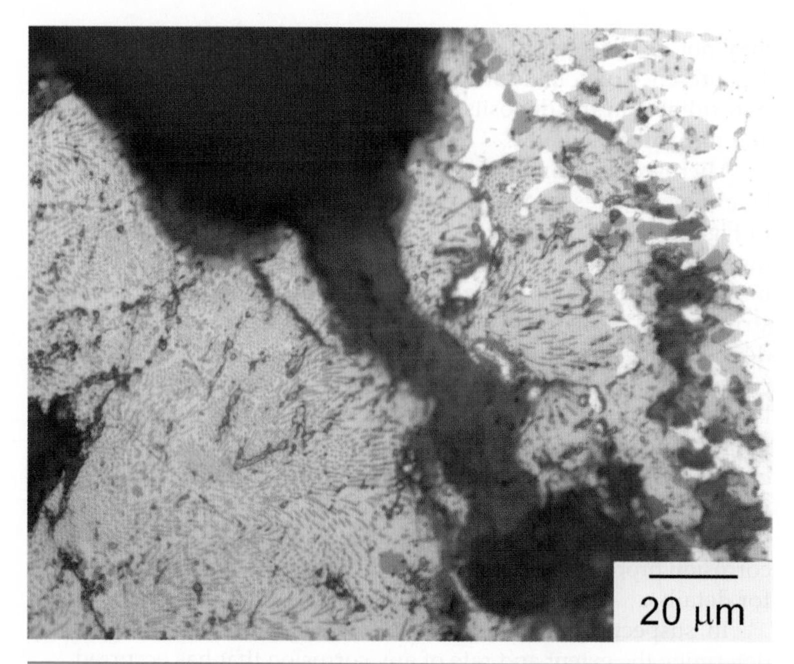

FIGURE 20.8 Eutectic phases in the corrosion products overlying a corroded surface in a waste heat boiler. Unetched.

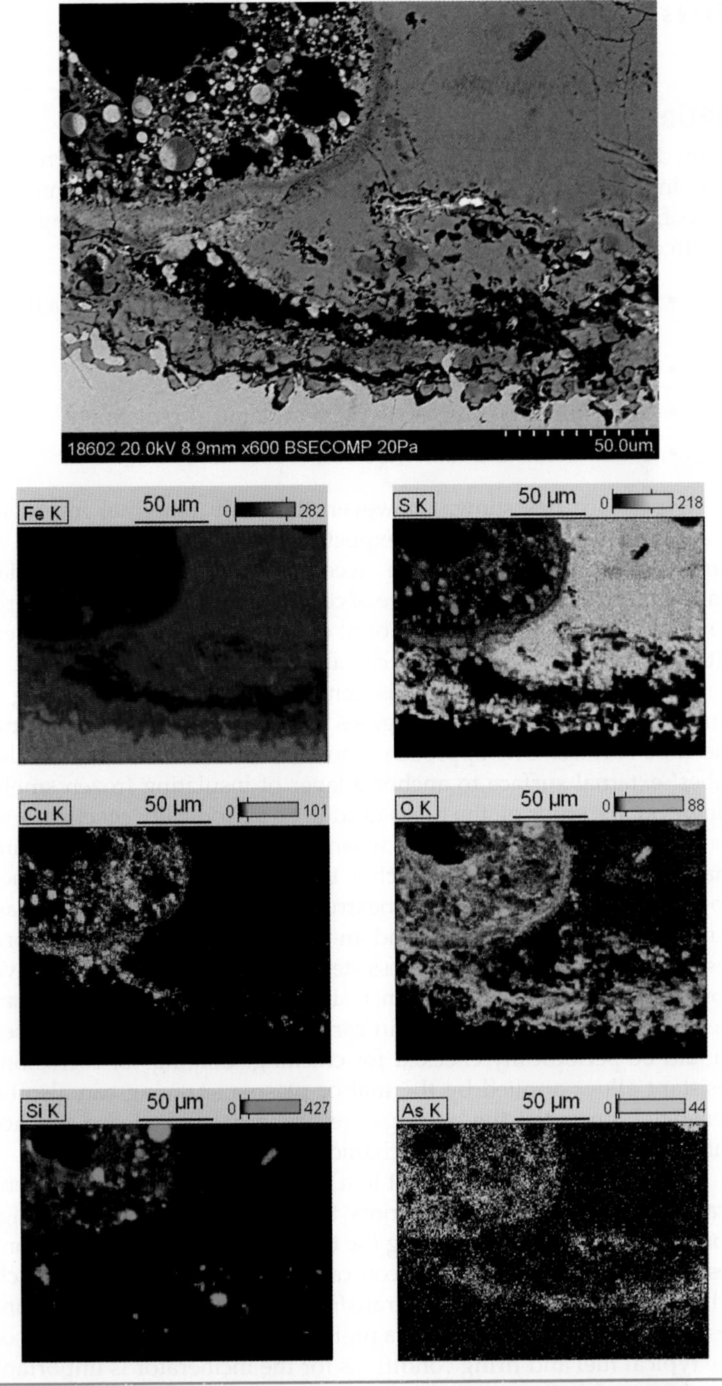

FIGURE 20.9 Map showing the elemental distribution in the ash deposits and corrosion products on the external surface in a waste heat boiler. (The SEM gray image is at the top, and element maps that show the distribution and relative concentration of some elements are below. Many elements were present in the ash deposits, and not all element maps are shown, for clarity.)

Elimination

Remedial measures to prevent or limit furnace waterwall corrosion may include changes in combustion, where possible, or implementation of barriers between the metal surface and the corrosive slag.

In coal-fired boilers, changes in combustion may include

- Improvement of burning efficiency by grinding coal to a finer, more uniform size
- Balancing of fuel supply to individual burners
- Adjustment of burners to prevent flame impingement
- An increase and redistribution of secondary air

Experience has shown, however, that only marginal improvement in combustion can be expected from these corrections. A furnace modification may be necessary to achieve substantial improvement. Changing the type of coal or blending the existing type with one containing a lower amount of aggressive species can lessen attack. Coal washing (cleaning) can also reduce aggressive species.

Attack in recovery boilers is generally controlled by creating a barrier between reduced sulfur gases in the furnace environment and the tube surface. One approach is to weld cylindrical studs on the tubes' external surface to anchor a layer of insulating frozen smelt. However, the studs on the tubes need to be periodically checked, and the tubes need to be restudded when they have wasted to the point that they do not effectively anchor the frozen smelt layer. Another approach is to use composite (coextruded) tubing that consists of an inner plain carbon steel tube and an outer clad layer of a corrosion-resistant alloy. Austenitic stainless steel alloys, such as type 304, have appreciably improved corrosion resistance to reduced sulfur gases (see Fig. 20.4) compared to plain carbon steel. However, clad tubes should be periodically checked for cracking. Coatings of corrosion-resistant alloys applied by thermal or plasma spraying can also be beneficial, provided the coatings are properly applied and are not subjected to spalling due to mechanical or thermal stresses.

The aggressive environment in a refuse-fired boiler furnace generally requires substantial barriers to protect carbon steel tubing. Some methods include studding the tubes and applying a corrosion-resistant refractory, such as silicon carbide. However, this approach will appreciably reduce heat transfer. Application of a corrosion-resistant weld overlay has been a preferred method. Consideration of the typical fuel and firing conditions for the incinerator is important in selecting the alloy for the overlay. In many cases, alloy 625 is used as an overlay to provide improved corrosion resistance.

Ultrasonic thickness surveys are used to determine the extent and severity of existing damage. If damage is minor, patch welding of affected areas may be satisfactory. If metal loss is severe, installation of thicker tubes or corrosion-resistant alloy tubes, use of thermal spray coatings, cladding or welding overlays of tubes, and the use of shields may be economically justifiable.

Case History 20.1

Industry:	Utility
Specimen location:	Roof tube
Orientation of specimen:	Horizontal
Years in service:	35
Drum pressure:	2150 psi (14.8 MPa)
Tube specifications:	3-in. (7.6-cm) outer diameter, studded
Fuel:	Coal, 13% ash

Six tube failures of the type illustrated in Fig. 20.10 occurred over several weeks. Only roof tubes had failed.

Visual examination of the external surface of one of the tubes revealed a deep, longitudinal fissure at the base of a stud (Fig. 20.10). The surface was covered with a hard, tenacious, light-colored slag. Chemical analysis of the fireside slag revealed 42% sulfur and 18% sodium. The pH of a 1% slurry of the deposit was 2.9.

Microstructural examinations revealed no thermal alteration of the tube metal. However, deep, intergranular fissures filled with a complex sulfate eutectic were observed to originate on the external surface.

Visual and microstructural examinations, coupled with chemical analysis of the slag, revealed that the cracking apparent in Fig. 20.10 resulted from penetration of molten sodium pyrosulfate along grain boundary pathways of the metal during boiler operation. Stresses imposed by normal internal pressure acted

Figure 20.10 Longitudinal fissure adjacent to a stud; fissure was caused by the combined effects of stresses imposed by internal pressure and the presence of a molten phase.

synergistically with the molten slag to produce the intergranular penetration. Microstructural examinations revealed that the attack was localized to a small region around the primary crack.

The presence of sodium pyrosulfates indicated reducing conditions or incomplete combustion in the firebox, possibly from insufficient oxygen in the burner zone or unsatisfactory grinding of the coal. Insufficient oxygen in the burner zone may be caused by insufficient excess air.

Case History 20.2

Industry:	Pulp and paper
Specimen location:	Adjacent to primary air port, recovery boiler
Orientation of specimen:	Vertical
Years in service:	10
Tube specifications:	3-in. (7.6-cm) outer diameter, stainless steel clad, carbon steel tube
Fuel:	Black liquor

Figure 20.11 shows metal loss from the external surface near a longitudinally oriented fin. Metal loss was confined to an elliptical region centered on the fin. The stainless steel cladding and some of the underlying carbon steel were corroded.

The corroded region was covered with a layer of brown corrosion product and deposits. Chemical analysis of the material indicated that it contained 52% iron, 24% sodium, and 14% carbonate.

Visual and microstructural evidence, coupled with analysis of the corrosion products, indicated that metal loss was caused by exposure of the metal to a molten sodium salt. The fusion temperature of this salt may have been depressed by the presence of the carbonate.

FIGURE 20.11 Metal loss around a fin resulting from exposure of a stainless steel–clad tube to a molten sodium salt.

Rates of metal loss from corrosion of this type vary depending on metal temperature and furnace design. Corrosion rates of 30 mpy (0.76 mm/y) have been reported.

Mitigation of this problem required redesign of tube openings so that seals are tight to flue gas. In addition, crevices where corrosive substances can concentrate must be eliminated.

Repair of damage is most successful when weld overlays of high-nickel stainless steels and thermal spray coatings are used.

Case History 20.3

Industry:	Pulp and paper
Specimen location:	Wall tube, recovery boiler
Orientation of specimen:	Vertical
Years in service:	13
Drum pressure:	875 psi (6 MPa)
Tube specifications:	2½-in. (6.4-cm) outer diameter
Fuel:	Black liquor

The corrosion apparent in Figs. 20.12 and 20.13 occurred over a very small area of the wall at a position 40 ft (12.2 m) above the floor and 4 ft (1.2 m) above oil guns. No failure of this type had occurred previously.

The sample had been water-washed before removal, which reportedly may have removed a layer of frozen smelt. The fire side of the boiler was water-washed twice per year.

The 3/8-in. (1.0-cm) perforation shown in Fig. 20.12 was centered in a region of severe external corrosion along the crown of the tube (Fig. 20.13). The corrosion produced deep, broad depressions, which gave the surface a rolling contour. Microstructural examinations revealed no evidence of overheating.

The localization of the corrosion in the region of the oil guns suggested that a reducing environment was created locally during the operation of the guns. This environment produced a corrosive low-melting molten phase, or possibly substantial amounts of corrosive reduced sulfur gases. In this case, corrosion would occur only during oilgun use, which was intermittent. However, the accumulated corrosion over 13 years of service was apparently sufficient to result in failure.

FIGURE 20.12 Perforation in a region of severe metal loss.

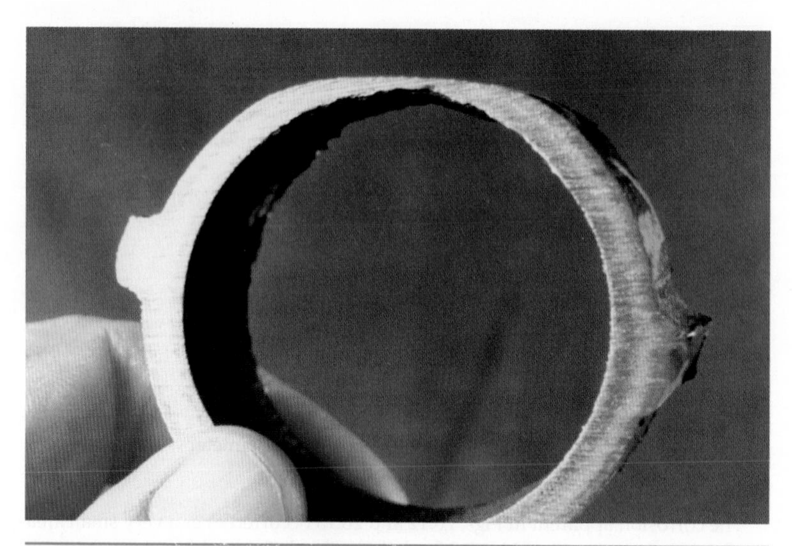

FIGURE 20.13 Severe metal loss along crown of tube resulting from exposure to a molten phase.

Case History 20.4

Industry:	Utility
Specimen location:	Wall tube
Orientation of specimen:	Slanted at bottom of furnace
Years in service:	18
Drum pressure:	2600 psi (17.9 MPa)
Tube specifications:	2¾-in. (7.0-cm) outer diameter, plain carbon steel (SA-210 A1)
Fuel:	Coal

A section of tubing from the lower rear waterwall of the coal-fired boiler contained a failure. Reportedly, numerous tubes in the area experienced damage. The hot-side external surface was covered by thick, tenacious deposit layers, as shown in Fig. 20.14. Analysis of the material indicated the presence of some unburned carbon and volatile sulfur. X-ray fluorescence indicated the material consisted primarily of iron with minor amounts (5 to 15%) of silicon, germanium, aluminum, and lead. Cross sections through the tube indicated nearly 50% wall loss on the hot side from the external surface. Removal of the deposit and corrosion product layers revealed an irregular surface contour with many narrow, circumferentially oriented fissures (Fig. 20.6). Microscopic examination showed many wedge-shaped fissures lined with oxide and sulfide layers, similar to the corrosion product layers elsewhere on the external surface. The fissures were caused by corrosion fatigue cracking; one of the fissures penetrated the tube wall.

Corrosion fatigue cracking was caused by cyclic bending stresses due to thermal expansion and contraction of the tube from fluctuations in the metal temperature. This case illustrates the significant contribution of sulfidation attack on waterwall tubing to the propagation of corrosion fatigue cracks. For more details on corrosion fatigue cracking, see Chap. 12.

FIGURE 20.14 Thick layers of ash and corrosion products covering a corroded waterwall tube in a coal-fired boiler.

Case History 20.5

Industry:	Utility
Specimen location:	Wall tube
Orientation of specimen:	Vertical
Years in service:	25
Drum pressure:	2600 psi (17.9 MPa)
Tube specifications:	3-in. (7.6-cm) outer diameter, plain carbon steel
Fuel:	Combination of coal and refuse-derived fuel (RDF)

Nondestructive testing during boiler inspections revealed significant tube wall thinning. Dimensional measurements on a section of waterwall tubing removed from the boiler revealed severe wastage from the hot-side external surface (Fig. 20.15). The attacked surface had an irregular contour with smoothened features that was covered by brown and black corrosion product layers. Bulk deposit analysis indicated appreciable levels of unburned carbon and some volatile sulfur. Iron was the main component of the deposit; some low levels of sodium and potassium were also present. SEM-EDS analysis indicated localized areas that contained up to 19% sulfur, 8% potassium, 4% sodium, and 3% chlorine, suggesting that at least some of the wastage was caused by molten pyrosulfates. Elevated chloride levels in the deposit layers, probably introduced primarily from the refuse-derived fuel, contributed to the corrosive conditions and promoted wastage.

Wastage in this area of the boiler was a chronic problem. Pad weld overlays were applied to the tube surfaces periodically to increase the overall tube-wall thickness. A cross section through a repaired region indicated attack of a plain carbon steel weld overlay. However, a nickel-based overlay (containing 59% nickel, 22% chromium, 10% iron, and 9% molybdenum) experienced minimal attack, as shown in Fig. 20.16. This highlights that weld overlays of corrosion-resistant alloys can be useful in some cases to reduce attack from molten pyrosulfates.

FIGURE 20.15 Severe wastage from the hot-side external surface of a waterwall tube.

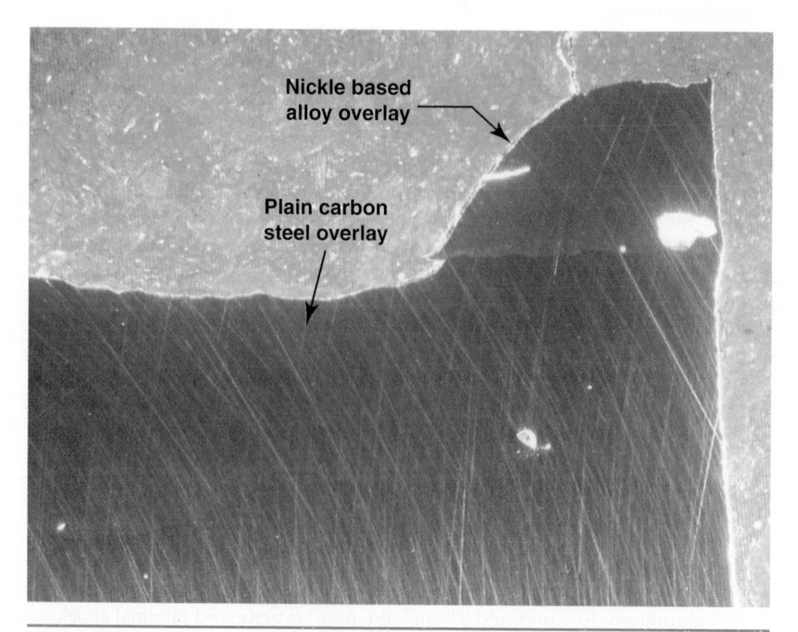

FIGURE 20.16 Wastage of a plain carbon steel overlay and minimal attack of a nickel-based alloy overlay. Unetched.

Case History 20.6

Industry:	Mining
Specimen location:	Radiant zone of a flash boiler
Orientation of specimen:	Slanted
Years in service:	5
Drum pressure:	1000 psi (6.9 MPa)
Tube specifications:	1½-in. (3.8-cm) outer diameter, plain carbon steel (SA-210A)
Fuel:	Waste heat gases and ash solids

The waste heat gases introduced to the boiler contained 17% SO_2, and the solids contained 30% sulfur, 22% iron, 12% silica, and 3% calcium, along with metallic elements such as zinc, arsenic, copper, and lead. A section from the waterwall of the combustion chamber contained a short, narrow rupture. The external surface in this region was covered by black deposit and corrosion product layers. The composition of the material was generally consistent with the compositions of the waste heat gases and solids, with the addition of a significant concentration of copper. The very irregular surface contour contained smoothened depressions (Fig. 20.17). Eutectic phases were identified in the corrosion products on the surface (Fig. 20.8). SEM-EDS element mapping indicated the eutectics primarily consisted of iron sulfides and iron oxides (Fig. 20.9). The eutectic phases penetrated along the grain boundaries. The microstructure did not show evidence of overheating.

External surface wastage was primarily caused by molten phases that formed on the metal surface. The rupture was apparently related to an intergranular fissure that penetrated the remaining tube wall in a severely wasted area. Application of a barrier layer was recommended to prevent interaction of the extremely aggressive ash deposits and the metal surface.

FIGURE 20.17 Severely wasted surface by the rupture containing smooth, rounded depressions typical of attack due to a molten phase. (The red marks identify locations where metallographic sections were taken.)

References

1. I. G. Wright, V. Nagarajan, and H. H. Krause, "Mechanisms of Fireside Corrosion by Chlorine and Sulfur in Refuse-Firing," Paper 201, *Corrosion 93*, NACE International, Houston TX, 1993.
2. H. T. Tran, "Recovery Boiler Corrosion," *Kraft Recovery Boilers*, T. N. Adams (Ed.), TAPPI Press, Atlanta, Ga., 1997, p. 298.
3. *Boiler Tube Failure Metallurgical Guide,* vol. 1: *Technical Report*, TR-102433-1, Electric Power Research Institute, Palo Alto CA, October 1993, pp. 3–155 to 3–158.
4. Tran, "Recovery Boiler Corrosion."
5. S. C. Kung, "Fireside Corrosion in Coal- and Oil-Fired Boilers," *Metals Handbook,* vol. 13B, ASM International, Materials Park, Ohio, 2002, p. 477.
6. Tran, "Recovery Boiler Corrosion."
7. D. Singbeil, L. Frederick, N. Stead, J. Colwell, and G. Fonder, "Testing the Effects of Operating Conditions on Corrosion of Water Wall Materials in Kraft Recovery Boilers," *1996 TAPPI Engineering Conference Proceedings*, Book 2, pp. 647–680.
8. D. N. French, *Metallurgical Failures in Fossil Fired Boilers*, 2d ed., John Wiley & Sons, New York, 1993, pp. 357–358.

CHAPTER 21

High-Temperature Fireside Corrosion

General Description

This chapter presents fireside corrosion mechanisms that can result in attack or accelerated wastage rates on high-temperature components in the boiler. Because wastage is caused by the development of aggressive ash on the metal surfaces, it is sometimes referred to as *slag attack*. The specific corrosion mechanisms associated with ash deposits depend on the type of fuel being burned in the unit.

In general, the aggressive elements in the fuel can react in the ash to form low melting point compounds, which cause dissolution and fluxing of the normally protective oxide on the metal surface. The ash that forms on metal surfaces depends on many factors, but the fuel type determines the species that cause attack. For instance, vanadium, sodium, and sulfur are the primary constituents responsible for attack in oil-fired boilers. The term *oil-ash corrosion* is given to the mechanism involving the vanadium compounds. Similarly, attack associated with complex alkali trisulfates due to sodium, potassium, and sulfur in coal is called *coal-ash corrosion*. In refuse-fired boilers, chlorine is the element generally responsible for fireside attack.

A wide variety of reactions can be associated with high-temperature fireside attack. Factors such as fuel type, environment, and temperature determine the reactions that can occur. Some reactions that can result in formation of aggressive ashes are discussed below, classified by fuel type.

Oil-Fired Boilers (Oil-Ash Corrosion)

Oil-ash corrosion occurs when molten slag containing vanadium compounds forms on the tube fireside surface according to the following sequence:[1]

1. Vanadium compounds and sodium compounds in the fuel are oxidized in the flame to V_2O_5 and Na_2O.

2. Ash particles adhere to metal surfaces, with Na_2O acting as a binding agent.
3. V_2O_5 and Na_2O and/or Na_2SO_4 react in the ash, forming a liquid (eutectic).

Although multiple mechanisms have been presented to describe the role of the molten vanadium salts on corrosion,[2] the primary role of the molten vanadate compounds is to flux the normally protective magnetite layer on metal surfaces, exposing the underlying metal to rapid oxidation.

Coal-Fired Boilers (Coal-Ash Corrosion)

During coal combustion, minerals in the coal are exposed to high temperatures, causing release of volatile alkali metal compounds and sulfur oxides. Coal-ash corrosion occurs when fly ash deposits on metal surfaces in the temperature range from 1050 to 1350°F (566 to 732°C). With time, alkali compounds (such as K_2SO_4 and Na_2SO_4) and sulfur compounds (SO_3) condense on the fly ash and react with it to form complex alkali trisulfates, such as $K_3Fe(SO_4)_3$ and $Na_3Fe(SO_4)_3$, at the metal/deposit interface. The presence of both potassium and sodium in the deposit results in lowered melting points for the complex sulfate. The molten slag fluxes the protective iron oxide layer covering the tube, exposing the metal beneath to accelerated oxidation.

The development of even small amounts of complex alkali trisulfates can lead to extensive attack, as the reactions on the metal surface and within the ash deposit allow for the corrosive species to recycle.[3] The molten trisulfates are believed to react with iron to form iron oxide (magnetite), iron sulfides, and alkali sulfates. Under oxidizing conditions, the iron sulfide can be oxidized to form sulfur trioxide, which can further react with the alkali sulfates to re-form the lower melting trisulfates.

Coal-ash corrosion can become an issue when the fuel supply or fuel type is changed, firing parameters are altered, or metal temperatures increase to allow for the production of an aggressive ash.

Refuse-Fired Boilers

Refuse used as fuel in waste-to-energy boilers (incinerators) contains a wide variety of materials that release constituents that can be very corrosive when burned. Municipal solid waste (MSW) is typically high in paper content and also contains appreciable amounts of plastics and metals.[4] Mass burning units burn the refuse in an unprocessed state. In some units, refuse-derived fuel (RDF) is burned. RDF is processed by separation to remove materials that can be reclaimed, and then the remaining material is prepared in various forms for burning.

Corrosion in refuse-fired boilers is usually related to chlorine. A major source of chlorine in the fuel is plastic material such as polyvinyl chloride (PVC). After combustion, chlorine can form hydrochloric acid (HCl), or it can remain as molecular chlorine in the flue gas. On high-temperature components of the boiler, corrosion can be caused by both liquid and vapor phases.[5] Many mechanisms have been proposed to describe the corrosion reactions associated with burning refuse, the details of which are presented elsewhere.[6] In general, the chlorine can react with other species in the fuel, such as lead and zinc, to form low melting point phases. Molten chloride compounds can flux the normally protective oxide that forms on the tube surfaces.

Numerous vapor-phase reactions can promote corrosion. Such reactions are generally influenced by the volatility or vapor pressure of the species that form. For instance, at superheater temperatures above 750°F (about 400°C), ferrous chloride ($FeCl_2$) in the ash deposits can react to form ferric chloride ($FeCl_3$). Ferric chloride is readily volatile from the surface. It will evaporate appreciably from the surface and result in dramatically increased corrosion rates in cases where metal temperatures exceed 900°F (482°C) and flue gas temperatures exceed about 1500°F (816°C).[7] The vapor pressures of alkali and heavy metal chlorides increase with increasing temperatures. Significant high-temperature corrosive attack can occur at high vapor pressures (typically above 10^{-4} atm).[8]

Black Liquor (Recovery) Boilers

The main mechanism responsible for fireside corrosion of superheater and reheater tubing in recovery boilers is the development of molten deposits from sulfidation/oxidation reactions. The deposits that form on the tube surfaces typically contain appreciable amounts of sodium carbonate (Na_2CO_3) and sodium sulfate (Na_2SO_4), in addition to some sodium sulfide (Na_2S).[9] Chloride, which is often present in the deposits, can dramatically depress the melting temperature of some deposit constituents. The melting temperature is about 1160°F (626°C) for a mixture of Na_2CO_3, Na_2SO_4, and NaCl.[10] Potassium present in the deposit can further lower the melting temperature. Corrosion is caused by the fluxing effect of the molten phase on the normally protective iron oxide layer.

Locations

Because slag attack mechanisms are generally liquid-phase corrosion phenomena, metal temperatures over 1000°F (538°C) are typically required for corrosion, especially for oil-ash and coal-ash mechanisms. Wastage can occur at much lower temperatures in refuse-fired boilers.

Attack predominately occurs in superheater and reheater sections of the boiler, especially utility boilers that operate at higher pressures and have high steam outlet temperatures. It may affect the tubes, which are cooled, or it may affect support and attachment equipment, which have higher surface temperatures than tubes during service. If corrosion occurs, it is typically greatest in tubes having the highest steam temperatures. The highest corrosion rates are generally found on outlet tubes of radiant superheater or reheater platens.

Corrosion is generally more severe on the windward sides of tubing with respect to flue gas flow for a number of different reasons. First, metal temperatures on the windward side of the tube are generally higher than on the leeward side. For coal-ash corrosion, the buildup of ash deposits can result in wastage flats along the sides of the tube on the windward sides, generally at the 2 o'clock and 10 o'clock positions, with the 12 o'clock position facing the flue gas path. In recovery boilers, attack on the windward side has been related to the reducing conditions that are generally present on this side. Corrosion around the tube's circumference is more variable for refuse-fired boilers.

In recovery boilers, superheater tubes that are subjected to deposition due to carryover from the smelt bed are more susceptible to attack. Such attack generally occurs in the lower superheater regions, on the leading edges of the tubes. Tubes that are subjected to radiant heating also are more prone to corrosive attack.[11]

Oil-ash corrosion can also be an issue in refinery and petrochemical plant equipment (e.g., furnaces) in which residual fuel oils are burned.

Critical Factors

Critical factors affecting corrosion include the type and composition of fuel, firing practices, and metal temperatures. Metal temperatures will depend on the operating pressure of the boiler, the steam outlet temperature, and the accumulation of thermally insulating material on the internal surface. For superheater and reheater tubing, this material generally consists of thermally formed iron oxide, although deposits from boiler water carryover or contaminated attemperation water can also contribute. Figure 21.1 indicates that as the thickness of internal oxide layer (scale) increases, the metal temperature also increases, since the internal surface oxide layer reduces heat transfer from the tube into the steam. Hence, in older units, which may have established relatively thick oxide layers on the internal surface, the metal temperature will increase and may exceed temperatures at which ash constituents become molten. In this case, the rate of tube-wall loss will be higher than that due to thermal oxidation alone, due to contributions from both thermal oxidation and corrosion. Generally, if the ratio of

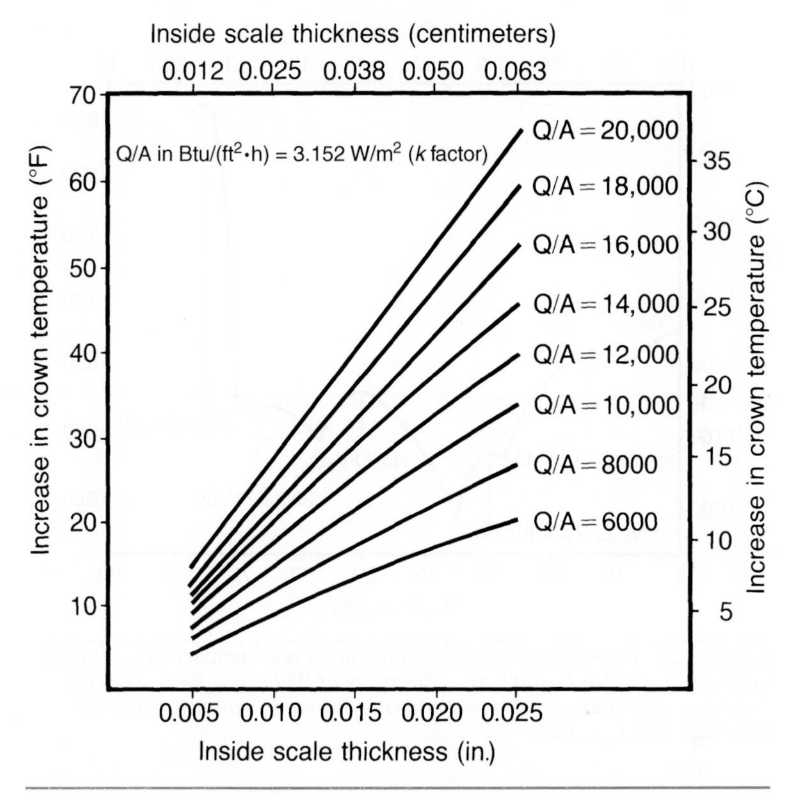

FIGURE 21.1 Relationship between increase in metal temperature and thickness of internal oxide scale for steam-cooled tubes. (*Courtesy of John Wiley & Sons, Metallurgical Failures in Fossil Fired Boilers, by David N. French, New York, 1983.*)

the wall loss to the steamside oxide layer thickness is less than 3:1, oxidation may be the primary wall loss mechanism. However, a ratio greater than 5:1 can indicate an appreciable contribution from ash corrosion mechanisms.[12]

Other critical factors specific to some different fuels and mechanisms are addressed below.

Oil-Fired Boilers (Oil-Ash Corrosion)

A number of factors are required to form a corrosive slag for oil-fired units:

1. High levels of vanadium, sodium, sulfur, or a combination of these elements in the oil
2. Inordinate amounts of excess air available for the formation of V_2O_5
3. Metal temperatures in excess of 1100°F (593°C)

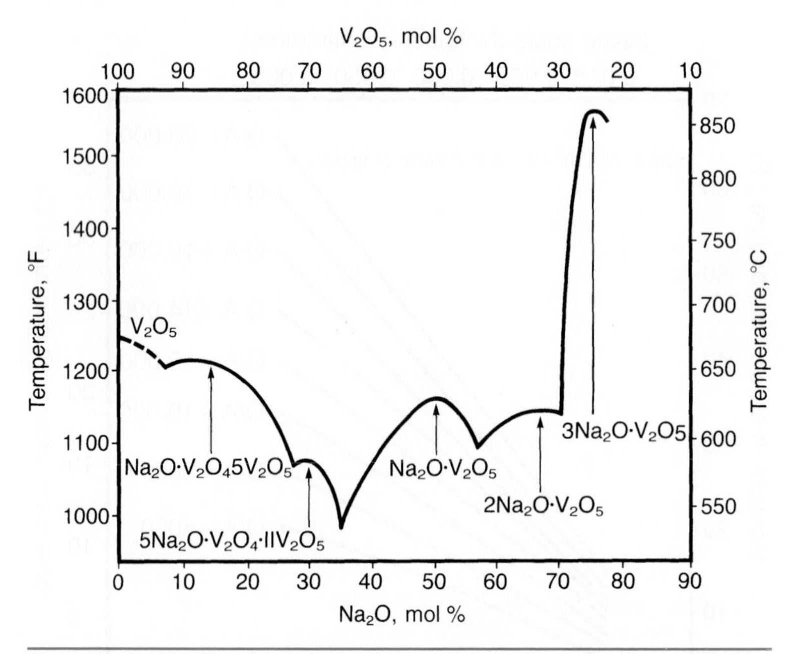

FIGURE 21.2 Relationship between temperature and compositions of liquid forms involving Na_2O and V_2O_5. (*Courtesy of William T. Reid, External Corrosion and Deposits: Boilers and Gas Turbines, American Elsevier, New York, 1971, p. 137.*)

As the temperature increases, the range of compositions of $Na_2O \cdot V_2O_5$ that form liquids expands considerably (Fig. 21.2). In addition, the diagram shows that the lowest melting point is achieved at a V_2O_5 to Na_2O ratio of about 2:1.

Corrosion rates will increase with increasing metal temperatures. Corrosion rates of 30 mpy (0.76 mm/y) have been reported. In austenitic alloys, increasing flue gas temperatures can result in substantial increases in metal loss rates.[13] However, the effect of flue gas temperature is not as great for ferritic alloys.

Coal-Fired Boilers (Coal-Ash Corrosion)

The critical factors affecting coal-ash corrosion include the use of a coal that produces an aggressive ash and metal temperatures that allow for the formation of complex alkali metal sulfates. Coal-ash corrosion can occur with any bituminous coal but is more probable with coals containing more than 3.5% sulfur and 0.25% chlorine.

The melting point of the complex alkali trisulfates varies appreciably with the ratio of sodium to potassium. The lowest melting point, approximately 1025°F (552°C), occurs at a Na:K ratio of about 2:3.[14]

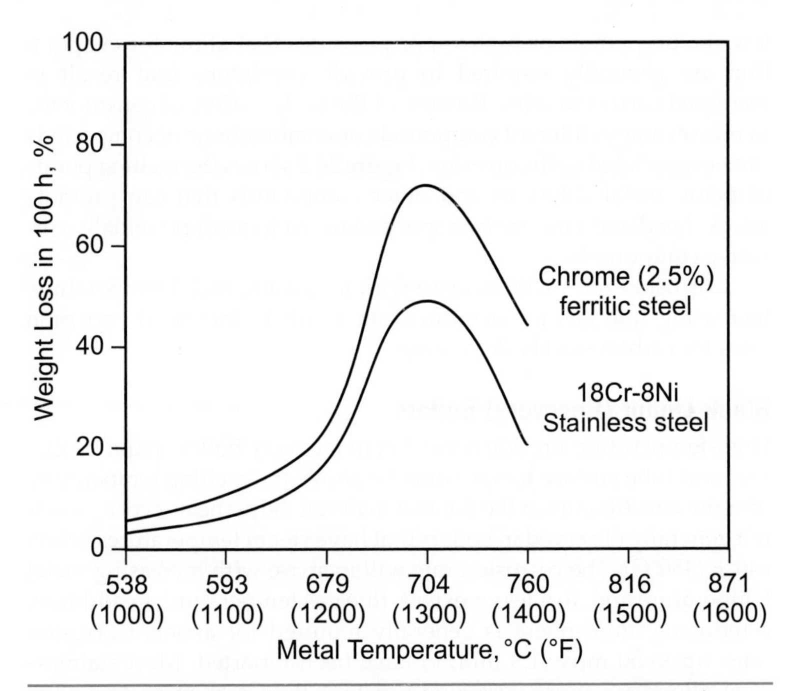

FIGURE 21.3 Bell-shaped curve relationship between temperature and corrosion rate that is associated with attack from molten complex alkali trisulfates for coal-ash corrosion. Note that the corrosion rate for the austenitic alloys is lower than that for the ferritic alloys, but the temperature range where attack occurs is still the same. (*Courtesy of The Babcock & Wilcox Company, Steam/Its Generation and Use, 41st edition.*)

Coal-ash corrosion generally occurs on metal surfaces in the temperature range from about 1025 to 1350°F (552 to 732°C). The corrosion rate is a nonlinear function of temperature, and it generally follows a bell-shaped curve (Fig. 21.3). The corrosion rate starts to increase at temperatures where molten phases form and reaches a maximum between 1250 and 1350°F (677 and 732°C). At higher temperatures, the corrosion rate decreases rapidly because of thermal decomposition of the corrosive species. This relationship between corrosion rate and temperature has been demonstrated for both ferritic and austenitic alloys.[15] Although the temperature range in which corrosion occurs is the same for ferritic and austenitic alloys, the corrosion rates for different alloys can vary appreciably. The corrosion rates are generally lower for austenitic stainless steel alloys than for low-alloy ferritic steels.

Refuse-Fired Boilers

Critical factors associated with high-temperature fireside corrosion in refuse-fired boilers primarily include the presence of appreciable concentrations of chlorine. Additionally, aggressive species having low melting points or high vapor pressures that allow for evaporation are generally required to provide conditions that result in increased corrosion rates. Because of the wide variety of constituents in refuse, many different compounds or combinations of compounds can be associated with corrosion. Figure 20.2 shows the melting points of many metal chlorides and other compounds that can promote attack. Lead and zinc are two species known to form potentially corrosive compounds.

Corrosion rates will increase with increasing metal temperature. Increasing flue gas temperatures can result in increased corrosion rates for carbon steel boiler tubing.

Black Liquor (Recovery) Boilers

High-temperature fireside corrosion in recovery boilers requires that the local tube surface temperature be above the melting temperature of some constituents in the deposit material. Superheater corrosion is not generally observed in boilers that have steam temperatures below 840°F (450°C). The corrosion rate will increase with increasing metal temperature and, to a lesser extent, flue gas temperature. In addition, a reducing atmosphere is generally required for attack. Corrosion rates up to 50 mpy (1.3 mm/y) have been reported. Most stainless steel alloys are more corrosion-resistant than carbon or low-alloy steels.[16]

Identification

All high-temperature fireside corrosion mechanisms will produce general metal loss that causes tube-wall thinning. Wastage is often most severe along one side of the tube. The surfaces will be covered by ash that varies in appearance with the type of fuel used. The underlying surface will generally have a very irregular contour with a smoothened appearance, as shown in Fig. 21.4.

Corrosion of the external surface reduces the wall thickness and load-bearing area of the tube. This reduction in load-bearing area results in an increase in hoop stress generated by internal pressure through the thinned region. The combined influence of increased stress level and high metal temperatures typically results in eventual final failure by creep (stress rupture). An example of a tube failure caused by oil-ash corrosion and mild long-term overheating is shown in Fig. 21.5.

Examination of high-temperature components that have experienced fireside corrosion should include chemical analysis of the ash deposits and corrosion products removed from attacked areas. Analysis of the fuel can be beneficial to determine what corrosive constituents may be present. A determination of ash fusion temperatures may also prove valuable.

FIGURE 21.4 Irregular surface contour with smoothened features on the external surface of a severely corroded tube from a superheater of a mass-fired refuse incinerator. The deposits were removed by bead blasting.

FIGURE 21.5 Narrow rupture located in a longitudinally oriented band of deep wastage that reduced the tube-wall thickness by over 50% due to oil-ash corrosion. The rupture was caused by the combination of corrosive metal loss and long-term overheating.

The analysis of the material covering attacked areas should include x-ray fluorescence, x-ray diffraction, and CHNS (carbon-hydrogen-nitrogen-sulfur). The presence of unburned carbon indicates incomplete combustion and reducing conditions. Volatile sulfur released during heating may be associated with alkali trisulfates, as these compounds decompose if heated to high temperatures. The amount and ratio of sodium and potassium are of special interest in coal-fired boilers, while vanadium and sodium are of interest for oil-fired boilers. Many species can be associated with corrosion in refuse-fired incinerators, although chlorine is frequently associated with attack.

Metallographic analysis will determine the morphology of ash deposits and corrosion products. Microscopic examination can reveal eutectics or globular phases in the corrosion products adjacent to the metal surface. This will confirm that the presence of a liquid phase caused attack (Fig. 21.6). Attack will generally create an irregular surface profile that may have some rounded features (Fig. 21.7). In many cases, attack by liquid eutectics in the ash will penetrate along the grain boundaries in the microstructure (Fig. 21.8).

Chemistries of the different constituents in the ash deposits and corrosion products on the external surface of corroded tubing can be

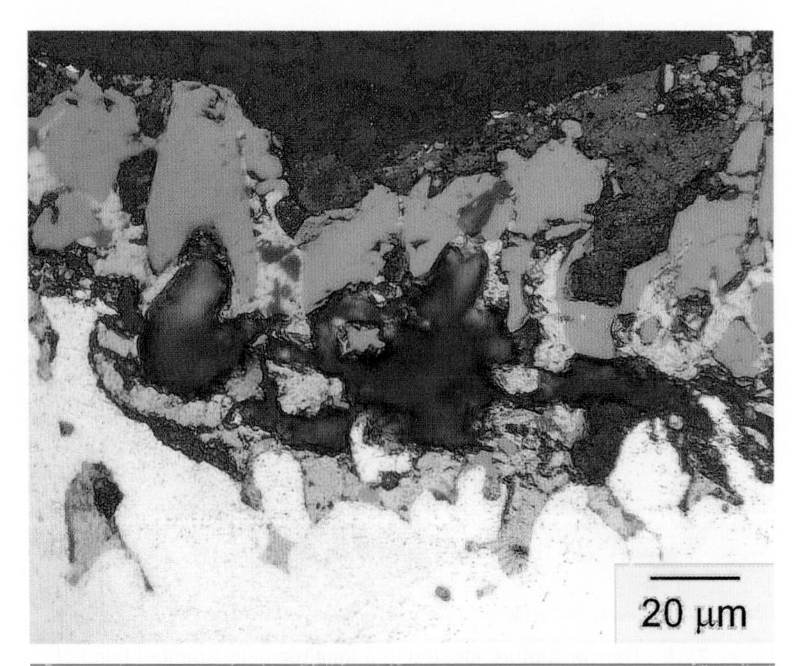

FIGURE 21.6 Microscopic examination, showing globular and eutectic phases at a corroded area on the external tube surface in a boiler that burned No. 6 fuel oil. Unetched.

FIGURE 21.7 Metallographic cross section on the external surface of a superheater tube from a refuse-fired boiler, showing an irregular surface profile containing stratified deposit and corrosion product layers. Sodium chloride was identified in the material, along with other chloride-containing compounds. Unetched.

FIGURE 21.8 Metallographic cross section, showing deep intergranular penetration of molten phases from the ash on the surface of a superheater from a recovery boiler. Note the stratified sulfide layers in the overlying deposit and corrosion products. Unetched.

determined on cross sections through the layers. Spot analysis or element mapping using a scanning electron microscope (SEM) equipped for energy-dispersive x-ray spectroscopy (EDS) is a very useful technique for examination of the chemistry of the constituents on a microscopic level. Element maps of deposit and corrosion product layers covering the external surface on a type 304 stainless steel superheater tube from a coal-fired boiler are shown in Fig. 21.9. Note the layers with elevated levels of potassium and sulfur in the ash deposits slightly away from the metal surface. Vanadium oxides are typically identified in corrosion products overlying areas where oil-ash corrosion occurred, as shown in Fig. 21.10. Low melting point phases containing chlorine can be observed in the corrosion products covering attacked surfaces in refuse-fired boilers, as illustrated in Case History 21.5.

Some features related to attack from specific fuels due to high-temperature fireside corrosion are highlighted separately below.

Oil (Oil-Ash Corrosion)

Figure 21.11 illustrates the appearance of oil-ash corrosion on a low-alloy steel tube. A section of a stainless steel reheater tube that has suffered oil-ash corrosion is shown in Fig. 21.12. Figure 21.13 illustrates the deterioration of a stainless steel tube at an attachment. The attachment, which protruded into the gas stream, acted as a heat-transfer fin, causing metal temperatures at its base to increase. Severe oil-ash corrosion occurred wherever the metal temperature exceeded 1100°F (593°C). Evidence of overheating generally accompanies oil-ash corrosion.

Coal (Coal-Ash Corrosion)

Coal-ash corrosion is identified by slag buildup on the tube surface and associated metal loss. Austenitic stainless steel tubes may exhibit a pockmarked surface appearance. Low-alloy or carbon steel tubes typically have a pair of flat zones of metal loss that are located on the windward side of the tube at orientations of 30° to 45° from the crown of the hot side, as shown in Fig. 21.14. Ultrasonic thickness measurements taken at these locations should indicate whether significant metal loss has occurred. The underlying corroded surfaces typically have a roughened appearance with smoothened features. Evidence of overheating of carbon steel and low-alloy tubing is generally present in attacked areas.

Corrosion is almost always associated with a sintered or slag-type deposit that is strongly bonded to the metal surface. This deposit consists of three distinct layers, as illustrated in Fig. 21.15. The outer layer is a bulky layer of porous fly ash. The intermediate layer

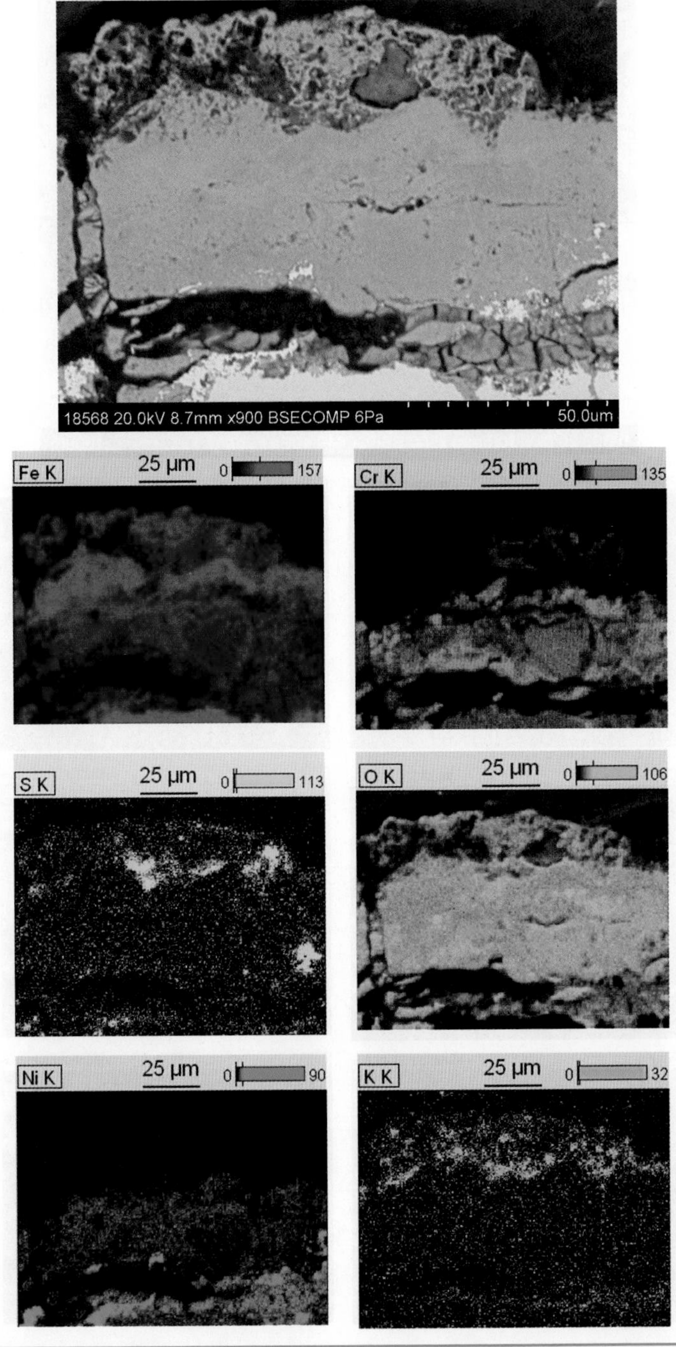

FIGURE 21.9 Element maps, showing elevated potassium and sulfur levels in the deposit layers on the external surface of a stainless steel superheater tube in a coal-fired boiler. (The SEM gray image is at the top, and element maps that show the distribution and relative concentration of some elements are below. Many elements were present in the ash deposits, and not all element maps are shown, for clarity.)

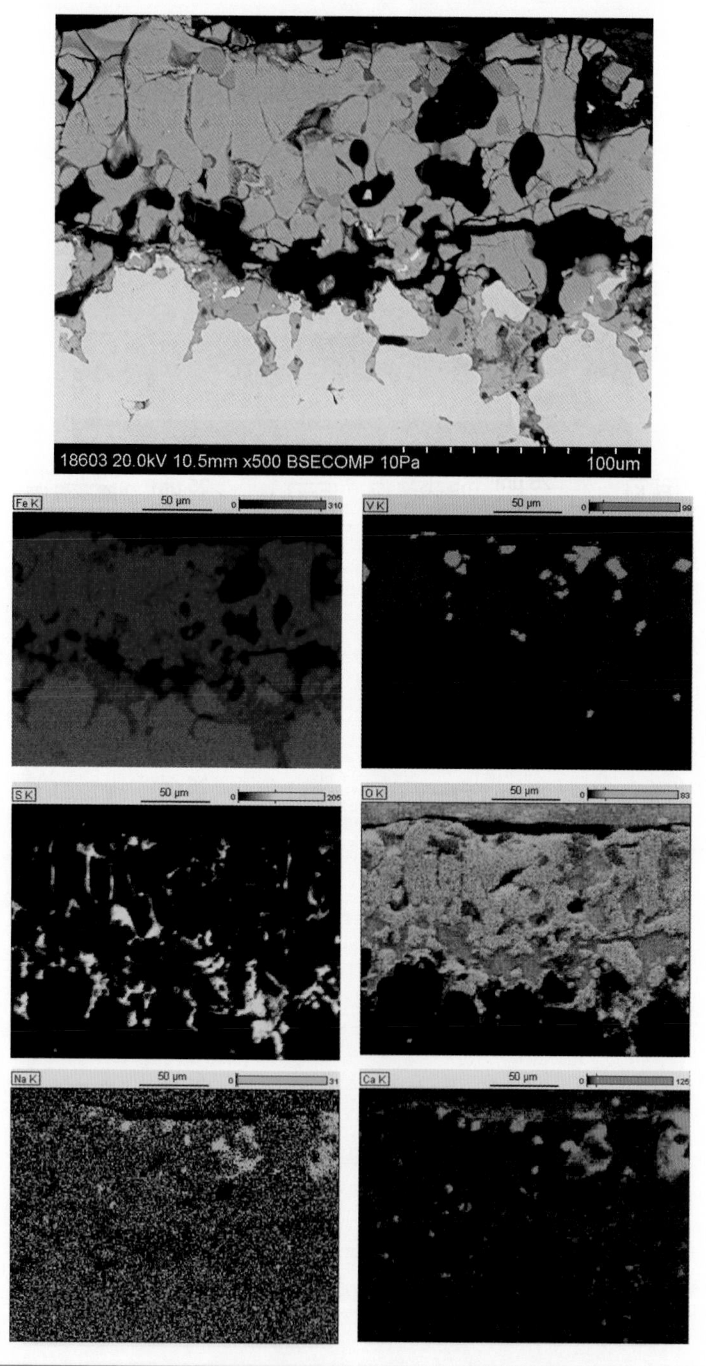

FIGURE 21.10 Element maps, showing globular phases containing vanadium and sulfur compounds in a carbon steel boiler tube that experienced oil-ash corrosion at high temperatures. (The SEM gray image is at the top, and element maps that show the distribution and relative concentration of some elements are below. Many elements were present in the ash deposits, and not all element maps are shown, for clarity.)

Figure 21.11 Severe wall thinning along one side of a superheater tube due to oil-ash corrosion.

Figure 21.12 Oil-ash corrosion on a stainless steel reheater tube. (*Courtesy of John Wiley & Sons, Metallurgical Failures in Fossil Fired Boilers, by David N. French, New York, 1983.*)

Figure 21.13 High-temperature corrosion at the base of an attachment (top of photograph) on a stainless steel reheater tube. (*Courtesy of Electric Power Research Institute.*)

FIGURE 21.14 Wastage flats on the external surface of a reheater tube from a coal-fired boiler.

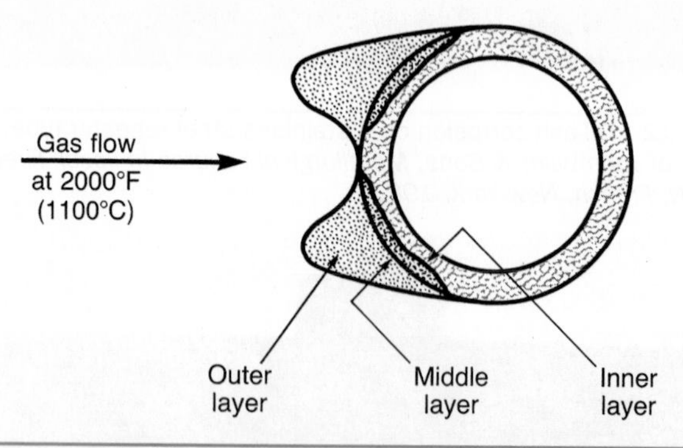

FIGURE 21.15 Schematic showing the layers of deposit and corrosion products associated with coal-ash corrosion.[17] *(Courtesy of The Babcock & Wilcox Company, Steam/Its Generation and Use, 41st edition.)*

consists of whitish, water-soluble alkali sulfates, which are responsible for the corrosive attack. This layer is typically 1/32 to ¼ in. (0.8 to 6.4 mm) thick. The innermost layer is generally thin and composed of glassy black iron oxides and sulfides. This layer is seldom thicker than 1/8 in. (3.2 mm). Components in the layers produce an acidic solution in water.

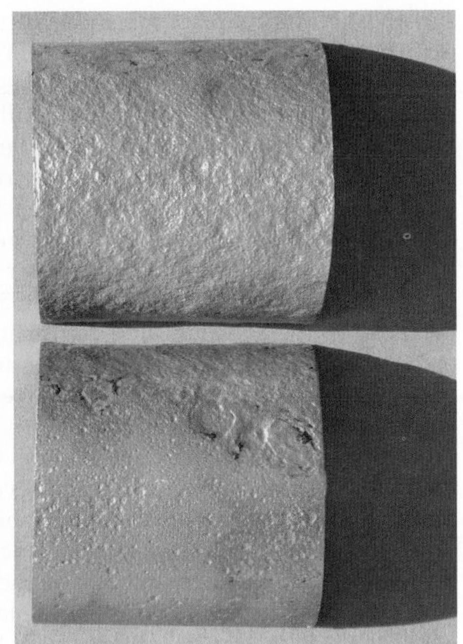

FIGURE 21.16 External surface of a corroded low-alloy superheater tube shown before and after cleaning. The tube was removed from a boiler that incinerated wood and other debris from demolished buildings. The deposits on the surface contained high levels of sulfur and lead.

Refuse-Fired Boilers

The external surfaces of corroded tubing will be covered by layers of ash, the appearance of which can vary appreciably. Analysis of the material on the surface typically indicates the presence of chlorine and sulfur. Metallic elements such as lead, zinc, and copper are often present. The ash layers usually cover corrosion products that are generally hard and overlie an irregular surface contour with smoothened features, as shown in Fig. 21.16. Microscopic examination often reveals significant stratification of the deposited ash and corrosion product layers (see Fig. 21.7). Because of the low melting point of some species that can cause accelerated wastage, thermal alteration of the microstructure (overheating) may not accompany attack.

Elimination

Reducing high-temperature fireside corrosion requires limiting the factors that promote attack, such as the formation of aggressive constituents in the ash and elevated metal temperatures. Corrosion of superheater and reheater tubes due to slag with melting temperatures in the 1100 to 1300°F (593 to 704°C) range was largely responsible for deviation from the trend in the early 1960s toward higher steam temperatures. Practically all utility boiler installations are now designed for maximum steam temperatures in the 1000 to 1025°F (538 to 552°C) range. However, the development of ultra-supercritical units to further increase plant efficiency and reduce emissions is expected to have temperatures that may approach 1380°F (750°C).

In general, superheater and reheater metal temperatures should be prevented from exceeding 1025°F (552°C). Methods of lowering the metal temperatures include decreasing steam temperatures, periodic cleaning of drainable superheater and reheater sections to prevent buildup of thick internal oxide layers, and redesign of affected areas to reduce heat-transfer rates.

Annual tube-wall thickness surveys using ultrasonic testing are useful for determining the extent and severity of attack, as the results can give early warning of impending problems. If corrosion is not severe, economical solutions may include periodic tube replacement, specifying of thicker tube walls, use of thermal spray coatings, or pad welding. Substitution of corrosion-resistant alloys, or possibly composite (clad) tubes, can be considered for locations where corrosion is severe.

Some specific remediation techniques are highlighted below for the different fuels.

Oil-Fired Boilers (Oil-Ash Corrosion)

Elimination of oil-ash corrosion is accomplished by controlling the critical factors that govern it. First, fuels containing very low quantities of vanadium, sodium, and sulfur should be specified where possible. Typical analyses of fuel oils are reported elsewhere.[18,19]

If fuel quality cannot be controlled, then recommendation of a fuel treatment additive to prevent the formation of low melting eutectics may be necessary. The use of magnesium compounds has proved to be economically successful in mitigating problems of oil-ash corrosion. Magnesium reacts with vanadium pentoxide to form a complex ($3MgO \cdot V_2O_5$) whose fusion temperature is significantly above that attained in most superheater and reheater sections. The effects of metal temperature, flue gas temperature, and fuel additives on weight loss of austenitic stainless steel alloys in an oil-fired unit are illustrated in Fig. 21.17. Other additives that increase the melting point of the ash constituents include calcium and nickel-based compounds.

Regarding firing practice, the boiler should be fired with low excess air to retard V_2O_5 formation.

Common alloys used for boiler tubing are not considered immune to oil-ash corrosion. However, resistance to oil-ash corrosion generally increases with increasing chromium content of the alloy.[20]

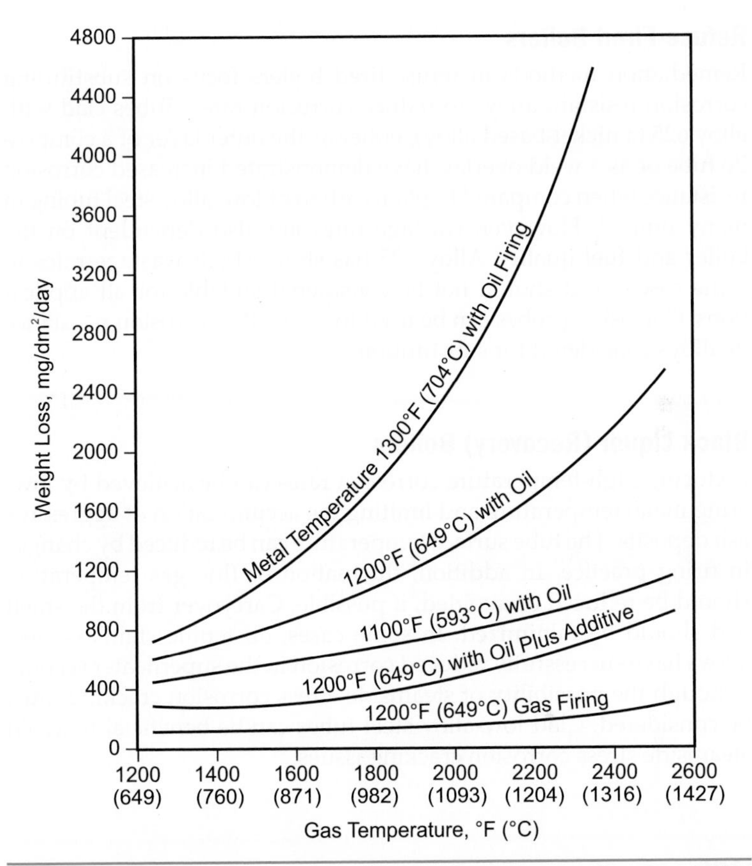

FIGURE 21.17 Effect of gas and metal temperatures on corrosion of austenitic stainless steel alloys in an oil-fired unit. Note the significant decrease in weight loss at a metal temperature of 1200°F (649°C) with the use of a fuel additive.[21] (*Courtesy of The Babcock & Wilcox Company, Steam/Its Generation and Use, 41st edition.*)

Coal-Fired Boilers (Coal-Ash Corrosion)

In view of the critical factors that govern coal-ash corrosion, it may be valuable to blend coals to reduce the percentage of corrosive constituents, such as sulfur and alkali metal species. Details on the properties of typical coals are presented elsewhere.[22,23] Coal washing can also be used to reduce sulfur and alkali contents in the coal. Some fuel additives, such as limestone, have been suggested to combat coal-ash corrosion.

Where coal-ash corrosion is severe, replacing or cladding of tubes with a resistant alloy may be required. Austenitic stainless steels generally have lower corrosion rates than low-alloy steels (see Fig. 21.3). The use of stainless steel heat shields over affected areas has been shown to extend tube life.

Refuse-Fired Boilers

Remediation methods in refuse-fired boilers focus on substituting corrosion-resistant alloys to reduce corrosion rates. Tubes clad with alloy 625 (a nickel-based alloy), either as the outer layer of a composite tube or as a weld overlay, have demonstrated increased corrosion resistance when compared to plain carbon or low-alloy steel tubing in many units.[24] However, wastage rates are also dependent on the boiler and fuel quality. Alloy 625 has shown high wastage rates in some cases, so it should not be considered suitable for all applications. Corrosion probes can be used to assess the corrosion resistance of alloys considered for substitution.

Black Liquor (Recovery) Boilers

Reducing high-temperature corrosion rates can be achieved by lowering metal temperature and limiting the accumulation of aggressive ash deposits. The tube surface temperature can be reduced by changes in firing practice. In addition, fluctuation of flue gas temperature should be reduced or avoided, if possible. Carryover from the smelt bed should be minimized. In some cases, austenitic stainless steel alloys have successfully reduced corrosion in the superheater section, although the possibility of steamside stress corrosion cracking must be considered. Clad low-alloy steel tubes can be beneficial to avoid steamside stress corrosion cracking issues.

Cautions

Ruptured superheater and reheater tubing that experience significant high-temperature fireside corrosion will generally also show evidence of overheating. Determining the individual contributions of corrosion and overheating to the failure often requires evaluation of the degree of metal loss from both internal and external surfaces. Chemical analysis and microscopic examination of material on the external surface are usually necessary to determine the most likely fireside corrosion mechanism.

Periodic removal of the ash from the surface of the tube, such as by soot blowing, can influence metal loss rates. In some cases, erosion due to soot blowing can contribute to wastage on the surface.

Related Problems

See also Chap. 3, Long-Term Overheating, and Chap. 24, Fireside Erosion.

Case History 21.1

Industry:	Utility
Specimen location:	Superheater
Orientation of specimen:	Vertical
Drum pressure:	1900 psi (13.1 MPa)
Steam temperature:	1050°F (566°C)
Tube specifications:	2-in. (5.1-cm) outer diameter, low-alloy steel (SA-213 T22)
Fuel:	No. 6 fuel oil

A superheater tube experienced a wide, fish-mouth rupture that was believed to be solely caused by overheating. However, cross sections showed severe wastage from the external surface along the side of the rupture (Fig. 21.18). The tube-wall thickness was reduced by slightly over 50% along this side. Tube diameter measurements indicated metal loss primarily occurred on the external surface. The wasted surface had an irregular appearance, as shown in Fig. 21.19. Comparison of the depth of metal loss to the steamside oxide thickness indicated a ratio of over 7, suggesting that the corrosive contribution to the external surface metal loss was considerable.

Microstructural evidence indicated that metal temperatures along the ruptured side ranged from 1100 to 1250°F (593 to 677°C) for a period of months to possibly years, well within the range where oil-ash corrosion could occur. Analysis of the external surface corrosion products by SEM-EDS indicated significant concentrations (greater than 10 wt%) of sulfur, sodium, vanadium, and iron, typical constituents in deposits and corrosion products associated with oil-ash corrosion. Fuel treatment additives were suggested to limit attack caused by oil-ash corrosion.

FIGURE 21.18 Severe wastage along the ruptured side on the external surface of a superheater tube from an oil-fired boiler.

FIGURE 21.19 Portion of a low-alloy superheater tube that experienced severe metal loss due to oil-ash corrosion on the ruptured side, shown before and after cleaning by bead blasting. Note the irregular, smoothened surface contour of the ruptured side (bottom).

Case History 21.2

Industry:	Utility
Specimen location:	Primary superheater
Orientation of specimen:	Vertical
Years in service:	10
Drum pressure:	2500 psi (17.2 MPa)
Steam temperature:	1000°F (537°C)
Tube specifications:	1½-in. (3.8-cm) outer diameter, low-alloy steel (SA-213 T22)
Fuel:	Periods with natural gas and No. 6 fuel oil

Natural gas was used as a fuel in the boiler for the majority of service for the superheater section. However, for 5 years prior to the occurrence of failures, No. 6 fuel oil was burned for extended periods. A magnesium-based treatment additive was used in an attempt to avoid oil-ash corrosion. Numerous failures occurred in the primary superheater within 1 to 2 years after fuel oil was used as the primary fuel source. The failures, in the form of narrow ruptures (see Fig. 21.5), were mostly localized to tubing near the outlet header of the superheater. Microscopic examination indicated metal temperatures ranged from 1100 to 1350°F (593 to 732°C) for extended periods, probably months to possibly years.

Analysis of the material on the surface indicated 60% vanadium, along with 6% magnesium from the fuel additive. However, more importantly, x-ray diffraction indicated the presence of bannermanite (NaV_6O_{15}). This sodium-vanadium compound has a melting temperature of slightly over 1200°F (649°C), as shown in Fig. 21.2 for $Na_2O{\cdot}V_2O_4{\cdot}5V_2O_5$. Even though a fuel additive was used, appreciable oil-ash corrosion occurred and contributed to the failure by long-term overheating. This case history illustrates that it is important to optimize the fuel treatment additives so they are effective in limiting the formation of low melting point compounds. Many factors need to be considered for fuel additives to be effective, including the analysis of the fuel and its variability, the type of additive, the amount of additive used, and the combustion parameters to ensure preferred reactions in the flue gas.

Case History 21.3

Industry:	Utility
Specimen location:	Secondary superheater outlet
Orientation of specimen:	Vertical
Years in service:	21
Drum pressure:	2500 psi (17.2 MPa)
Steam temperature:	1050°F (566°C)
Tube specifications:	2-1/8-in. (5.4-cm) outer diameter, austenitic stainless steel (SA-213-321)
Fuel:	Eastern coal, 10% ash, 7% S

The boiler from which the tubes shown in Figs. 21.20 and 21.21 were removed experienced chronic fireside corrosion problems in the outlet pendants of the secondary superheater. Ten separate failures occurred over a 3-year period. The boiler was operated continuously.

The tube submitted for analysis did not fail, but had sustained an l-in.- (2.5-cm-) wide band of severe corrosion along one side of its external surface. Measurement indicated a reduction in tube wall thickness of 0.085 in. (2.2 mm), about 30% of the tube-wall thickness relative to measurements away from obvious areas of wastage.

Microstructural examinations revealed solidified eutectics of slag material covering the external surface in the corroded zone. SEM-EDS chemical analysis

FIGURE 21.20 Contour of external surface in corroded zone. Note pockmarked surface appearance characteristic of attack on stainless steel.

FIGURE 21.21 Transverse profile of tube, showing metal loss.

of material covering the corroded region revealed large concentrations of iron, sulfur, and potassium, as well as components of fly ash.

Evidence from visual, microstructural, and chemical analyses identified the wastage as coal-ash corrosion resulting from the formation of complex alkali sulfates at the metal surface. Changing or blending the fuel was recommended to reduce the formation of aggressive ash. More frequent soot blowing was recommended to remove slag deposits.

Case History 21.4

Industry:	Utility
Specimen location:	Superheater tube
Orientation of specimen:	Vertical
Drum pressure:	2600 psi (17.9 MPa)
Steam temperature:	Unknown
Tube specifications:	2-in. (5.1-cm) outer diameter, SA-213 TP347 welded to SA-213 T22
Fuel:	Coal, unspecified

A wide, thin-lipped rupture occurred in a superheater tube at a location where a low-alloy steel tube segment was welded to an austenitic stainless steel tube segment, as shown in Fig. 21.22. The rupture was located within a flat wasted region on one side of the low-alloy steel segment. The surface contour in this region had a generally smoothened appearance with distinct depressions and edges (Fig. 21.23). Bulk analysis of the material covering the wasted region indicated a substantial concentration of sulfur (over 30%) and minor concentrations (5% or less) of sodium and potassium. Localized analysis of material in the corroded region by SEM-EDS indicated areas of elevated sodium and potassium concentrations in conjunction with high sulfur levels. Metallographic analysis of the low-alloy steel indicated that metal temperatures ranged between 1100 and 1250°F (593 and 677°C) for extended periods. The ash deposit analysis and metal temperatures were consistent with coal-ash corrosion.

Figure 21.22 Wide rupture localized to a wastage flat on the low-alloy steel side of the section.

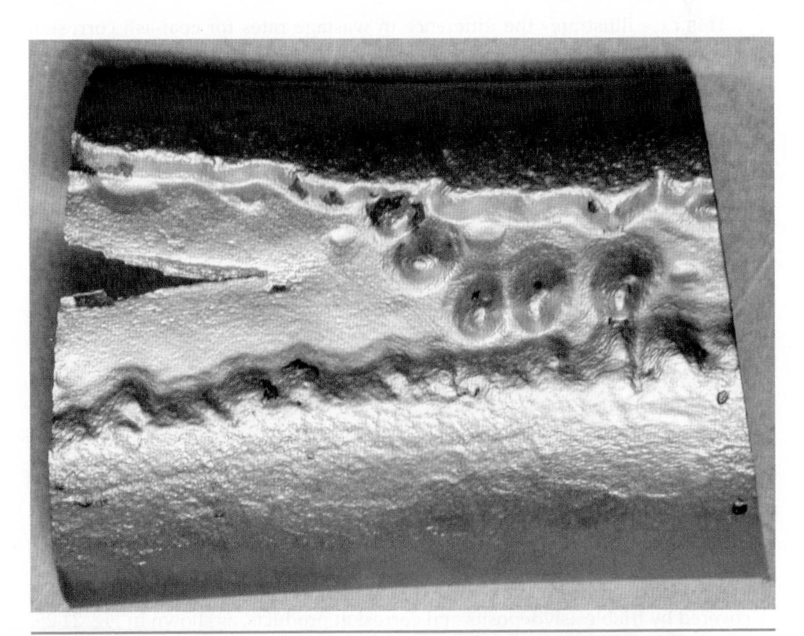

Figure 21.23 Close-up of the wastage flat at the end of the rupture. Smoothened features along the edges and depressions are revealed after cleaning.

FIGURE 21.24 Ring sections comparing the wastage profile on the tube segments. Note the deep localized wastage on the low-alloy steel section (on the right) compared to the austenitic stainless steel section (on the left).

This case illustrates the difference in wastage rates for coal-ash corrosion between ferritic low-alloy steel and austenitic stainless steel, assuming the same service life of the segments. Cross sections through the tube segments revealed much deeper metal loss on the low-alloy steel, even at locations away from the rupture (Fig. 21.24). However, both alloys exhibited deep attack, illustrating that austenitic stainless steels are also susceptible to coal-ash corrosion. Because attack was localized to one small area of the boiler, the installation of shielding for vulnerable tubes was suggested to avoid direct contact with the ash deposits.

Case History 21.5

Industry:	Utility
Specimen location:	Superheater
Orientation of specimen:	Vertical
Years in service:	Unknown
Drum pressure:	650 psi (4.5 MPa)
Steam temperature:	850°F (454°C)
Tube specifications:	2-in. (5.1-cm) outer diameter, SA-213 T22 low-alloy steel
Fuel:	Refuse-derived fuel (RDF)

The external surface of a severely thinned section of superheater tubing was covered by friable ash deposits and corrosion products, as shown in Fig. 21.25. External surface metal loss occurred along the entire circumference of the tube, but tended to be deeper along two sides (Fig. 21.26). Measurements indicated that the tube-wall thickness was reduced by more than 85% at one location.

FIGURE 21.25 Ash deposits covering an irregular surface contour on a superheater tube from a refuse-derived fuel-fired boiler.

FIGURE 21.26 Ring section, showing severe metal loss in most places on the external surface along the tube circumference.

The surface contour was very irregular with smoothened features. No evidence of overheating was observed.

Bulk analysis of the material covering the external surface identified high levels of iron (~50%) along with minor concentrations (5 to 15%) of calcium, sulfur, and chlorine. X-ray diffraction indicated the presence of sodium chloride, iron oxides, and calcium compounds.

Element mapping of the deposit and corrosion product layers by SEM-EDS indicated a distinct layer of a sodium/potassium chloride salt, as shown in Fig. 21.27. Corrosion was caused by a very aggressive ash that formed on the tube surfaces. Sodium and potassium chloride can react with other chloride compounds to form mixtures that have melting points considerably less than normal operating temperatures for superheater tubing. Tubing clad with a more corrosion-resistant alloy was suggested to limit the severity of the attack.

Case History 21.6

Industry:	Utility
Specimen location:	Superheater
Orientation of specimen:	Vertical
Years in service:	Unknown
Drum pressure:	680 psi (4.7 MPa)
Steam temperature:	850°F (454°C)
Tube specifications:	2-in. (5.1-cm) outer diameter, SA-213 T22 low-alloy steel
Fuel:	Refuse

A tube from the outlet of the primary superheater bank experienced external surface metal loss that reduced the tube-wall thickness by over 85%. The incinerator was a mass-burning unit that fired municipal solid waste (MSW). Metal loss was deepest in areas on the leeward side of the tube. The surfaces were covered by patches of white ash deposits overlying red and black corrosion product layers (Fig. 21.28). Analysis of material covering the surfaces in multiple locations in the superheater section revealed consistently high sulfur and calcium levels, with variable amounts of lead, chlorine, sodium, and zinc. The underlying surface had an irregular contour. Microscopic analysis revealed stratified corrosion products. Evidence of molten phases was not observed. The metal did not experience overheating.

Wastage was predominately caused by gas-phase corrosion. Gaseous hydrochloric acid can react to form chloride species that degrade the normally protective oxide on the surface. Attack was most severe near the outlet of the superheater, where metal temperatures were highest. Substitution of Incoloy 825 (a nickel-based alloy), in areas of the superheater where low-alloy steel experienced considerable wastage, proved effective at reducing wastage rates. A section of Incoloy 825 tubing was found to have 25% wall loss after more than 18 years of service, as shown in Fig. 21.29.

Case History 21.7

Industry:	Pulp and paper
Specimen location:	Superheater
Orientation of specimen:	Vertical
Years in service:	7
Drum pressure:	1250 psi (8.6 MPa)
Tube specifications:	2¼-in. (5.7-cm) outer diameter, low-alloy steel
Fuel:	Black liquor

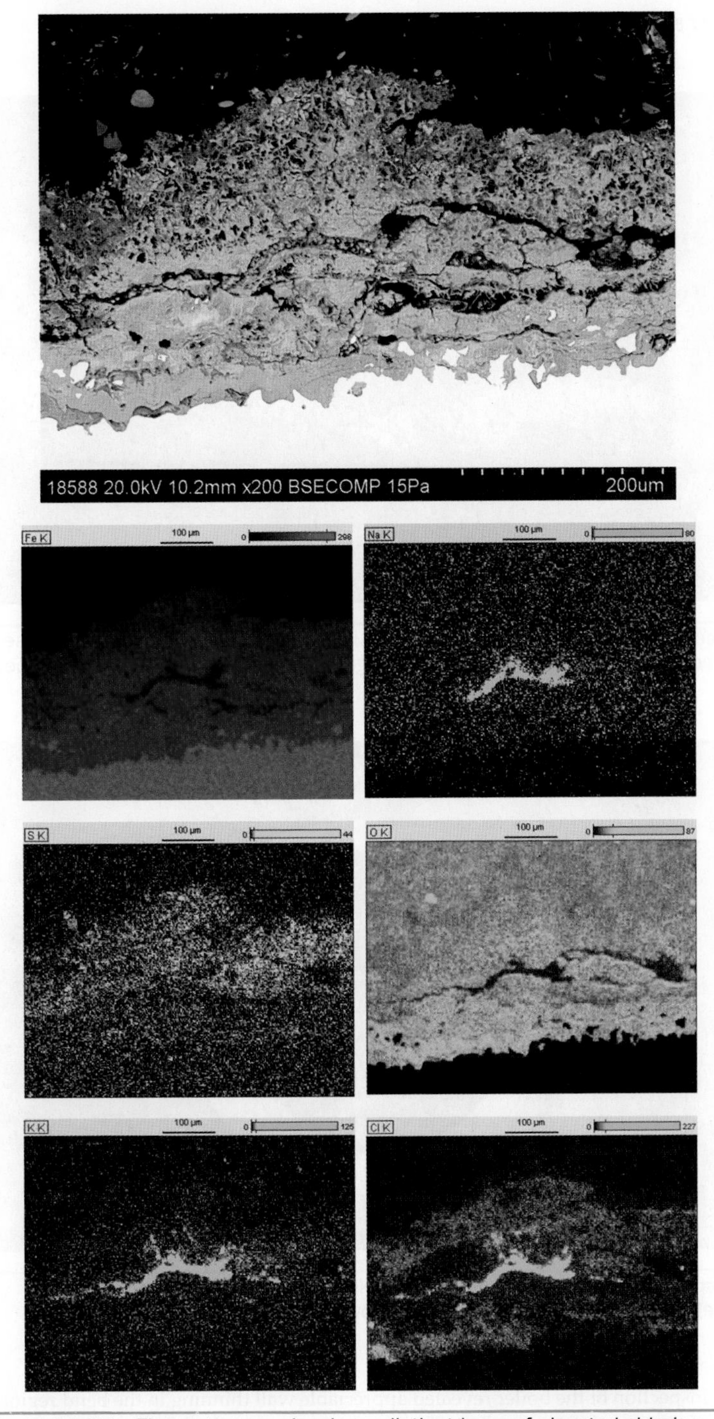

FIGURE 21.27 Element maps showing a distinct layer of elevated chlorine, sodium, and potassium in the corrosion products on the external surface. (Many elements were present in the ash deposits, and not all element maps are shown, for clarity.)

FIGURE 21.28 Spotty white ash deposits on a superheater tube in a refuse-fired boiler. Note the significant wastage in areas along the side of the tube.

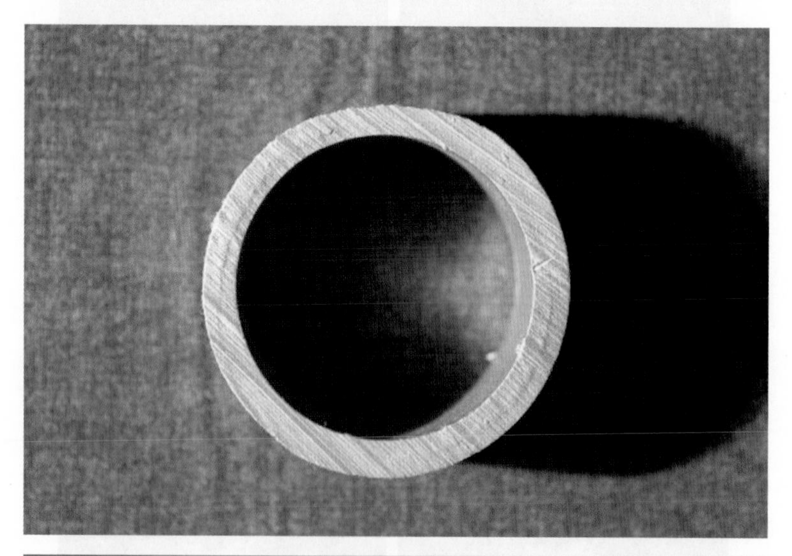

FIGURE 21.29 Ring section of an Incoloy 825 tube, showing only minor metal loss from the external surface over many years of service.

Inspection of the boiler revealed appreciable wall thinning in the bend regions of the superheater pendant. The surface in this area had an irregular contour and contained moderately deep depressions, as shown in Fig. 21.30. The material covering the surface consisted predominately of iron with minor levels of sodium, sulfur, chlorine, and potassium. Microscopic examination of the

FIGURE 21.30 Moderately deep, wide depressions superimposed on an irregular contour on the external surface of a superheater tube from a recovery boiler.

material overlying the attacked areas indicated stratified corrosion product layers containing distinct sulfide layers (Fig. 21.8). Moderately deep intergranular penetration extended into the metal. SEM-EDS analysis indicated localized areas with concentrations of chlorine and potassium exceeding 7 and 5%, respectively.

Moderately deep metal loss and intergranular penetration were primarily caused by a sulfidation-oxidation reaction resulting from aggressive molten deposits. Melting temperatures of constituents in the aggressive ash were lowered considerably by chlorine and potassium in the deposits.

Case History 21.8

Industry:	Pulp and paper
Specimen location:	Bottom bend of the primary superheater
Orientation of specimen:	Horizontal
Years in service:	2
Drum pressure:	600 psi (4.1 MPa)
Tube specifications:	Composite tubing, type SA-213 T22 clad with type 310 austenitic stainless steel
Fuel:	Black liquor

The primary superheater of the boiler experienced multiple failures over a few years. Composite tubing with austenitic stainless steel cladding was used to limit external surface wastage from the aggressive deposits and conditions at this location. The failures, which were longitudinally oriented ruptures, were

FIGURE 21.31 Bottom bend of a composite superheater tube that experienced multiple ruptures. Note the thick ash deposits in places along the surface.

localized to the bottom bends of the superheater (Fig. 21.31). This location was subjected to considerable carryover of particles from the smelt bed. Chemical analysis of the ash deposits indicated they consisted primarily of calcium, sulfur, magnesium, and manganese with minor amounts (4 to 6%) of sodium, chlorine, and potassium.

Although the austenitic stainless steel cladding layer is expected to be more resistant to high-temperature fireside corrosion at this location, the superheater was subjected to periodic events of boiler water carryover. The internal surface deposits that collected at the bottom bends raised tube metal temperatures significantly, exceeding 1450°F (788°C) in places near the rupture and 1175°F (635°C) at other locations. At these elevated temperatures, the clad layer experienced appreciable metal loss in the form of shallow general wastage and moderately deep intergranular attack (Fig. 21.32). The clad layer was completely penetrated in places. Intergranular corrosion penetrated along grain boundary phases, as shown in Fig. 21.33. This is typical of sulfidation of type 310 stainless steel, where the sulfur tends to preferentially react with chromium to form sulfides. SEM-EDS element mapping confirmed that particles in the attacked region contained elevated levels of chromium and sulfur. Although the failures in the superheater section were predominately caused by overheating conditions, this case illustrates the strong effect that metal temperature can have in increasing metal loss, even for more corrosion-resistant alloys.

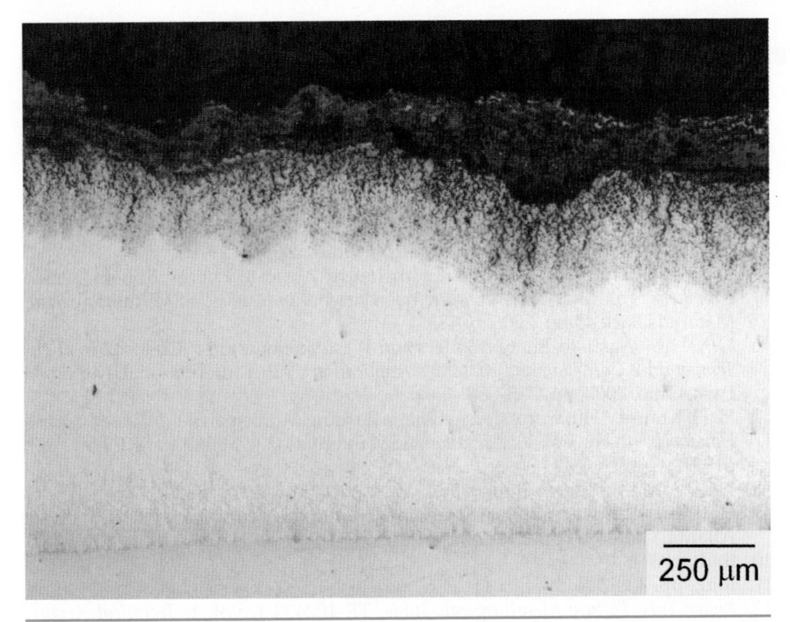

FIGURE 21.32 Wastage and intergranular penetration along the austenitic stainless steel clad layer on a recovery boiler superheater tube. Unetched.

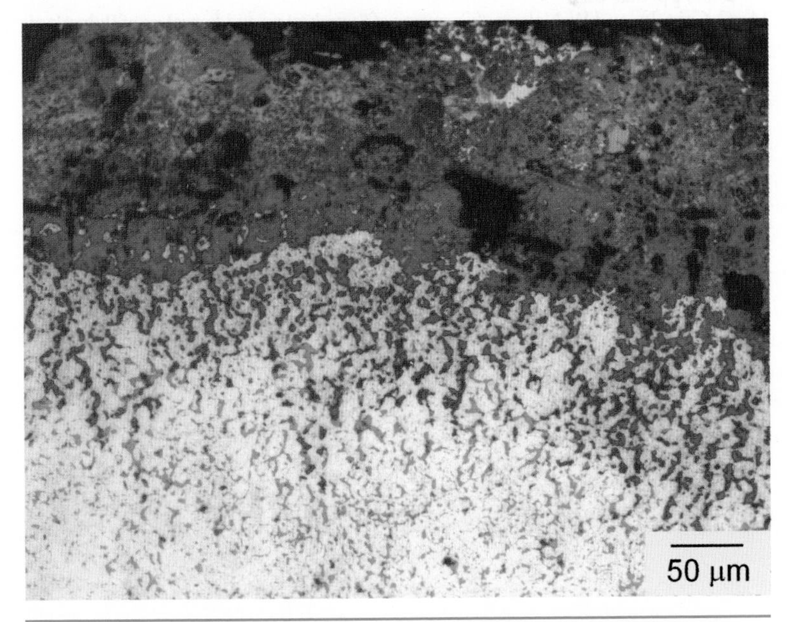

FIGURE 21.33 Higher magnification, showing sulfide in the deposit and corrosion product layers and preferential attack along chromium-rich grain boundary phases. Unetched.

References

1. D. N. French, *Metallurgical Failures in Fossil Fired Boilers*, 2d ed., John Wiley & Sons, New York, 1993, p. 362.
2. William T. Reid, *External Corrosion and Deposits: Boilers and Gas Turbines*, American Elsevier, New York, 1971, pp. 135–138.
3. Ibid., p. 131.
4. S. C. Stultz and J. B. Kitto (Eds.), *Steam: Its Generation and Use*, 41st ed., Babcock & Wilcox Company, Barberton, Ohio, 2005, p. 29-1.
5. G. Y. Lai, "High-Temperature Corrosion in Waste-to-Energy Boilers," *ASM Handbook*, vol. 13C: *Corrosion: Environments and Industries*, ASM International, Materials Park, Ohio, 2006, p. 483.
6. G. Y. Lai, "Waste-to-Energy Boilers and Waste Incinerators," Chap. 12 in *High-Temperature Corrosion and Materials Applications*, ASM International, Materials Park, Ohio, 2007, pp. 335–358.
7. H. H. Krause, "Hot Corrosion in Boilers Burning Municipal Solid Waste," *Metals Handbook*, 9th ed., vol. 13: *Corrosion*, ASM International, Metals Park, Ohio, 1987, p. 997.
8. Lai, "Waste-to-Energy Boilers and Waste Incinerators," p. 343.
9. H. Tran, "Recovery Boiler Corrosion", *Kraft Recovery Boilers*, TAPPI Press, Atlanta, Ga., 1997, pp. 296–298.
10. Ibid., p. 311.
11. Ibid., pp. 308–314.
12. *Boiler Tube Failure Metallurgical Guide*, TR-102433-1, vol. 1: *Technical Report*, Electric Power Research Institute, Palo Alto CA, October 1993, pp. 3–99.
13. G. Y. Lai, "Oil-Fired Boilers and Furnaces," Chap. 11 in *High-Temperature Corrosion and Materials Applications*, ASM International, Materials Park, Ohio, 2007, pp. 325–326.
14. C. Cain, Jr., and W. Nelson, "*Corrosion of Superheaters and Reheaters of Pulverized-Coal-Fired Boilers,*" *Journal of Engineering. Power, Trans. ASME*, October 1961, p. 486.
15. Stultz and Kitto, *Steam: Its Generation and Use*, pp. 21–21.
16. Tran, "Recovery Boiler Corrosion," pp. 308–309.
17. Stultz and Kitto, *Steam: Its Generation and Use*, pp. 21–21.
18. J. G. Singer (Ed.), *Combustion Fossil Power Systems*, 3d ed., Combustion Engineering, Inc., Windsor, Conn., 1981, pp. 2–27 to 2–32.
19. Stultz and Kitto, *Steam: Its Generation and Use*, pp. 9–12 to 9–16.
20. Lai, "Oil-Fired Boilers and Furnaces," p. 328.
21. Stultz and Kitto, *Steam: Its Generation and Use*, pp. 21–25.
22. Singer, *Combustion Fossil Power Systems*, pp. 2–1 to 2–19.
23. Stultz and Kitto, *Steam: Its Generation and Use*, pp. 9–5 to 9–11.
24. Lai, "Waste-to-Energy Boiler and Waste Incinerators," p. 339.

CHAPTER 22

Cold-End Corrosion during Service

General Description

Cold-end corrosion refers to a fireside corrosion mechanism that can occur in the relatively low-temperature region of the boiler. It is also termed *acid dew point corrosion*, as it is caused by conditions that allow acidic vapors in the flue gas to condense on metal surfaces and cause corrosion. Because commonly used fuels for boilers such as oil and coal typically contain sulfur, condensation of sulfuric acid is most commonly responsible for cold-end corrosion. The reactions associated with the formation of sulfuric acid vapors in the flue gas are discussed below.

In most combustion systems, flue gas temperatures can range from 3000°F (1650°C) in the flame to 250°F (121°C) or less at the stack. This temperature drop can cause numerous chemical and physical changes in the constituents of the flue gas. Among the most troublesome changes is the reaction between water vapor and sulfur trioxide to form sulfuric acid as the flue gas cools.

In general, the problem is associated with the combustion of fuel that contains sulfur or sulfur compounds. Sulfur in the fuel is oxidized to sulfur trioxide in the following sequence. First, oxygen combines with sulfur from the fuel to form sulfur dioxide:

$$S + O_2 \rightarrow SO_2 \tag{22.1}$$

A small fraction (1 to 5%) of the sulfur dioxide produced is additionally oxidized to sulfur trioxide by direct reaction with atomic oxygen in the flame:

$$SO_2 + O \rightarrow SO_3 \tag{22.2}$$

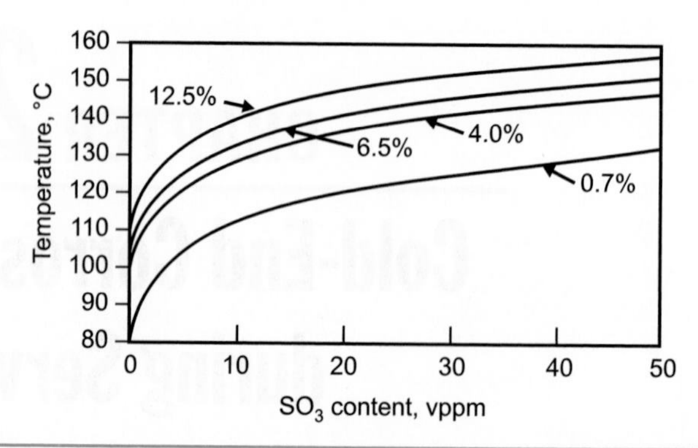

FIGURE 22.1 Effect of flue gas SO_3 concentration on the sulfuric acid dew point for various flue gas–water vapor percentages. The dew point increases with both increasing SO_3 and water vapor concentrations in the flue gas.[1] (*Reprinted with permission of ASM International. All rights reserved. www.asminternational.org*)

Catalytic oxidation can promote SO_3 formation if ferric oxide, vanadium pentoxide, or nickel is present in the flue gas:

$$SO_2 + 1/2\ O_2 + \text{catalyst} \rightarrow SO_3 \tag{22.3}$$

Sulfur trioxide and water can react to produce sulfuric acid:

$$SO_3 + H_2O \rightarrow H_2SO_4 \tag{22.4}$$

The quantity of sulfur trioxide and water vapor in the flue gas affects the sulfuric acid dew point temperature (the temperature at which sulfuric acid first condenses). The graph in Fig. 22.1 illustrates the general relationship between sulfur trioxide concentration, water vapor content, and dew point. Dew point temperatures increase as a function of increasing sulfur trioxide concentration and water vapor content of the flue gas.

Sulfuric acid may condense on a relatively cool surface. Thus, corrosion may occur wherever metal temperatures are less than the sulfuric acid dew point. Sulfuric acid will corrode steel according to the following reaction:

$$H_2SO_4 + Fe \rightarrow FeSO_4 + H_2 \uparrow \tag{22.5}$$

Note that it is the temperature of the metal, not the temperature of the flue gas, that is critical. Even when the flue gas temperature is well above the acid dew point, corrosion is a distinct possibility wherever metal temperatures are less than the acid dew point.

Locations

Cold-end corrosion can occur at locations where the metal temperature drops below the dew point of acid gases in the flue gas. The primary gas of concern is sulfuric acid. Most problems caused by cold-end corrosion occur on relatively low-temperature boiler components. Economizer tubing is susceptible, and attack will generally be most severe at the inlet end of the economizer, where water and metal temperatures are lowest. Some other components in the flue gas path that are susceptible to cold-end corrosion are air preheaters, induced-draft fans, flue gas scrubbers, electrostatic precipitators, ducting, and stacks.

Critical Factors

The critical factors that influence cold-end corrosion include both the concentration of sulfur trioxide and water vapor in the flue gas and the presence of metal surface temperatures below the sulfuric acid dew point. The acid dew point increases as the sulfur trioxide concentration and the water vapor content in the flue gas increase.

The amount of sulfur trioxide produced is dependent on the fuel and firing parameters of the boiler. The type of fuel and its sulfur level are important parameters. The severity of cold-end corrosion is most extreme with "sour," or high-sulfur, gaseous fuels. The sulfuric acid dew point in oil-fired units varies appreciably with the sulfur content of the oil, which can be less than 1% for lighter oils to 3.5% for some residual oils. The sulfuric acid dew point typically ranges from 240 to 300°F (116 to 149°C), depending on the oil. Cold-end corrosion is not normally an issue in natural gas-fired units, as natural gas is typically very low in sulfur content. However, the combined burning of sulfur-containing fuel oil with natural gas can be more problematic than use of oil alone, because of the high water vapor content that results from burning of natural gas. Even though many coals have appreciable sulfur contents, coal-fired boilers are generally less susceptible to cold-end corrosion than oil-fired boilers since coal ash contains alkaline species that can neutralize some of the acid that forms.

Another major factor that affects the concentration of sulfur trioxide in the flue gas is the amount of excess air. Figure 22.2 shows the dramatic effect of increasing corrosion rates with high levels of excess air. Maintaining excess air levels to below 1 to 2% will generally significantly decrease cold-end attack from condensed sulfuric acid. However, even periodic conditions of elevated excess air can significantly increase corrosion.

The amount of water vapor present in the flue gas is dependent on many factors. Sources include the moisture content in the fuel, fuel combustion, leaks in boiler tubes, and steam from soot blowing. The sulfuric acid dew point increases with increasing water content, as shown in Fig. 22.1.

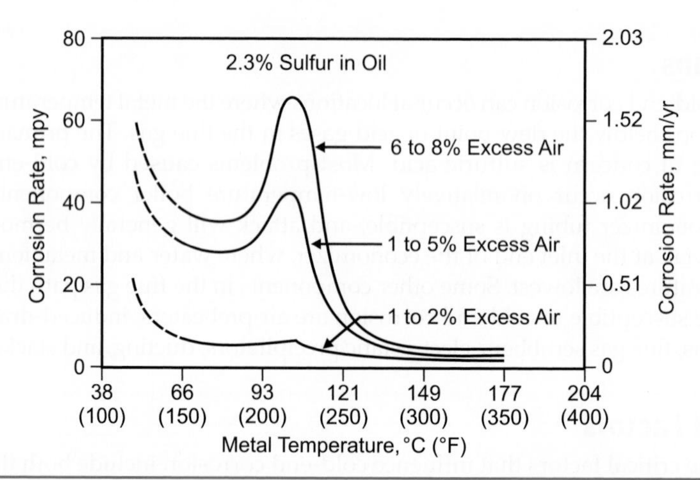

FIGURE 22.2 Corrosion rate of carbon steel due to condensing sulfuric acid vapors as a function of metal temperature, showing the effect of excess air in the flue gas.[2] (*Courtesy of The Babcock & Wilcox Company, Steam/Its Generation and Use, 41st edition.*)

Figure 22.2 also shows that corrosion rates of steel tubing due to condensed sulfuric acid are generally greatest about 40 to 100°F (4 to 38°C) below the acid dew point, because the rate of acid deposition tends to peak in this range.

Although sulfuric acid is by far the most common acid gas responsible for cold-end corrosion, other gaseous compounds in the flue gas, such as HCl, NO_2, and CO_2, can produce acidic solutions on metal surfaces in some cases. Hydrochloric acid may be more of a concern in waste incinerators because of the chloride content of the fuel. However, because these species have much lower dew points than sulfuric acid, attack related to these acids is rare.

Identification

The corroded surfaces are typically covered by rust-colored layers of deposits and corrosion products. The layers are often very thin, and the black oxide layer (magnetite) that typically forms on boiler tube surfaces may be absent if corrosion was active shortly before inspection. Analysis of the material covering the surface will often reveal appreciable levels of volatile sulfur in cases where condensed sulfuric acid was responsible for the metal loss. In addition, pH measurements of a 1% slurry can often highlight the aggressiveness of the acidic conditions.

Cold-end corrosion typically produces an irregular surface contour having a pockmarked or orange peel type of appearance, as shown in Fig. 22.3. Distinct undercut pits typical of acid attack may

FIGURE 22.3 Pockmarked contour of economizer tube exposed to condensed sulfuric acid.

be present. In many cases, attack produces general, smooth, featureless metal loss. Attack due to low-pH conditions often highlights microstructural features, such as sulfide inclusion stringers. Shallow and parallel striations corresponding to the features may be observed in attacked areas. This characteristic is shown in Fig. 22.4, which

FIGURE 22.4 Shallow preferential attack along the seam weld of a tube in an area with otherwise smooth wastage. Attack was caused by condensed sulfuric acid vapors.

FIGURE 22.5 Thin-edged perforation located in a deep metal loss area caused by cold-end corrosion on an economizer tube.

highlights preferential attack in places along the seam weld of the tube. On externally finned tubes, the attack will generally be more severe on the tube surface and at the fin base due to lower temperatures at these locations.

Failures in service will often have the form of thin-lipped, longitudinally oriented ruptures. Ruptures can occur when stresses developed due to internal pressurization exceed the load-carrying capacity of the thinned tube wall. Thin-walled perforations can also occur in areas of severe metal loss, as shown in Fig. 22.5.

Elimination

Elimination of cold-end corrosion is achieved by gaining control of the critical factors that govern it. The approaches to limit attack are often focused on reducing the level of sulfur trioxide in flue gas, lowering excess quantities of water vapor in the flue gas, and increasing the surface temperature of components above the sulfuric acid dew point.

Lowering the level of sulfur trioxide in flue gas can be achieved by specifying fuels with lower sulfur contents, reducing the amount of excess air for firing, and limiting air infiltration. The amount of excess air should generally be maintained below 1 to 2% to avoid cold-end corrosion.[3]

The presence of excessive quantities of water vapor in the flue gas can be avoided by specifying fuel with low moisture content, reducing the amount of soot blowing, preventing tube leaks, and limiting operation of the boiler with active leaks.

Operating the economizer at higher temperatures, if possible, can increase metal temperatures to avoid cold-end corrosion. If this cannot be achieved, then substantial design changes are often required to eliminate metal surface temperatures below the sulfuric acid dew point. In some cases, external heaters can be used to raise metal temperatures. Insulating ductwork and components in the stack subjected to attack can be beneficial. However, detailed design changes are beyond the scope of this book.

Fuel additives can provide a chemical solution to control cold-end corrosion. This approach is generally suitable only for some specific fuel types, such as oil. The additives, which are often magnesium-based compounds, react with sulfur trioxide preferentially to limit the formation of sulfuric acid.

Periodic water washing, when done properly, can remove aggressive deposits on the metal surfaces. This is generally a process that helps reduce corrosion during idle times. However, in boilers equipped with multiple air preheaters, it is possible to isolate and perform an on-line washing of individual preheaters.

Substitution of corrosion-resistant materials can be considered if other methods are not effective at reducing or eliminating attack.

Cautions

Metal surfaces that have suffered cold-end corrosion may be covered with deposits and/or corrosion products, making visual identification difficult. In addition, cold-end corrosion can produce smooth, uniform, featureless metal loss, such that the attacked surfaces closely resemble the original unaffected surface contour. Thickness measurements using ultrasonic techniques serve as a nondestructive means of evaluation. Visual comparison of surface profiles and ring sections cut from tubing may also reveal the reduction in thickness caused by cold-end corrosion.

The visual features of cold-end corrosion during service are very similar to those of dew point corrosion during idle times. However, review of the operational history of the boiler, consideration of the metal temperatures of affected components during service, and observation of subtle features associated with the wastage pattern can indicate if attack predominately occurred during service or idle time.

Related Problems

See Chap. 23, Dew Point Corrosion during Idle Periods.

Case History 22.1

Industry:	Pulp and paper
Specimen location:	Economizer, recovery boiler
Orientation of specimen:	Vertical
Years in service:	20
Drum pressure:	600 psi (4.1 MPa)
Tube specifications:	2½-in. (6.4-cm) outer diameter
Fuel:	Black liquor

Figure 22.3 is a close-up photograph of the external surface of an economizer tube that sustained general corrosion. Note the pockmarked contour. Corrosion occurred in regions of the economizer where sulfuric acid condensed from the flue gas. Such condensation, if it occurs, generally affects the tubes at the inlet end of the economizer, as metal temperatures are lower on the end where the feedwater enters.

Case History 22.2

Industry:	Sugar
Specimen location:	Economizer
Orientation of specimen:	Horizontal
Years in service:	20
Tube specifications:	1-7/8-in. (4.8-cm) outer diameter
Fuel:	No. 6 fuel oil (1.9% S)

The corrosion and perforations apparent in Fig. 22.6 represent a chronic problem in this boiler. The external surface, including both the tube and fin surfaces,

FIGURE 22.6 Corrosion and perforations of a finned economizer tube resulting from exposure to condensed sulfuric acid. Note detachment of fins from tube wall.

exhibited smooth, general metal loss. In many cases, severe corrosion along the attachment line had caused the fins to separate from the tube wall. This deeper metal loss along the fin base demonstrates the increased attack associated with cold-end corrosion on the cooler surfaces.

The deposits and corrosion products covering the corroded surfaces were analyzed by x-ray diffraction and identified as hydrated iron sulfate. The pH of a 1% slurry of the material was measured at 2.3. The corrosion was caused by condensation of sulfuric acid on the cold tube surfaces during boiler operation.

Case History 22.3

Industry:	Utility
Specimen location:	Economizer
Orientation of specimen:	Horizontal
Tube specifications:	2-in. (5.1-cm) outer diameter, SA-178A carbon steel
Years in service:	15
Boiler pressure:	1150 psi (7.9 MPa)
Fuel:	Refuse

Recurrent failures occurred in the economizer section of the boiler, which incinerated scrap tires. The failures were in the form of thin-lipped ruptures, shown in Fig. 22.7, and small, thin-edged perforations. The external surface experienced general metal loss that reduced the tube-wall thickness by over 50% at most locations, relative to the nominal wall thickness of the tube. The external surface contour was irregular and contained small pits and shallow grooves along the tube weld seam, as shown in Fig. 22.8.

FIGURE 22.7 Wide, thin-lipped rupture of an economizer tube in a refuse-fired boiler. The tube experienced deep, general metal loss on the external surface caused by cold-end corrosion.

FIGURE 22.8 External surface after cleaning, showing an irregular contour with some small pits and a groove along the tube seam weld.

Analysis of the material on the external surface indicated that it consisted primarily of iron and zinc. The presence of high levels of zinc was likely related to metal contained in the tire scrap, such as galvanized wires. The analysis also indicated 19% volatile sulfur, and about 17% of the sample was attributed to water. The combination of these species is strong evidence that the attack was caused by sulfuric acid that condensed on the external surface. In addition, the boiler was reportedly operated continuously except for annual inspections, so the contribution to metal loss from attack during idle times was probably minor. As the fuel source could not be altered, firing practices were changed to reduce excess air. Excessive soot blowing was also minimized in order to limit moisture in the flue gas.

Case History 22.4

Industry:	Utility
Specimen location:	Economizer from a heat recovery steam generator (HRSG)
Orientation of specimen:	Vertical
Tube specifications:	2-in. (5.1-cm) outer diameter
Years in service:	4
Fuel:	Natural gas and fuel oil

The vertical economizer tubes in the unit experienced deep, localized metal loss. Metal loss was localized to a circumferential band adjacent to a plate through which the tubes passed to enter the boiler (Fig. 22.9). Many small, thin-edged perforations were found in the deep metal loss bands (Fig. 22.10). Close examination revealed many shallow parallel striations on the smoothened surface contour. These are features that are typically associated with acid attack.

FIGURE 22.9 Localized band of severe metal loss at a location where the tube passed through a bottom plate in the boiler.

Attack was caused by cold-end corrosion. Sulfuric acid that condensed on the tube surfaces collected in a region at the bottom of the tubes adjacent to the plate. The acid caused the circumferential band of deep wastage, which resulted in the failures. This case illustrates the attack possible in an HRSG economizer section when cofiring the unit with supplemental sulfur-containing oil in addition to natural gas. Avoiding cofiring with sulfur-containing oils would limit cold-end corrosion. If oil must be used as a supplemental fuel source, then fuel additives could be considered.

Case History 22.5

Industry:	Pulp and paper
Specimen location:	Economizer from a recovery boiler
Orientation of specimen:	Varies, 90° bend
Tube specifications:	2-in. (5.1-cm) outer diameter
Years in service:	30
Boiler pressure:	1250 psi (8.6 MPa)
Fuel:	Black liquor

A tube immediately downstream from the inlet header to the economizer experienced a wide, thin-lipped rupture (Fig. 22.11). The rupture was caused by severe wall thinning from metal loss on the external surface, which reduced the tube wall thickness by as much as 80% at the rupture edge. Rupture occurred

FIGURE 22.10 Shallow, parallel striations superimposed on the deeply wasted surface in the circumferential band of metal loss.

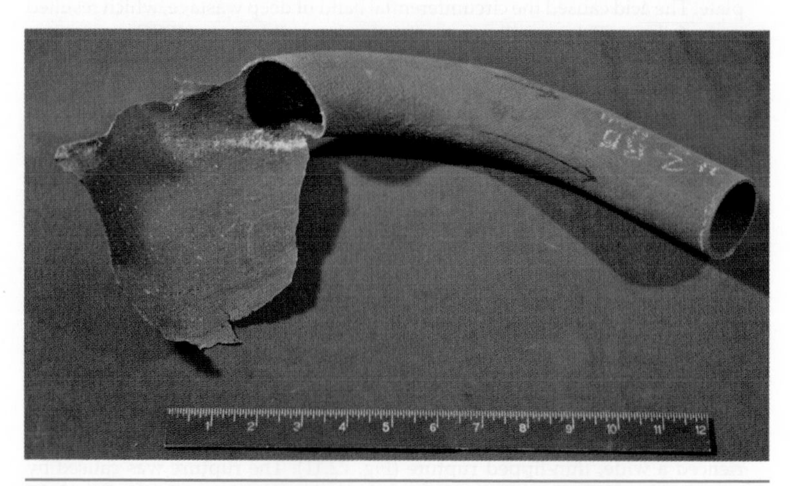

FIGURE 22.11 Wide rupture with thinned edges at the inlet end of economizer tubing in a recovery boiler.

FIGURE 22.12 External surface of recovery boiler economizer tubing, showing many discrete pits scattered on an irregular contour.

when the tube wall was so severely thinned that it could not support the stresses generated from internal pressure. The external surface was covered by a very thin oxide layer that was easily removed from the surface to reveal bare metal. Spot tests of material on the surface indicated that it contained sulfates and produced an acidic solution when mixed with water. The surface also contained scattered pits, a characteristic of acid attack shown in Fig. 22.12.

Attack was primarily caused by cold-end corrosion during service and was active shortly before the rupture. The localized attack at the inlet end of the economizer highlights the importance of metal temperature for cold-end corrosion to occur. Attack only occurred where metal temperatures were below the sulfuric acid dew point for the flue gas. The reported temperature of the water at the economizer inlet was about 265°F (129°C), a temperature that can be below the sulfuric acid dew point if the concentrations of water vapor and sulfur trioxide in the flue gas are high. The boiler operational history was reviewed to check for possible changes in firing practice and feedwater inlet temperatures that may have promoted the attack.

Case History 22.6

Industry:	Refining
Specimen location:	Economizer
Orientation of specimen:	Varies, 180° bend
Tube specifications:	2-in. (5.1-cm) outer diameter, 0.125-in. (3.1-mm) wall thickness
Years in service:	25
Boiler pressure:	600 psi (4.1 MPa)
Fuel:	No. 6 fuel oil

A boiler inspection revealed tube-wall thinning as a general problem occurring in all return bends of the economizer section. The tube wall thickness was reduced

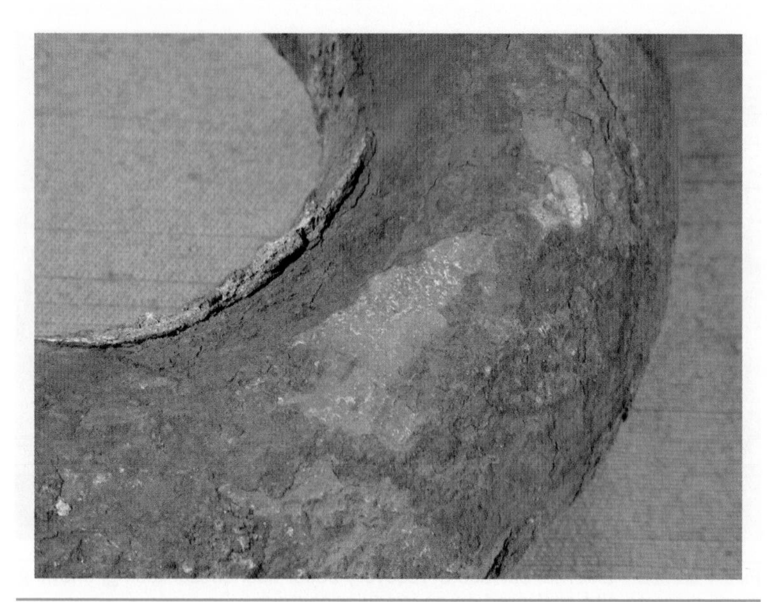

FIGURE 22.13 External surface along the side of an economizer tube bend, showing an area of essentially bare tube metal with an irregular contour.

to as little as 0.028 in. (0.71 mm) in places. In thinned areas on the tube bends, the external surface contour was irregular (Fig. 22.13). Close examination revealed a pockmarked morphology consisting of intersecting depressions (Fig. 22.14). Bare metal surfaces indicated that attack was actively occurring shortly before removing the section from service. Elsewhere, especially along the intrados of

FIGURE 22.14 Close-up, showing a pockmarked appearance on the external surface of the tube.

the tube bend, the surface was covered by layers of reddish-brown deposits and corrosion products. X-ray fluorescence of the material indicated a high sulfur concentration (14 wt%). In addition, ion chromatography of the water-soluble species in the material indicated appreciable sulfate levels.

The characteristics of the metal loss and the analysis of the material on the external surface indicated wastage due to cold-end corrosion. The reported operating temperatures of the tube bends were in the range of 255 to 354°F (124 to 179°C), which is well within the typical range of sulfuric acid dew points. Flue gases passed into the economizer box that enclosed the tube bends, allowing for condensation of sulfuric acid on the cooler metal surfaces.

References

1. W. Cox, "Components Susceptible to Dew-Point Corrosion," *ASM Handbook,* vol. 13C: *Corrosion: Environments and Industries*, S. D. Cramer and B. S. Covino, Jr. (Eds.), ASM International, Materials Park, Ohio, 2006, p. 491.
2. J. B. Kitto and S. C. Stultz (Eds.), *Steam: Its Generation and Use*, 41st ed., Babcock and Wilcox Company, Barberton, Ohio, 2005, pp. 21–27.
3. Ibid.

CHAPTER 23

Dew Point Corrosion during Idle Periods

General Description

Dew point corrosion during idle periods refers to corrosive attack that can occur on the fire side when the unit is out of service. As the boiler cools, the metal temperature may drop below the dew point for water, allowing moisture to form on tube surfaces. Although the presence of water containing dissolved oxygen can cause some metal loss on the surface (see Chap. 9, Oxygen Corrosion), wastage is generally superficial as long as the moisture does not remain on the metal surfaces for extended periods. However, when the moisture hydrates acid-producing deposits, a low-pH solution forms that is capable of causing severe attack during idle periods. Dew point corrosion due to the presence of aggressive deposits can result in significant corrosion of metal surfaces in relatively short periods. Corrosion rates as high as 500 mpy (12.7 mm/y) have been measured.

Because appreciable corrosion is associated with acid-producing deposits, deepest attack is generally present at locations of deposition. Figure 23.1 illustrates an ash-covered tube with dew point corrosion before the deposits were removed. Figure 23.2 illustrates the irregular contour of the corroded tube surface after the deposits were removed.

Aggressive deposits that accumulate on metal surfaces are typically sulfurous. Sulfur is generally present in most fuels, such as coal, oil, and black liquor, at various concentrations. Thus, the ash deposits from burning these fuels contain sulfurous species. Other components in the ash that can accelerate corrosive attack when moisture is present include chlorides. Chlorides can form in the combustion products of some fuels, especially from refuse that is burned in municipal solid waste (MSW) and refuse-derived fuel (RDF) units. However, sulfurous deposits are most commonly associated with dew point attack.

FIGURE 23.1 Ash deposits covering the fireside surface of a boiler tube.

FIGURE 23.2 Generally wasted surface having an irregular contour, revealed after removal of the deposit and corrosion product layers.

Locations

Dew point corrosion during idle periods can occur in any locations in the boiler where external surfaces are covered with acid-forming deposits. Areas where deposits and moisture tend to collect (such as low points in the system) are more susceptible.

Critical Factors

Two factors are required for accelerated attack to occur:

1. The reduction of metal temperatures within the boiler to below the dew point of water
2. The presence of ash deposits that contain aggressive acid-forming species

The species that typically cause attack contain sulfur from the fuel source. The aggressive species in the ash deposits form a low-pH solution when liquid water is present.

Identification

Deposits generally cover the affected surfaces, unless they are removed by water washing. Once deposits are removed, simple visual inspection should disclose the occurrence of dew point corrosion (Fig. 23.2). Typically, metal attack will be confined to surfaces that were covered with the aggressive ash deposits. The attack often produces well-defined regions of metal loss with distinct edges. Islands of metal may be left relatively intact.

The underlying attacked areas are often bare-surfaced or covered by a very thin layer of orange flash rust. The surfaces often contain many pits and depressions. Pits and depressions can intersect and produce general metal loss over wide areas. Smoothened surface contours may be present, especially in places where low-pH solutions flowed across the metal surface.

Analysis of ash deposits covering the metal surfaces is often beneficial to identify the aggressive constituents responsible for attack. The analysis will generally indicate sulfur-containing material from the fuel. The aggressiveness of the deposits can be determined by measuring the pH of a 1% solution.

Tube failures are generally in the form of thin-edged perforations centered in regions of severe wall thinning, as illustrated in Fig. 23.3. Thin-lipped ruptures can occur in some cases if the tube is pressurized, especially during start-up of the unit after an extended idle period.

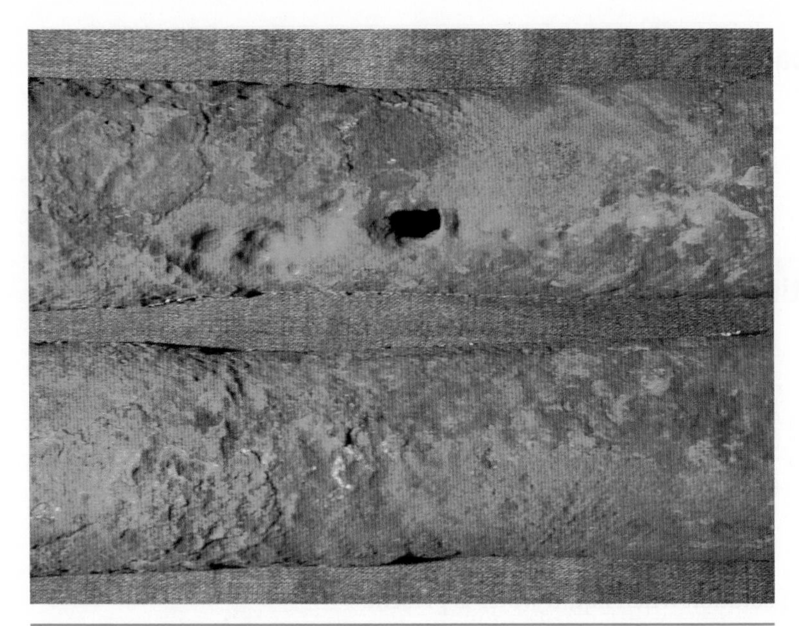

Figure 23.3 Thin-edged perforation centered in a region of deep metal loss caused by dew point corrosion on the fire side of a tube.

Elimination

Dew point corrosion may be controlled by altering the critical factors that cause the corrosion. Two critical factors govern dew point corrosion: (1) the presence of ash deposits containing sulfurous or other aggressive species that form a low-pH solution when combined with water and (2) condensation of moisture as fireside metal temperatures are reduced below the dew point of water.

To reduce the effect of sulfurous ash deposits, the following steps may be taken:

- Specify fuels with lower sulfur content. This reduces or eliminates the formation of corrosive ash.
- Remove fireside deposition from metal surfaces immediately after boiler shutdown by using high-pressure water sprays. This should be followed by a lime wash to completely neutralize remaining acidic substances. Allow the metal surfaces to air-dry.
- Apply corrosion inhibitors or hydrophobic coatings, such as light oils, after the surfaces are cleaned and dried.

It is generally not possible to maintain metal temperatures above the water dew point during idle times, unless the boiler is kept in a hot, standby condition by using low firing rates. In some cases, containers of unslaked lime or other desiccant materials can be placed in the firebox during shutdown to help keep the air dry. This desiccant should be renewed periodically.

Cautions

Corrosion may not be visually apparent until deposits are removed. Because attack is caused by low-pH conditions, dew point corrosion during idle periods and cold-end corrosion during service can appear similar, especially on economizer and air heater tubing. In addition, attacked surfaces for both mechanisms will generally be covered by acid-producing, sulfurous material. However, the irregular pattern of metal loss typical of dew point corrosion may be helpful in distinguishing it from cold-end corrosion. Moreover, consideration of boiler operation, such as amount of time in service and combustion parameters, can indicate which mechanism is more probable.

Localized wastage can sometimes be confused with metal loss due to erosion damage. Observation of flow-oriented features and consideration of the surrounding environment (e.g., the presence of soot blowers or nearby leaking tubes) can indicate that erosion mechanisms are more probable. In some cases, attacked areas may exhibit evidence of contributions from both erosion and dew point corrosion.

Related Problems

See Chap. 22, Cold-End Corrosion during Service, and Chap. 24, Fireside Erosion.

Case History 23.1

Industry:	Pulp and paper
Specimen location:	Screen tube, recovery boiler
Orientation of specimen:	Vertical
Years in service:	10
Drum pressure:	600 psi (4.1 MPa)
Tube specifications:	2-in. (5.1-cm) outer diameter
Fuel:	Black liquor

Figure 23.4 illustrates the appearance of the external surface of a screen tube that experienced dew point corrosion. General metal loss produced an irregular surface contour. Microstructural examinations revealed that the external surface was covered with a dense layer of iron oxide beneath a layer of iron sulfide. This corrosion occurred during idle periods when moisture, possibly from inadequate water washing, combined with sulfur-containing deposits on the external surface to form an acidic solution.

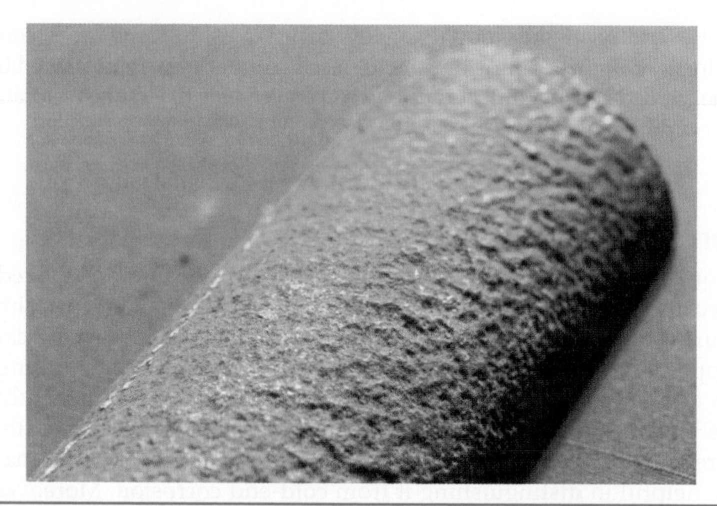

FIGURE 23.4 Irregular contour of the external surface of a screen tube after exposure to moisture and corrosive deposits during idle periods.

Case History 23.2

Industry:	Pulp and paper
Specimen location:	Front row of economizer, recovery boiler
Orientation of specimen:	Bend, slanted
Years in service:	11
Drum pressure:	600 psi (4.1 MPa)
Tube specifications:	2-in. (5.1-cm) outer diameter
Fuel:	Black liquor

The economizer tube shown in Fig. 23.5 had an irregular, pebble like surface contour covered with nonprotective corrosion products and deposits. The appearance of the surface was characteristic of dew point corrosion that occurs during

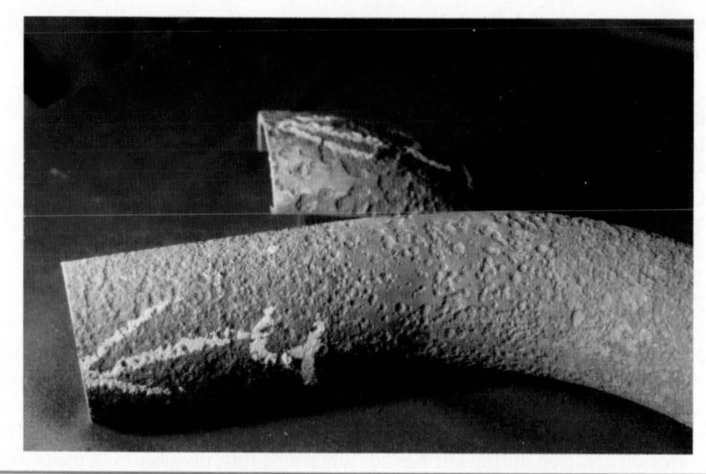

FIGURE 23.5 External surface of economizer tubing from a recovery boiler following dew point corrosion.

idle periods, in which acid-producing salts combined with atmospheric moisture to produce a corrosive solution.

Case History 23.3

Industry:	Pulp and paper
Specimen location:	Wall tube, 2 ft (0.6 m) below roof
Orientation of specimen:	Vertical
Years in service:	12
Drum pressure:	600 psi (4.1 MPa)
Tube specifications:	3-in. (7.6-cm) outer diameter
Fuel:	Black liquor

The metal loss shown in Fig. 23.6 was found in the upper furnace area of the boiler around openings. All metal loss occurred under refractory. The boiler was water-washed twice annually and was not dried after washing. Measurements revealed a 16% reduction in tube-wall thickness.

The corrosion occurred as a result of sulfur-containing deposits on the metal surface that were not removed during water washes. Corrosion of this type is not uncommon in areas where water can collect, such as beneath refractory. Corrosion rates of up to 50 mpy (1.27 mm/y) have been reported in similar circumstances.

Case History 23.4

Industry:	Steel
Specimen location:	Downcomer 12 in. (0.3 m) from steam drum
Orientation of specimen:	Slanted 45°
Tube specifications:	2½-in. (6.4-cm) outer diameter
Fuel:	High-sulfur fuel oil

FIGURE 23.6 Idle-time corrosion occurring in a region that was covered by refractory in service.

FIGURE 23.7 Grooving and perforation of a slanted tube resulting from corrosive water dripping from above.

The tube illustrated in Fig. 23.7 was one of several similar tubes discovered during a fireside inspection. The tube was positioned at a 45° angle in the boiler. A massive, deep groove containing a perforation formed in the vertical direction. It was observed that the groove was directly aligned with water dripping from the economizer section above.

An inspection of the economizer section revealed another series of grooved tubes. The grooves in these tubes were aligned with water dripping from the air preheater section above the economizer.

An inspection of the bottom of the air preheater section, which had recently been water-washed, revealed an accumulation of wet, dripping deposits. Water dripping from the deposits had a pH of 2. Analysis of the deposits by x-ray fluorescence disclosed a high concentration of sulfur.

Incomplete washing of highly corrosive deposits in the air preheater furnished a source of very corrosive water, which corroded tubes in the path of the dripping water. Tubes in both the economizer and the boiler proper were affected.

From a corrosion engineering perspective, this case underscores the importance of recognizing boiler systems as a continuum. Environmental conditions existing in one part of the system can have a direct, adverse impact on another part of the system that is physically separate and apparently unconnected.

Case History 23.5

Industry:	Chemical processing
Specimen location:	Tubesheet and tubes in a firetube boiler
Orientation of specimen:	Vertical tubesheet, horizontal tubes
Tube specifications:	2½-in. (6.4-cm) outer diameter tubing, carbon steel
Years in service:	2
Boiler pressure:	150 psi (1.0 MPa)
Fuel:	Waste gases from an incinerator

The incinerator burned various fuel sources and operated intermittently with weekend shutdowns. The tubes of the boiler were cleaned monthly by hydroblasting. Boiler tube failures were occurring routinely, with attack localized

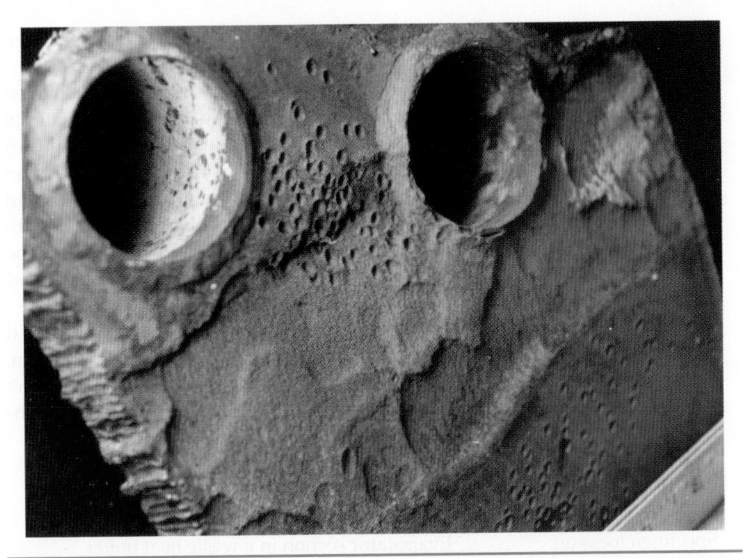

FIGURE 23.8 Areas of moderately deep wastage on the tubesheet surface, with some preferential attack of the welds joining the tubes to the tubesheet.

to areas near the inlet tubesheet. Figure 23.8 shows areas of deep general wastage on the inlet tubesheet. Close examination revealed parallel striations in the wasted areas in line with the apparent rolling direction of the plate. These are features associated with acid attack.

Tubes in the unit contained small perforations located in areas with severe pitting on the internal (fireside) surface, as shown in Fig. 23.9. The areas with

FIGURE 23.9 Cluster of deep pits containing a perforation located in an area of shallow metal loss at the bottom of a firetube.

deep pitting were present on the bottom sides of the tubes. The pitted areas were surrounded by shallow wastage having distinct edges.

The internal surface of the tubes and tubesheet surface were generally covered by thin brown-orange layers of deposits and oxide. Analysis of the material on the surface indicated it consisted primarily of iron, with minor amounts of sodium, silicon, phosphorus, sulfur, and chlorine. The deposits produced an acidic solution when moisture formed on the surfaces during idle periods. In addition, residual water from monthly cleaning by hydroblasting reacted with the deposits and created a low-pH solution that collected in areas on the bottoms of the tubes. The acidic conditions that formed during idle times caused the severe attack along the tube bottoms and tubesheet.

Attack on the fireside surfaces was reduced by running the boiler at low load on the weekends to maintain temperatures above the dew point. In addition, the periodic cleanings increased in frequency and incorporated a neutralizing solution to limit the amount of aggressive, acid-producing deposits that accumulated on the fireside surfaces.

Case History 23.6

Industry:	Coke processing
Specimen location:	Evaporator section in a waste heat boiler
Orientation of specimen:	Vertical tubes
Tube specifications:	2-in. (5.1-cm) outer diameter, SA-178A carbon steel
Years in service:	11
Boiler pressure:	800 psi (5.5 MPa)
Fuel:	Coke battery exhaust gas

Multiple units at the facility experienced many tube failures, in both the economizer and evaporator sections. The evaporator tube failures were in the form of thin-edged, irregular perforations (Fig. 23.10). The perforations occurred in areas

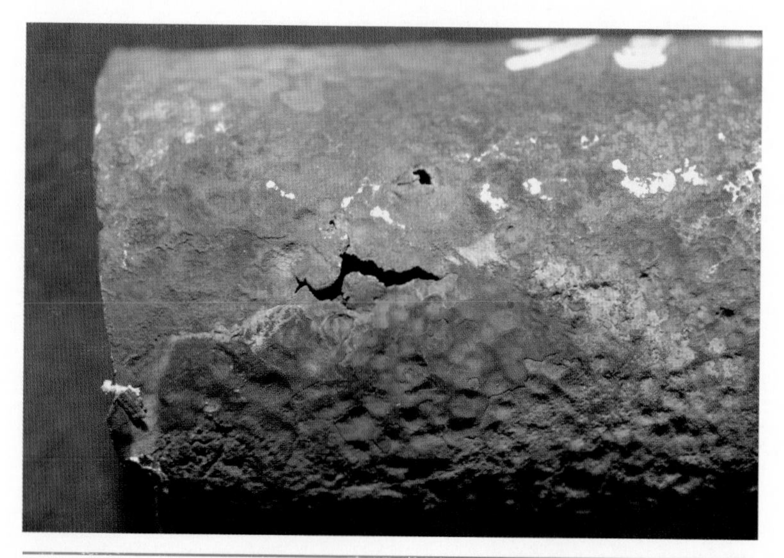

FIGURE 23.10 Thin-edged failure in an area of deep, smoothened, dimpled metal loss.

FIGURE 23.11 Area on the outer bend of a tube where the fins were completely consumed.

of deep, smoothened, dimpled metal loss. Metal loss in other locations resulted in complete wastage of the fins along one side of the tube, as illustrated in Fig. 23.11. Attack in the evaporator section was more severe in regions at the bottom of the tubes near the inlet header for the section.

The external surfaces were covered by thin deposit and corrosion product layers. Analysis of the layers, using a scanning electron microscope equipped for energy-dispersive (x-ray) spectroscopy (SEM-EDS), revealed major concentrations of iron and sulfur. Spot tests indicated the deposited material produced an acidic solution when mixed with water. The attack was caused by the sulfurous deposits forming low-pH solutions during idle time, when moisture was present on the external surfaces of the tubes. In addition to condensation forming at temperatures below the dew point of water, soot blowing and tube leaks may have contributed to the presence of liquid water on the fire side. The boiler experienced many idle periods due to multiple economizer failures. More frequent washing to remove aggressive deposits on the tube surfaces, as well as subsequent neutralization, was recommended during shutdowns to avoid dew point attack.

Case History 23.7

Industry:	Food
Specimen location:	Air heater tubes
Orientation of specimen:	Vertical
Tube specifications:	2½-in. (6.4-cm) outer diameter
Years in service:	3
Fuel:	Mixture of wood and bagasse, with fuel oil on start-up

Flue gas flowed through the tube sides of air heater tubes to heat combustion air that flowed across the shell sides. Severe metal loss that resulted in thin-walled failures occurred near the top of the tubesheet, as shown in Fig. 23.12. Examination

FIGURE 23.12 Thin-edged perforations on one end of a section from an air heater tube; note that metal loss on the external surface was not significant.

of the tube indicated that wall thinning was caused by deep, general metal loss, primarily on the internal surface (Fig. 23.13). Close examination revealed adherent corrosion product layers in places on the internal surface. Analysis of these layers by x-ray fluorescence indicated they consisted primarily of iron. X-ray diffraction determined the iron was in the form of iron oxides and iron oxide hydroxides. Minor amounts of calcium, silicon, and sulfur were also present in the deposited layers on the internal surface. The pH of a 1% slurry was mildly acidic.

The characteristics of the wastage and deposits on the internal surface indicate attack occurred primarily during idle time. Acid-forming sulfurous deposits accelerated attack when moisture formed on the surfaces. Attack during service was improbable as the reported gas temperatures were sufficiently above the sulfuric acid dew point. Proper water-washing techniques were suggested to limit attack.

Case History 23.8

Industry:	Utility
Specimen location:	Economizer tube in a heat recovery steam generator (HRSG)
Orientation of specimen:	Horizontal
Tube specifications:	2-in. (5.1-cm) outer diameter, externally finned
Years in service:	6
Boiler pressure:	875 psi (6.0 MPa)
Fuel:	Natural gas and No. 2 fuel oil

The HRSG unit experienced many failures over a 2-week period. Inspection indicated areas with severe wastage along one side of the horizontally oriented

FIGURE 23.13 Severe, general metal loss on the internal surface near the failed end.

tubes in the economizer section (Fig. 23.14). The tubes contained thin-lipped ruptures in places (Fig. 23.15). The fins were almost completely wasted in most regions along the failed side. Tube wall thickness measurements highlighted gradual metal loss along the circumference at locations away from the failure. The external surfaces were covered by thin, orange-brown layers of deposits and

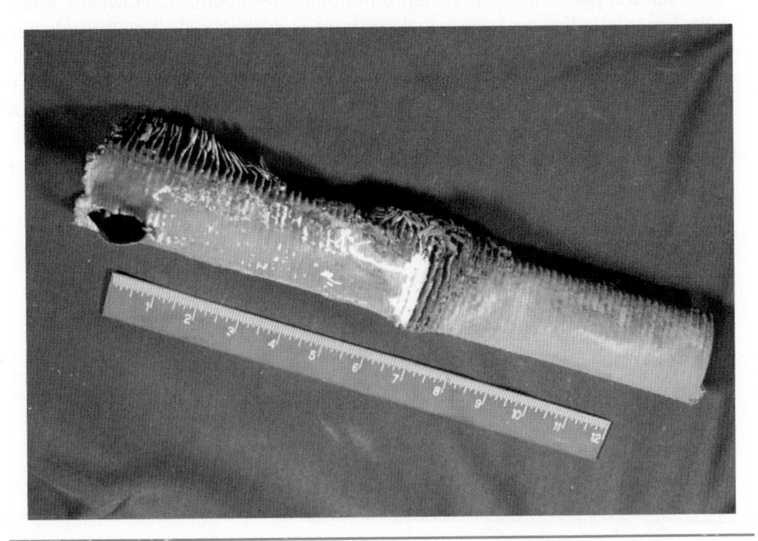

FIGURE 23.14 Economizer section of tubing in an HRSG that experienced severe wastage along one side of the tube.

FIGURE 23.15 Close-up of the thin-lipped rupture along the side with deep metal loss.

iron oxides. Analysis of the material covering the surface indicated it consisted predominately of iron with elevated levels of sulfur as sulfate. The pH of a 1% solution was 1.7.

The metal loss on the section was primarily caused by dew point corrosion during idle times, as review of the boiler operation indicated it was intermittently in service and subjected to long idle periods. Normally, HRSGs that only burn natural gas as a fuel experience minimal fireside attack. However, when supplemental sulfur-containing fuels are burned, such as fuel oil, acid-producing sulfurous deposits typically form on tube surfaces. These deposits can result in dramatically accelerated corrosion rates when moisture forms on the surface during idle times. Attack could have been avoided if only fuels with low sulfur content were used. However, once aggressive deposits form on the fireside surfaces, they should be removed after shutdown of the unit by following proper cleaning and neutralization procedures.

CHAPTER 24

Fireside Erosion

Introduction

Most erosion-related failures occur on the fire side of boilers. Erosive damage can be attributed to a number of different sources, such as particles in fly ash, falling slag, soot blowing, steam cutting, and mechanical rubbing. Details on the mechanism of erosive metal loss on the water and steam sides have been presented previously in Chap. 18. Because most erosion-related failures in the boiler occur on the fire side, some specific causes of fireside erosion are discussed in detail in this chapter.

Hard-Particle (Fly Ash and Coal) Erosion

General Description

Hard particles, either directly from the fuel or from the ash that forms after combustion, can cause erosive damage to boiler components. *Fly ash erosion* refers to damage caused by ash particulate entrained in high-speed flue gases striking metal surfaces. Coal particle erosion is caused by high-velocity coal particles that impinge on tube walls. Although these terms typically refer to attack from coal, other fuels that contain hard particles can cause similar metal loss. Boilers burning wood, bagasse, refuse, and other waste materials are particularly likely to suffer erosion, since they may contain significant amounts of abrasive materials such as sand, dirt, and cement dust that will be entrained in combustion gases.

Wastage can be attributed to two mechanisms. One common mechanism is impingement of particles on the surface, which causes abrasive wear and direct metal loss. In another mechanism, the oxide that normally covers the boiler tube surface is repeatedly removed by abrasive particles and subsequently re-formed, allowing for an accelerated oxidation rate.

For fly ash erosion, major accelerating factors are high flue gas velocity and large concentrations of entrained abrasive components in the fly ash. These factors accelerate metal loss by increasing both

Type of Firing or Fuel	Baffle Arrangement	
	Multipass	Single-Pass
Pulverized coal (PC)	75	75*
Spreader stoker	50	60
Chain-grate stoker, anthracite	60	75
Chain-grate stoker, coke breeze	60	75
Chain-grate stoker, bituminous	100	100
Underfeed stoker	75	100
Blast furnace gas	75	100
Cyclone furnace	—	100
Wood or Other Waste Fuels Containing:		
Sand	50	60
Cement dust	—	45
Bagasse	60	75

*For PC units burning fuels having more than 30% ash on a dry basis, limit the maximum velocity through the free-flow area to 65 ft/s (19.8 m/s). For PC units burning coals producing fly ash with known high-abrasive tendencies, such as Korean or central Indian coals, limit the maximum velocity through free-flow area to 45 ft/s (13.7 m/s).

Note: 1 ft/s = 0.3048 m/s

Source: Courtesy Babcock & Wilcox, *Steam: Its Generation and Use*, 39th ed., pp. 15-9, Babcock & Wilcox, New York, 1978.

TABLE 24.1 Design Gas Velocity (ft/s) through Net Free-Flow Area in Tube Banks to Prevent Flue Dust Erosion

the amount of kinetic energy per impact and the number of impacts per unit time in a given area. The tendency for fly ash erosion depends on the type of firing and the fuel. Damage can occur when the local gas flow velocity is excessive. Table 24.1 shows maximum design gas velocities for some various types of firing and fuel.

Fly ash erosion is common in boilers fired with overfeed stokers, which allow considerable amounts of ash to enter the gas stream more readily. Those boilers using an overfire air system may reduce particulate entrained in the flue gas. Partial suspension burning also causes greater amounts of particulate matter to enter the gas stream. Thus, collectors are generally used. When collectors malfunction, damage increases. Fluidized-bed and other special-purpose boilers sometimes suffer severe erosive damage. Incinerators suffer similar fireside problems.

For coal particle impingement, attack generally occurs in tubes lining the cyclone burner. Erosion becomes pronounced when the refractory covering the tubes or the wear liners is damaged or worn, exposing unprotected tubes. The high-velocity coal particles, moving at speeds up to 300 ft/s (92 m/s) (in utility service), impinge on tube walls and cause rapid wear.

Locations

Damage due to fly ash erosion frequently occurs in economizer, superheater, reheater, and roof tubing, although other tubes may be affected. Since fly ash is usually more erosive when particle temperatures are lower, economizers are particularly susceptible to erosion. Inlet areas of reheaters are common wastage sites due to higher gas velocities and swirling flow (eddying) that may occur at these locations. Areas in the superheater where slagging is pronounced are common problem regions, since gas flow velocity increases through the narrow channels between slag buildup. Any location where channeling or eddying of gases occurs is susceptible to wastage. Erosion is usually localized and frequently is restricted to regions such as gaps between tube rows, banks, and duct walls.

Water-cooled tubes that line the cylindrical combustion chamber of cyclone-type coal burners are common attack sites for coal particle erosion. Incompletely burned coal particles can also accelerate fly ash erosion in superheaters and wall tubing.

Critical Factors

Erosive metal loss increases as gas flow velocity, ash concentration, and particle hardness increase. Of these factors, gas flow velocity and ash concentration are most important (Fig. 24.1). Erosive metal loss also depends on the angle of impingement between gas flow and the metal surface. Impingement angles between 15° and 35° generally produce maximum erosion rates.[1]

The amount of ash varies considerably between the different grades of coal.[2] Size, hardness, and composition of particulate matter also influence erosive metal loss. Particles larger than about 0.001 in. (0.025 mm) are more erosive because of high particle kinetic energy. In addition, coals that contain high concentrations of aluminum, silicon, and iron compounds are more erosive because of high particle hardness. As temperature increases, erosive metal loss decreases because particles become softer.

The degradation of the refractory and wear liners in combustion chambers contributes to erosion by coal particles in cyclone burners.

Identification

Erosion may be localized or general. Fly ash erosion frequently causes smoothly polished surfaces. In other cases, irregular flow marks and grooves are produced by high-velocity swirling flow around slag encrustations, hangers, brackets, and the like. (See Fig. 24.2.)

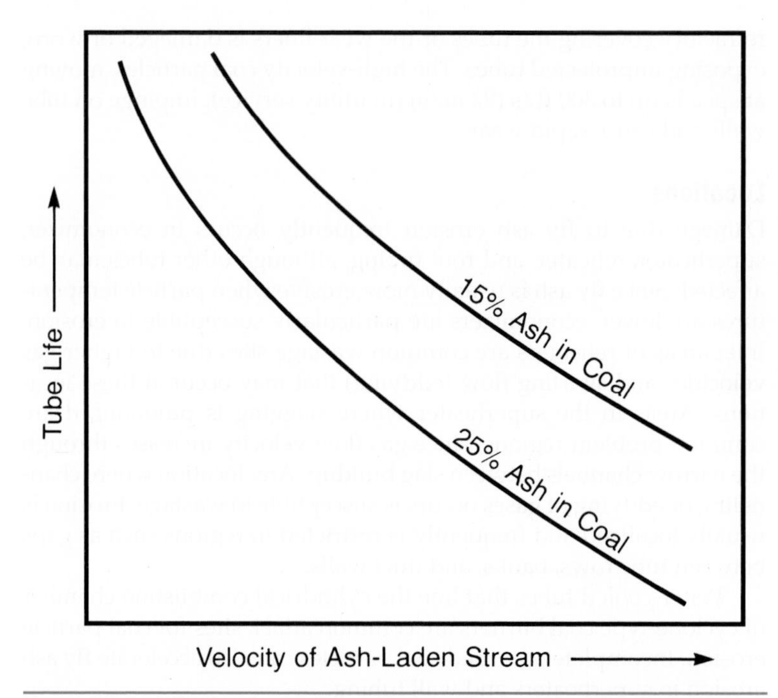

FIGURE 24.1 Effects of velocity and ash content on the life of a tube. Increases in the number and velocity of the impacting ash particles can result in reduced service life. (*Courtesy of Electric Power Research Institute, EPRI Manual for Investigation and Correction of Boiler Tube Failures, EPRI, 1985. Originally in E. Raask, "Tube Erosion by Ash Impaction" WEAR, 13: 1969, pp. 301–315.*)

FIGURE 24.2 Deep grooves cut into the hot side of a waterwall tube. Grooving was caused by channeling of furnace gases containing entrained fly ash.

In extreme cases, tube-wall thinning can cause rupture. When a rupture occurs, it is usually thin-edged. Close inspection of the rupture site often reveals metal loss on the external surface and no overheating or corrosion.

Fireside oxide and deposit layers are usually much thinner on wasted surfaces, allowing discrimination from coal-ash or oil-ash corrosion, in which deposits and corrosion product accumulations are extensive. In addition, the surface contour is smooth rather than rough, as might be expected due to cold-end corrosion.

Metallographic examination of wasted areas will reveal a very smooth profile that is not covered by deposit or oxide layers if erosion was recently active. Microstructural features, such as banded pearlite colonies and inclusion stringers, will intersect the wasted surface. Microscopic examination of thinned surfaces caused by impingement of high-velocity particles will generally indicate narrow zones of plastic deformation, as shown in Fig. 24.3.

Discrimination between fly ash erosion and other forms of fireside erosion requires detailed information about the boiler. Locations in the boiler where erosion occurs, positions of deflecting baffles, amount of entrained fly ash, and local gas velocity are important clues regarding erosive wastage.

In most cases, diagnosis of coal particle erosion is simple. Damage closely resembles fly ash erosion but occurs at or near a cyclone

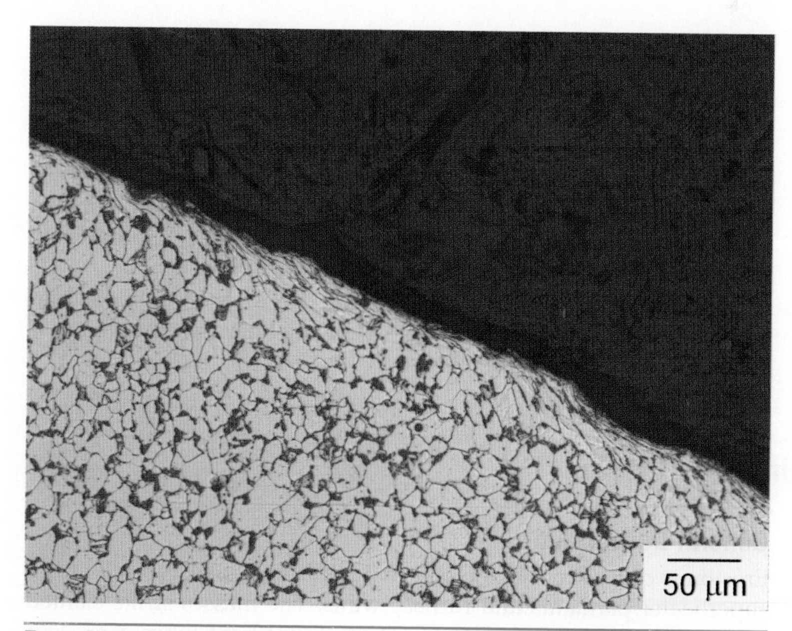

FIGURE 24.3 Narrow zone of grain deformation present along the external surface in a metal loss region caused by hard-particle impingement. Nital etch.

burner and damages the refractory or wear liners. Inspection usually reveals damaged areas and the associated spalled refractory.

Elimination

Decreasing fly ash erosion requires a system operations approach. This includes making sure all baffles, collectors, refractory linings, and other fireside components are in good working order. In extreme cases, redesign of boiler components may be required.

Of course, reducing the amount and velocity of fly ash will also limit damage. In extreme cases, fuels less prone to produce erosive ash may have to be used. High load and excess air increase flue gas velocity and thus increase erosion damage.

Slagging promotes fly ash erosion by channeling gases and increasing swirling flow. Appropriate fuel additives and soot blowing can reduce slagging. Baffles have been used to distribute flue gases, and consequently, fly ash, more evenly. However, where gas flow is horizontal through tube banks, baffling is generally absent. Slag fences have been used to prevent larger slag pieces from entering horizontal tube banks. Shielding and erosion-resistant coatings are beneficially used in certain erosion-prone locations.

Frequent inspection and periodic maintenance of the refractory and wear liners will eliminate most damage in combustion chambers due to coal particle impingement.

Cautions

Hard-particle erosion resembles other forms of erosion damage, including soot blower erosion, impingement from nearby leaking tubes, and falling-slag erosion. Attack due to coal-ash and oil-ash corrosion can have similar features to erosive damage, such as wastage flats. Cold-end corrosion also can be confused with erosion by fly ash. Location of failures and knowledge of boiler operation are important in making a correct diagnosis.

Not all eroded tubes need to be replaced. However, tubes should always be inspected to determine wall thickness after damage is discovered.

Falling-Slag Erosion

General Description

Erosion is caused by slag masses that fall from above, typically from superheater pendants and furnace walls. The masses strike slanted walls below and cause wastage. Dented and bent tubes can occur from the impact of large slag chunks.

Locations

This damage is rare and usually is confined to slanted tube walls near the bottom of large boilers, which direct ash into the ash hopper. Most damage occurs near side walls where greater amounts of slag tend to accumulate, since slag from side walls is more likely to strike these areas.

Critical Factors

The damage rate is controlled by the amount of slag per unit area and the tendency of this slag to fall. Slag formation is favored in a furnace containing a large wall area; having a low or high flue gas velocity; burning a high-slagging coal containing high sodium and chlorine concentrations; having low flue gas temperature; or cycling thermally and/or running high tube temperatures. Such slag can be easily shed.

Identification

Falling-slag erosion produces flat spots on slanted tubes. If large slag chunks falling from great heights strike surfaces, tubes can be dented and deformed. Microscopic examination often reveals plastic deformation associated with this damage.

Elimination

Reduction of slagging decreases damage. This may necessitate a change of coal type to one with a low tendency to form slag. The higher the fusion temperature of the ash, the lower the slagging rate will be. Use of fuel-additive chemicals may also reduce slagging potential. Structural alterations, such as increasing tube-wall thickness and the use of weld overlays or wear bars, have been used to extend tube life.

Cautions

Falling-slag erosion can resemble erosion caused by fly ash, coal dust, soot blowers, and steam cutting. However, the widespread attack on slanted tubes near ash hoppers is usually definitive.

Related Problems

See the discussion under "Hard-Particle Erosion" in this chapter.

Soot Blower Erosion

General Description

Soot blowers are used to limit slag accumulation on tubes, which can result in fouling of the surfaces and subsequent channeling of the flue gases. These conditions can promote erosion from hard particles entrained in the flue gas and possibly falling slag. Although soot blowers can be used to reduce the tendency for fly ash and falling-slag

FIGURE 24.4 Tube rupture caused by a misdirected soot blower. Rupture edges are thin and ragged due to tearing of the thinned steel in the eroded zone.

erosion, excessive use can result in erosive damage if they are not operated properly. In fact, perhaps the most common cause of erosion in boilers is from soot blower operation.

Soot blowers are designed to direct flow tangentially to or between tube surfaces (see Fig. S1.14 in the Introduction to Boilers and Steam Systems in Sec. 1). A misdirected soot blower allows a high-velocity jet of steam, or air carrying condensed water droplets, to impinge directly upon tube surfaces. Physical abrasion and accelerated oxidation cause metal loss. Damage can be accelerated by fly ash entrained in the high-velocity jet stream directed against the tube surface. This is sometimes referred to as soot blower–enhanced ash erosion. In this case, the ash in the flue gas supplies hard particles, and the soot blower increases the local velocity to promote erosive metal loss. Erosive thinning often leads to tube rupture, as shown in Fig. 24.4.

Locations

As the term *soot blower erosion* implies, damage occurs near or in the direct path of soot blower discharge. Superheater tubing is usually attacked. Common damage locations include tubes along the path of retractable soot blowers, particularly those tubes nearest wall entrances. Other damage locations include furnace corners opposite wall blowers. Platens in the convection section are often targets, as are any tubes near malfunctioning soot blowers.

Critical Factors

Soot blower erosion is caused by improper blower alignment or malfunction of the system. Entrainment of either condensed water or fly ash in the blower gas stream also accelerates erosion. Raising the blowing pressure increases gas velocity and thus promotes damage by entrained fly ash. Improper alignment and operation are the most common sources of damage.

Identification

Attacked surfaces are locally wasted; usually producing longitudinally aligned flattened zones, bordered on both sides by shoulders of unattacked metal. When viewed in transverse cross section, the tube appears to have been "planed" along its length (Fig. 24.5). Grooving and more irregular wastage will be present if eddies and gas channeling are pronounced.

Metal surfaces will be smooth or have smoothly undulating contours. Only a thin, dark oxide layer, or no oxide layer at all, will be present if attack is fresh (Fig. 24.6). If attack is old or intermittent (as is often the case with soot blower erosion), oxide and deposit layers will cover thinned surfaces. However, the oxide and deposit layers will usually be much thinner than on adjacent unattacked surfaces. Close visual observation of wasted surfaces will often show very shallow wavelike striations (Figs. 24.6 and 24.7). The striations will be aligned perpendicular to steam flow across the surface.

Metal loss is usually present at the opening of the soot blower valve and continues along the blower path, decreasing in severity as the distance from the blower increases. Frequently, the blower is misaligned, is frozen in position, or is blowing wet steam because of water entrainment.

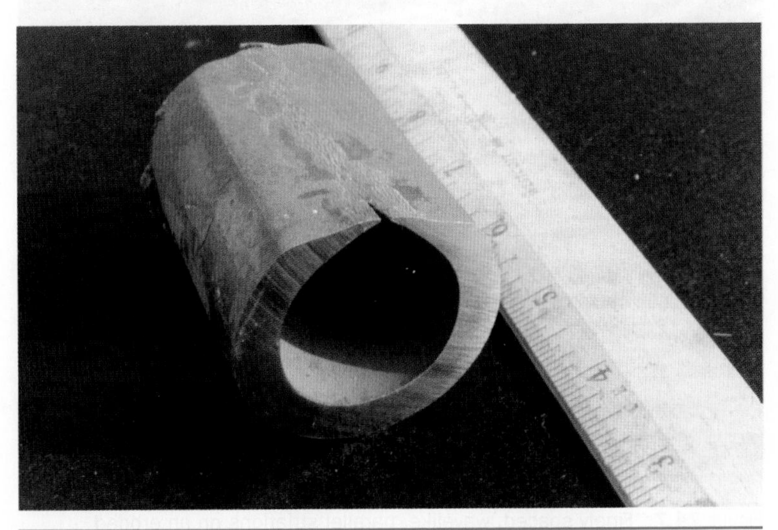

FIGURE 24.5 Superheater section thinned by soot blower erosion. Note the flattened surface. The tube bulged slightly and ruptured on the thinned side. Also note how the eroded surface color is similar to that of unattacked regions, indicating intermittent attack.

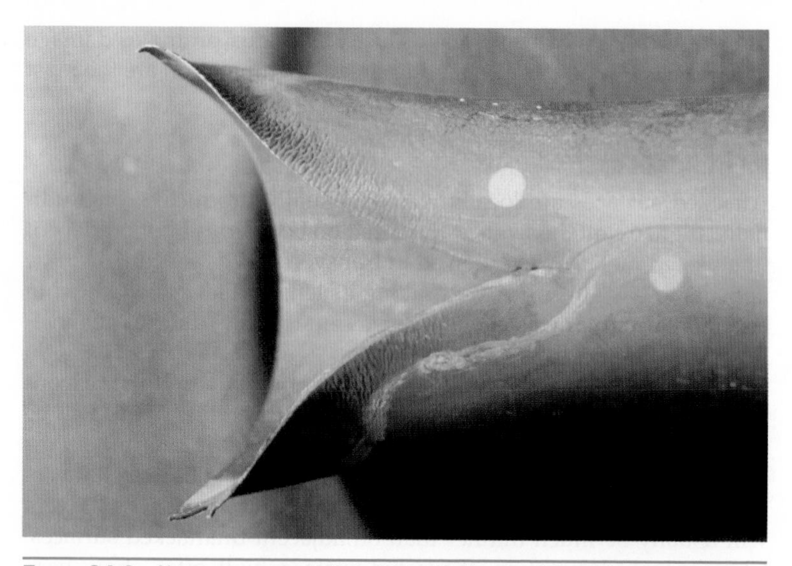

FIGURE 24.6 Numerous parallel surface ripples on the external surface at the edges of a rupture in a superheater tube. The light circles are regions where acid drops removed the thin oxide layer during laboratory testing.

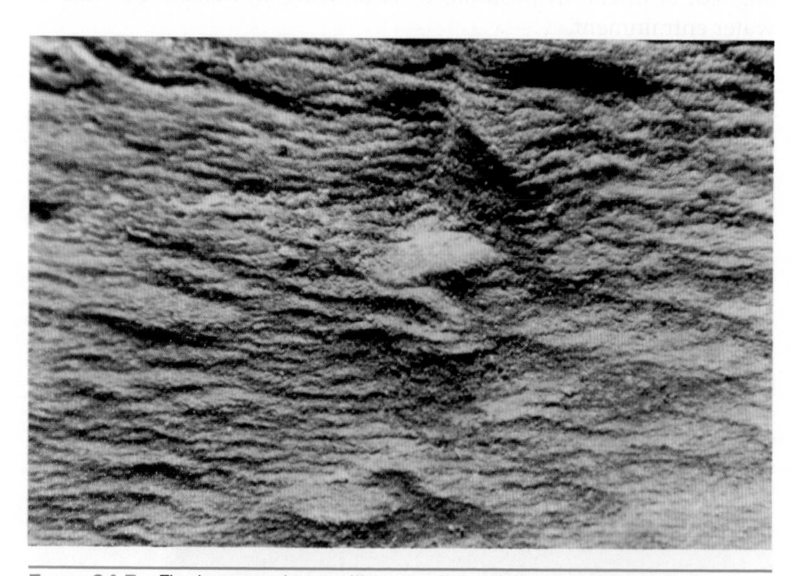

FIGURE 24.7 Finely spaced, wavelike, parallel striations on an eroded fireside external surface of a superheater tube. The striations resemble waves viewed from the air and are aligned perpendicular to the direction of steam impingement. (Compare with Fig. 24.6.)

Bulging will usually be absent in internally pressurized tubes suffering erosion. Rupture is usually longitudinal but can follow along grooves in the wasted contour, whatever their orientation. Rupture edges will be thin and ragged (Fig. 24.4).

Elimination

Proper soot blower operation, alignment, and function are required to reduce erosion. Periodic inspection of nozzle position and alignment is recommended. Elimination of moisture in blowing steam can be accomplished by allowing sufficient steam warm-up and providing for adequate drainage of steam supply lines.

Metallizing, welding, and plasma spray techniques can increase wall thickness in affected areas and thereby prolong tube life. Coating processes, however, can slightly reduce heat transfer locally and do not remove the root cause of the problem.

Cautions

Attack by entrained particulate in the flue gas resembles soot blower erosion. Location of damage in the direct path of the soot blower stream is necessary for diagnosis of failure.

Soot blower damage can resemble oil-ash or coal-ash corrosion. If heavy deposit accumulations are present atop thinned superheater and reheater tube walls, damage may be due to oil-ash or coal-ash corrosion, even though these tubes may exhibit other characteristics of erosion.

Related Problems

See also Chap. 21, High-Temperature Fireside Corrosion, and the sections entitled "Hard-Particle Erosion," "Falling-Slag Erosion," and "Steam Cutting from Adjacent Tube Failures" in this chapter.

Steam Cutting from Adjacent Tube Failures

General Description

Damage is produced by escaping high-velocity fluids from a pressurized tube leak, causing the nearby tubes to be steam-cut. The wastage mechanism is essentially the same as soot blower attack. However, attack is usually more localized. Wastage occurs rapidly.

Locations

Tubes affected by steam cutting can occur in any part of the boiler. A nearby failed tube is always present. Usually damage is highly localized and is worst in line of sight with the nearby failed tube. Cutting is most severe when internal steam pressures, and consequently temperatures, are high. Thus, superheaters and reheaters are often severely attacked. Occasionally, if the original failure is not detected promptly, a single failure leads to chain reactions involving multiple tube breaches.

Critical Factors

Pressure of escaping fluids and proximity of leaking tubes dictate the damage potential. Wastage rates increase as pressures (and consequently temperatures) increase and as distance decreases.

Identification

Wasted surfaces sometimes resemble those produced by soot blower erosion (see the section entitled "Soot Blower Erosion"). Surfaces close to nearby leaking tubes are likely to be smoothly undulating and grooved (Figs. 24.8 and 24.9). The irregularity of wastage increases as distance to the erosion source decreases. Freshly attacked surfaces contain almost no oxide or deposition. Surfaces are usually shallowly pebbled or striated and have an undulating contour. Often, the damaged area will be in line with a failure and the direction of steam from the leak.

Elimination

Since steam cutting is caused by other unpredictable tube failures, and such failures can occur anywhere in the boiler, it is not practical to design or plan for the damage. The only reasonable way to reduce the incidence of failure is to reduce the likelihood of other failures.

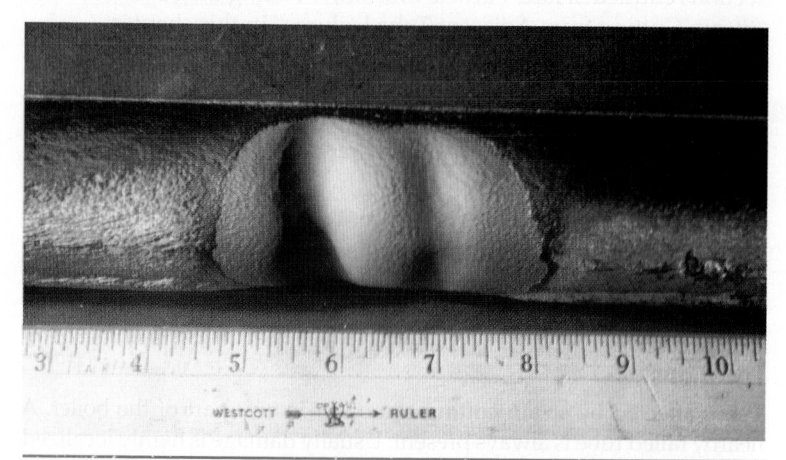

FIGURE 24.8 Utility superheater tube cut by steam leaking from an adjacent tube that failed at a manufacturing defect. Note the smoothly undulating surface and removal of red and black oxides and deposits in the wasted area.

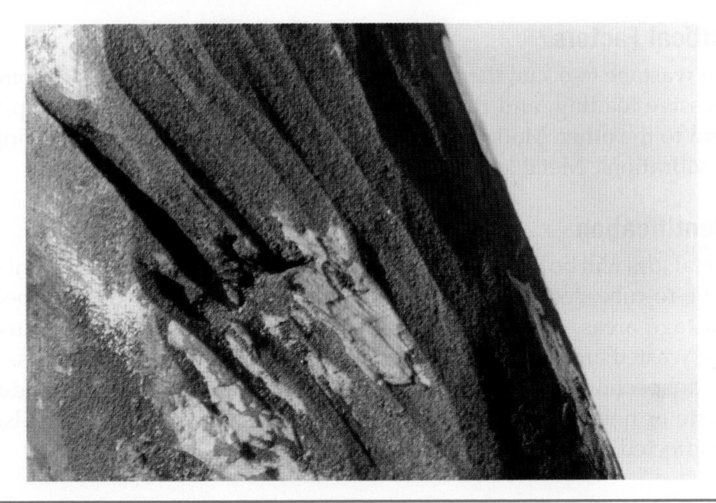

FIGURE 24.9 Grooving on pendant U-bend superheater leg caused by steam leaking through a corrosion fatigue crack at the U bend. Grooves were rusted after removal from boiler.

Cautions

Steam cutting is usually obvious because of associated failure or failures. However, this wastage can sometimes be confused with damage from other forms of fireside erosion. Also, once the steam-cut tube begins leaking, escaping fluids from subsequently damaged tubes can modify the nearby failure that was the primary cause of damage.

Related Problems

See also the sections entitled "Soot Blower Erosion," "Hard-Particle Erosion," and "Falling-Slag Erosion" in this chapter.

Mechanical Rubbing

General Description

Movement of contacting surfaces can result in erosive metal loss due to mechanical rubbing. The damage mechanism involves wastage at contacting points due to relative motion of the surfaces across one another. The motion at contact points can result in removal of the protective oxide layers and accelerated oxidation or abrasive wear of the metal in severe cases.

Locations

Metal loss will occur at contact points, such as touching tubes or locations of brackets on tubes. Relative movement of the surfaces is often related to insufficient or malfunctioning tube supports.

Critical Factors

For wastage to occur, the components need to be in contact with compressive loading, and there must be relative motion of one component to the other. Motion of the components may be related to sliding or vibrations. Metal loss is often due to insufficient tube support.

Identification

Metal loss will be localized to the contact points between components. Tube-to-tube contacting surfaces will generally have oval-shaped wastage areas. These eroded areas have concave surface contours. The wasted areas will generally be free of oxide and deposit layers.

Inspection of the boiler may reveal misaligned tubing or inadequate or nonfunctioning tube supports. Other nearby tubes can also be affected if they have similar contact points.

Elimination

Metal loss can be prevented by eliminating contact points between components, such as realigning tubing. If the components must touch, the motion of the components across the surfaces must be abated. This may require repairing nonfunctioning tube supports or adding supports to control motion.

Case History 24.1

Industry:	Utility
Specimen location:	Corner tube in a water wall
Specimen orientation:	Vertical
Years in service:	19
Drum pressure:	2250 psi (15.5 MPa)
Tube specifications:	3-in. (7.6-cm) outer diameter
Fuel:	Pulverized coal

A rupture occurred at a repair weld. Welding overlays on fireside surfaces were used to increase lost wall thickness to about 50% of the specified wall thickness. External surface metal loss was caused by erosion due to the impingement of fly ash entrained in flue gas. Deep grooves were cut in the tube wall at membrane gaps (Fig. 24.2). This boiler had a history of slagging problems. High solids and silica concentration in the fuel accelerated erosion.

Case History 24.2

Industry:	Utility
Specimen location:	Division wall in a circulating fluidized-bed boiler
Specimen orientation:	Vertical
Years in service:	1½, but only operating about ½ year because of numerous maintenance shutdowns
Drum pressure:	1275 psi (8.8 MPa), ~1700°F (927°C) external surface temperature
Tube specifications:	3-in. (7.6-cm) outer diameter, mild steel
Fuel:	Furnace bed of sand, lime, and wood chips

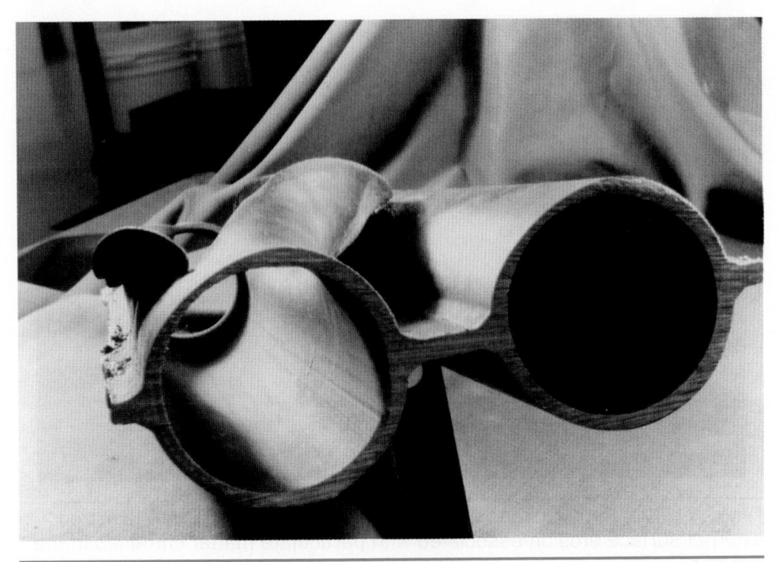

FIGURE 24.10 Ruptured division wall tube from fluidized-bed boiler. Metal loss was caused by sand abrasion in the bed.

The section sustained a massive longitudinal rupture in a zone of severe external surface metal loss (Fig. 24.10). Failure was caused by external surface erosion. Metallographic examination revealed that hard particulate matter (sand) was entrained in bed gases. Impingement of sand against the tube surface substantially reduced wall thickness, resulting in severe thinning. Rupture occurred when hoop stresses due to internal pressure exceeded the yield strength of the thinned tube.

The forced-draft fan maintained a 60- to 70-in. (152- to 178-cm) pressure head of water to fluidize the bed. No refractory or other surface protective devices such as studs or wear bars were present at the failure site.

Case History 24.3

Industry:	Pulp and paper
Specimen location:	Economizer
Specimen orientation:	Curved, predominantly vertical
Years in service:	8
Drum pressure:	1200 psi (8.3 MPa)
Tube specifications:	2-in. (5.1-cm) outer diameter, mild steel
Fuel:	Black liquor

Substantial metal loss occurred on the external surface at the inner curvature of a bend in an economizer tube. This metal loss produced a ragged, thin-lipped rupture (Fig. 24.11). Away from the rupture, surfaces were relatively unattacked.

The rupture was caused by erosion. Normal internal pressures could no longer be contained at the eroded site, and the tube ruptured. Erosion was caused by impingement of hard particulate matter entrained in flue gases. There was evidence of mild cold-end corrosion on all external surfaces.

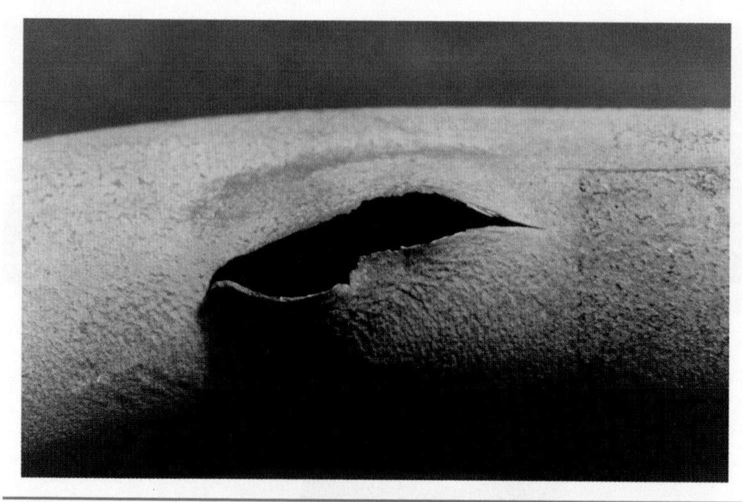

FIGURE 24.11 Rupture in an economizer tube caused by severe localized external surface erosion associated with particulate matter entrained in the flue gas.

Case History 24.4

Industry:	Refining
Specimen location:	Flue gas cooler used to preheat boiler feedwater
Specimen orientation:	Vertical
Years in service:	4
Drum pressure:	150 psi (1.0 MPa)
Tube specifications:	2-in. (5.1-cm) outer diameter

A flue gas cooler tube, used to preheat boiler feedwater, contained helical grooves on the internal surface (Fig. 24.12). Some grooves penetrated the

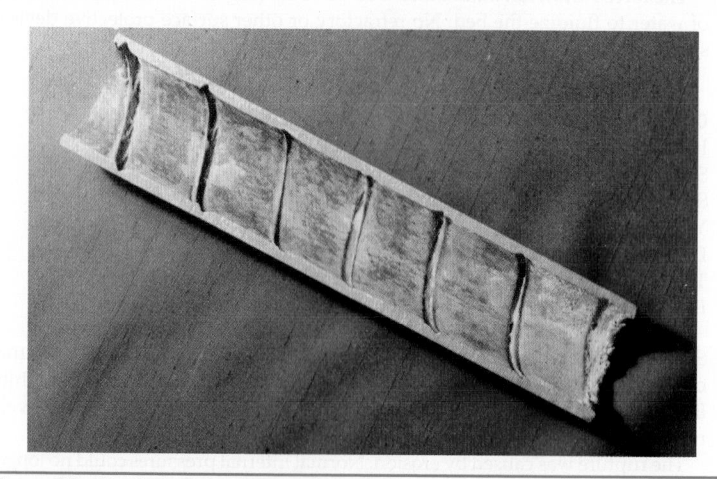

FIGURE 24.12 Spiral grooves cut into internal surface of a flue gas cooler by erosive gas flow. A spiral, stainless steel coil present inside the tube caused increased localized turbulence.

tube wall, producing helical cracks, while other grooves went only one-third of the way through the wall. Grooves were undercut in the direction of gas flow.

Erosion occurred along a helical coil inside the tube. The helix was used to increase turbulence and eliminate deposits. High-velocity flue gas flow became sufficiently turbulent at the helix to cause erosion failures.

Case History 24.5

Industry:	Metal processing
Specimen location:	Tube in bed section of a circulating fluidized-bed boiler
Specimen orientation:	Horizontal
Years in service:	0.75
Drum pressure:	1280 psi (8.8 MPa)
Tube specifications:	2½-in. (6.4-cm) outer diameter, mild steel

A tube in the bed of a circulating fluidized-bed (CFB) boiler developed multiple ruptures within a year of service, as shown in Fig. 24.13. The ruptures were thin-lipped and located in an area of deep tube-wall loss from the external surface. The localized metal loss area had a very smooth surface contour covered by a very thin oxide layer. Microscopic analysis revealed a narrow zone of grain deformation in the area of metal loss (Fig. 24.3). The localized metal loss and deformation of the microstructure indicated that wastage was caused by hard-particle impingement.

The failure occurred near a location of a damaged tuyere (air injector) in the bed. The damaged air injector resulted in excessive flow velocities that caused the localized metal loss.

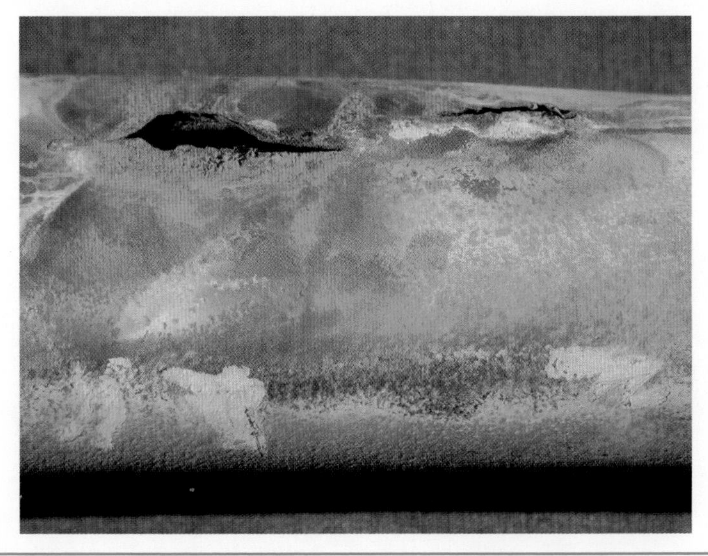

FIGURE 24.13 Multiple thin-edged ruptures in a localized area of metal loss on a tube in a bed of a circulating fluidized-bed boiler. Wastage was caused by hard-particle impingement.

Case History 24.6

Industry:	Utility
Specimen location:	Economizer
Specimen orientation:	Horizontal
Years in service:	4
Drum pressure:	1350 psi (9.3 MPa)
Tube specifications:	2-in. (5.1-cm) outer diameter, mild steel
Fuel:	Unknown

The economizer section experienced multiple failures. Failures were in the form of thin-lipped ruptures. The ruptures were located in longitudinally oriented wastage flats. These flats had very smooth contours with distinct edges, as shown in Fig. 24.14.

The characteristics of the wastage are typical of soot blower erosion. The boiler tube failures were occurring near soot blowers that were not operating properly.

Case History 24.7

Industry:	Utility
Specimen location:	Superheater
Specimen orientation:	Vertical
Years in service:	5
Drum pressure:	2400 psi (16.5 MPa)
Tube specifications:	2½-in. (6.4-cm) outer diameter, SA-213
Fuel:	Pulverized coal

A series of failures of superheater tubes occurred almost 2 years after extensive retubing. Nearby tubes contained forming defects present as deep fissures that ultimately penetrated to failure due to corrosion fatigue cracking. These fissures opened in service, causing steam cutting of adjacent tubes (Fig. 24.8).

FIGURE 24.14 Thin-lipped rupture that occurred in a wastage flat caused by soot blower erosion. Note the distinct shoulder along the edge of smooth surface wastage.

It is remarkable that failures similar to this one occurred almost 2 years earlier and for essentially the same reason. In spite of the extensive retubing, some defective tubes were missed. Tubes containing deep fissures remained in service for at least 2 years before failure occurred. When these defective tubes finally failed, extensive steam cutting of nearby tubes resulted.

Case History 24.8

Industry:	Utility
Specimen location:	Primary superheater pendant
Specimen orientation:	Vertical
Years in service:	25
Drum pressure:	1800 psi (12.4 MPa)
Tube specifications:	2-3/8-in. (6.0-cm) outer diameter
Fuel:	Coal

A superheater tube failed at a pendant U bend. Failure was caused by corrosion fatigue cracking. Leakage was relatively slight before opposite legs were steam-cut (Fig. 24.9). More tubes had to be replaced because of chain reaction failures associated with steam cutting than because of the precipitating corrosion fatigue.

Case History 24.9

Industry:	Utility
Specimen location:	Front convection wall
Specimen orientation:	Vertical
Years in service:	8
Drum pressure:	2500 psi (17.2 MPa)
Tube specifications:	2-3/8-in. (7.0-cm) outer diameter, SA-210, grade A1
Fuel:	Coal

A tube failed at a weld support. Escaping water cut away the support and perforated the adjacent tube (Fig. 24.15).

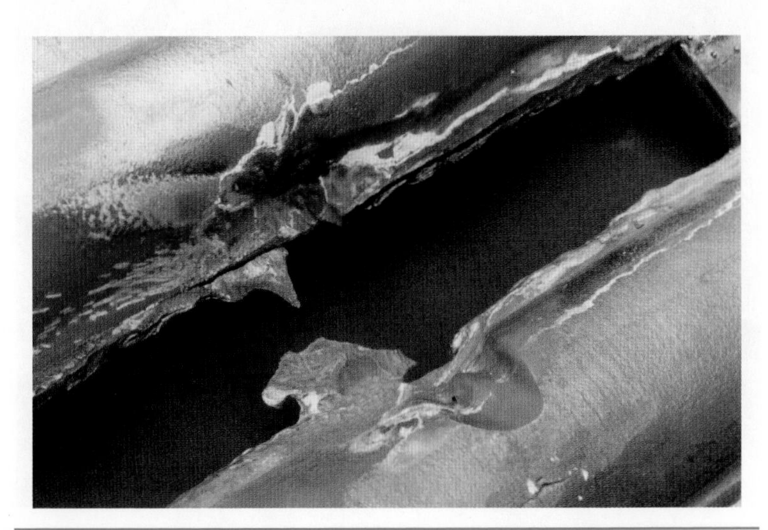

FIGURE 24.15 Steam cutting of utility convection tubes. The original failure occurred at a welded support bracket and was due to corrosion fatigue. The adjacent tube was breached by escaping steam from the original failure.

The question that naturally arises in multiple failures is, Which failure occurred first? If erosion is severe, identifying evidence of the original failure can be entirely eliminated. Luckily, other nearby supports were included with the received section. Careful visual inspection revealed small corrosion fatigue cracks at poorly fused support welds. Upon sectioning and microscopic examination, small fissures were located near the original failure; these fissures were almost identical to (but deeper than) those found at adjacent braces. Hence, the original failure was likely caused by corrosion fatigue at a weld support.

Case History 24.10

Industry:	Mining—flash furnace of a smelter
Specimen location:	Convection section in a waste heat boiler
Specimen orientation:	Vertical
Years in service:	26
Drum pressure:	800 psi (5.5 MPa)
Tube specifications:	1.5-in. (3.8-cm) outer diameter, mild steel
Fuel:	Waste heat from a smelting furnace

Two tube segments attached by a bracket were removed from the generating bank. One of the tube segments was firmly attached to the bracket, while the other was loose. The loose tube segment from the generating bank had a uniformly flattened tube wall only within the portion that contacted the bracket, as shown in Fig. 24.16. A small perforation occurred where external surface metal loss was deepest. The external surface within the flattened portion was fairly smooth and covered with thin layers of deposits and corrosion products. Away from the bracket location, no significant external surface metal loss occurred.

FIGURE 24.16 Severely thinned generating bank tube due to external surface metal loss at a bracket contact location. Note the flush fit of the tube with the bracket.

The localization of the wastage indicates that metal loss was primarily caused by wear resulting from rubbing or sliding contact of the external surface of the tube across the internal surface of the bracket. Other locations of tube-bracket connections were recommended to be inspected, as loose connections would also be susceptible to mechanical wear. At loose connections, welds could be used to firmly attach the tubing to the bracket and stop relative movement that resulted in rubbing of the components.

References

1. *Boiler Tube Failure Metallurgical Guide*, vol. 1: *Technical Report*, TR-102433-1, Electric Power Research Institute, Palo Alto, CA, October 1993, pp. 3–127.
2. J. B. Kitto and S. C. Stultz (Eds.), *Steam: Its Generation and Use*, 41st ed., Babcock & Wilcox Company, Barberton, Ohio, 2005, pp. 9–10 to 9–11.

SECTION 3

Material Defects

General Description

Failures resulting from manufacturing defects are relatively rare. In fact, manufacturing defects account for less than 1% of total failures examined. Perhaps the most common defects occur in weld seams and circumferential welds. Welding defects will be described in greater detail in Chap. 25, Welding Defects.

In rare instances, seamless tubes may contain defects that are similar in appearance to weld seam defects. These defects are generally a consequence of internal voids in the casting from which the tube was formed. Voids that are not closed and metallurgically bonded during tube manufacturing will result in a defect.

Forming defects may be produced in tubing during manufacturing and/or installation. Such forming defects may produce tube cross sections that are "out of round," i.e., not circular. Tubes with irregular cross sections will not necessarily cause or promote damage. However, excessive residual stresses imparted by forming may contribute to failure mechanisms. See Chap. 11, Stress-Assisted Corrosion; Chap. 12, Corrosion Fatigue Cracking; and Chap. 13, Stress Corrosion Cracking.

Locations

Defects that affect serviceability are typically found in tubes. Defective material can be inadvertently placed in any tubed region of the boiler.

Critical Factors

Installation of construction materials having significant defects may result from insufficient adherence to specified fabrication and quality control practices on the part of the manufacturers of tubing, components, and/or boilers.

Identification

Identification of welding defects is presented in Chap. 25. Defects that are produced in seamless tubes due to forming from a casting containing internal voids will appear similar to a seam defect, as shown in Fig. 25.6. They may also be in the form of straight or gently curving cracks, which may run longitudinally for some distance (Fig. S3.1).

Tubes that have irregular cross sections may be identified by inspection using oblique lighting after surfaces are cleaned. In some cases, irregularities may be felt by hand. An irregular cross section is most obvious when the tube is cut in cross section (Fig. S3.2). Certainly, if such defects are found prior to start-up of the boiler, it is ensured that they formed during tube manufacturing or installation. However, following boiler operation, simple visual examination cannot definitely distinguish such defects from bulges or deformation caused by overheating. Consequently, metallographic examination is required to determine the cause for the irregular cross section.

FIGURE S3.1 Defect in seamless tube that was formed from a casting that contained internal voids.

FIGURE S3.2 Tube forming defect that produced an egg-shaped cross section.

Elimination

Generally, once construction is completed, there is no economical way of detecting specific manufacturing defects. If they exist and are serious, their presence is generally revealed by failure of the defective component.

If failure of a component results from a manufacturing defect, it may be advisable to survey similar components for evidence of defects. This usually requires the use of nondestructive testing techniques such as ultrasonic testing.

Cautions

Thick-walled ruptures caused by overheating are sometimes incorrectly diagnosed in the field as material defects. Confirmation by metallographic examination is required.

Related Problems

See also Chap. 2, Short-Term Overheating, and Chap. 3, Long-Term Overheating.

CHAPTER 25

Welding Defects

General Description

Welding is defined as joining two or more pieces of material by applying heat, pressure, or both, with or without filler material, to produce a localized union through fusion or recrystallization across the interface.[1]

In an ideal weld, the metallurgical bond provides a smooth, uninterrupted, microstructural transition across the weldment. The weldment should be free of significant porosity and nonmetallic inclusions and should form smoothly flowing surface contours with the section being joined. The weldment should not contain excessive residual welding stresses both within the weld metal and surrounding heat-affected zones.

A complete listing of possible weld defects is well beyond the scope of this guide. Rather, only the more common weld failures will be presented.

Locations

A large, utility-size boiler may contain more than 50,000 welds. In addition, longitudinal seam welds are present in some grades of tubing. Each weld is a possible defect site. Largely as a result of the rigorous code requirements for pressure vessels set forth by the American Society of Mechanical Engineers (ASME), weld failures account for a very small percentage of boiler failures. Because of the relatively severe environments present in superheater and reheater sections, many of the weld failures tend to occur in these areas.

Critical Factors

All weld defects represent a departure from one or more of the desirable features described in the section entitled "General Description" in this chapter. However, it should be realized that welds are not perfect. In consideration of this fact, all major welding codes allow for welding defects, but set limitations on the severity of the defect. An acceptable weld is not one that is defect-free, but one in which existing defects allow satisfactory service. Several different types of weld defects may be produced. The most common defects are described below.

Porosity

Identification

Porosity refers to the entrapment of gas bubbles or pores in the weld metal resulting either from decreased solubility of a gas as the molten weld metal cools or from chemical reactions that occur within the weld metal. Porosity is usually not apparent from the surfaces of the weldment. However, in some cases pores may be exposed at the surfaces. Porosity must be identified through metallographic analysis. Figure 25.1 shows a metallographic cross section of a porous weld. Internal porosity may also be produced by shrinkage of the weld metal during solidification (also known as *shrinkage cavities*) depending upon factors such as the size of the weld. The distribution of pores within the weldment can be classified as uniformly scattered porosity, cluster porosity, or linear porosity. Porosity at and near surfaces will have a significant effect on the integrity of the weld.

Elimination

Porosity can be limited by using clean, dry materials and by maintaining proper weld current and arc length.

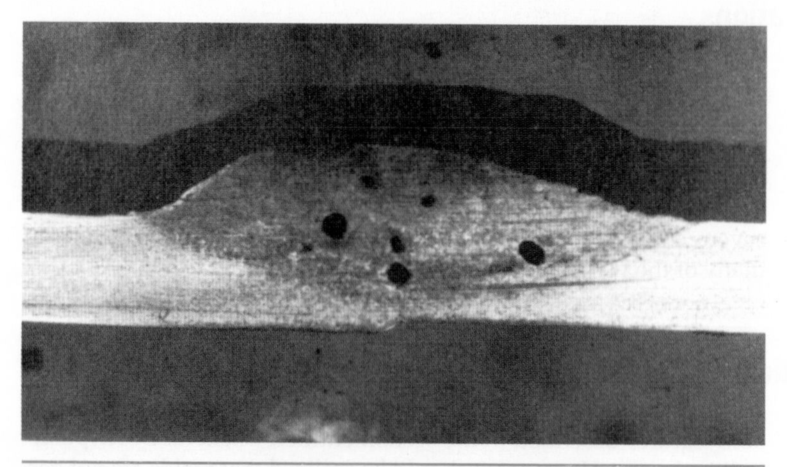

Figure 25.1 Cross section through weld, showing severe porosity. This weld had been in low-pressure service for 40 years without failure. (*Reprinted with permission from Helmut Thielsch, Defects and Failures in Pressure Vessels and Piping, New York, Van Nostrand Reinhold, 1965.*)

Slag Inclusions

Identification

The term *slag inclusion* refers to nonmetallic solids trapped in the weld deposit or between the weld metal and base metal (Fig. 25.2). Slag forms by a high-temperature chemical reaction, which may occur during the welding process due to poor welding practice and/or contamination of the weldment. These inclusions may be present as isolated particles or as continuous or interrupted bands. Slag inclusions are not visible unless they emerge at a surface, so they must be identified through metallographic analysis. Figure 25.3 is a metallographic cross section of a linear slag inclusion located at the weld metal/base metal interface.

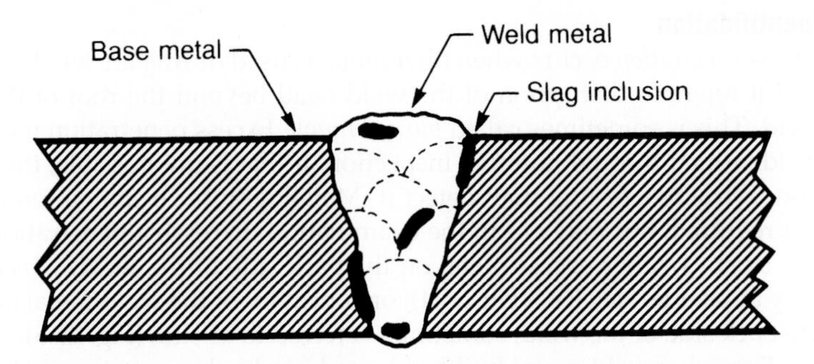

FIGURE 25.2 Schematic of possible slag defect locations within a weld.

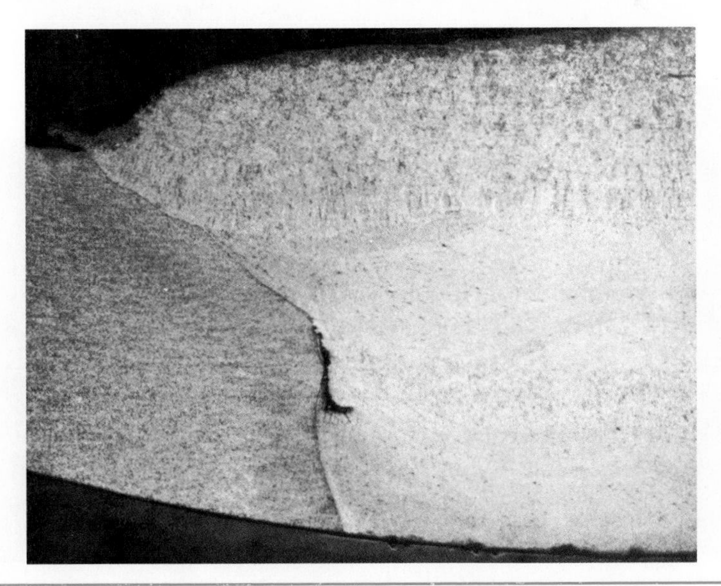

FIGURE 25.3 Cross section through a linear slag inclusion. This weld had been in 650 psi (4.5 MPa) service for 25 years. (*Reprinted with permission from Helmut Thielsch, Defects and Failures in Pressure Vessels and Piping, New York, Van Nostrand Reinhold, 1965.*)

Service failures are generally associated with surface-lying slag inclusions, or inclusions that are of such size that they significantly reduce the cross-sectional area of the wall, decreasing mechanical strength.

Elimination

The number and size of slag inclusions can be minimized by maintaining weld metal at low viscosity, preventing rapid solidification, and maintaining sufficiently high weld metal temperatures. Weld metal contamination must also be avoided.

Excess Penetration

Identification

Excess penetration occurs when filler metal is used during arc welding, and it refers to disruption of the weld bead beyond the root of the weld. This is sometimes called *burn-through*. Excess penetration may be identified through internal inspection. If that is not possible, then nondestructive testing may detect it. Metallographic examination is not required, but may aid in the examination. Excessive penetration can exist as either excess metal on the back side of the weld, which may appear as "icicles" (Fig. 25.4), or concavity of the weld metal on the back side of the weld, sometimes referred to as sink (Fig. 25.5).

Excessive weld metal buildup is undesirable because this material can disrupt boiler water or steam flow, possibly causing localized

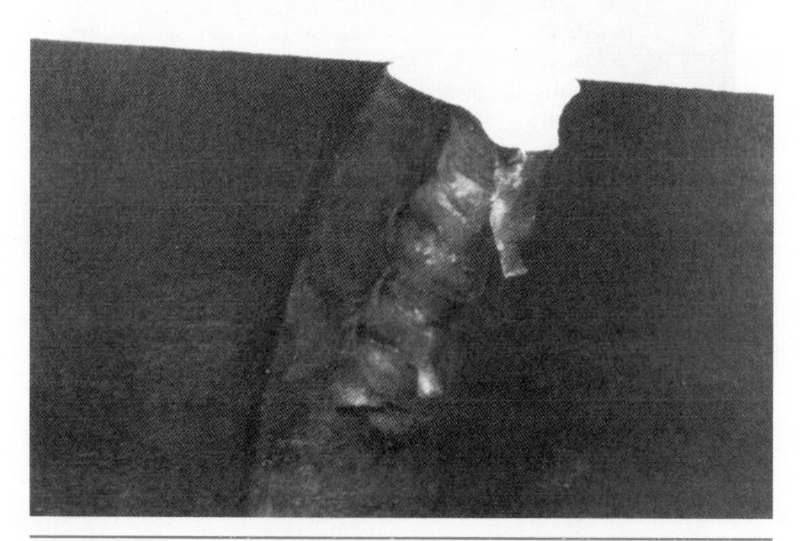

FIGURE 25.4 Burn-through resulting in icicles on the underside of a butt weld. (*Reprinted with permission from Helmut Thielsch, Defects and Failures in Pressure Vessels and Piping, New York, Van Nostrand Reinhold, 1965.*)

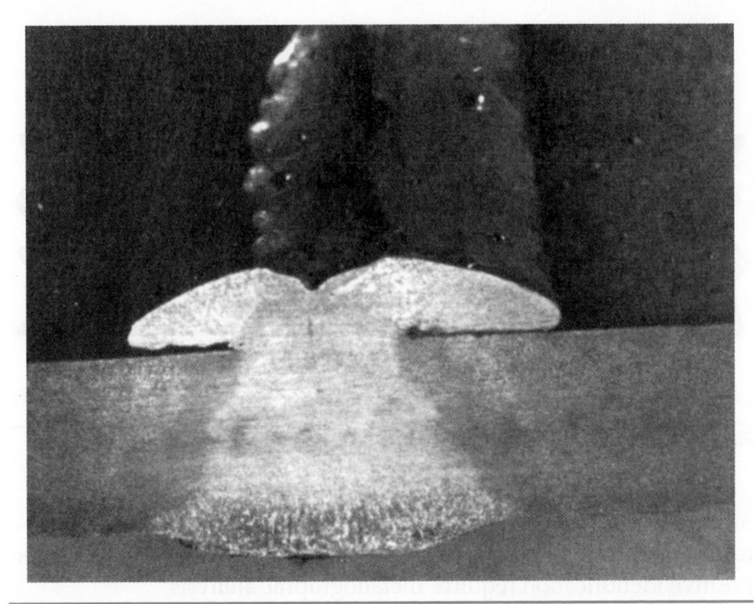

FIGURE 25.5 A burn-through cavity formed on the backing ring of a tube weld. (*Reprinted with permission from Helmut Thielsch, Defects and Failures in Pressure Vessels and Piping, New York, Van Nostrand Reinhold, 1965.*)

deposition, under-deposit corrosion, overheating downstream of the defect in a water tube, or localized overheating downstream of the defect in a superheater or reheater tube.

Concavity, if severe, can cause root-pass cracking. Such cracks may not be revealed by nondestructive testing (NDT). Concavity can also substantially reduce fatigue life and may promote initiation of corrosion fatigue failures since they provide sites of stress concentration.

Elimination

Excess penetration is frequently caused by improper welding techniques, poor joint preparation, and poor joint alignment.

Incomplete Fusion

Identification

Incomplete fusion, as the term implies, refers to incomplete melting and attachment of the mating components within a weld joint. It can occur in welds that do not use filler metals, such as electric resistance welds, used to manufacture seam welds. In such cases, incomplete fusion is characterized by a straight, V-shaped continuous or discontinuous groove along the seam. Incomplete fusion may also occur in arc welds between individual weld beads, between the base metal and weld

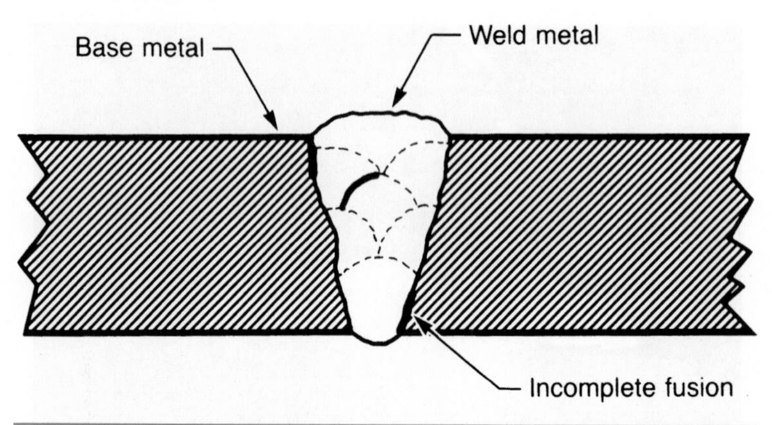

FIGURE 25.6 Schematic of incomplete fusion defect in weld.

metal, or at any point in the welding groove (Fig. 25.6). Nondestructive testing may indicate the possibility of incomplete fusion. However, positive identification requires metallographic analysis.

Failures resulting from incomplete fusion of internal surfaces of the weld are infrequent, unless it is severe. Incomplete fusion to 10% or more of the wall thickness is considered to be critical. Due to stress concentrating effects, incomplete fusion at surfaces is more critical than internal incomplete fusion, and it can lead to failure by mechanical fatigue, corrosion fatigue, and stress corrosion cracking. Crevices that are exposed to the waterside surfaces are susceptible to corrosion, especially during chemical cleaning due to concentrating mechanisms for acids and entrapment of acids.

Elimination

Incomplete fusion can be eliminated by increasing weld current or reducing weld speed. Incomplete fusion can also be caused by failure to remove or flux oxides and nonmetallic materials adhering to the surfaces to be welded.

Related Problems

See the section titled "Inadequate Joint Penetration" in this chapter.

Undercutting

Identification

The term *undercutting* refers to the creation of a continuous or intermittent groove melted into the base metal either at the interface of the weld metal and surface (toe of the weld) (Fig. 25.7) or at the bottom or base (root) of the weld. Nondestructive testing may indicate the possibility of appreciable undercutting. Metallographic analysis is required to positively diagnose subsurface undercutting. Undercutting at exposed

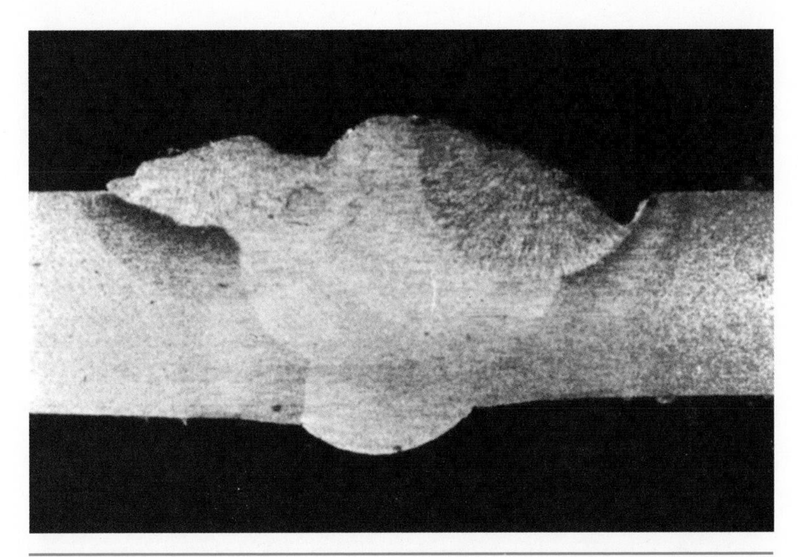

FIGURE 25.7 Cross section of welds containing severe undercut. (*Reprinted with permission from Helmut Thielsch, Defects and Failures in Pressure Vessels and Piping, New York, Van Nostrand Reinhold, 1965.*)

surfaces may be identified visually. Depending upon depth and sharpness, undercutting may provide points of stress concentration that can promote cracking by either fatigue or corrosion fatigue mechanisms.

Elimination

Serious undercutting may be repaired either by grinding or by depositing additional weld metal. Undercutting is generally caused by using excessive welding currents for a particular electrode or by maintaining an arc that is too long.

Inadequate Joint Penetration

Identification

In some cases weld filler metal incompletely penetrates the thickness of the joint (Fig. 25.8). Nondestructive testing may indicate the possibility of inadequate joint penetration. However, positive identification requires metallographic analysis. This usually applies to the initial weld pass or to passes made from one or both sides of the joint. On double-welded joints, the defect may occur within the wall thickness (Fig. 25.9).

Inadequate joint penetration is one of the most serious welding defects. It has caused failures in both pressure vessels and tube welds. Failures by mechanical fatigue, corrosion fatigue, stress corrosion cracking, and various corrosion mechanisms, particularly acid corrosion, have been associated with this defect.

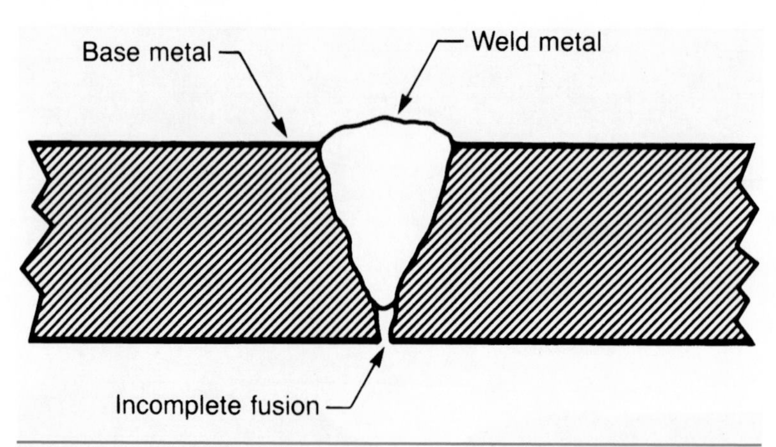

FIGURE 25.8 Schematic of inadequate joint penetration in weld.

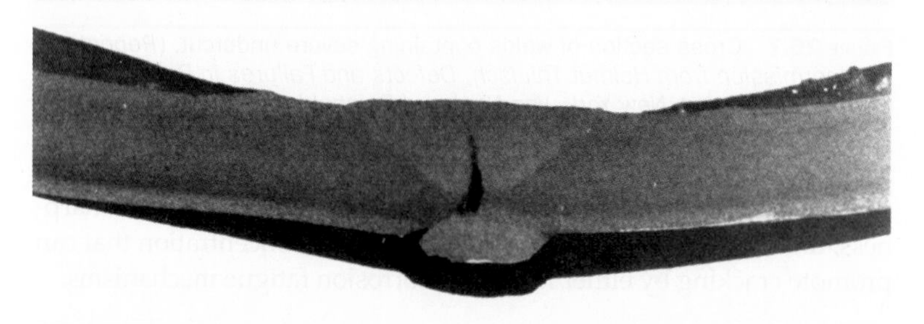

FIGURE 25.9 Cross section of circumferential butt weld. (*Reprinted with permission from Helmut Thielsch, Defects and Failures in Pressure Vessels and Piping, New York, Van Nostrand Reinhold, 1965.*)

Elimination

Inadequate joint penetration is generally caused by unsatisfactory groove design, the use of an electrode that is too large, excessive weld travel rate, or insufficient welding current.

Related Problems

See the section titled "Incomplete Fusion" in this chapter.

Cracking

Identification

Nondestructive testing may indicate the possibility of cracking. However, positive identification requires metallographic analysis. Cracks may appear as linear openings at the metal surface. They can be wide, but frequently are tight. Such cracks are typically thick-walled and exhibit very little, if any, plastic deformation. Cracking of weld metal

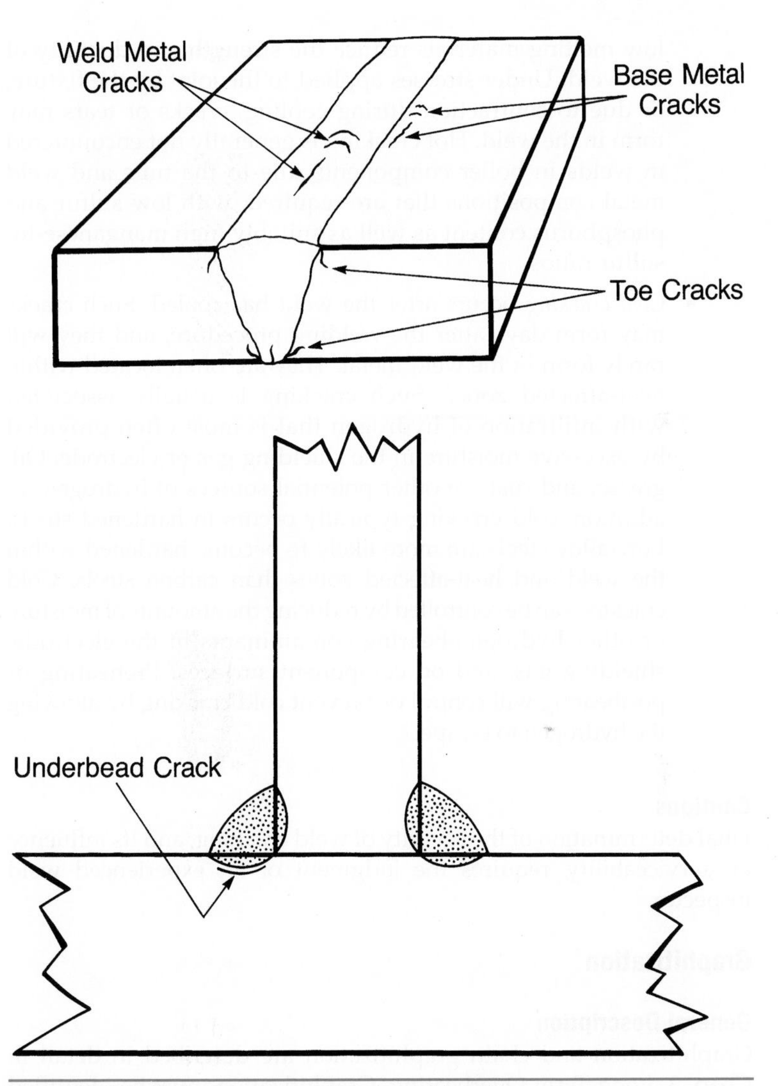

FIGURE 25.10 Typical crack locations in weldments.

can be critical. Base-metal cracking (toe cracking, under-bead cracking) is also critical and has caused service failures. Cracking may be either transverse or longitudinal (Fig. 25.10).

Cracks may form in weldments due to the welding process through two primary mechanisms:

- *Hot cracking of the weld metal* occurs immediately upon solidification. Such cracks are primarily and most commonly caused by the presence of excessive amounts of low melting sulfides and phosphides in the weld metal. These low melting materials reduce the strength and ductility of the weld. Under stresses applied to the joint by the fixture, or due to contraction during cooling, cracks or tears may form in the weld. Hot cracking is generally not encountered in welds in boiler components due to the tube and weld metal compositions that are required, with low sulfur and phosphorus content as well as suitably high manganese-to-sulfur ratios.

- *Cold cracking* occurs after the weld has cooled. Such cracks may form days after the welding procedure, and they will rarely form in the weld metal. They are often located within heat-affected zones. Such cracking is usually associated with infiltration of hydrogen that is most often provided by excessive moisture in the shielding gas or electrode. Oil, grease, and rust are other potential sources of hydrogen. In addition, cold cracking typically occurs in hardened steels. Low-alloy steels are more likely to become hardened within the weld and heat-affected zones than carbon steels. Cold cracking can be controlled by reducing the amounts of moisture or other hydrogen-bearing contaminants in the electrode, shielding gas, and on component surfaces. Preheating or postheating will control or prevent cold cracking by allowing the hydrogen to escape.

Cautions

Final determination of the severity of weld cracking, and its influence on serviceability, requires the judgment of an experienced weld inspector.

Graphitization

General Description

Graphitization and chain graphitization are described in detail in Chap. 3, Long-Term Overheating. Graphitization must be identified through metallographic examination.

Graphitization in welded steels may result in the formation of continuous chains of graphite that preferentially align along the low-temperature edge of the heat-affected zone in the base metal. Cracks may readily propagate through continuous chains or surfaces of the very brittle, weak graphite.

Locations

Plain carbon and some low-alloy steels containing $< 0.5\%$ chromium can be susceptible to graphitization at weld heat-affected zones. Such steels may be used in low-temperature sections of the superheater and reheater. Failures are more frequent in steam piping than in boiler tubing.

Critical Factors

The critical factors leading to potential failure resulting from graphitization are the use of a susceptible material, welding of susceptible material, and exposure of the weldment to temperatures above the range of 800 to 1200°F (427 to 649°C) for a prolonged period.

Identification

Fractures tend to occur within the heat-affected zones approximately 1/16 in. (1.6 mm) from the weld interface. A nondestructive technique for confirming the occurrence of graphitization does not exist. Graphitization must be identified through metallographic examination. Graphite nodule chain formation will have an appearance similar to that in Fig. 25.11.

Elimination

Avoiding prolonged exposure of welded metal to the temperature range over which graphitization occurs is the best method of eliminating graphitization in suspect material. In the equipment design phase, specifying material that contains appropriate concentrations of chromium and molybdenum to control graphite formation can prevent failures by graphitization.

Welding Stresses

Residual stress is defined as a stress in the metal that is not supplied by externally applied forces. As such, residual stresses are imparted to

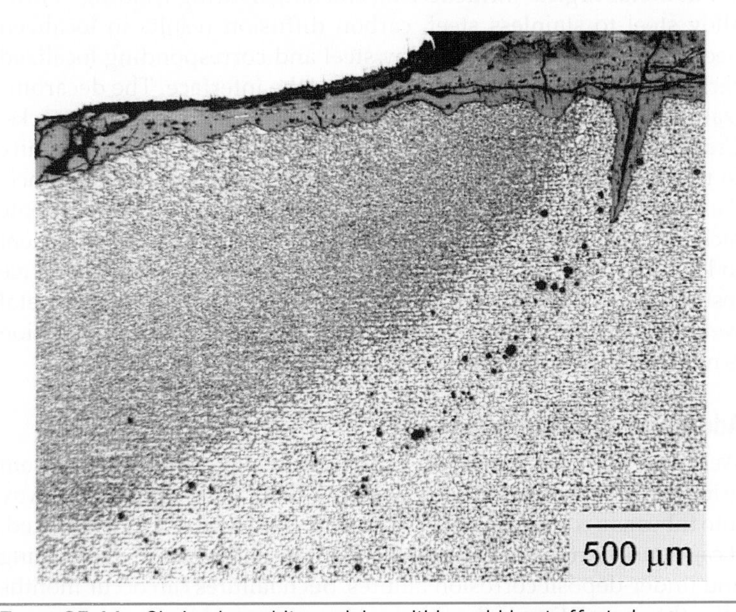

FIGURE 25.11 Chained graphite nodules within weld heat-affected zone.

the metal by various forming operations. Residual stresses are produced at welds primarily by thermal expansion and contraction between the hot weld metal and the much cooler base metal. Weld metal shrinkage during solidification may also provide residual stresses. Residual stress may reach the yield strength of the metal itself. Corrosion and cracking will tend to occur more readily in locations of high residual stress. Cracks may form within the weld or base metal under high residual stress in some cases. For more in-depth discussions concerning the effects of residual stress see Chap. 13, Stress Corrosion Cracking; Chap. 12, Corrosion Fatigue Cracking; and Chap. 11, Stress-Assisted Corrosion. Unfortunately, conventional nondestructive testing cannot detect high residual stresses. Residual stresses may be reduced through use of appropriate preheating, postheating, and stress relief annealing, particularly for low-alloy steels. Welding of thick components will almost certainly require preheating.

Dissimilar Metal Welds

In some locations in certain boiler designs, metals with substantially different compositions, such as low-alloy steels and stainless steels, are joined by welding. Due to different thermal expansion coefficients and thermal conductivities of the different alloys, significant stresses may be imparted to the weld joint. Under extreme conditions, fatigue or corrosion fatigue cracks may form due to the application of thermally induced stresses. There is much debate about metallurgical influences on cracking. During welding of low-alloy steel to stainless steel, carbon diffusion results in localized decarburization of the low-alloy steel and corresponding localized carburization of the stainless steel near the interface. The decarburization in the low-alloy steel has been suspected to initiate cracks. Cracks form within the decarburized base metal in close proximity to the weld metal due to the loss of strength and creep resistance. Austenitic stainless steel weld metal may be replaced with appropriate nickel alloys, which are more similar in thermal expansion coefficient to low-alloy steel and have a lower carbon solubility. Nondestructive testing may be conducted on a periodic basis at dissimilar metal welds to determine if cracking occurred. Metallographic examination is required to determine the exact cause of cracking.

Additional Considerations

Welding debris such as weld spatter, shavings, filings, chips from grinding tube ends, and even welding tools have found their way into tubes as a consequence of tube repair. If this debris is not removed, it can cause partial blockage of coolant flow and result in overheating and under-deposit corrosion failures. Such failures can occur months after the completion of the repair.

Internal surface defects and defects deep within the weld metal are usually inaccessible for repair, and these would require removal and rewelding.

Backing rings have been used at circumferential welds in boilers in the past. Their use in newer installations is rare. Weld defects may occur at backing rings if they are improperly installed or designed, and due to improper welding technique. Their use may promote deposition or overheating downstream of the weld, due to intrusion that causes flow disruption. In addition, aggressive agents may concentrate within gaps between backing rings and tube surfaces to cause corrosion.

Case History 25.1

Industry:	Utility
Specimen location:	Superheater outlet header
Specimen orientation:	Slanted
Years in service:	7
Water treatment program:	Ammonia, hydrazine
Drum pressure:	3800 psi (26.2 MPa)
Tube specifications:	2¼-in. (5.7-cm) outer diameter, SA-213 T22

The tube section shown in Figs. 25.12 and 25.13 had been welded to the finishing superheater outlet header in the penthouse. The specimen shows a brittle fracture at the weld, which apparently disconnected from the header after the failure. Similar fractures had not occurred previously, but several additional cracked and leaking welds were discovered upon the inspection associated with this failure. The boiler was base-loaded and in continuous service except for yearly maintenance outages.

Visual examination of the weld and the fracture face revealed that large areas of the weld root were unfused. Radial striations on the fracture face originated at the unfused region.

The specimen had failed at the weld that joined the tube to the header. Cracks initiated at the weld root, where incomplete fusion had occurred, and propagated from this region by a corrosion fatigue mechanism. The unfused portion

FIGURE 25.12 Cross section of tube showing weld metal (top of tube).

FIGURE 25.13 Fracture surface, showing tube wall (inner ring) and fractured weld metal (outer ring).

of the weld formed sharp crevices, which acted as stress concentration sites that locally elevated normal stresses.

Cyclic stressing in this region of the boiler is frequently caused by differences in rates and directions of thermal expansion between the header and the tubes. The thermal expansion and contraction stresses are often associated with start-up and shutdown.

Case History 25.2

Industry:	Utility
Specimen location:	Screen tube
Specimen orientation:	45° slant
Years in service:	20
Water treatment program:	Coordinated phosphate
Drum pressure:	1500 psi (10.3 MPa)
Tube specifications:	3-in. (7.6-cm) outer diameter
Fuel:	Coal

A "window" section had been cut out of the hot side of the tube illustrated in Fig. 25.14 and a new section welded into place from the external surface. Cross sections cut through the welded window revealed deep fissures associated with the welds (Fig. 25.15). The appearance of this defect suggests that entrapped slag, dirt, or flux prevented the formation of a sound metallurgical bond. Defects of this type can lead to through-wall fatigue cracking or corrosion fatigue due to the stress-concentrating effects of the fissures.

Case History 25.3

Industry:	Utility
Specimen location:	Economizer outlet header
Specimen orientation:	Horizontal
Years in service:	2
Water treatment program:	Equilibrium phosphate
Drum pressure:	2600 psi (17.9 MPa)
Tube specifications:	2½-in. (6.4-cm) outer diameter

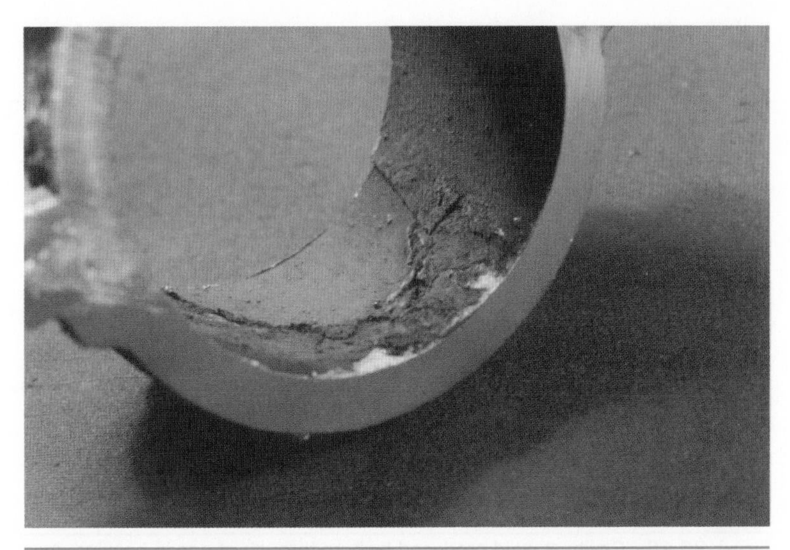

FIGURE 25.14 Internal surface of tube, showing oval-shaped replacement window weld into place.

FIGURE 25.15 Cross section through window weld showing deep fissure in weld metal.

A backing ring was used at a circumferential weld (Fig. 25.16). There was a significant gap between the longitudinally saw-cut ends of the ring. Within this gap, excessive weld metal penetration occurred, the backing ring cut through, and weld metal flowed out onto the adjacent part of the ring. A deep cavity was produced in the weld at the gap location, the base of which was close to the external surface. A perforation occurred at the cavity after only 2 years

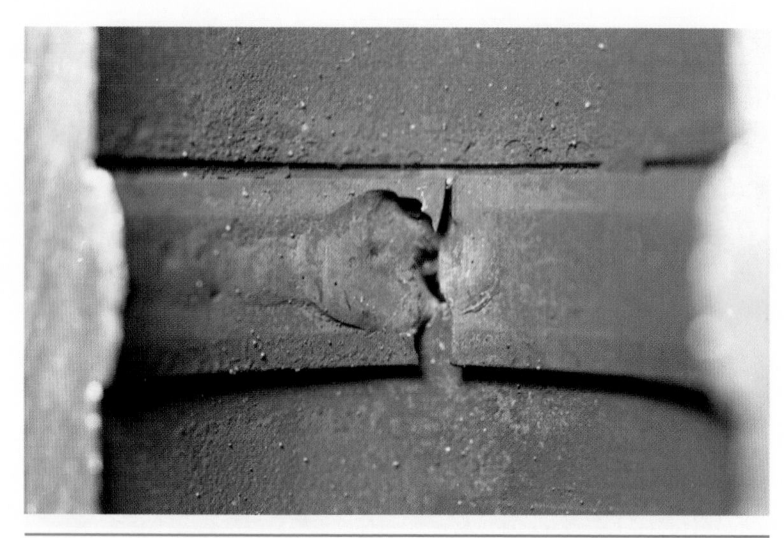

FIGURE 25.16 Cavity at weld produced by burn-through at gap in backing ring.

of service. The cause of final perforation could not be accurately determined, because erosion by escaping steam removed evidence that defined the cause of damage responsible for the final failure. It is possible that the damage was caused by corrosion due to concentration of aggressive agents within the cavity, or due to a cracking mechanism. Nondestructive testing would have identified the deep cavity that formed at this weld, which would have prompted repairs that may have prevented a failure. Backing rings are rarely used at welds in boilers due to problems such as those described in this case history, flow disruption, and crevices that form between the ring and tube wall. Crevices beneath backing rings may promote corrosion, particularly during cleaning (see Chap. 10, Corrosion during Cleaning).

Case History 25.4

Industry:	Utility
Specimen location:	Screen wall
Specimen orientation:	Vertical
Years in service:	9
Water treatment program:	Coordinated phosphate
Drum pressure:	900 psi (6.2 MPa)
Tube specifications:	2-in. (5.1-cm) outer diameter
Fuel:	Coal

Figure 25.17 illustrates a massive fracture along the weld bead attaching the membrane to the tube wall. Microstructural examinations revealed an aligned chain of graphite nodules in the heat-affected zone immediately adjacent to the weld bead (Fig. 25.11). The fracture occurred along this chain. The proximity and alignment of the graphite nodules provided a weak plane where a crack could readily propagate due to stresses imposed by internal pressure. The tube section had been exposed to metal temperatures in excess of 850°F (454°C) for a very long time.

FIGURE 25.17 Massive fracture along weld bead resulting from weld-related graphitization. (*Courtesy of National Association of Corrosion Engineers.*)

Case History 25.5

Industry:	Pharmaceutical
Specimen location:	Wall tube
Specimen orientation:	Vertical
Years in service:	0 (New)
Drum pressure:	900 psi (6.2 MPa)
Tube specifications:	3¼-in. (8.3-cm) outer diameter

A tube leak occurred at a weld seam in a newly installed boiler during a hydrotest. The failure was not available for analysis, but a section adjacent to it revealed deep grooves along the seam on the internal surface (Fig. 25.18). Metallographic examination at the seam revealed incomplete penetration of weld metal to the

FIGURE 25.18 Deep grooves formed along weld seam on internal surface of a new wall tube.

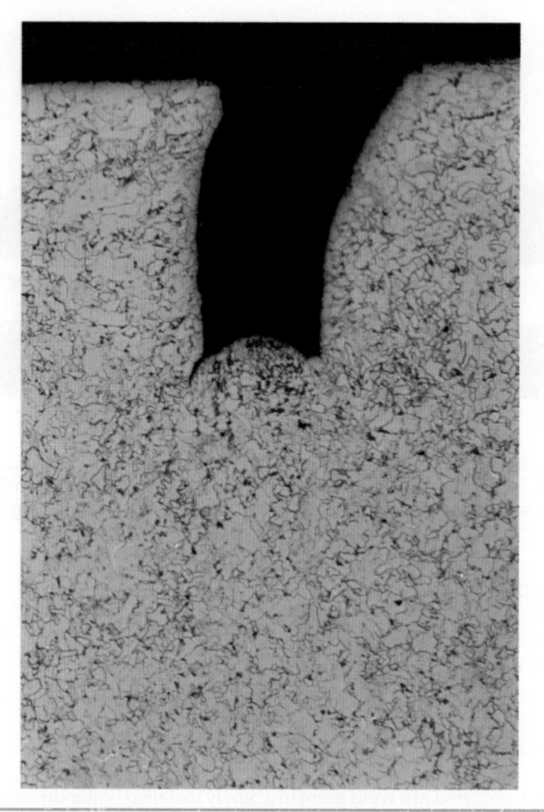

FIGURE 25.19 Metallographic cross section at the seam reveals incomplete penetration of weld metal to the internal surface.

internal surface (Fig. 25.19). It is probable that incomplete penetration and/or incomplete fusion was responsible for the failure. A partially open seam serves as a location for stress concentration and is susceptible to corrosion fatigue crack initiation and growth. Corrosion may occur due to concentration of aggressive agents, particularly mineral acids, within the crevice as well. Tubes containing significant seam defects should be removed and replaced.

Case History 25.6

Industry:	Utility (HRSG)
Specimen location:	Superheater welded to header
Specimen orientation:	Vertical
Years in service:	5
Water treatment program:	Coordinated phosphate
Drum pressure:	900 psi (6.2 MPa)
Tube specifications:	2-in. (5.1-cm) outer diameter
Fuel:	Natural gas

The HRSG was operated on an intermittent basis, ranging from 15 to 30 on/off cycles per year. An austenitic stainless steel superheater tube was butt-welded to a low-alloy steel outlet header with a 309 stainless steel filler metal. The tube separated from the header at the base of the weld (Fig. 25.20). The

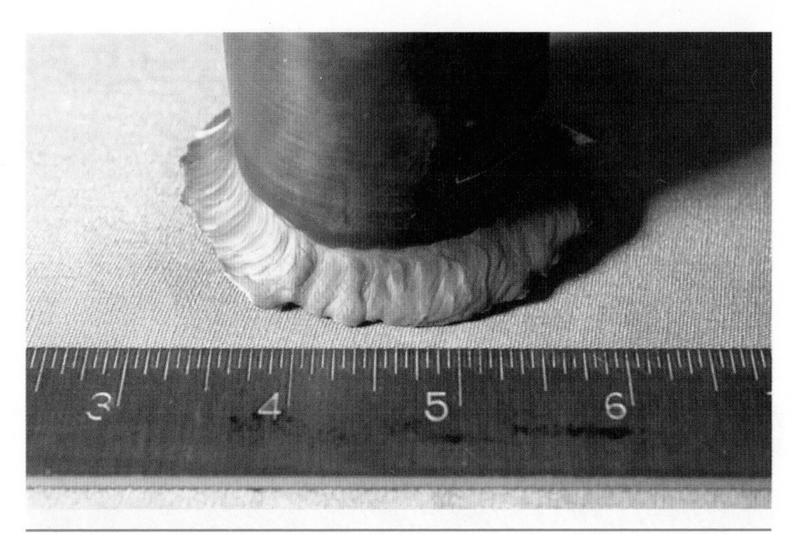

FIGURE 25.20 Stainless steel tube separated from the header at the base of the weld.

welds to stainless steel bonds were sound, but the base of the weld that contacted the low-alloy steel header had a pebbly contour and was covered with a layer of dull gray corrosion products and deposits (Fig. 25.21). Metallographic examination revealed that thin layers of decarburized low-alloy steel from the header remained on the fracture surfaces (Fig. 25.22), indicating that fracture occurred within the low-alloy base metal. Failure was caused in part by

FIGURE 25.21 Base of the weld that contacted the low-alloy steel header has a pebbly contour.

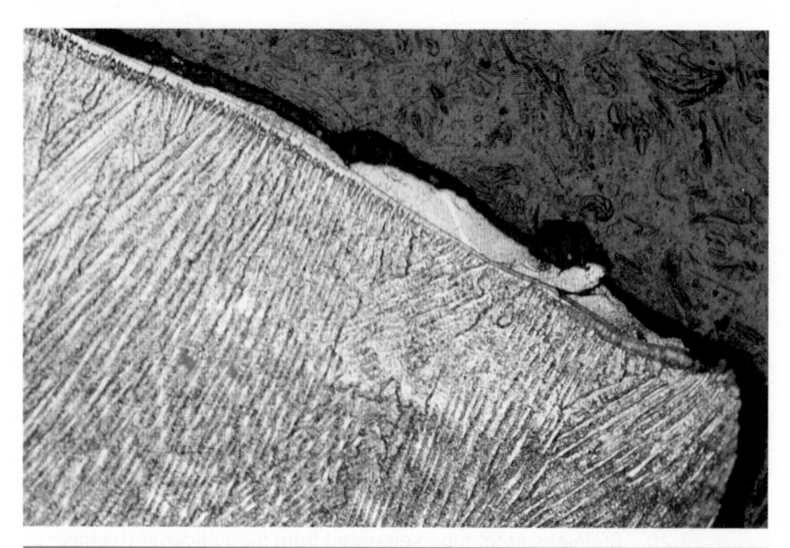

FIGURE 25.22 Metallographic cross section through fracture surface of the weld exhibits thin layers of decarburized low-alloy steel from the header.

corrosion fatigue cracking due to the application of cyclic tensile stresses. These stresses were partially supplied by differential thermal expansion and contraction between the stainless steel weld metal and low-alloy steel header. The numerous on/off cycles contributed significantly to crack growth. The tube was also inadequately supported. Decarburization occurred within the low-alloy steel base metal close to the weld, due to migration of carbon into the weld metal during welding. Decarburization substantially reduced the strength of the steel directly adjacent to the weld, allowing corrosion fatigue cracks to initiate and propagate.

Austenitic stainless steel weld metal may be replaced with appropriate nickel alloys, which have thermal expansion coefficients more similar to those of low-alloy steel, as well as lower carbon solubility.

Nondestructive testing may be conducted on a periodic basis at dissimilar metal welds to determine if cracking occurred and its extent.

Case History 25.7

Industry:	Pulp and paper
Specimen location:	Screen tube, recovery boiler
Specimen orientation:	Vertical
Years in service:	11
Water treatment program:	Coordinated phosphate
Drum pressure:	1200 psi (8.3 MPa)
Tube specifications:	2½-in. (6.4-cm) outer diameter
Fuel:	Black liquor

The longitudinal tube rupture apparent in Fig. 25.23 is straight and oriented in the longitudinal direction. V-shaped, straight seam contours were clearly visible along the length of the tube on the internal surface inline with the rupture.

FIGURE 25.23 Straight rupture that occurred at a seam defect. Erosion occurred on the external surface of the adjacent tube due to steam impingement from the rupture.

Examination of a metallographic cross section taken inline with the rupture revealed a straight, wedge-shaped fissure that formed due to incomplete fusion of the resistance welded seam during original tube manufacturing. The unfused weld seam penetrated 40% of the tube-wall thickness (Fig. 25.24). An acid cleaning of the boiler caused the cavernous metal loss apparent near

FIGURE 25.24 Metallographic cross section at a location nearby and inline with the rupture. The straight, wedge-shaped fissure was formed due to incomplete fusion of the resistance seam weld during original tube manufacturing. The wide, undercut cavity near the internal surface resulted from acid corrosion during chemical cleaning.

the internal surface. External surface metal loss on the adjacent tube due to erosion from steam and water escaping from the through-wall rupture is apparent in Fig. 25.23.

Reference

1. *ASM Metals Handbook*, vol. 1, 8th ed., 1990, ASM International, Materials Park, Ohio, p. 40.

Glossary

acid A species having the tendency to lose a proton (H^+). pH values of 0 or greater, but less than 7, are considered to be acidic.

air preheater A component usually used in high-pressure boilers that heats incoming air to be used for combustion with flue gases. The preheated air increases combustion efficiency and reduces fuel use.

alkaline Synonym for *basic*. A base is a substance that tends to accept a proton (H^+). pH values above 7 and up to 14 are considered to be alkaline (basic).

alloy steel Steel with additions of specified quantities of alloying elements (typically other than carbon) that produce desired changes in mechanical or physical properties.

all-volatile treatment (AVT) Boiler treatment programs in which the chemicals added do not contribute any dissolved solids to the boiler water.

amines Derivatives of ammonia in which one or more hydrogen atoms has been replaced by a functional group. A variety of organic amines are used to control corrosion within steam and condensate systems. Neutralizing amines control acid corrosion, while filming amines produce a water-repellent film that controls oxygen corrosion and assists in control of acid corrosion.

amphoteric Capable of reacting chemically either as an acid or as a base. In reference to certain metals, signifies their propensity to corrode at both high and low pH.

anneal To heat-treat such that the microstructure and consequently the properties of a material are altered. Often conducted to relieve residual stresses that result from welding, or to soften a metal that has been cold worked.

anode In a corrosion cell, the area over which corrosion occurs and metal ions enter solution; oxidation is the principal reaction.

applied stress Stress that results from the application of an external load, e.g., due to internal pressurization, thermal expansion and contraction, vibrations, etc.

ash fusion temperatures Temperatures at which samples of ash in a cone form show evidence of softening when heated; the analysis provides information on the ash melting characteristics.

attemperator Apparatus for reducing and controlling the temperature of superheated steam.

austenite A face-centered cubic solid solution of carbon or other elements in gamma iron.

austenitic stainless steel A nonmagnetic stainless steel that has alloying additions, generally nickel, to stabilize the face-centered cubic austenite phase.

AVT See *all-volatile treatment.*

backing In welding, a material placed under or behind a joint to enhance the quality of the weld at the root. The backing may be a metal ring or strip; a pass of weld metal; or a nonmetal such as carbon, granular flux, or a protective gas.

baffle A component that is used to direct flow of fluids. It may also be used to provide tube support in shell-and-tube heat exchangers.

bagasse Dry fibrous material that remains from sugarcane after extraction of juice.

base metal (1) In welding, the metal to be welded. (2) After welding, the part of the metal that was not melted.

belly plate A plate in the steam drum that directs the steam-water mixture from the risers to the remainder of the steam separation equipment.

biomass Material produced by living organisms, typically plant matter, that can be used for combustion.

black liquor The liquid material remaining from pulpwood cooking in the soda or sulfate papermaking process.

blowdown In connection with boilers, the process of discharging a significant portion of the aqueous solution in order to remove accumulated dissolved and suspended solids and other impurities.

borescope An optical device (e.g., an optical fiber) used to inspect an otherwise inaccessible space, such as the internal surface of a boiler tube.

brass An alloy consisting primarily of copper (> 50%), with varying proportions of zinc. Smaller amounts of other elements may also be added.

British thermal unit (Btu) Standard unit used as a measure of the heating value of a fuel. It is the amount of energy needed to heat 1 lb (0.454 kg) of water 1°F (0.56°C) at 39°F (3.89°C).

brittle fracture Material separation accompanied by little or no macroscopic plastic deformation.

bronze An alloy consisting primarily of copper, with varying combinations and proportions of elements including tin, zinc, manganese, silicon, nickel, aluminum, or iron.

Btu See *British thermal unit.*

buckets Components that are installed at the perimeter of disks mounted on shafts of turbines, also named blades. Steam is injected from a nozzle to impinge on the buckets (blades) to drive the turbine.

buckstay Boiler component that provides horizontal reinforcement and support to waterwall tubes.

butt weld A weld that attaches two mating components of similar geometry at their ends. For a cylindrical tube or pipe it may be called a circumferential weld.

carryover Transport of boiler water in steam to components downstream of the steam drum. Vaporous carryover of metals or compounds may occur at high pressure.

carryunder Migration of steam into the downcomer tubes from the steam drum, and possibly farther into the mud drum.

cast iron An iron-based alloy with a greater carbon content than steels. Most commercial cast iron alloys contain 3.0 to 4.5 wt% carbon and 1 to 3 wt% silicon.

cathode In a corrosion cell, the area over which reduction is the principal reaction. It is usually an area that is not attacked.

caustic cracking Caustic stress corrosion cracking; a form of stress corrosion cracking most frequently encountered in carbon steels or some iron-chromium-nickel alloys (such as stainless steels) that are exposed to concentrated hydroxide solutions at temperatures of 200 to 250°C (392 to 482°F).

caustic embrittlement An obsolete historical term for caustic stress corrosion cracking.

cementite A compound of iron and carbon known chemically as iron carbide and having the approximate chemical formula Fe_3C.

chelating agent Also called a ligand, complexing agent, chelant, and sequestrant. Chelating agents are water-soluble organic molecules that form coordinate covalent bonds with calcium, metals, and other ions.

circumferential weld See *butt weld.*

cladding A layer of metal of different composition than the substrate metal that it covers. This layer is typically used to control corrosion. For tubes it is typically applied by welding or coextrusion.

coal washing Preparation method that removes low grades of coal from raw coal, typically to reduce particulate loading. Also called coal cleaning.

coke Solid, carbonaceous substance remaining after destructive distillation of coal. Often it is used as blast furnace fuel for smelting of ore in the production of iron and other metals. Petroleum coke is the solid residue that remains after the cracking process during oil refining.

cold side The side of a wall tube that faces away from the firebox. In other types of tubes in a boiler, it can be used to describe the side that experience much lower heat flux than on the opposite, or hot, side.

cold work Plastic deformation of a metal below its recrystallization temperature.

complexing agent A component of an internal chemical treatment program for a boiler that combines with dissolved solids to form a stable complex or unreactive compound to control deposition. They may also complex with metals under certain conditions.

composite tubing Tubing comprised of an outer tube metallurgically bonded to an inner tube, usually formed by coextrusion. Also referred to as coextruded or cladded tubing.

condensate Liquid water that is produced when steam cools below the saturation temperature.

condensate polishing A process used to remove impurities from condensate prior to use for boiler feedwater.

convective heat transfer Transfer of heat from a higher-temperature substance to a lower-temperature substance that occurs due to mass motion of a fluid, e.g., air or water. In boilers, convection is a less significant type of heat transfer than radiation (see *radiant heat transfer*).

coring Compositional inhomogeneities (differences) throughout a metal resulting from solidification of the molten metal.

corrosion Deterioration of a substance, typically a metal, due to a reaction with its environment.

corrosion inhibitors Compounds that may be used on the fireside and in steam and condensate systems to control corrosion. Corrosion inhibitors may be used in some low-pressure boilers and hot water systems that do not employ suitable deaeration and oxygen scavenging.

corrosion product A substance that results from a corrosion process; a solid material based on the oxidized metal or metals that have been corroded.

crack A thick-walled, brittle fracture that occurs without appreciable plastic deformation.

creep Plastic flow or deformation of metals at stresses lower than the yield stress, occurring at elevated temperatures over long periods.

creep rupture See *stress rupture*.

crystallographic defects Imperfections in the crystal lattice of a material that increase its stored energy.

cupronickel Alloys used for feedwater heater and condenser tubes that contain copper concentrations ranging from 70 to 90%, with the balance consisting of nickel. Typical compositions are 90% copper-10% nickel (90-10) and 70% copper-30% nickel (70-30).

cyclone furnace Burner system for pulverized coal that consists of a horizontal cylindrical barrel that is attached to the side of a boiler with air introduced tangentially to produce a vortex.

deaerator A device that is used to remove dissolved oxygen and other noncondensable gases from boiler feedwater. Several different types are used.

deformation twins A special type of crystallographic defect produced by mirror-image positions of the atoms in the crystal lattice on both sides of the boundary. They occur in some metals (such as iron) due to applied mechanical shear forces.

demineralized water Essentially pure water produced by pretreatment processes that remove virtually all dissolved solids.

dendritic microstructure A crystalline structure that has a pattern similar to tree branches and forms during solidification. The branches are called dendrites. The term *interdendritic* refers to locations between the dendrites. Dendritic microstructures can be found in cast components and weld beads.

denickelification Type of dealloying in which nickel is selectively removed from a nickel-containing alloy, most commonly copper-nickel alloys (cupronickels).

deposit weight The mass of material covering the waterside surface of a tube per unit area. Typical units are grams per square foot and milligrams per square centimeter.

depression Term used for metal loss that occurs within a small, localized area where the depth of metal loss is smaller than the width of the area exposed. (Compare to *pit*.)

dew point The temperature to which a vapor species must be cooled for condensation to occur.

dezincification Type of dealloying in which zinc is preferentially removed from a zinc-containing alloy, most commonly copper-zinc alloys containing less than 85% copper.

diffusion Transport of mass that occurs by atomic motion.

dissolved solids Compounds (generally inorganic salts) present in solution in water, often including ions of calcium, magnesium, silicon, and others that are typically dissolved in natural waters, as well as metal ions and inorganic treatment chemicals.

downcomer Boiler tubes in which fluid flow is away from the steam drum. Such tubes receive little or no heat input.

duct burners Rows of burners in the flue gas path used for supplemental heating, typically in heat recovery steam generators.

ductile fracture Separation characterized by tearing of metal accompanied by appreciable plastic deformation and expenditure of considerable energy.

ductility The ability of a material to deform plastically without fracturing.

duplex alloy An alloy that consists of two distinct phases (see *phase*).

dye penetrant A type of nondestructive testing that utilizes a dye to detect defects, such as tight cracks that are not observable by unaided visual examination alone.

economizer A heat-exchange device, usually consisting of tubes, for increasing boiler feedwater temperature by recovery of heat from gases leaving the boiler.

eddy current A type of nondestructive testing that utilizes electromagnetic inductance to detect flaws (e.g., pits and cracks) in an electronically conductive material. It may also be used to measure material thickness and distinguish between different metals.

EDS See *energy dispersive (x-ray) spectroscopy.*

electrolyte A nonmetallic electrical conductor in which current is carried by the motion of ions. For corrosion, the electrolyte is generally an aqueous solution.

electrostatic precipitator Device that uses an induced electrostatic charge to remove particles from a flowing gas stream, typically to remove particulate matter from flue gas before it is exhausted to the atmosphere.

element mapping An analytical technique, typically conducted using a scanning electron microscope with an energy dispersive x-ray spectrometer, that shows the distribution and relative amounts of elements across an examined area.

energy dispersive (x-ray) spectroscopy (EDS) Analysis of characteristic x-rays emitted from a material due to its interaction with an electron beam. EDS produces information about the concentrations of specific elements in the material.

erosion Destruction of metals or other materials by the abrasive action of moving fluids, usually accelerated by the presence of solid particles or matter in suspension. When corrosion occurs simultaneously, the term *erosion-corrosion* is often used.

etchant A chemical solution used to selectively attack a metal to reveal microstructural details.

eutectic A reversible reaction in which a liquid solution is converted into two or more distinct solid phases.

evaporator tube Terminology that refers to a steam-generating tube, typically in heat recovery steam generator units.

excess air The amount of air above that theoretically required for complete combustion of the fuel source. Some excess air is required because the mixing of air and fuel is not perfect.

exfoliation (1) In boiler tubes, removal of oxide layers in the form of flakes or chips from a metal surface. (2) In high-pressure utility boiler feedwater heaters, a form of corrosion that produces distinct layers of oxide material on the condensing steamside surfaces of cupronickel tubes.

extrados The exterior curve of a bend or arch.

failure A general term used to imply that a part in service (1) has become completely inoperable, (2) is still operable but is incapable of satisfactorily performing its intended function, or (3) has deteriorated seriously, to the point that it has become unreliable or unsafe for continued use.

fatigue Damage caused by cyclic or fluctuating tensile stresses having a maximum value less than the tensile strength of the material. It results in thick-walled, brittle-type fractures.

fatigue limit Stress below which failure by fatigue apparently will not occur for any number of cycles.

feedwater Pretreated and deoxygenated (in most boilers) water that is supplied to the boiler during the initial fill and as makeup. (See *makeup water.*)

feedwater heater A heat exchanger that is used to preheat boiler water prior to introduction to the boiler to increase efficiency. These exchangers are primarily used for high-pressure utility boilers.

ferrite A body-centered cubic solid solution of one or more elements in alpha iron.

ferritic stainless steel A magnetic stainless steel that is a solid solution of (alpha) ferrite with a chromium content that typically varies from 11.5 to 27%.

ferrule Insert that is used on the tubeside at the inlet end of a tube used in a shell-and-tube heat exchanger.

finite element analysis A method for mathematically modeling any number of properties (thermal, mechanical, etc.) of a system utilizing a complex system of points called nodes, which form a grid called a mesh. Examples include thermal properties of fluids on the fireside of a boiler, or stress levels in a component.

fins Thin sheets of metal that are attached perpendicular to the external surface of a tube in parallel and in close proximity, usually in a spiral pattern, to provide greater surface area for heat transfer.

fireside The side of a tube in a boiler that faces the flue gases.

firetube A tube in a boiler that uses a shell-and-tube design (specifically, a firetube boiler). Flue gases flow through the tubeside, while boiler water flows across the shellside.

fissure A cracklike penetration that goes partially through a component.

flash rust Iron oxide, typically orange, brown, or red hematite, that forms on carbon, alloy steel, and cast iron surfaces due to exposure to residual water that is oxygenated by air during idle periods.

flue gas The gaseous products of combustion that flow through the fireside of a boiler and are eventually released to the atmosphere as exhaust.

flue gas scrubber Module that is used to reduce concentrations of undesirable pollutants from the flue gas emissions (sulfur dioxide, acid gases, etc.), mostly by utilizing wet processes.

fluidized bed A type of reactor in which small solid particles are suspended in an upward-flowing stream of fluid. In fluidized-bed boilers, solid fuel particles are suspended in upward-blowing jets of air during combustion, promoting effective chemical reactions and heat transfer.

flux (1) A substance that is capable of dissolving and removing metal oxides and other undesirable materials and is able to prevent oxidation. (2) The rate of transfer of fluid, particles, or energy across a given surface.

forging A process in which metal is heated and shaped by plastic deformation by application of compressive force.

fuel treatment additives Chemicals that are mixed with fuels to limit the formation of potentially corrosive species by reacting to produce less aggressive compounds.

galvanic Associated with the terms *galvanic cell, galvanic corrosion*, and *galvanic series.* Galvanic corrosion occurs when a galvanic cell is produced by dissimilar conductors, which may consist of dissimilar metals, or similar metals exposed to dissimilar electrolytes. The galvanic series is a chart that lists dissimilar metals. Contacting metals that are widely separated on the list are susceptible to producing galvanic corrosion.

gas porosity Fine holes or pores within a metal that are caused by entrapped gas or by evolution of dissolved gas during solidification.

generating bank In many industrial boiler designs, a steam drum and a mud drum that are connected by riser and downcomer tubes. Steam is generated in the bank (rows) of riser tubes.

grain An individual crystal in a polycrystalline metal or alloy.

grain boundary An interface between grains that corresponds to the transition from one crystallographic orientation to another.

grain growth An overall increase in grain size of a metal due to elevated-temperature exposure.

gram loading See *deposit weight*.

graphite nodules A spherical form of carbon that can form in plain carbon steel due to prolonged elevated-temperature exposure.

graphitization A metallurgical term describing the formation of graphite in iron or steel, usually from decomposition of iron carbide at elevated temperatures. Graphitization should not be confused with graphitic corrosion.

hard-facing Application of a harder material to a base metal for the purpose of wear resistance or to replace metal lost during service.

hardness (material) Resistance of a material to deformation, as measured by surface indentation or abrasion.

hardness (water) Concentration of multivalent cations in water, most commonly calcium and magnesium.

header A reservoir, usually a larger-diameter pipe or tube that collects fluid from several tubes upstream to supply components downstream.

heat flux Heat-transfer rate per unit area (see *flux*).

heat recovery steam generator (HRSG) Boiler system that uses a hot gas stream from a gas-fired turbine to produce steam that can be used in a process or to produce electricity in a steam turbine.

heat-affected zone (HAZ) In welding, that portion of the base metal that was not melted during welding, but whose microstructure and mechanical properties were altered by the heat.

heat treatment Controlled heating and cooling of metal to change its physical and mechanical properties while keeping its shape constant.

hematite A form of iron oxide, Fe_2O_3, that is gray to bright red. The presence of significant quantities of hematite in waterside corrosion products is often associated with oxygen corrosion.

high-temperature transformation products A general term that indicates microstructural constituents that form from heating above the lower critical transformation temperature and then cooling.

hoop stress The circumferentially oriented stress that develops in a cylindrical component, such as a tube, either due to internal or external pressurization or from residual stresses that often develop from fabrication.

hot side The side of a wall tube that faces the firebox. In other types of boiler tubes, it can refer to the side that experiences much higher heat flux than on the opposite, or cold, side such as the side facing the flue gas flow or radiant section of the furnace.

HRSG See *heat recovery steam generator.*

hydroblasting Technique for cleaning that uses a high-velocity stream of pressurized water to remove material from surfaces.

hydrolysis A double decomposition reaction in which a compound chemically reacts with water.

impingement To strike, e.g., by a rapidly flowing fluid stream upon a solid surface.

incinerator General term used for a furnace that burns waste such as trash.

inclusions Particles of foreign material in a metallic matrix. The particles are usually compounds (such as oxides, sulfides or silicates), but may be any substance that is foreign to (and essentially insoluble in) the matrix.

inhibited brass Brass that contains a small concentration of arsenic, phosphorus, or antimony to prevent the occurrence of dealloying.

intergranular Occurring between crystals or grains. Also, intercrystalline.

intergranular corrosion Corrosion occurring preferentially along grain boundaries, usually with slight or negligible attack on the adjacent grains.

intergranular oxidation Preferential formation of oxide that penetrates along grain boundaries, typically as small particles.

intrados The interior curve of a bend or arch.

iron (II) testing Analytical methods capable of measuring iron in the ferrous, Fe(II), oxidation state. This testing can be useful to identify elevated concentrations of soluble corrosion products that are released by some corrosion mechanisms, such as flow-accelerated corrosion.

laning The intentional or unintentional formation of a bypass or short circuit for flue gases resulting in redistribution of heat-transfer rates.

Larson-Miller parameter (LMP) Value from a parametric relation that incorporates the effects of time and temperature. It can be used to predict conditions that may result in stress rupture.

layup A practice followed during idle periods to control or prevent corrosion damage that may be caused by exposure to atmospheric conditions.

LMP See *Larson-Miller parameter.*

load The demand for steam for use in power generation or for industrial processes.

low alloy steel Steel that contains relatively low alloy additions (typically between 2 to 10% total alloy content). For boiler tubing, these steels typically contain chromium and molybdenum additions that provide increased resistance to oxidation and creep deformation.

low-NO_x burner Combustion system that typically uses staged firing to reduce nitrogen oxide emissions.

lower critical transformation temperature The equilibrium temperature A_1 that represents the boundary between the ferrite and cementite phase field and the ferrite and austenite or austenite and cementite phase fields in the iron-carbon phase diagram. In carbon steel, it is used as the boundary temperature between mild and severe overheating.

magnetic particle inspection Nondestructive testing method that utilizes magnetic fields and small magnetic particles, such as iron filings, to detect defects in a ferromagnetic component.

magnetite A magnetic form of iron oxide, Fe_3O_4. Magnetite is dark gray to black and forms a protective layer on iron surfaces.

makeup water Feedwater that supplies the boiler due to steam consumption or losses from returned condensate.

mandrel A tool that is used on the tubeside during tube manufacturing to maintain proper shape.

martensite For steel alloys, a supersaturated solid solution of carbon in iron characterized by microstructure having an a needlelike morphology. It is produced by quenching (rapid cooling) of austenite, preventing diffusion from occurring. It is very hard and brittle.

martensitic stainless steel A stainless steel alloy that is hardened to achieve appropriate mechanical properties by quenching and heat treatment. The alloy typically contains about 12% chromium and no nickel, and it is used for applications where erosion or wear resistance is desired, such as for turbine components.

matrix The principal phase in which other constituents are distributed.

membrane A strip of metal that longitudinally connects the external surfaces of adjacent waterwall tubes by welding to provide a gastight seal. The membrane metal typically has a composition similar to that of the tube steel.

microstructure The appearance and distribution of grains and phases in a metal determined by microscopic examination of a polished specimen. Etching is often used to reveal structural details of the metal.

mild steel Carbon steel having a maximum carbon content of approximately 0.25% that also contains 0.4 to 0.7% manganese and 0.1 to 0.5% silicon.

monel A nickel-copper alloy that is used for feedwater heater tubing.

morphology The form and structure of something.

MSW See *municipal solid waste.*

mud drum The lower closed-ended cylindrical vessel (drum) within the generating bank. Riser and downcomer tubes are connected to the drum on opposite sides. Water flows downward from the steam drum into the mud drum via downcomer tubes and toward the steam drum in riser tubes. Sediments tend to collect in this drum.

municipal solid waste (MSW) Refuse generated from an urban setting that incorporates household and commercial garbage.

NDT See *nondestructive testing.*

Neumann bands Deformation twins specifically formed by shock loading in metals with a body-centered cubic crystal structure, such as iron.

nital Etchant used for some steels and irons to highlight boundaries between ferrite grains and other microstructural features. The solution consists of a few percent of nitric acid (typically 1–5%) in alcohol (usually ethanol or methanol).

noncondensable gas Gas or gases that are removed in the deaerator and that may be present within steam and condensate lines. These gases can cause corrosion if not adequately removed. The most significant noncondensable gases in boiler, steam, and condensate systems include oxygen and carbon dioxide.

nondestructive testing (NDT) Inspection methods that can detect flaws such as pits, cracks, and metal loss without significant alteration of the component to be inspected.

nose arch In field-erected boilers, a portion of the waterwall that bends into the firebox in order to direct flue gas flow and shield convection surfaces (e.g., the superheater) from radiant heat. Also it is called the bull nose.

once-through steam generator (OTSG) Boiler unit that converts water to a steam-water mixture in one pass, without any recirculation of the fluids.

ORP See *oxidation-reduction potential.*

OT See *oxygenated treatment.*

OTSG See *once-through steam generator.*

oxidation limit A temperature or range of temperatures at which the rate of conversion from metal to oxide becomes excessive.

oxidation-reduction potential (ORP) A value representing the net sum of contributions from reducing agents and oxidizers in a liquid to the overall solution potential. The value depends on temperature and the reference electrode used for measurement.

oxidizing Condition that allows for the reaction where elements or ions lose electrons, resulting in an increase in positive valence of the specie.

oxygen scavenger A type of treatment chemical that reacts with dissolved oxygen in boiler feedwater.

oxygenated treatment (OT) Treatment program that is used in some high-pressure utility boilers with all steel construction. Maintaining specific dissolved-oxygen concentration produces a corrosion-resistant iron oxide layer on the steel surfaces. This treatment requires the use of ultrahigh purity makeup water and returned condensate.

package boiler A small boiler that is commonly shop-assembled, often used in industrial applications.

pad weld A weld that covers an area on the external surface of a tube to replace lost wall thickness or to provide erosion resistance.

parting Synonym for *dealloying.*

pearlite A microstructural aggregate consisting of alternate lamellae (layers) of ferrite and cementite.

pendant A design for superheater, economizer, and reheater tubes in which the tubes are mostly vertically oriented.

penetration In welding, the distance from the original surface of the base metal to that point at which fusion ceased.

pH The negative logarithm of the hydrogen ion activity; it denotes the degree of acidity or basicity of a solution. At 25°C (77°F), 7.0 is the neutral value. Decreasing values below 7.0 indicate increasing acidity; increasing values above 7.0 indicate increasing basicity.

phase A physically and chemically homogeneous and distinct portion of a material system.

phase diagram A graphical representation of phase fields and phase reactions in a system, generally under equilibrium conditions. For solid materials, phase boundaries are functions of temperature and composition. The iron-carbon phase diagram is useful to understand the possible phase transformations that can occur when heating plain carbon steel boiler tubing.

phase transformation Reaction that occurs, typically on heating or cooling, and results in a change in phase or phases.

picral Etchant used for some steels and irons to highlight boundaries between ferrite and cementite with superior resolution to Nital, but with less definition of ferrite grain boundaries. The solution consists of a few percent of picric acid (typically 4%) in alcohol (usually ethanol or methanol).

pit Term used for metal loss that occurs within a small, localized area where the depth of metal loss is greater than the width of the exposed area. (Compare to *depression.*)

plain carbon steel Steel containing carbon up to about 2% and only residual quantities of other elements except those added for deoxidation.

plastic deformation Deformation that is permanent after release of an applied load due to crystallographic displacements that are not recovered.

platen A design for superheater, economizer, and reheater tubes in which the tubes are mostly horizontally oriented.

ppm Parts per million. 1 ppm equals 1 mg/L of water. This is a common term used to measure the concentrations of various components in boiler water, steam, and condensate.

pretreatment Various mechanical and chemical processes used to remove suspended and dissolved solids to make feedwater suitable for use in a boiler. Pretreatment may consist of demineralization, reverse osmosis (RO), softening, condensate polishing, and others.

primary, secondary, tertiary air Terms used to identify the sources of air in the different stages of combustion in a boiler. Primary air is used for initial combustion; secondary air and tertiary air are used to complete combustion.

quench In metallurgical terms, to rapidly cool the metal from elevated temperatures.

radiant heat transfer Transfer of heat (thermal energy) caused by electromagnetic radiation emitted from a substance due to its temperature. In boilers, radiation is the most significant type of heat transfer.

RDF See *refuse-derived fuel.*

recovery boiler A steam-generating unit whose furnace burns black liquor from the Kraft pulping process to recover the cooking chemicals as smelt.

reducing Condition that allows for the reaction in which elements or ions gain electrons, so that their valence is reduced.

re-formed pearlite In previously normalized carbon steel, recrystallized pearlite formed during slow cooling from temperatures above the lower critical temperature. Re-formed pearlite colonies are different in size and shape from the original pearlite colonies.

refractory Material that is applied to fireside surfaces to provide insulation to control or prevent overheating.

refuse-derived fuel (RDF) A fuel source that is prepared from garbage after separation of components to remove material of value.

reheater A boiler component consisting of a series of tubes that is used to increase the temperature of steam extracted from a turbine. This reheated steam is sent to a lower-pressure turbine.

residual stress Stress that remains in a material even when it is free from external forces or thermal gradients. Residual stresses often result from welding, casting, cold forming, and other manufacturing processes.

rifling Shallow, rectangular grooves that are formed in a spiraling pattern on the internal surface of a tube used in a watertube boiler. The grooves are designed to promote swirling flow.

riser Boiler tubes in which fluid flow is toward the steam drum. These tubes typically contain a steam-water mixture.

rolling A term used to describe mechanical deformation of tube ends into holes in drum walls and tubesheets to impart a tight, leak-resistant seal.

root crack A crack in either a weld or the heat-affected zone at the root of a weld.

root of joint In welding, the portion of a weld joint where the base metal components are closest to one another before welding.

root of weld The points at which the weld metal intersects the base metal surfaces either nearest to or coincident with the root of joint.

SAGD See *steam-assisted gravity drainage.*

saturated steam temperature The boiling temperature of water, as determined by its pressure. Saturated steam refers to steam that is in equilibrium with liquid water.

scale (1) Waterside deposits consisting of calcium, magnesium, silicon, and other compounds that form from dissolved solids. In many cases, scale forms due to contamination of the boiler water, such as due to condenser cooling water leaks. (2) A term sometimes used to refer to the oxide formed on a metal surface at high temperatures. However, *thermal oxidation* is a more appropriate term.

scaling temperature See *oxidation limit.*

scanning electron microscopy (SEM) Analytical technique using a microscope that utilizes an electron beam to produce a variety of signals at the surface of solid materials. It may generate visual images as well as information about chemical composition and crystal structure. See also *energy dispersive (x-ray) spectroscopy.*

Schikorr reaction A chemical reaction describing the decomposition of ferrous hydroxide to magnetite and hydrogen. It is instrumental in the reaction of bare steel with water and/or steam when dissolved-oxygen concentration is very low or zero. The resulting magnetite layer acts as a protective coating on surfaces contacting water and/or steam.

screen tube A tube within a set or bank of tubes that control the amount of heat that reaches the superheater and reheater tubes located behind them.

seam The location where a sheet or plate of steel is formed into a tube shape and is joined at the longitudinally oriented mating ends by a weld.

seam welding Making a longitudinal weld in sheet metal or tubing.

SEM See *scanning electron microscopy*.

sensitization In austenitic stainless steels, precipitation of chromium-rich carbides at grain boundaries within the approximate metal temperature range of 1020 to 1560°F (550 to 850°C). Metal adjacent to the grain boundaries has a decreased corrosion resistance due to localized chromium depletion. Sensitization may be prevented through use of low-carbon grades.

shell and tube A heat exchanger that consists of a series of parallel tubes which are fastened into round holes drilled through disk-shaped tubesheets. The assembly is encased in a vessel that separates the tubeside and shellside environments.

shellside The external surface of a tube or pipe in a shell-and-tube unit.

shielding Regarding boilers, barriers that are typically used to limit heat flux or exposure to erosive or corrosive conditions.

slag (fireside) In boiler furnaces, the noncombustible ash that has reached fusion temperatures and accumulates on fireside surfaces.

slagging The tendency to form slag deposits.

smelt In recovery boilers, the molten salt mixture that develops from combustion of black liquor, which accumulates in a bed at the bottom of the furnace.

soot blower A device used in the fireside of a boiler consisting of a tube containing a port or ports that periodically issue steam that impinges upon tube surfaces to remove fireside deposits.

sour Regarding boiler fuels, those that contain appreciable amounts of sulfur compounds, such as hydrogen sulfide in a gaseous fuel.

spalling The breaking off of chips of material from a surface.

spheroidal carbides Iron or alloy carbides that are essentially spherical.

spheroidization The thermal transformation of lamellar cementite in pearlite colonies to a rounded, globular morphology.

spray coating A layer of a hard material that is applied to a surface to control fireside erosion using a thermal or plasma spraying process.

stainless steel Ferrous (iron-based) alloys that contain a minimum of 11% chromium. The addition of a sufficient amount of chromium results in the formation of a chromium-based oxide layer that provides for improved corrosion resistance in many environments compared to plain carbon or low-alloy steels.

standby When a boiler is operated at low load when steam demand is low. Such operation is used as an alternative to shutdown.

steam-assisted gravity drainage (SAGD) Enhanced oil recovery technique that uses steam stimulation in horizontal wells by injection of steam into a top well and collection of steam and water in the bottom well.

steam blanket When departure from nucleate boiling (DNB) occurs, particularly in horizontally oriented or slanted tubes, the steam space (blanket) produced by steam/water stratification along the top side.

steam drum The upper closed-ended cylindrical vessel (drum) within the generating bank. Riser and downcomer tubes are connected to the drum on opposite sides. Water flows downward from the steam drum via downcomer tubes and toward the steam drum in riser tubes. Steam collects within this drum where steam separators are housed.

steam quality The mass fraction of steam in a steam-water mixture, often given as percent steam by weight.

steam separator A device or series of devices used to mechanically separate entrained water from steam.

stoker-chain grate A device for conveying solid fuel across a furnace such that the grate acts as a burning platform.

stress Generally, load (measured in units of force) divided by the cross-sectional area. Common units are pounds per square inch (psi) or pascals (Pa). See also *applied stress* and *residual stress.*

stress concentrator or stress raiser A location or feature where applied tensile stress will be amplified, usually on the surface (e.g., pits, depressions, fissures, weld toes, sharp angles on machined surfaces, etc.) or internal (e.g., casting porosity or weld porosity).

stress relief Heat treatment to reduce residual stresses arising from processes such as welding, cold forming, and casting.

stress rupture A type of fracture that results from the accumulation of creep damage, typically characterized by interconnecting creep voids or creep cracks. Also called creep rupture.

striations Small grooves or channels, often parallel to one another.

stringers Elongated inclusions or impurities in a metal.

studded tubing Tubing with short, cylindrical pieces of metal (studs) welded to the external surface. In recovery boilers, studs aid in retention of a frozen layer of protective smelt in some areas of the boiler.

sulfidation The reaction of a metal or alloy with a sulfur-containing species to produce a sulfide-based corrosion product that forms on or beneath the surface. In boilers, this is a typical fireside corrosion mechanism in environments with reduced sulfur gases.

superheater Boiler component that is supplied saturated steam by the steam drum and that is composed of rows of tubes to increase the temperature to become superheated, mostly for use in turbines.

surface condenser A shell-and-tube heat exchanger used to condense steam on the shellside surfaces of the tubes. Cooling water flows through the tubes to promote condensation.

suspended solids Very small particles of solid material that are in permanent suspension, i.e., will not settle out if left undisturbed. These may contain calcium, magnesium, silicon, and other elements that are typically found in natural waters, as well as metals and possibly components of inorganic treatment chemicals.

swaging Forming of a metal by forcing it to conform to a die, usually with cold work. Examples related to boiler systems include tube flaring or rolling and fabrication of seamless tubing using a mandrel.

synergism Cooperative action of discrete factors such that the total effect is greater than the sum of the effects taken independently.

TDS See *total dissolved solids.*

tensile strength The maximum load-carrying capability of a material, as determined by a tensile test. It is the stress calculated from the ratio of the maximum load to original cross-sectional area. Also it is called ultimate strength.

thermal conductivity A measurement of the ability of a material to conduct heat for a unit distance within a unit of time.

thermal expansion and contraction Changes in the dimensions of a material due to changes in temperature, since atomic vibrations increase with increasing temperature. Expansion occurs on heating and contraction on cooling.

thermal expansion coefficient Fractional change in a material's length, divided by the change in temperature.

thermal fatigue Specific case of corrosion fatigue caused by alternating heating of metal and quenching by liquid water, e.g., due to repeated formation and collapse of pockets or layers of steam along a boiler tube surface.

thermal oxidation Conversion of the metal to oxide at elevated temperatures, typically above the oxidation limit.

thermally formed oxide Layer produced when the metal or alloy is converted to oxide at elevated temperatures.

total dissolved solids (TDS) A measurement in parts per million of the amount of dissolved solids in the water (see *dissolved solids*). The amount of TDS must be properly controlled.

transgranular Occurring through or across crystals or grains; also intracrystalline or transcrystalline.

tube bore (1) A circular opening within a tubesheet, drum wall, or header where a tube is inserted and housed. (2) The cylindrical space within the tubeside.

tube crown For a membrane attached waterwall tube, this is the portion of a tube wall located midway between the membranes, which extends farthest into the firebox.

tube wall The material between the inside diameter and outside diameter of a tube at any location.

tubercle Localized, structured iron oxide corrosion products in the form of a knoblike mound.

tubesheet A thick plate that contains drilled holes that house tubes. Tubes may be rolled, swaged, or welded to maintain a tight mechanical connection and seal within the holes.

tubeside The internal surface of a tube.

tunnel pits Areas of localized metal loss often observed along chloride stress corrosion cracks in austenitic stainless steel, caused by preferential attack in some areas due to compositional inhomogeneities within the metal.

turbidity A unit of measurement used to describe the relative amount of suspended material in water based on water clarity. A common unit of measurement is the nephelometric turbidity unit (NTU), which is determined by several different methods.

turbine A device that employs a series of disks, mounted perpendicular on a shaft, that house blades or buckets at their outer edges. Steam is injected onto the blades or buckets to rotate the turbine. Energy supplied by turbine rotation can generate electricity, drive pumps, blow air, and perform other functions.

turbulence In fluids, departure from laminar (smooth) flow.

tuyere A pipe or nozzle used to inject air.

ultrasonic testing A nondestructive test in which an ultrasonic beam is applied to sound-conductive materials having elastic properties. The test is used to determine wall thickness and can locate inhomogeneities or structural discontinuities within the material.

under-bead crack A subsurface crack in the base metal near a weld.

under-deposit corrosion A type of waterside corrosion that is promoted by the concentration of aggressive agents beneath deposit layers. Specific types of under-deposit corrosion and associated damage in boilers include caustic and alkaline salt corrosion, acid corrosion, and hydrogen damage.

undercut In weldments, a groove melted into the base metal adjacent to the toe of a weld that is left unfilled.

undercut pit A pit that has significant lateral propagation of the pit beneath the surface. Such pit morphology may indicate acid corrosion.

upper critical transformation temperature The equilibrium temperature A_3 that represents the boundary between the austenite phase field and the ferrite and austenite phase field in the iron-carbon phase diagram.

volatile sulfur Regarding fireside deposits, sulfur that is released due to vaporization on heating.

waterside The surface of a tube or component in a boiler that contacts water and/or steam.

watertube Boiler water flows through the tubeside, while flue gases flow across the shellside. A boiler containing watertubes is named a watertube boiler.

waterwalls The walls that enclose many boiler types, composed of a series of parallel, vertically oriented watertubes. Either the watertubes are connected by membranes, or gaps between are sealed by refractory.

wear liners Replaceable erosion-resistant material at the burner end of a cyclone furnace in the region where pulverized coal is introduced.

weld Two or more pieces of metal that are joined by heat, pressure, or both with or without filler metal. The joint is produced by fusion or recrystallization across the interface.

weld bead A deposit of filler metal from a single welding pass.

weld metal The portion of a weld that is melted during welding.

weld overlay A layer produced on a metal surface by application of multiple weld beads.

welding current The current flowing through a welding circuit during the making of a weld.

weldment A component with base metal constituents that are joined by welding.

windbox A chamber attached to the boiler through which air for combustion is introduced to the furnace.

wustite Form of iron oxide (with the approximate formula FeO) that generally forms at high temperature.

x-ray examination A nondestructive technique utilizing x-ray radiation to detect flaws and differences in thickness in a component.

yellow metal A copper-based alloy.

yield strength The stress required to produce a specified, usually small, amount of plastic strain (elongation).

Bibliography

2007 ASME Boiler & Pressure Vessel Coe, Section II —Materials (includes Addenda for 2008), American Society of Mechanical Engineers, New York, NY, 2008.

Andrews, K. W., "Empirical Formulae for the Calculation of Some Transformation Temperatures," *J. Iron Steel Inst.*, 1965, vol. 203.

ASM Committee on Failures of Pressure Vessels, Boilers and Pressure Piping, "Failures of Boilers and Related Steam-Power-Plant Equipment," *ASM Metals Handbook: Failure Analysis and Prevention*, 8th ed., vol. 10, ASM International, Metals Park, Ohio, 1975.

ASM Metals Handbook: Corrosion, 9th ed., vol. 13, American Society for Metals, Metals Park, Ohio, 1987.

ASM Metals Handbook: Failure Analysis and Prevention, 8th ed., vol. 10, American Society for Metals, Metals Park, Ohio, 1975.

ASM Metals Handbook: Properties and Selection of Metals, Definitions Relating to Metals and Metalworking, vol. 1, American Society for Metals, Metals Park, Ohio, 1961.

Avallone, E. A., and Baumeister, T., III, *Marks' Standard Handbook for Mechanical Engineers*, 10th ed., McGraw-Hill, New York, 1996.

Babcock & Wilcox, *Plant Service Bulletin 29A*, Barberton, Ohio, 2003.

Banweg, A., "Flow Accelerated Corrosion in Industrial Boiler Feed Water Systems, International Water Conference," Pittsburgh, Pa., October 18–21, 1999.

Barer, R. D., and Peters, B. F., *Why Metals Fail*, Gordon and Breach Science Publishers, New York, 1970.

Beavers, J. A., Agrawal, A. K., and Berry, W. E., *Corrosion-Related Failures in Feedwater Heaters*, CS-3184, Electric Power Research Institute, Palo Alto, CA, 1983.

Bignold, G. J., Garbett, K., Garnsey, R., and Woolsey, I. S., "Erosion Corrosion of Mild Steel in Ammoniated Water," *Proceedings of the 8th International Congress on Metallic Corrosion*, Mainz, 1981.

Bouchacourt, M., EDF Internal Report, (1982), Ref.: HT-PVD. XXX MAT/T.42).

Cain, C., Jr., and Nelson, W., "Corrosion of Superheaters and Reheaters of Pulverized-Coal-Fired Boilers," *Journal of Engineering, Trans. ASME*, Oct. 1961.

Cameron, D. W., and Hoeppner, D. W., "Fatigue Properties in Engineering," *ASM Handbook: Fatigue and Fracture*, vol. 19, ASM International, Metals Park, Ohio, 1996.

Chen, T. Y., and Godfrey, M. R., "Monitoring Corrosion in Boiler Systems with Colorimetric Tests for Ferrous and Total Iron," *Corrosion*, 1995, vol. 51, no. 10.

Chen, T. Y., Moccarri, A. A., and Macdonald, D. D., "Development of Controlled Hydrodynamic Techniques for Corrosion Testing," *Corrosion*, 1992, vol. 48, no. 3.

Ciaraldi, S. W., "Stress-Corrosion Cracking of Carbon and Low-Alloy Steels (Yield Strengths Less than 1241 MPa)," in *Stress Corrosion Cracking—Material Performance and Evaluation*, R. H. Jones (Ed.), ASM International, Materials Park, Ohio, 1992.

Cohen, A., "Corrosion of Copper and Copper Alloys," *ASM Handbook: Corrosion: Materials*, vol. 13B, ASM International, Materials Park, Ohio, 2005.

Corcoran, S. G., "Effects of Metallurgical Variables on Dealloying Corrosion," *ASM Handbook: Corrosion: Fundamentals, Testing, and Protection*, vol. 13A, ASM International, Metals Park, Ohio, 2003.

Cox, W., "Components Susceptible to Dew-Point Corrosion," *ASM Handbook: Corrosion: Environments and Industries*, vol. 13C, S. D. Cramer and B. S. Covino, Jr. (Eds.), ASM International, Materials Park, Ohio, 2006.

Desch, P. B., Dillon, J. J., and Vrijhoeven, S. H. M., "Case Histories of Failures in Heat Recovery Steam Generating Systems," Paper 03488, National Association of Corrosion Engineers, Corrosion 2003, San Diego, CA, March 2003.

———, Dillon, J. J., and Vrijhoeven, S. H. M., "Case Histories of Stress Assisted Corrosion in Boilers," Paper 04516, National Association of Corrosion Engineers, Corrosion 2004, New Orleans, LA, March 2004.

———, Dillon, J. J., and Vrijhoeven, S. H. M., "Case Histories of Stress-Assisted Corrosion in Boilers," *Power Plant Chemistry*, 2005.

———, Dillon, J. J., and Vrijhoeven, S. H. M., "The Metal Doesn't Lie—Investigative Boiler Failure Analysis," Paper 06457, National Association of Corrosion Engineers, Corrosion 2006, San Diego, CA, March 2006.

———, Dillon, J. J., and Vrijhoeven, S. H. M., "Stress Related Boiler Failures," Paper 07444, National Association of Corrosion Engineers, Corrosion 2007, Nashville TN, March 2007.

Dooley, R. B., and Chexal, V. K., "Flow-Accelerated Corrosion," Paper 347, NACE CORROSION 99.

Duplex Stainless Steels—Fighting Corrosion Worldwide (brochure), Sandvik Steel, Sandviken, Sweden.

During, E. D. D., *Corrosion Atlas*, vol. 1, Elsevier Science Publishing Company, New York, 1988.

Electric Power Research Institute, Final Report CS-3945, *Manual for Investigation and Correction of Boiler Tube Failures*, Palo Alto, CA, April 1985.

———, TR-102433-1, *Boiler Tube Failure Metallurgical Guide*, vol. 1: *Technical Report*, Palo Alto, CA, October 1993.

———, TR-102433-2, *Boiler Tube Failure Metallurgical Guide*, vol. 2: *Appendices, Appendix C, Elevated Temperature Properties*, Palo Alto, CA, October 1993.

———, TR-102433-2, *Boiler Tube Failure Metallurgical Guide*, vol. 2: *Appendices, Appendix D, Microstructural Catalogs*, Palo Alto, CA, October 1993.

———, TR-106611, *Flow-Accelerated Corrosion in Power Plants*, Pleasant Hill, Calif., 1996.

———, TR-1008082, *Guidelines for Controlling Flow-Accelerated Corrosion in Fossil and Combined Cycle Plants*, Palo Alto, CA, 2005.

———, TR-1004503, *Heat Recovery Steam Generator Tube Failure Manual*, Palo Alto, CA, 2002.

———, *HRSG Tube Failure Manual*, Palo Alto, CA, 2002.

———, Program *on Technology Innovation: Oxide Growth and Exfoliation on Alloys Exposed to Steam*, TR-1013666, Palo Alto, CA, 2007.

Elliott, P., "Materials Selection for Corrosion Control," *ASM Handbook: Corrosion: Fundamentals, Testing, and Protection*, vol. 13A, ASM International, Materials Park, Ohio, 2003.

Evans, U. R., *The Corrosion and Oxidation of Metals, First Supplementary Volume*, Edward Arnold, Ltd., London, 1968.

———, *Corrosion and Oxidation of Metals: Scientific Principles and Practical Applications*, Edward Arnold, Ltd., London, 1960.

Flynn, D. J., *The Nalco Water Handbook*, 3d ed., McGraw-Hill, New York, 2009.

Fontana, M. G., and Greene, N. D., *Corrosion Engineering*, 2d ed., McGraw-Hill, New York, 1978.

French, D. N., "Failures of Boilers and Related Equipment," *ASM Handbook: Failure Analysis and Prevention*, 9th ed., vol. 11, ASM International, Materials Park, Ohio, 2002.

———, *Metallurgical Failures in Fossil Fired Boilers*, 2d ed., John Wiley & Sons, New York, 1993.

Frey, D., "Chemical Cleaning of Boilers," 54th Annual Meeting, International Water Conference, Pittsburgh, Pa., 1993.

Hammitt, F. G., and Heymann, F. J., "Liquid-Erosion Failures," *ASM Handbook: Failure Analysis and Prevention*, 8th ed., vol. 10, ASM International, Metals Park, Ohio, 1986.

Heifner, S., and Thurston, E. P., 2005, "Collecting and Using Boiler Tube Deposit Loading Data," Paper 05442, National Association of Corrosion Engineers, NACE International, Houston TX, 2005.

Herro, H. M., and Port, R. D., *The Nalco Guide to Cooling Water Systems Failure Analysis*, McGraw-Hill, New York, 1993.

Heymann, F. J., "Liquid Impingement Erosion," *ASM Metals Handbook: Friction, Wear and Lubrication Technology*, vol. 18, American Society for Metals, Metals Park, Ohio, 1992.

Jones, R. H., and Ricker, R. E., "Mechanisms of Stress-Corrosion Cracking," in *Stress-Corrosion Cracking: Materials Performance and Evaluation*, R. H. Jones (Ed.), ASM International, Materials Park, Ohio, 1992.

Kitto, J. B., and Stultz, S. C., *Steam: Its Generation and Use*, 41st ed., Babcock & Wilcox Company, Barberton, Ohio, 2005.

Krause, H. H., "Hot Corrosion in Boilers Burning Municipal Solid Waste," *Metals Handbook: Corrosion*, 9th ed., vol. 13, ASM International, Metals Park, Ohio, 1987.

Krauss, G., *Principles of Heat Treatment of Steel*, American Society for Metals, Metals Park, Ohio, 1980.

Kung, S. G., "Fireside Corrosion in Coal- and Oil-Fired Boilers," *ASM Metals Handbook*: vol. 13B Corrosion: Materials, ASM International, Materials Park, Ohio, 2002.

Lai, G. Y., "Coal-Fired Boilers," Chap. 10 in *High-Temperature Corrosion and Materials Applications*, ASM International, Materials Park, Ohio, 2007.

———, "High-Temperature Corrosion in Waste-to-Energy Boilers," *ASM Handbook: Corrosion: Environments and Industries*, vol. 13C, ASM International, Materials Park, Ohio, 2006.

———, "Oil-Fired Boilers and Furnaces," Chap. 11 in *High-Temperature Corrosion and Materials Applications*, ASM International, Materials Park, Ohio, 2007.

———, "Waste-to-Energy Boilers and Waste Incinerators," Chap. 12 in *High-Temperature Corrosion and Materials Applications*, ASM International, Materials Park, Ohio, 2007.

Lawson, F. R., and Miller, J., "A Time-Temperature Relationship for Rupture and Creep Stresses," *Trans. AIME*, July 1952.

Logan, H. L., *The Stress Corrosion of Metals*, John Wiley & Sons, New York, 1966.

Manual for Investigation and Correction of Boiler Tube Failures, Electric Power Research Institute, Palo Alto, CA.

McCall, J. L., and French, P. M. (Eds.), *Metallography in Failure Analysis*, Plenum Press, New York, 1978.

McCoy, J. M., *The Chemical Treatment of Boiler Water*, Chemical Publishing Co., New York, 1981.

Muldoon, T. J., "Stress Corrosion Cracking (SCC) of 304SS Tubes at Outlet of the Desuperheating Zone in Feedwater Heaters," *Feedwater Heater Technology Symposium*, Electric Power and Research Institute, 1004121, Palo Alto, CA, 2004.

Patterson, W. R., *Designing for Automotive Corrosion Prevention, Proceedings*, Nov. 8–10, 1978, Society for Automotive Engineers, Troy, Mich., p. 78.

Pawel, S. J., Willoughby, A. W., Longmire, H. F., and Singh, P. M., "An Experience with Detection and Assessment of SAC in a Recovery Boiler," Paper 04517, National Association of Corrosion Engineers, 2004.

Phull, B., "Evaluating Corrosion Fatigue," *ASM Handbook: Corrosion: Fundamentals, Testing, and Protection*, vol. 13A, ASM International, Metals Park, Ohio, 2003.

Port, R. D., "Flow Accelerated Corrosion," Paper 721, *NACE Corrosion 98, NACE International,* San Diego, 1998.

Postlethwaite, J., and Nesic, S., "Erosion-Corrosion in Single and Multiphase Flow," in *Uhlig's Corrosion Handbook,* 2d ed., R. Winston Revie (Ed.), John Wiley & Sons, New York, 2000.

Rehn, I. M., Apblett, W. R., Jr., and Stringer, J., "Controlling Steamside Oxide Exfoliation in Utility Boiler Superheaters and Reheaters," *Materials Performance,* June 1981.

Reid, W. T., *External Corrosion and Deposits-Boilers and Gas Turbines,* American Elsevier Publishing Co., Inc., New York, 1971.

Samuels, L. E., *Microscopy of Carbon Steels,* ASM International, Materials Park, Ohio, 1999.

Schoch, W., and Spahn, H., "On the Role of Stress Induced Corrosion and Corrosion Fatigue in the Formation of Cracks in Water Wetted Boiler Components," *Corrosion Fatigue: Chemistry, Mechanisms, and Microstructures,* National Association of Corrosion Engineers, Houston, TX, 1972.

Shreir, L. L. (Ed.), *Corrosion,* George Newness, Ltd., Tower House, London, 1963.

Singbeil, D., Frederick, L., Stead, N., Colwell, J., and Fonder, G., "Testing the Effects of Operating Conditions on Corrosion of Water Wall Materials in Kraft Recovery Boilers,"*1996 TAPPI Engineering Conference Proceedings,* Book Singer, J. G. (Ed.), *Combustion Fossil Power Systems,* 3d ed., Combustion Engineering, Inc., Windsor, Conn., 1981.

Speller, F. N., *Corrosion/Causes and Prevention,* McGraw-Hill, New York, 1951.

Stafford, S. W., and Mueller, W. H., "Failure Analysis of Stress-Corrosion Cracking," in *Stress Corrosion Cracking—Material Performance and Evaluation,* R. H. Jones (Ed.), ASM International, Materials Park, Ohio, 1992.

Stultz, S. C., and Kitto, J. B., *Steam: Its Generation and Use,* 41st ed., Babcock & Wilcox Company, Barberton, Ohio, 2005.

Thielsch, H., *Defects and Failures in Pressure Vessels and Piping,* Reinhold Publishing Corp., New York, 1965.

Tran, H. T., "Recovery Boiler Corrosion," in *Kraft Recovery Boilers,* T. N. Adams (Ed.), TAPPI Press, Atlanta, Ga., 1997.

Uhlig, H. H. (Ed.), *The Corrosion Handbook,* John Wiley & Sons, New York, 1948.

Unterweiser, Ed, *Case Histories in Failure Analysis,* American Society for Metals, Metals Park, Ohio, 1979.

Vrijhoeven, S. H. M., Desch, P. B., and Dillon, J. J., "Case Histories of Unusual Boiler Failures," Paper 05431, National Association of Corrosion Engineers, 2005.

Warke, W. R., "Stress-Corrosion Cracking," in *ASM Handbook: Failure Analysis and Prevention,* vol. 11, ASM International, Materials Park, Ohio, 2002.

Wright, I. G., Nagarajan, V., and Krause, H. H., "Mechanisms of Fireside Corrosion by Chlorine and Sulfur in Refuse-Firing," Paper 201, *Corrosion 93,* NACE International, Houston, Tex., 1993.

Wyatt, L. M., and Germmill, N. G., "Experience with Power Generating Steam Plant and Its Bearing on Future Developments," *Proceedings of the Joint International Conference on Creep,* New York, NY, 1963.

Zeuthen, A. W., *Heating, Piping, and Air Conditioning,* 1970, vol. 42, no. 1, p. 152.

Index

Index

A

B

C

D

G

H

N

O

T

U

X

Y